Stereo Vision

Edited by
Dr. Asim Bhatti

Stereo Vision

Edited by Asim Bhatti

Republished by InTech

Janeza Trdine 9, 51000 Rijeka, Croatia

Technical Editor Teodora Smiljanic

Cover Designer Vladimir Šijan

Additional hard copies can be obtained from orders@intechopen.com

Stereo Vision, Edited by Asim Bhatti
p. cm.
ISBN 978-953-7619-22-0

For EU product safety concerns:
IN TECH d.o.o., Prolaz Marije Krucifikse Kozulić 3, 51000 Rijeka, Croatia,
info@intechopen.com or visit our website at intechopen.com.

Meet the editor

Dr. Bhatti is affiliated with Centre for Intelligent Systems Research, Deakin University, Australia. He has been actively involved in Research and development in the areas of Computer Vision and Image/Signal processing as well as Haptics and human machine interfaces. Dr. Bhatti has authored two books on the theory and applications of wavelets and multiwavelets analysis in signal and image process and has edited one book on stereo vision. Dr. Bhatti has published over 45 peer reviewed research papers and is awarded a worldwide patent on haptics. He has been part of number of successful competitive research grants and R&D funding through industrial collaboration.

Preface

Computer vision has gained enormous research interest over the last two decades with exponentially growing focus on stereo vision. Stereo vision has long been studied and lot of research involving new discoveries, techniques and applications has been reported and published over the years. Although, these published texts serve as fine introductions and references to the core mathematical ideas, however, cannot hope to keep pace with the vast and diverse outpouring of new research activities. In contrast, this volume has accumulated the most recent advances and trends of stereo vision research from around the globe. The goal of this book is to provide an insight of the current research trends and advances in the field of stereo vision. Furthermore, to provide a particularly good way for experts in one aspect of the field to learn about advances made by their colleagues with different research interests.

It is quite understandable that visual information in 3D view possesses more information about the objects in the scene than its counterpart 2D view. During the image formation process of the cameras, explicit depth information about the scenes is lost. In many applications, such as *industrial assembly and inspection, robot obstacle detection and path planning, autonomous vehicle navigation of unfamiliar environments, image based object modelling, surveillance and security, medical image analysis, and human-computer interaction*, one of the most critical tasks is the recovery and estimation of depth information. Therefore, depth information has to be inferred implicitly from the 2D views of the scenes. There are a number of techniques available in the literature for depth estimation; however the most widely used techniques are based on stereo vision. Stereo vision techniques are mainly inspired by the human visual system, where two perspective views of eyes lead to slight offset of objects (disparities) in the two monocular views, leading to the phenomenon of depth perception. In spite of the fact that research on stereo vision spans over almost two decades; the performance of the stereo vision systems could not be compared with the human visual system. Stereo vision has long been studied and a number of techniques have been reported. Depth by stereo is achieved by estimating the pixel correspondences of similar features in the stereo perspective views originated from the same 3D scene. However, finding correct corresponding points is still an ill posed problem. Some of the factors that make the correspondence problem difficult are its inherent ambiguity, occlusions, photometric, radial and geometric distortions. A number of algorithms have been proposed to address some of the aforementioned problems in stereo vision however it, relatively, is still an open problem.

To address aforementioned issues, the presented book consists of topics from theoretical aspects to the applications of stereo vision. The book consists of 20 chapters, each highlighting different aspects of the research in stereo vision. The book can be categorized into the spectrum of three broad interests that includes the theoretical aspects of the stereo vision, algorithm development for robust disparity (consequently depth) estimation and the applications of stereo vision in different domain. Generally, understanding theoretical aspects and the algorithm development in solving for the robust solutions goes hand in hand. Similarly, the algorithm development and the relevant applications are also tightly coupled as generally algorithms are customized to achieve optimum performance for specific applications which is hard to achieve, otherwise. For instance; chapters 18, 9, 20, 15, 19, 2, 17, 1 and 7 fall into the category of first two interests that involves the discussion of theoretical aspects as well as the development of algorithms in reference to stereo vision. Particularly, chapter 20 describes Graph Cut whereas chapter 2 describes multiresolution analysis based algorithms that fall into two famous classes of stereo vision algorithms, i.e. global optimization algorithm and multi-resolution based hierarchical algorithms, respectively. Furthermore, chapters 6, 8, 10, 11-14 and 16 present applications of stereo vision and current trends. The applications that have been covered in this book includes driver assistance system, robotic bin-picking, human computer interaction (HMI), head detection and tracking, 3D underwater mosaicking, moving target tracking, robot navigation, pose estimation, smart homes and 3D avatar communication. Chapters 5 and 14 present hardware based approaches regarding the development of stereo vision systems that has direct commercial applications.

In summary this book comprehensively covers almost all aspects of stereo vision and highlights the current knowledge trends as well as corresponding solutions. In addition reader can find topics from defining knowledge gaps to the state of the art algorithms as well as current application trends of stereo vision to the development of intelligent hardware modules and smart cameras. It would not be an exaggeration if this book is considered to be one of the most comprehensive books published in reference to the current research in the field of stereo vision. Research topics covered in this book makes it equally essential and important for students and early career researchers as well as senior academics linked with computer vision.

Editor

Dr. Asim Bhatti
Centre for Intelligent Systems Research
Deakin University
Vic 3217, Australia

Contents

1

Calibration and Sensitivity Analysis of a Stereo Vision-Based Driver Assistance System

András Bódis-Szomorú, Tamás Dabóczi and Zoltán Fazekas
Computer and Automation Research Institute &
Budapest University of Technology and Economics
Hungary

1. Introduction

As safety and efficiency issues of transportation - hand-in-hand with the intelligent vehicle concept – have gathered ground in the last few years in the automotive research community and have penetrated into the automotive industry, vision-based applications have become increasingly important in driver assistance systems (Kastrinaki et al., 2003; Bertozzi et al., 2002). The primary targets of the safety and efficiency improvements are intelligent cruise control (e.g. vehicle following), lane keeping and lane departure warning systems, assistance in lane changing, avoidance of collision against vehicles, obstacles and pedestrians, vision enhancement and traffic sign recognition and signalling. Our focus of interest here is stereo machine vision used in the context of lane departure warning and lane keeping assistance. The primary purposes of our vision system are to determine the vehicle's position and orientation within the current lane, and the shape of the visible portion of the actual lane on structured roads and highways. That is, we face a somewhat simplified simultaneous localization and mapping (SLAM) problem here, with the assumption of a structured man-built environment and a limited mapping requirement. The localization is achieved through the 3D reconstruction of the lane's geometry from the acquired images.

Fig. 1. Images acquired with our wide-baseline stereo vision system in a typical highway scene. The overlaid epipolar lines in the left image correspond to the marked points in the right image.

In monocular lane detection systems, reconstruction usually relies on a number of assumptions concerning the scene geometry and the vehicle motion, such as flat road and constant pitch and roll angles, which are not always valid (as also stated in Kastrinaki et al., 2003; Marita et al., 2006). Stereo vision provides an effective and reliable means to extract correct range information without the need of relying on doubtful assumptions (Hartley & Zisserman, 2006). Nevertheless, there are some stereo systems that still make use of such assumptions while providing a higher level of robustness compared to monocular systems. For example, certain systems use inverse perspective mapping (IPM) to remap both the left and the right images to the road's plane, i.e. to generate a "bird's view", where lane detection is performed while stereo disparity can be exploited in vehicle detection by analysing the difference between the remapped images (Bertozzi & Broggi, 1998). Another solution is to use a stereo pair in standard configuration (parallel optical axes and image coordinate axes) and the Helmholtz shear equation to relate the precomputed sparse stereo disparity to the distances measured in the road's plane and also to classify the detected 3D points to "near-road" and "off-road" points (Weber et al., 1995). Some algorithms perform a sparse stereo reconstruction of the 3D scene through the classical stereo processing steps of feature detection, correlation-based matching to solve the correspondence problem and triangulation (Nedevschi et al., 2005). Stereo matching is performed by making use of the constraints arising from the epipolar geometry, that is, the search regions are simplifed to epipolar lines (Hartley & Zisserman, 2006), as shown in Figure 1. In the standard arrangement, the correspondence problem simplifies to a search along corresponding scanlines, thus, such a setup is more suitable for real-time applications (Weber et al., 1995). Even if the cameras are in general configuration it is possible to virtually rotate the cameras and rectify the images with a suitable homography (Hartley & Zisserman, 2006). However, this transformation is time-consuming and when sub-pixel accuracy can be achieved in feature extraction, image warping can jeopardize the overall accuracy (Nedevschi et al., 2006). This consideration is increasingly valid for lane detection systems where the scene's depth can easily reach 60 to 100 meters.

Even if up-to-date tracking methods (involving dynamical scene and ego-motion models) are used to make the reconstruction more robust, the precision and even the feasibility of the reconstruction using any of the techniques mentioned above is seriously affected by the precision of the camera parameters that are typically determined prelimarily with camera calibration. IPM techiques (and also the one based on the Helmholtz shear equation) may fail when the relative orientations of cameras with respect to the road's plane are not precisely determined (Weber et al., 1995; Broggi et al., 2001). Since epipolar lines and the rays at the triangulation step are highly dependent on the camera parameters, both stereo matching and 3D reconstruction may prove unsatisfactory (Marita et al., 2006). These facts highlight the importance of an accurate calibration and motivate a thorough sensitivity analysis in the design of such safety-critical systems.

In this chapter, we investigate the effects of parameter uncertainties on the quality of 3D geometrical reconstruction. We propose an off-line and far-range camera calibration method for stereo vision systems in a general confguration. Due to the high depth range, we restrict our analysis to perspective cameras. The cameras are calibrated individually by using fairly common planar patterns, and camera poses with respect to the road are computed from a specifc quasi-planar marker arrangement. Since a precise far-range arrangement might be difficult and costly to set up, we put up with an inexpensive and less precise arrangement, and formulate the maximum likelihood estimate of the camera parameters for it. We demonstrate how our method overperforms the widely used reprojection error minimization method when significant errors are present in the marker arrangement. By applying the presented methods to real images, we give an estimate on the precision of 3D lane boundary reconstruction.

The chapter is organized as follows. We outline our preliminary lane detection algorithm for stereovision in Section 2. Next, a brief overview is given of existing calibration methods and their applicability in the field in Section 3. Then, in Section 4, we present an intrinsic camera calibration and a single-view pose estimation method. The proposed stereo calibration method and the overall sensitivity analysis, including epipolar line uncertainty are derived in Section 5. In Section 6, evaluation of the methods based on real data is presented. Finally, Section 7 concludes the chapter.

2. The lane detection algorithm

In the current section, we suppose that our wide-baseline system with two forward-looking grayscale cameras fixed to the side mirrors of a host vehicle (we refer to Figure 1) is already calibrated and the camera poses are known. Then we return to the problem of calibration in details in Section 3. The outline of the algorithm is depicted in Figure 2.

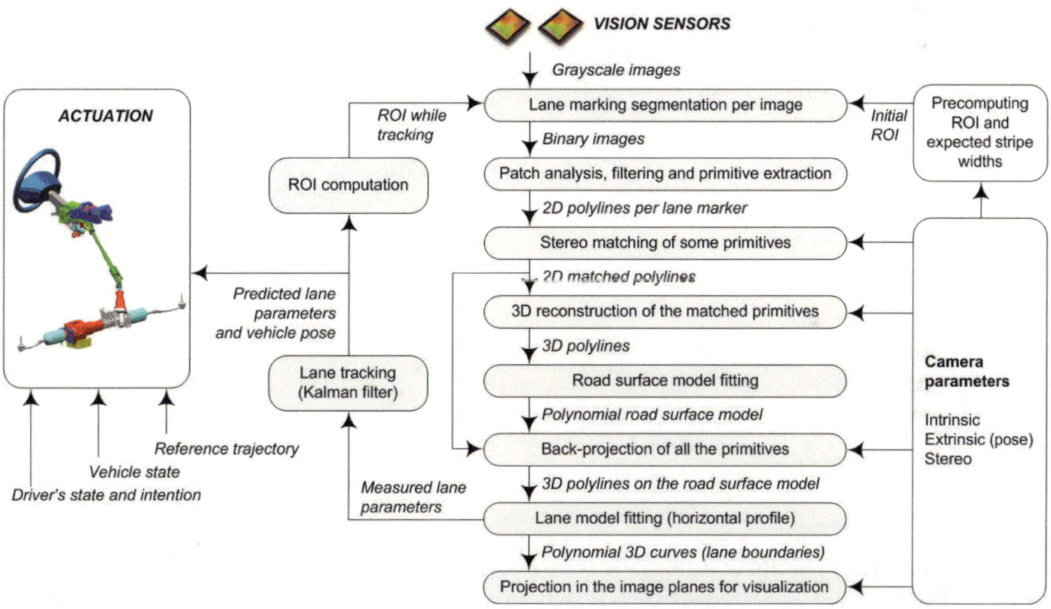

Fig. 2. Outline of a lane keeping assistant system based on the lane boundary detection and a stereo lane reconstruction algorithm. The algorithm highly depends on the camera parameters. ROI stands for Regions-of-Interest.

In order to minimize the resource requirements of the algorithm, we avoid using dense stereo reconstruction and image rectification. In sparse methods, interesting features are defined and extracted from the images by a feature detector.

2.1 Lane marking extraction

The presented algorithm currently relies on the presence of lane markings. We have developed a lane marking detector that is capable of extracting perspectively distorted bright stripes of approximately known metrical width over a darker background. The lane marking extraction is performed with 1-dimensional templates of precomputed width per scanline within precomputed regions-of-interest (ROI). The ROI computation requires the a priori knowledge of the camera parameters as the ROI boxes in the images are determined from rectangular regions specified metrically in the ground plane (which does not exactly match the road's surface). In the current study, we focus on the pure lane extraction algorithm working on independent successive frames without considering temporal relations and only using the known camera parameters as a priori information. Tracking of the lane boundary curves over time (e.g. with a Kalman filter) can be added easily to this framework. It gives more robustness and stability to the estimated parameters. Solutions for an elaborate dynamical modeling of the problem exist in the literature (Eidehall & Gustafsson, 2004; Bertozzi et al., 2002). Thus, the presented algorithm can be considered as an initialization stage for lane (and object) tracking (see Figure 2).

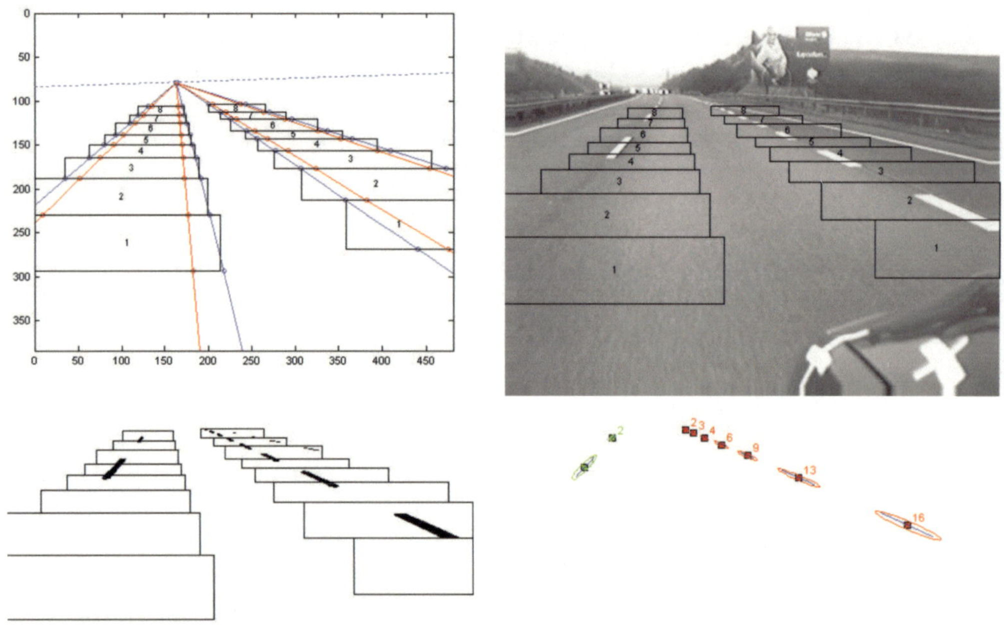

Fig. 3. Intermediate results of automatic lane marking extraction for the image of the left camera. The ROI and expected stripe width computation (top-left), ROI boxes over the input image (top-right), segmentation results (bottom-left), patch analysis results (bottom-right).

The first part of the algorithm is performed independently and simultaneously in the two images. Intermediate results are shown in Figure 3. Lane marking segmentation consists of a lane marking enhancement step and a binarization step. Binarization uses automatic threshold computation. This is followed by a patch analysis stage where several properties of the identified patches are determined, for example, patch area and eccentricity of the ellipse with equivalent second central moments (the ellipse is only a means to measure elongatedness of the patches). These properties are then used to filter the detected patches: small and more or less circular patches are removed. Next, the ridge of each patch is determined along scanlines. After this step, the identified lane markings are represented as chains of points (primitives). If two primitives overlap in the horizontal direction, the external one is removed in order to avoid ambiguities at feature matching and to avoid the detection of the boundary of a potential exit lane (slip road) next to the current lane. The points grouped into these chains are radially undistorted before proceeding.

2.2 Lane reconstruction

The second part of the algorithm uses stereo information. The primitives are matched by cycling through the points of each primitive belonging to either boundary of the lane in the left image and then by computing the intersection of the corresponding epipolar lines with the primitives belonging to the same boundary in the right image. Presently, we describe the primitives as polylines. An alternative solution is to fit low-order models (e.g. line or polynomial curve segments) to the ridge points for each lane marking, and compute the intersection of the epipolar lines with these models. The matched points identifying lane boundary sections detected in both images are then triangulated into 3-space. The reconstructed 3D points belonging to the lane's boundary are used to find the *road*'s surface. The road surface model used is a second-order polynomial in z, which is the longitudinal coordinate of the vehicle reference frame and linear in the lateral coordinate x. The x-axis of the vehicle reference frame is defined to point from the right to the left. As we use right-handed reference frames, the remaining axis y points upwards. The road surface model explicitely contains the vehicle's roll angle φ and pitch angle ϑ measured with respect to the road, the vertical road curvature c_V and the vertical distance h between the road's surface and the origin of the vehicle's reference frame (the height coordinate):

$$y(x,z) = h + \varphi x + \vartheta z + \frac{c_v}{2} z^2 . \quad (1)$$

This model is currently fitted to the triangulated 3D boundary-points in the least-squares (LS) sense. An alternative method would be to use a robust fitting method, e.g. RANSAC (Fischler & Bolles, 1981).

Mono systems usually rely on the assumption $y(x,z) = 0$, i.e. the vehicle motion is simplified to a planar motion, the road's surface is modeled as a constant plane in the vehicle's reference frame or equivalently, vehicle pitch and roll angle, as well as, the vertical curvature of the road is neglected. This may cause instabilities in the next step, when a lane model is fitted to the back-projected feature points as depicted in Figure 4.

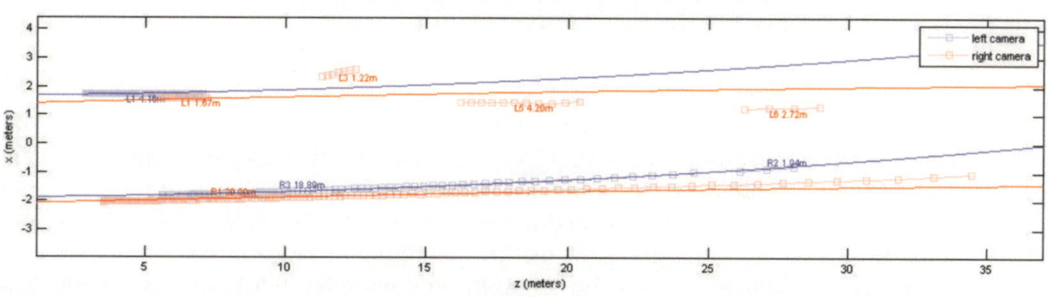

Fig. 4. An example of divergence present at lane model fitting when the left and right cameras are not interpreted as a stereo pair and instead the $y(x,z) = 0$ assumption is used.

In Figure 4, the point chains - shown as connected squares - represent the identified and reprojected lane markings while the continous double-curves represent the fitted polynomial lane model. The left and right-side chains converge due to an unmodeled pitching. It should be noted that some mono systems are able to estimate vehicle pitching by optimizing the pitch angle until the reprojected primitives become parallel. In Figure 4, at the left-side boundary, an outlier segment is present that is resulted from an imperfect segmentation.

In our stereo approach, as soon as the road's surface is found, all the detected points are projected onto it, including those that were ruled out at stereo matching. Then a double-polynomial lane boundary model is fitted to the 3D data as shown in Figure 5. Some of the roads are designed using constantly varying curvature (e.g. in Europe) while others include straight and circular segments (e.g. in the United States). Corresponding to a constantly varying curvature, clothoid lane models may be used (Nedevschi et al., 2005, Kastrinaki et al., 2003) (European case), but we experienced that the LS-fitting is relatively unstable when the polynomial order is higher than two. Again, a robust model fitting method such as the RANSAC could be used to account for some outliers and make the detection more stable. Figure 6 shows some outputs of the discussed lane geometry reconstruction algorithm.

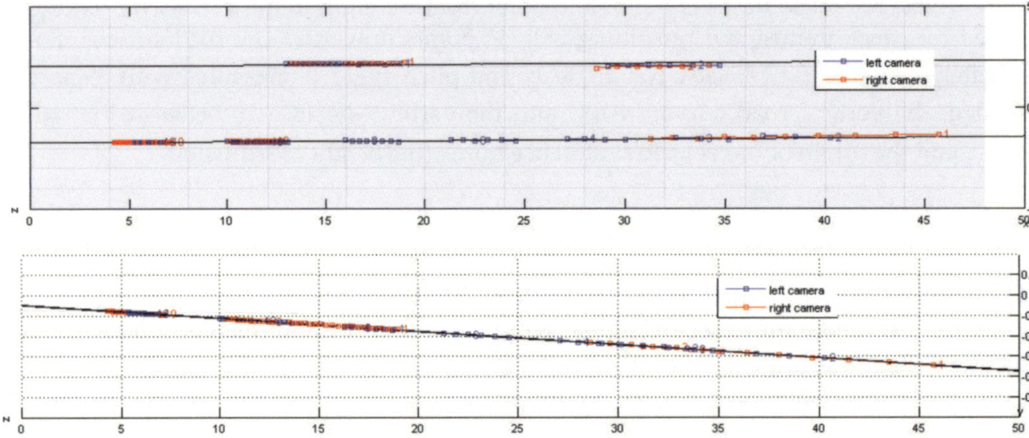

Fig. 5. Example of polynomial lane model fitting that followed the road surface detection based on stereo data. The image shown in Figure 3 and its right pair were used as input. There is a slight vehicle pitching (note the different scales on the axes) as shown in the side view of the road surface (bottom) but the lane profile (top) shows that the reprojected primitives are parallel, just like the true lane boundaries.

2.3 On reconstruction errors

Reconstruction errors can have multiple sources. An imperfect segmentation causing outliers can disturb stereo feature matching, road surface model fitting and horizontal lane profile model fitting, the effect on the latter two being more severe. In most of the cases, the effects caused by outliers can be moderated or even eliminated by using robust algorithms at the model fitting stage. Ambiguities at feature matching are rare in the discussed lane boundary detection algorithm since horizontally overlapping boundary segments are removed as a prevention (the epipolar lines are almost horizontal). However, feature matching becomes a difficult problem when the lane detection algorithm is extended by vehicle, obstacle or guard rail detection. In such cases, area-based matching techinques are popular and ambiguities may occur at repeating patterns. Also, imprecisions in feature matching cause errors in 3D reconstruction, especially if the 3D point is far from the vehicle. As mentioned earlier, the ROI computation, feature matching and 3D reconstruction require the knowledge of the camera parameters (see Figure 2) that are determined by calibration. Any imprecision in the camera parameters may jeopardize the whole procedure. In such a case, the computed epipolar lines do not exactly pass through the corresponding points at point feature matching. Therefore, in case of area-based matching, the correlation threshold may not be reached along an epipolar line and the point pair to match may be rejected. Globally, this results a decreased number of reconstructed feature points per snapshot. This may cause, e.g. missed obstacles if obstacles are searched based on a vicinity criterion of the reconstructed points. In the meantime, the mentioned threshold should be kept as high as possible to avoid false matches. Even if matches are accepted, their localization may be imprecise which, together with the imprecisely known camera parameters can cause signficant errors in the reconstruction by triangulation. Considering further processing stages, high reconstruction errors can affect the model fitting stage seriously (similarily to the case shown in Figure 4). Therefore, extra care is required at camera calibration.

Fig. 6. Some results of the discussed stereo reconstruction algorithm. The reconstructed 3D lane geometry is reprojected to the source images.

3. Camera calibration preliminaries

There are several common ways to calibrate a stereo rig. For example, it is possible to compute the fundamental matrix $\mathbf{F}$ from point correspondences (e.g. by using the well known normalized 8-point algorithm or the 7-point algorithm) without any knowledge of the scene or motion. It has been shown that the reconstruction based on this information only is possible up to a projective transformation (Hartley & Zisserman, 2006). The camera matrices determine $\mathbf{F}$ up to scale, but not vice-versa. If additional knowledge either of the motion or of the scene's geometry (e.g. parallel scene lines or planes) an affine or a metric reconstruction may be reached. This is still not enough to achieve a "ground truth" (true Euclidean) lane reconstruction. The reconstruction algorithm should not rely on such scene constraints as these might not always be available in a road scene. We can state that exact information on the vehicle's reference frame is required in such applications. $\mathbf{F}$ does not provide information on the 3D Euclidean reference frame but camera models do (Hartley & Zisserman, 2006). Consequently, we need to determine the two camera models first, only then can we proceed with computing $\mathbf{F}$ and the epipolar lines for stereo matching. In general form, the model of a single camera relating a 3D world point $\mathbf{W} \in \mathrm{R}^3$ (given in metrical coordinates) to its 2D image denoted by $\mathbf{I} \in \mathrm{R}^2$ is as follows:

$$\mathbf{I} = \Phi\,(\mathbf{p}_{\mathrm{in}}, \mathbf{p}_{\mathrm{ex}}, \mathbf{W}), \tag{2}$$

where $\mathbf{p}_{\mathrm{in}}$ is a vector formed of the intrinsic camera parameters and $\mathbf{p}_{\mathrm{ex}}$ is the 6-vector of the extrinsic camera parameters, the latter representing the 6-Degrees-of-Freedom (DoF) Euclidean transformation (i.e., a 3D rotation and a 3D translation) between camera and world reference frame. The mapping Φ may model certain non-linear effects (e.g., radial and tangential lens distortions), as well. These distortions can be - in certain cases - neglected, or can be removed from Φ by an adequate non-linear image warping step. The remaining Φ_{lin}, representing the pinhole camera model, is linear in a homogeneous representation:

$$\tilde{\mathbf{I}} = \mathbf{KR}[\,\mathbf{E}_{3x3} \,|\, -\mathbf{t}]\tilde{\mathbf{W}} \qquad \mathbf{K} = \begin{bmatrix} \alpha & \gamma & u_0 \\ 0 & \beta & v_0 \\ 0 & 0 & 1 \end{bmatrix} \tag{3}$$

where $\mathbf{E}_{3x3}$ is the identity matrix, $\tilde{\mathbf{I}} \in \mathrm{P}^2$ and $\tilde{\mathbf{W}} \in \mathrm{P}^3$ are homogeneous representations of the points and $\mathbf{K}$ is the camera calibration matrix incorporating the relative focal lengths α and β, the skew γ and the principal point $(u_0, v_0)^T$. These are all intrinsic parameters contained in $\mathbf{p}_{\mathrm{in}}$. $\mathbf{R}$ is the 3-DoF rotation matrix, and $\mathbf{t}$ is the camera position. $\mathbf{R}$ and $\mathbf{t}$ correspond to $\mathbf{p}_{\mathrm{ex}}$.

Calibration of a monocular camera consists of determining the camera parameters $\mathbf{p}_{\mathrm{in}}$ and $\mathbf{p}_{\mathrm{ex}}$ from properly measured $\mathbf{W}_i \mapsto \mathbf{I}_i$ point correspondences. We note here that for the intrinsic parameters, novel methods tend to use planar calibration patterns instead of 3D calibration objects (Malm & Heyden, 2003; Zhang, 2000; Sturm & Maybank 1999). A single planar arrangement does not provide enough information for estimating all the intrinsic paremeters, however, a solution for $\gamma = 0$ and known aspect ratio β / α exists (Tsai, 1987). Algorithms for calibrating from planar patterns are developed for the stereo case, as well

(Malm & Heyden, 2001), but those determine the relative pose between the cameras while the absolute poses are required in our application.

As to lane-related applications, Bellino et al. have used a single planar pattern and the aforementioned method for calibrating a monocular system used in heavy vehicles (Bellino et al, 2005). The authors used a fixed principal point and did not involve γ. They have investigated the reconstruction errors of two target points on the ground plane vs. the tilt angle of a calibration plane given with respect to the vertical position. The target points were placed up to 11.5 m from the camera and their positions were measured with a laser-based meter for validation. They found that the inclination of the plane had considerable effect on the quality of the result.

A more general intrinsic parameter estimation is given in (Hartley & Zisserman, 2006) for multiple planes with unknown orientations. Homographies (i.e., perspective 2D-to-2D mappings) $\mathbf{H}_j$ between the planes and the image are estimated first. Each $\mathbf{H}_j$ gives rise to two constraints on the image of the absolute conic (IAC) $\boldsymbol{\omega}$. The IAC is formulated as $\boldsymbol{\omega} = (\mathbf{K}\mathbf{K}^T)^{-1}$. The constraints are linear in the elements of $\boldsymbol{\omega}$, namely $\mathbf{h}_1^T\boldsymbol{\omega}\mathbf{h}_2 = 0$ and $\mathbf{h}_1^T\boldsymbol{\omega}\mathbf{h}_1 = \mathbf{h}_2^T\boldsymbol{\omega}\mathbf{h}_2$, where $\mathbf{h}_i$ is the i-th column of $\mathbf{H}_j$. Since $\boldsymbol{\omega}$ is the homogeneous representation of a conic, it has 5 DoF, so 3 planes suffice to estimate $\boldsymbol{\omega}$. $\mathbf{K}$ can be computed from $\boldsymbol{\omega}^{-1}$ by Cholesky-factorization (Hartley & Zisserman, 2006), or by using direct non-linear formulas (Zhang, 2000). In the literature, some comparisons can be found between the method of Tsai and the method of Zhang (Sun & Cooperstock, 2005; Zollner & Sablatnig, 2004). It turns out, that Zhang's model and his method overperforms the others with respect to residual errors and convergence. The price payed is the relatively high number of iterations, which is not a serious problem in case of off-line applications.

Having determined the intrinsic camera parameters, the rotation matrix $\mathbf{R}$ and the translation vector $\mathbf{t}$ need to be recovered. This is called pose estimation and a single general planar arrangement with at least 4 points suffices for a unique solution (Lepetit & Fua, 2005). A rough pose estimation can be performed by first estimating the homography between the world plane and the image and then by re-using the orthogonality constraints with known camera calibration matrix $\mathbf{K}$ (Malm & Heyden, 2003; Lepetit & Fua, 2005).

The planar pattern used in intrinsic calibration is unsuitable for far-range systems, because it minimizes errors for close-range (as also stated in Marita et al., 2006; Bellino et al., 2005). For this purpose, Broggi et al. used a medium-range grid painted on the ground (Broggi et al., 2005) while Marita et al. used vertical X-shaped markers placed on the ground in front of the vehicle in a distance up to 45 m (Marita et al., 2006). In the latter work, the intrinsic and the extrinsic parameters of each camera are computed separately. Extrinsic parameters are computed by minimizing the reprojection error in the image for the available control points. A constrainted Gauss-Newton minimization is used with the constraints $\mathbf{R}^T\mathbf{R} = \mathbf{I}$. The calibration is validated by comparing the 3D reconstruction errors of the control points to the actual 3D measurements. However, there is no information available about the accuracy of the control point setup. Alternatively, line features can also be used for pose estimation (Kumar & Hanson, 1994).

With the road/lane following application in mind, and making use of the calibration methods used in computer vision, we present here a calibration scheme and method that is optimal under some reasonable assumptions. It should be emphasized that the proposed method takes into consideration the errors present in the 3D setup.

4. Calibration of a single camera

4.1 Camera model

We use a pinhole camera - extended with a fifth-order radial distortion model and a tunable distortion centre (Hartley & Zisserman, 2006) - as our camera model:

$$\mathbf{D} = \begin{pmatrix} x_D \\ y_D \end{pmatrix} = (1 + d_1 r^2 + d_2 r^4) \begin{pmatrix} x_P - c_x \\ y_P - c_y \end{pmatrix} + \begin{pmatrix} c_x \\ c_y \end{pmatrix}, \tag{4}$$

where $(c_x, c_y)^T$ is the distortion center, d_1 and d_2 are the distortion coefficients, while $(x_P, y_P)^T$ is the point perspectively projected to the normalized image plane and $\mathbf{D}$ is the corresponding distorted point on this plane. Thus, the distortion model is applied between the projection step and the rasterization step, the latter being modeled with the homogenous transformation represented by the camera calibration matrix $\mathbf{K}$, like in equation (3). Lens distortion models involving tangential distortion are also available in the literature (Heikkila & Silvén, 1997). An explicit tangential distortion model is not required in our case, as the distortion center, with its two additional parameters, models imperfect alignment of the lenses and of the sensor. Therefore, we have, in total, the nine intrinsic parameters $\mathbf{p}_{in} = (\alpha, \beta, \gamma, u_0, v_0, d_1, d_2, c_x, c_y)^T$ for each camera.

4.2 Intrinsic calibration

Intrinsic calibration is performed separately for the two cameras. We used a hand-held checkerboard pattern shown in m different orientations to the camera. Alternatively, a pattern with circular patches could have been used (Heikkila & Silvén, 1997, Zollner & Sablatnig, 2004). Corners in the checkerboard pattern can be localized with sub-pixel accuracy with our corner detector based on a robust saddle-point search in small regions around the corners.

For an initial guess of $\mathbf{p}_{in}$, we used the method of Zhang, however we extended it with some optional constraints: $\omega_{12} = \omega_{21}$ (for $\gamma = 0$), $\omega_{11} = \omega_{22}\beta^2\alpha^{-2}$ (for a fixed aspect ratio, e.g. square pixels), and finally $\omega_{13} = -u_0\omega_{11}$ and $\omega_{23} = -v_0\omega_{22}$ (for a fixed principal point). The problem is usually overdetermined because of the great number of the control points available. Having carried out a homography estimation for each image, the solution for $\boldsymbol{\omega}$ is obtained with a simple SVD-based method that gives a least-squares solution while exactly fulfilling the constraints. $\mathbf{K}$ is determined using the formulae arising from $\boldsymbol{\omega}^{-1} = \mathbf{K}\mathbf{K}^T$.

Next, for each of the m views, the 6 extrinsic parameters $\mathbf{p}^*_{ex,j}$ ($j = 1, 2...m$) are determined, that is, the planar pose estimation problem is solved for each view. For this, we used the orthogonality constraints satisfied by $\mathbf{K}^{-1}\mathbf{H}_j$, where $\mathbf{H}_j$ denote the homography matrix estimated for the j-th view. Since measurements are noisy and orthogonality is not exactly satisfied in practice, some tolerances were used in the orthogonality test. In the representation of the rotation, we used the Rodrigues-vectors $\mathbf{r}_j = \varphi_j \mathbf{a}_j$, where $\mathbf{a}_j$ represents the rotation axis and $\varphi_j = \|\mathbf{r}_j\|$ is the rotation angle. This has several advantages over the rotation matrix-based representation (Lepetit & Fua, 2005).

As the set of parameters $\hat{\mathbf{p}}_{in} \in R^9$ and $\hat{\mathbf{p}}^*_{ex} = (\hat{\mathbf{p}}^{*T}_{ex,1},...,\hat{\mathbf{p}}^{*T}_{ex,m})^T \in R^{6m}$ determined earlier minimize an algebraic error, a refinement of the parameters is preferable by minimizing a geometrically meaningful error. A reasonable choice is to minimize the sum of the reprojection errors in all the m images for all the n feature points:

$$f(\mathbf{p}_{in}, \mathbf{p}^*_{ex}) = \sum_{i,j} d^2(\mathbf{I}^*_{ij}, \hat{\mathbf{I}}^*_{ij}) = \sum_{i,j} \| \mathbf{I}^*_{ij} - \hat{\mathbf{I}}^*_{ij} \|_2^2 = \| \mathbf{I}^* - \hat{\mathbf{I}}^* \|_2^2 , \tag{5}$$

where $\mathbf{I}^*_{ij}$ is the measured location of the i-th corner in the j-th image and $\hat{\mathbf{I}}^*_{ij}$ is the reprojection of the world point $\mathbf{W}^*_{ij}$ using the parametrized camera model (2). $\mathbf{I}^*$ is the measurement vector containing all the $\mathbf{I}^*_{ij}$'s and $\hat{\mathbf{I}}^*$ contains all the $\hat{\mathbf{I}}^*_{ij}$'s. d denotes Euclidean distance in the pixel reference frame. The cost function (5) can be minimized using a gradient-based iterative method. It can be shown, that (5) is a Maximum Likelihood (ML) cost function, provided the checkerboard pattern is precise, significant errors are due to the corner localization, and the errors have uniform Gaussian distribution all over the images. These assumptions make it possible to estimate the deviation σ of the detection noise in the image simply as the standard deviation of the residual errors $(\mathbf{I}^*_{ij} - \hat{\mathbf{I}}^*_{ij})$ evaluated at the optimum. In order to determine the quality of the calibration, the estimated noise deviation $\hat{\sigma}$ is then back-propagated to the camera parameters through a linearized variant of the camera model (2).

4.3 Optimal pose estimation per view

The relative orientations of the views with respect to the imaged checkerboards (incorporated in $\hat{\mathbf{p}}^*_{ex}$) are of no interest for us from the point of view of our application. Instead, for both cameras, we need an estimate of the pose $\hat{\mathbf{p}}_{ex}$ with respect to the vehicle's reference frame. To determine the poses, the vehicle with the mounted cameras is stopped over an open flat area and marker plates - with an X-shape on each - are placed in front of the vehicle (see Figure 7). Similar arrangements has already been used (Marita et al., 2006).

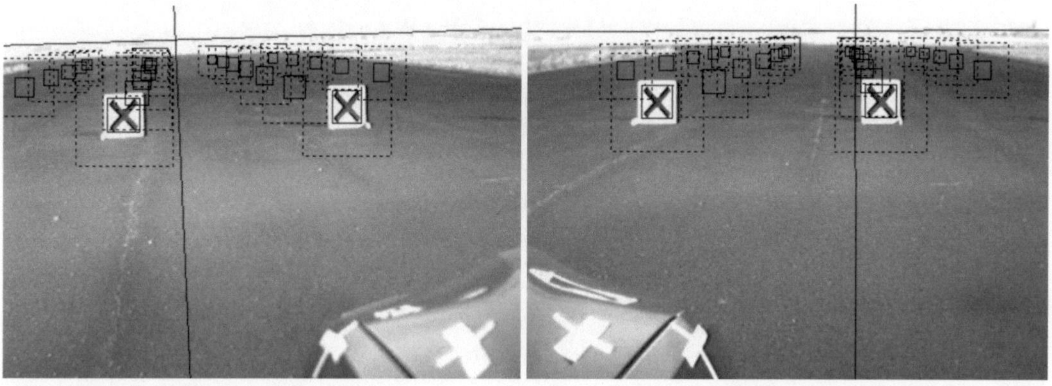

Fig. 7. An image pair of the calibration scene with overlayed marker detection and calibration results. The dashed boxes are the search regions for template matching, while the solid boxes correspond to the detected markers. The solid lines represent the vanishing lines of the world reference planes.

The feature points are the centres-of-gravity (CoG's) of the individual X-shapes. The 3D marker locations $\mathbf{W}_i$ were measured by a laser-based distance meter from two reference points. The locations of the reference points are measured with respect to the vehicle. $\mathbf{I}_i$'s are extracted from the images using normalized cross-correlation with an ideal X-shaped template. We derive the pose estimation formula for one camera first and based on that we derive it for the stereo case. The initialization of the algorithm is done with the pose estimation method based on the orthogonality criteria discussed in Section 4.2. Therefore, an initial guess is available for $\mathbf{p}_{ex}$, and this is to be fine-tuned by using a geometrically meaningful expression. The measurement of the 3D point locations and the detection of their image are two independent measurements modeled with two independent Gaussian distributions. If we have N markers and we introduce the vector of all the measured 3D coordinates $\mathbf{W} := (\mathbf{W}_1^T, ..., \mathbf{W}_N^T)^T$, and the vector of all the measured pixel coordinates $\mathbf{I} := (\mathbf{I}_1^T, ..., \mathbf{I}_N^T)^T$, then the likelihood functions for $\mathbf{W}$ and $\mathbf{I}$ are:

$$L(\mathbf{W} \mid \bar{\mathbf{W}}) = \frac{1}{\left(\sqrt{2\pi}\right)^{3N} \sqrt{\det(\mathbf{C}_W)}} \exp\left\{ -\frac{1}{2} \| \mathbf{W} - \bar{\mathbf{W}} \|_{\mathbf{C}W}^2 \right\}, \tag{6}$$

$$L(\mathbf{I} \mid \bar{\mathbf{I}}) = \frac{1}{\left(\sqrt{2\pi}\right)^{2N} \sqrt{\det(\mathbf{C}_I)}} \exp\left\{ -\frac{1}{2} \| \mathbf{I} - \bar{\mathbf{I}} \|_{\mathbf{C}I}^2 \right\}. \tag{7}$$

Here, $\| \mathbf{a} - \mathbf{b} \|_C^2 = (\mathbf{a} - \mathbf{b})^T \mathbf{C}^{-1} (\mathbf{a} - \mathbf{b})$ denotes the squared Mahalanobis distance between the vectors $\mathbf{a}$ and $\mathbf{b}$ with respect to the covariance matrix $\mathbf{C}$. $\mathbf{C}_I$ and $\mathbf{C}_W$ denote the covariance matrices of the measurement vectors $\mathbf{I}$ and $\mathbf{W}$, respectively. The likelihood function of all the measurements is the product of $L(\mathbf{I} \mid \bar{\mathbf{I}})$ and $L(\mathbf{W} \mid \bar{\mathbf{W}})$ because of the independence. Therefore, the Maximum Likelihood Estimate (MLE) for the expected 2D and 3D locations $\bar{\mathbf{I}}$ and $\bar{\mathbf{W}}$ of the feature points can be found by minimizing the function

$$g(\bar{\mathbf{I}}, \bar{\mathbf{W}}) = \| \mathbf{I} - \bar{\mathbf{I}} \|_{\mathbf{C}I}^2 + \| \mathbf{W} - \bar{\mathbf{W}} \|_{\mathbf{C}W}^2 . \tag{8}$$

$\bar{\mathbf{I}}$ and $\bar{\mathbf{W}}$ are related by the camera model (2), so $\bar{\mathbf{I}} = \Phi(\mathbf{p}_{in}, \mathbf{p}_{ex}, \bar{\mathbf{W}})$. Both $\mathbf{C}_I$ and $\mathbf{C}_W$ are block-diagonal provided the measurement of the control points are independent from each other. The blocks in the diagonals are $\mathbf{C}_{Ii}$ (the 2x2 covariance matrices of each markers' localization errors in the image) and $\mathbf{C}_{Wi}$ (the 3x3 covariance matrices of the markers' localization errors in 3-space), respectively. Thus, (8) can be rewritten as

$$g(\mathbf{p}_{ex}, \bar{\mathbf{W}}) = \sum_{i=1}^{N} \left\{ \| \mathbf{I}_i - \Phi(\mathbf{p}_{in}, \mathbf{p}_{ex}, \bar{\mathbf{W}}_i) \|_{\mathbf{C}Ii}^2 + \| \mathbf{W}_i - \bar{\mathbf{W}}_i \|_{\mathbf{C}Wi}^2 \right\}. \tag{9}$$

Several important conclusions can be drawn from equation (9). The first is that the covariance matrices of the measurements are involved, so this ML cost function can not be used when no uncertainty information is available of the measurements. In such cases the cost function (5) could be used for pose estimation, as well. The second one is that the expected 3D locations represented by $\bar{\mathbf{W}}_i$ are involved, therefore not only the searched 6 extrinsic parameters in $\mathbf{p}_{ex}$ are to be optimized, but the marker locations of the 3D calibration arrangement, as well. This adds extra 3N dimensions to the parameter space. Optionally, the 3N extra dimensions can be eliminated by approximating $\bar{\mathbf{W}}_i$ with the closest point to the measured 3D point $\mathbf{W}_i$ on the ray through the radially corrected image point $\mathbf{I}_i$.

5. Optimal stereo calibration and sensitivity analysis

5.1 Optimal two-view calibration and pose estimation

Up to this point we considered the two cameras independently. Clearly, it is possible to determine the extrinsic parameters of the left ($\mathbf{p}_{exL}$) and right cameras ($\mathbf{p}_{exR}$) independently (Marita et al., 2006). However, this model does not take into consideration that the 3D control point setup is common for the two views. Also, equation (9) requires the knowledge of the exact intrinsic parameters $\mathbf{p}_{in}$, however, only its estimation $\hat{\mathbf{p}}_{in}$ is available from intrinsic calibration. These problems can be solved by formulating the MLE for the overall problem that involves all the measurements including those available from checkerboard-based calibration and for both cameras. As a result, the cost function to minimize becomes slightly more complex:

$$\begin{aligned} h(\mathbf{p}_{inL},\mathbf{p}_{exL},\mathbf{p}_{inR},\mathbf{p}_{exR},\mathbf{p}^*_{exL},\mathbf{p}^*_{exR},\bar{\mathbf{W}}) = \frac{1}{\sigma_L^2}\|\mathbf{I}^*_L - \bar{\mathbf{I}}^*_L(\mathbf{p}_{inL},\mathbf{p}^*_{exL})\|_2^2 + \frac{1}{\sigma_R^2}\|\mathbf{I}^*_R - \bar{\mathbf{I}}^*_R(\mathbf{p}_{inR},\mathbf{p}^*_{exR})\|_2^2 + \\ + \|\mathbf{W} - \bar{\mathbf{W}}\|^2_{\mathbf{C}_W} + \|\mathbf{I}_L - \bar{\mathbf{I}}_L(\mathbf{p}_{inL},\mathbf{p}_{exL},\bar{\mathbf{W}})\|^2_{\mathbf{C}_{IL}} + \|\mathbf{I}_R - \bar{\mathbf{I}}_R(\mathbf{p}_{inR},\mathbf{p}_{exR},\bar{\mathbf{W}})\|^2_{\mathbf{C}_{IR}} \end{aligned}, \quad (10)$$

where the first two terms come from the intrinsic calibration performed independently for the left and right camera (see the right side of equation (5)), the last two terms represent the localization errors in the images in the X-marker-based calibration and the third term represents the errors in the world point locations in the X-marker-based pose estimation. Each measurement influences the solution weighted with the inverse of its uncertainty. The outline of the proposed algorithm is as follows.

1. Perform an intrinsic calibration independently for the two views by using planar patterns and by minimizing the cost function (5). Compute $\hat{\sigma}_L$ and $\hat{\sigma}_R$.
2. Set up a far-range planar arrangement of visible control points for stereo pose estimation (as suggested by Figure 7). Locate the markers and estimate the measurement covariances both for the images and for the 3D arrangement.
3. Initialize the yet unknown extrinsic parameters $\mathbf{p}_{exL}$ and $\mathbf{p}_{exR}$ with respect to the road by solving the pose estimation problem based on the orthogonality criteria independently for the two views. Initialize $\bar{\mathbf{W}}$ with the measured 3D locations and minimize cost function (10).

5.2 Sensitivity of the camera parameters

Supposing that the optimal solution of (10) has been found (this can be checked with a residual analysis together with a linearity test of the cost function h), the overall quality of the calibration can be characterized by performing a sensitivity analysis. First of all, the cost function (10) can be approximated as

$$h(\mathbf{q}) = \| (\mathbf{m} - \overline{\mathbf{m}}) - (\hat{\mathbf{m}} - \overline{\mathbf{m}}) \|^2_{\mathbf{C}_m} \approx \| (\mathbf{m} - \overline{\mathbf{m}}) - \mathbf{J}_m(\mathbf{q} - \overline{\mathbf{q}}) \|^2_{\mathbf{C}_m}, \quad (11)$$

where $\mathbf{q}$ is the vector of all the parameters to optimize, $\mathbf{m}$ is the vector of all the measurements $\mathbf{I}_L, \mathbf{I}_R, \mathbf{I}^*_L, \mathbf{I}^*_R$ and $\mathbf{W}$. $\mathbf{C}_m$ is the block-diagonal covariance matrix of all these measurements. $\hat{\mathbf{m}}$ contains, on the one hand, the image points $\hat{\mathbf{I}}_L, \hat{\mathbf{I}}_R, \hat{\mathbf{I}}^*_L, \hat{\mathbf{I}}^*_R$ reprojected using the camera model, on the other hand, the optimized world point coordinates $\hat{\mathbf{W}}$. $\overline{\mathbf{m}}$ represents the "ground thruth" values of the parameters that are always unknown and $\mathbf{J}_m$ is the analytically computed Jacobian matrix of the mapping $\mathbf{q} \mapsto \hat{\mathbf{m}}$ evaluated at the optimal parameter vector $\hat{\mathbf{q}} \approx \overline{\mathbf{q}}$.
The measurement uncertainty incorporated in $\mathbf{C}_m$ can be back-propagated to the parameters, as $\mathbf{C}_q = (\mathbf{J}_m^T \mathbf{C}_m^{-1} \mathbf{J}_m)^{-1}$. Since we are primarily interested in the uncertainty of the intrinsic parameters $\mathbf{p}_{inL}$ and $\mathbf{p}_{inR}$ and of the poses $\mathbf{p}_{exL}$ and $\mathbf{p}_{exR}$, the corresponding two 30x30 sub-matrix should be extracted from $\mathbf{C}_q$. It is important that nothing prevents the parameters of the two cameras to cross-correlate. Therefore, at the uncertainty estimation of any computation involving the parameters of both cameras, the 30x30 covariance matrix has to be considered instead of the two 15x15 blocks corresponding to the two cameras, independently.

5.3 Uncertainty of the epipolar lines

In the feature matching stage of the lane detection algorithm, we compute the intersection of epipolar lines and 2D primitives (polylines or curves). It is well known that the fundamental matrix required for epipolar line computation can be derived from the camera parameters (Hartley & Zisserman, 2006). We use the formulation

$$\mathbf{F} = \mathbf{K}_R^{-T} \mathbf{R}_R \mathbf{R}_L^T [\mathbf{R}_L (\mathbf{t}_R - \mathbf{t}_L)]_\times \mathbf{K}_L^{-1}, \quad (12)$$

where the indices L and R refer to the left camera and right camera, respectively. $[\cdot]_\times$ denotes the 3x3 matrix of rank 2 corresponding to the cross product operator, so that $\mathbf{a} \times \mathbf{b} = [\mathbf{a}]_\times \mathbf{b}$. Then the epipolar line $\mathbf{l}_R$ in the right image corresponding to a point $\tilde{\mathbf{I}}_L$ in the left image can be computed as $\mathbf{l}_R = \mathbf{F}\tilde{\mathbf{I}}_L$, where both $\mathbf{l}_R$ and $\tilde{\mathbf{I}}_L$ are homogeneous 3-vectors. The epipolar constraint $\mathbf{l}_R^T \tilde{\mathbf{I}}_R = 0$ assures that the line $\mathbf{l}_R$ passes through the point $\tilde{\mathbf{I}}_R$, which is the right image of the same world point, as shown in the left side of Figure 8.
Uncertainty in the camera parameters propagates to the derived fundamental matrix and to the epipolar lines. If the vector of the involved camera parameters is denoted by $\mathbf{p}$, then the formula (12) can be interpreted as a mapping $\mathbf{p} \mapsto \mathbf{f}$ where $\mathbf{f}$ is a vector formed from the elements of $\mathbf{F}$. If this mapping can be approximated by a linear mapping in the range of the noise, and Gaussian distributions are supposed just like in earlier derivations, then the forward propagation of covariance can be given as

$$\mathbf{C}_f = \mathbf{J}_f \mathbf{C}_p \mathbf{J}_f^T, \quad (13)$$

where $\mathbf{C}_p$ is the covariance matrix of $\mathbf{p}$, $\mathbf{C}_f$ is a covariance matrix associated to $\mathbf{F}$ and $\mathbf{J}_f$ is the Jacobian matrix of the mapping $\mathbf{p} \mapsto \mathbf{f}$ evaluated at the mean values $\hat{\mathbf{p}} \approx \overline{\mathbf{p}}$ and $\hat{\mathbf{f}} \approx \overline{\mathbf{f}}$ (hat over the letter denotes estimated values and bar denotes the unknown ground thruth). $\mathbf{J}_f$ is either available analytically as the partial derivatives of (12) with respect to the camera parameters or it can be approximated numerically. To avoid ambiguities, $\mathbf{F}$ is always normalized so that $\|\mathbf{F}\|=1$.

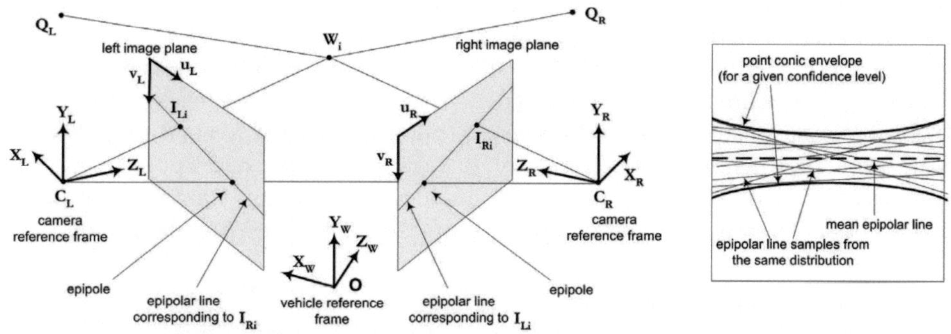

Fig. 8. Epipolar geometry of cameras in general configuration (left). The 3D point $\mathbf{W}_i$ given in the vehicle reference frame is "seen" as $\mathbf{I}_{Li}$ and $\mathbf{I}_{Ri}$ in the pixel reference frames. The nature of epipolar line uncertainty (right).

Next, the uncertainty present in the fundamental matrix can be forward-propagated to the epipolar lines very similarily, by linearizing the mapping $\mathbf{f} \mapsto \mathbf{l}_R / \|\mathbf{l}_R\|$. It is known that the set of epipolar line samples corresponding to a given confidence level form a line conic that is bounded by a point conic envelope of 5 DoF (Hartley & Zisserman, 2006). This is illustrated in the right side of Figure 8. The point conic can be analytically derived if the covariance matrix $\mathbf{C}_{lR}$ of the epipolar line $\mathbf{l}_R$ is known. The envelope conic for a given confidence level λ (99% for example) characterises very well the sensitivity of an epipolar line. However, it is more practical to use the distribution of the angle between the epipolar line and the u-axis of the image reference frame to characterize uncertainty. If $\mathbf{l}_R = (a,b,c)^T$, this angle can be expressed as

$$\Theta = \arctan(-a / b) . \tag{13}$$

Moreover, the mapping $\mathbf{l}_R \mapsto \Theta$ should also be linearized in order to propagate the covariance of the epipolar lines to Θ. As a result, the complete chain of the uncertainty propagation from the camera parameters to the angle Θ is $\mathbf{p} \mapsto \mathbf{F} \mapsto \mathbf{l}_R \mapsto \Theta$.

A practical verification of the uncertainty computations is done by Monte-Carlo simulations. A high number of samples $\{\mathbf{p}_k\}$ were generated of $\mathbf{p}$ with mean the estimated camera parameters $\hat{\mathbf{p}}$ and covariance matrix $\mathbf{C}_p$ that is extracted from the computed $\mathbf{C}_q$ matrix as described in Section 5.2. Then by selecting a point $\mathbf{I}_L$ in the left image, the corresponding epipolar line and Θ_k are computed for each sample $\mathbf{p}_k$. Finally, the standard deviation of the set $\{\Theta_k\}$ is calculated. This way, to any point $\mathbf{I}_L$ in the left image, a single value is associated that characterizes the uncertainty of the corresponding epipolar line. If the output of the computationally expensive Monte-Carlo approach coincides with the results received from the discussed cheaper linear approximation method, then the non-linear $\mathbf{p} \mapsto \mathbf{F} \mapsto \mathbf{l}_R \mapsto \Theta$ mapping is nearly linear in the range of the uncertainities around $\hat{\mathbf{p}}$.

5.4 Uncertainty in the reconstruction

The main reason why camera parameter uncertainties are studied is to predict reconstruction errors due to an imprecise knowledge of the camera parameters. The errors in stereo point reconstruction can be simulated with a Monte-Carlo method that is very similar to the one discussed in Section 5.3. The estimated camera parameters are perturbed corresponding to the estimated parameter covariance matrix and for each parameter set, several 3D points are reconstructed. As a result, point clouds are formed in 3-space that correspond to reconstruction errors. Similarily, one can go further, and apply further steps of those detailed in Section 2, e.g. road surface model fitting and lane model fitting for each generated set of parameters. Some results based on real data are presented in Section 6.

6. Evaluation on real images

6.1 Numerical results of the two-step camera calibration method

A setup with two analog 1/3" b&w CCD cameras with a resolution of 720x576 pixels and 8 mm lenses with 34° horizontal field-of-view were mounted on the side mirrors of a test vehicle. For various reasons, the acquired images were resized to a size of 480x384 pixels when the recorded videos were post-processed in order to remove interlacing effects. For intrinsic calibration, we used a checkerboard pattern with 11x7 control points and a square size of 3 cm. Images were taken in 16 different views (which is much more than required) and a constrainted minimization with $\gamma = 0$ has been carried out by using the Levenberg-Marquardt method. As a result, the estimated focal lengths are (777.6, 849.8) ± (3.3, 4.1) pixels for the left camera, and (776.1, 847.2) ± (3.0, 3.5) pixels for the right, the principal point is located at (215.7, 201.9) ± (10.4, 7.6) pixels in the image of the left camera and at (236.0, 168.7) ± (8.8, 6.9) pixels in that of the right one. The uncertainties given here correspond to the 99% confidence interval of the uncertainty in the parameter space that is calculated by back-propagating the standard deviation of the residual reprojection errors given in the image to the parameters, as mentoned in Section 4.2. The above uncertainties are only given for reference; clearly they do not completely characterise the uncertainty (cross-covariance information is also required). The standard deviations of the residual errors were 0.26 and 0.23 pixels for the left and right cameras, respectively. The distortion centers are located approximately at the principal points while the estimated radial distortion coefficients are (-0.505, 0.878) ± (0.044, 0.540) for the left camera and (-0.516, 0.957) ± (0.041, 0.540) for the right one. Note that the second coefficient is estimated with a relatively high uncertainty (alternatively, it could be forced to zero).

In the second step, 24 X-markers of size 50x50 cm were placed in front of the vehicle along four lines at a distance of 3 meters laterally and in a depth range from 10 to 40 meters. Single rows of markers were placed at a time to have a clear view on each, that is, to avoid masking due to the perspective effect (see Figure 7). Marker distances from two reference points were measured with a laser-based distance meter and the 3D locations were computed by triangulation. Also, errors in the reference point locations, distance measurements, non-ideal marker-placing and deviations with respect to the planar ground assumption were estimated and forward-propagated to the computed 3D locations resulting in an estimate of each $\mathbf{C}_{Wi}$ $(i = 1,2...N)$ (see Bodis et al., 2007, for more details). The resulted 99% covariance ellipsoids of the 3D measurements are plotted in Figure 9C for some markers. Using a quasi-Newton optimization with a termination criterion of 10^{-5} on the relative change in the parameter values, the minimization of (10) converged in 116 iterations. For comparison, we also minimized (5), like if there would be no covariance information. A comparison of the results provided by the two approaches is shown in Fig. 9A-B.

It is clear from the results (Figure 9A-B) that inappropriate modeling of the problem or the lack of covariance information misleads the optimization when the 3D marker locations are not known precisely. Naturally, the high reprojection errors shown in Figure 9A is undesirable because it is directly related to the corresponding 3D reconstruction errors.

Because camera skews were forced to be zero, we had 2x8 intrinsic parameters, 2x16x6 extrinsic parameters with respect to the checkerboards and 2x6 extrinsic parameters with respect to the road (or stopped vehicle). As to the measurements, we had 2x2x16x11x7 coordinates from the checkerboard corners, while 2x2x24 image coordinates and 3x24 world coordinates were available from the X-marker based measurement. In total, there were 220 parameters and 5096 measured values. In Figure 9C, we can see that the covariance ellipsoids of the estimated 3D locations are much smaller than those of the measured locations. This is what we expected from a regression-like problem.

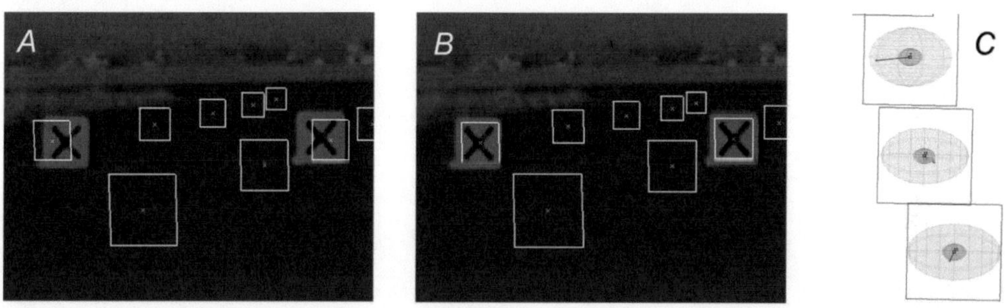

Fig. 9. A-B) Markers reprojected to one of the source images of the right camera using the camera parameters that resulted from the minimization of A) the cost function given in (5), B) the proposed cost function (10). C) 99% condence levels of the 3D location measurements for three 50x50 cm marker plates placed at 20, 25 and 35 m (outer ellipsoids) and 99% confidence level of the position estimate after optimization, from $\mathbf{C}_q$ (inner ellipsoids).

6.2 The sensitivity of point reconstruction

The evaluation of the calibration is performed by simulating the effects of parameter uncertainties represented by $\mathbf{C}_p$ on 3D point recunstruction by triangulation.

The reference points to reconstruct by Monte-Carlo simulations were chosen from the calibration scene: the estimated marker centers were shifted by 25 cm vertically to lie on the ground surface. The left and right images of these 3D points were computed by using the set of the optimal camera parameter estimates (see the top of Figure 10). Then, 1000 perturbed parameter sets were generated around the estimated parameters corresponding to $\mathbf{C}_p$. There are 2x14 estimated camera parameters that are to be considered here, zero skew being fixed, so $\mathbf{C}_p$ is a 28x28 matrix. A 3D subspace of the generated distribution is shown at the bottom of Figure 10. This corresponds to the 99% confidence level of the 3D location of the left camera given in the vehicle's reference frame. The sizes of this ellipsoid are ±14 laterally (X), ±35 cm in depth (Z) and ±10 cm in the direction perpendicular to the ground (Y). The ellipsoid is elongated in the longitudinal (Z) direction because the estimation of the camera positions in this direction is more sensitive to the uncertainty in 3D marker locations of the planar and far-range calibration scene. Interestingly, this does not mean that any 3D point can be reconstructed with a maximum precision of ±35 cm, because reconstruction quality is affected by the covariances between all the 28 parameters, as well.

After that the points were radially corrected in both images by using the perturbed radial distortion parameters in each experiment, they were reconstructed by using the simple mid-point triangulation method. The resulted 3D point clouds are shown in Figure 11 and their measured extent are plotted in Figure 12.

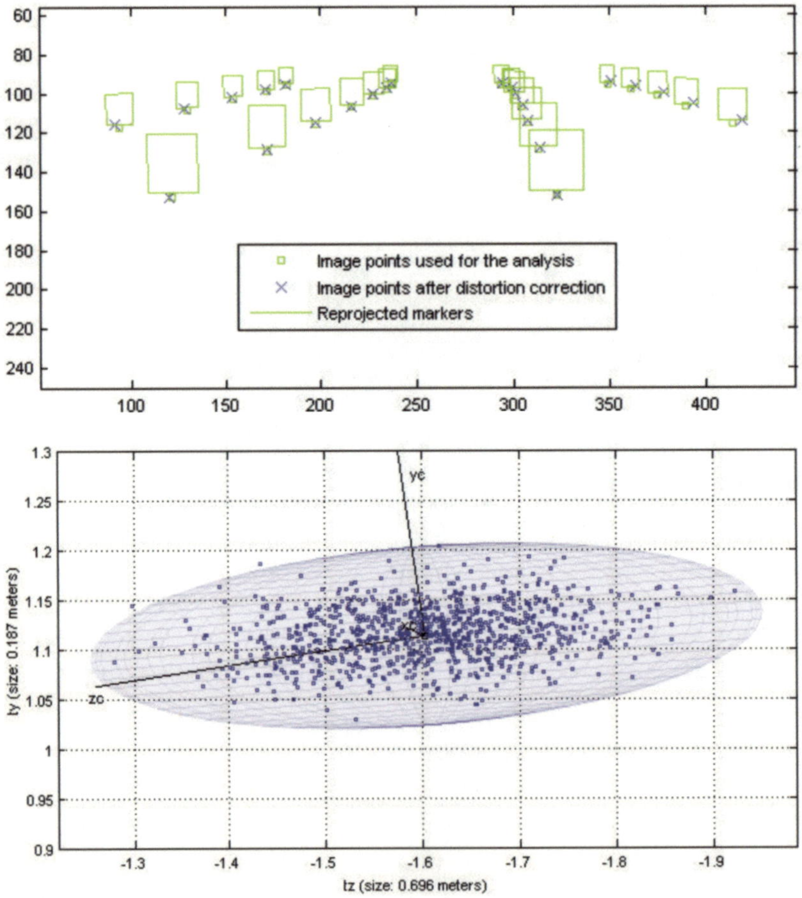

Fig. 10. The points to be reconstructed (triangulated) by Monte-Carlo simulations are the marker centers shifted vertically to the ground surface (top). 99% covariance ellipsoid of the left camera's position and verification of the generated noise in parameter space, 1000 experiments (bottom). 99.4% of the points fell inside the ellipsoid in this realization.

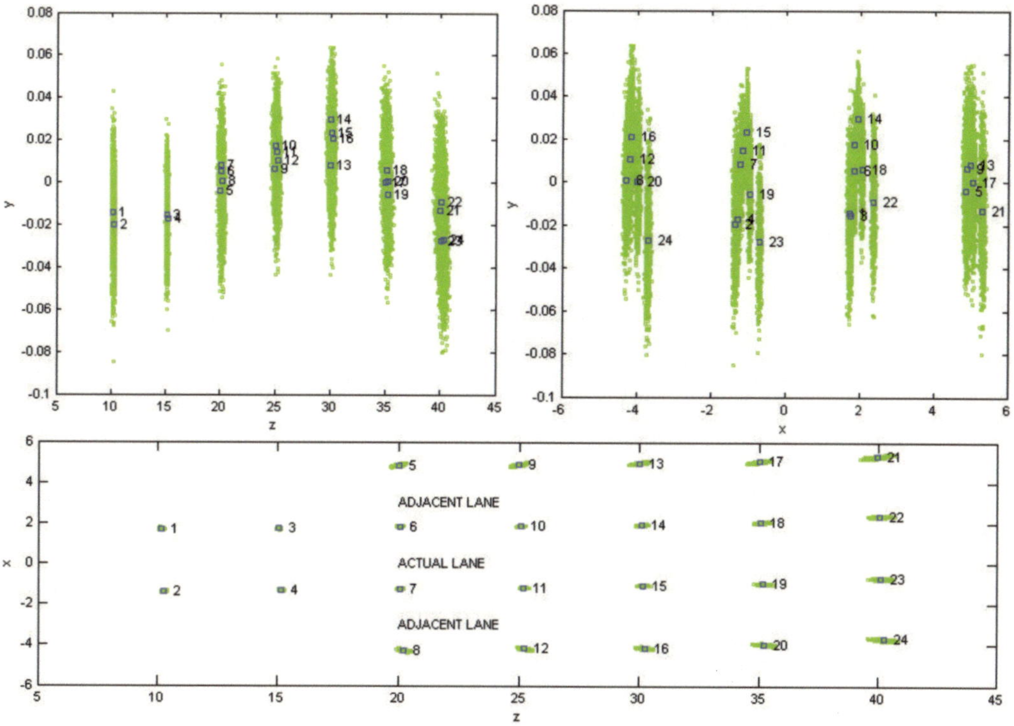

Fig. 11. Different views of the reconstructed 3D point clouds resulted from the Monte-Carlo simulations. YZ (top-left), XY (top-right) and XZ (bottom) view, where X is the lateral, Y is the height and Z is the longitudinal (depth) coordinate in meters. 1000 experiments were carried out.

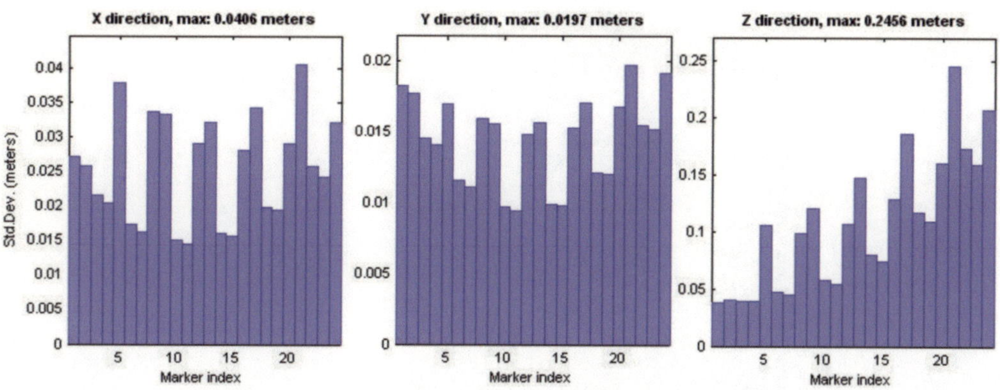

Fig. 12. Standard deviation (in meters) of the point clouds' sizes for each of the 24 reference points when only the optimal camera parameters are perturbed.

The largest deviation is 25 cm in the longitudinal (Z) direction at a distance of 40 meters from the car. This means that the "extent" (the 99% confidence interval) of the corresponding 3D point cloud is ±64 cm. In the direction perpendicular to the ground, the 99% confidence interval is ±5 cm and in the lateral direction, it is ±10 cm. We should emphasize that this only characterizes reconstruction errors due to uncertainty present in the camera parameters and not due to errors in stereo matching.

In order to simulate the effects of a random error (but not outliers) present at the stereo matching of point features, as well, random 2D Gaussian noise has been added to the image point locations in each experiment. We repeated the whole experiment with different standard deviations that ranged from 0 to 0.66 pixels in 0.11 pixels steps. In the case of a 2D Gaussian distribution, 0.33 pixels correspond to a 99% confidence interval of ±1 pixels while 0.66 corresponds to ±2 pixels. This is the simulated precision of the feature localization and stereo matching solution. The resulted point reconstruction errors are shown in Figure 13.

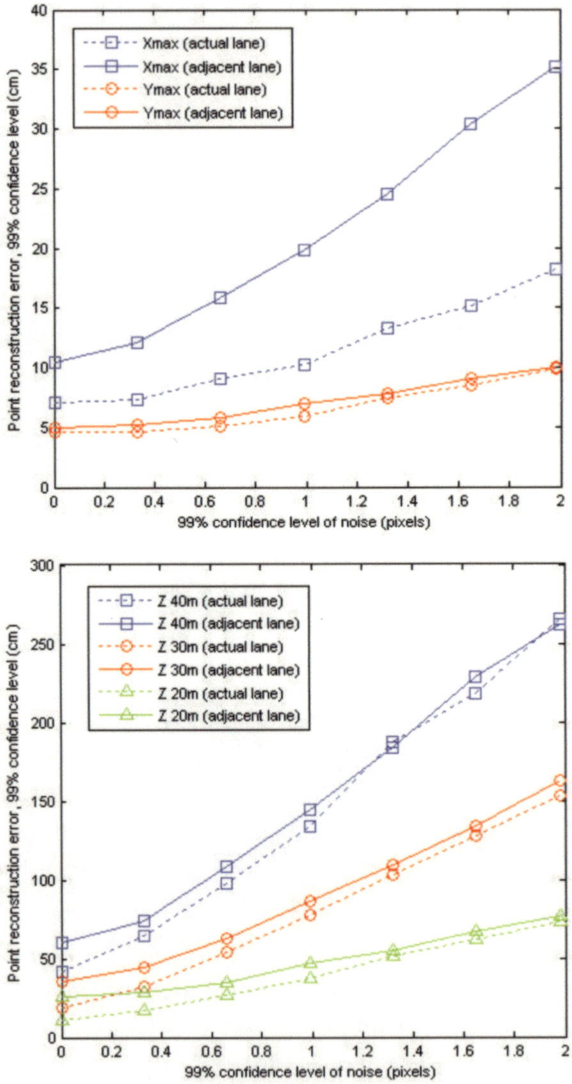

Fig. 13. Maximum reconstruction errors in the X and Y (top) and Z (bottom) directions vs. the simulated point feature localization and stereo matching errors in the image while the optimal camera parameters are perturbed corresponding to $\mathbf{C}_p$.

In Figure 13, both 3D and 2D errors are given as the 99% confidence interval of the corresponding distribution. It can be seen that the effects of the localization and feature matching noise in the image starts to dominate the uncertainty present in the camera parameters from ±0.5 pixels (Z plot). As we expected, the depth coordinate is the most sensitive one.

It should be noted that although parameter uncertainty is simulated as a random noise in order to measure the uncertainty of the parameter estimates with respect to the true parameters, the error of a single realized calibration remains constant when the calibrated system is on-line. In contrast, feature localization is realized in every acquired frame. However, the random perturbation is still valid, since we are interested in the deviation of the reconstructed features from the true ones.

6.3 Uncertainty of the epipolar lines

Next, the sensitivity of the epipolar lines was analysed as described in Section 5.3. The epipolar line uncertainty was characterized by the deviation of the line's angle with respect to its mean value. To every pixel center in the left image, the angle deviation of the corresponding epipolar line in the right image is associated. The resulting surface is shown in Figure 14.

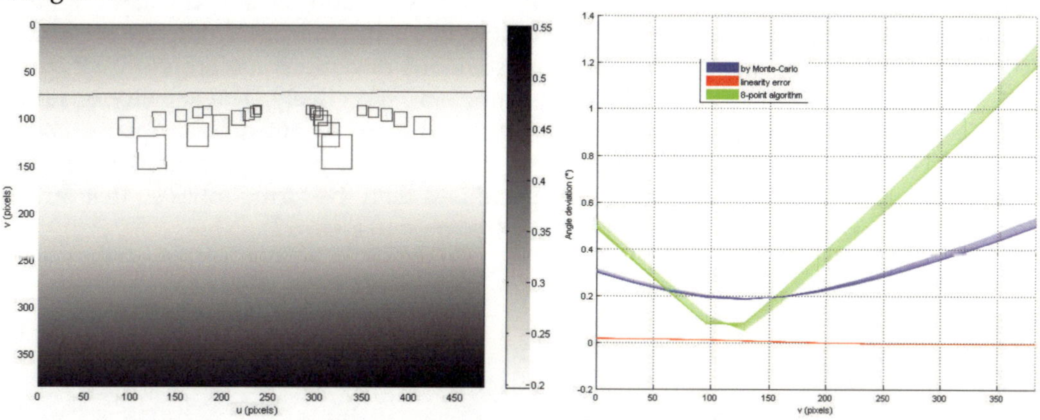

Fig. 14. Epipolar line uncertainties. Every pixel in the left image has an associated epipolar line uncertainty. The uncertainty is encoded in gray level values over the pixels of the left image (left side), while the side view of the resulting surface is compared with that of the eight-point algorithm (right side).

The angle deviation was not only computed by Monte-Carlo simulations but also with the linear covariance-propagation method detailed in Section 5.3. The difference between the two resulted surfaces is the linearity error surface that is also shown in the right side of Figure 14 (the side view of this error surface is a curve around zero degrees).

As a reference, we plotted the surface received from the eight-point algorithm used to determine the fundamental matrix. Since this method breaks down in the case of flat arrangements, we used the center and all the four corners of the markers in both images (five times more reference points than those used in the second step of the calibration procedure). We should also mention that the eight-point algorithm, or more generally the fundamental matrix, in itself does not suffice for the specific purpose because - as discussed in Section 3 - it does not provide Euclidean information about the camera poses with respect to the scene.

The uncertainty in the angle of epipolar lines does not significantly depend on the horizontal coordinate of the corresponding point (the epipolar lines are almost all horizontal). Although the objective function during optimization was not the uncertainty in the epipolar lines itself, it is clear from Figure 14 that this is minimal in the interesting zone. This is due to the specific arrangement (the marker locations and the horizon are overlayed for this purpose). The minimum of the angle deviation is around 0.2° and the maximum is 0.5°.

6.4 The sensitivity of road surface detection and lane model fitting

In order to see how the uncertainty in the camera parameters affect the quality of road surface reconstruction, we have used the Monte-Carlo technique, once again. In each of the 100 experiments, the feature (primitive or point chain) matching, the stereo point reconstruction and the road surface model were recomputed. The computations were performed for each frame over a 50 frames sequence, which corresponds to 2 seconds in real-time. The computed optimal camera parameters were perturbed corresponding to the estimated parameter covariance $\mathbf{C}_p$. The pitch angle, roll angle and height parameters resulted from the road surface model fitting are shown in Figure 15. Although there are some outliers (e.g. at frame indices 4, 10 and 49, that may correspond to surfaces with relatively high residual errors), the sensitivity of the estimation remains constant. In other words, the LS-fitting, in itself, is not very reliable in all circumstances, but the sensitivity estimation still remains stable over time. The standard deviations are around 0.11°, 0.36° and 5.4 cm, for the pitch, roll and height parameters respectively. The stability of fitting could be increased by using a robust fitting method or a weighted least-squares (WLS) method by giving more weight to the farther reconstructed points or primitives. This is because much more points constitue closer primitives than the farther ones so that farther points are not really involved in model shaping (we refer to Figure 5).

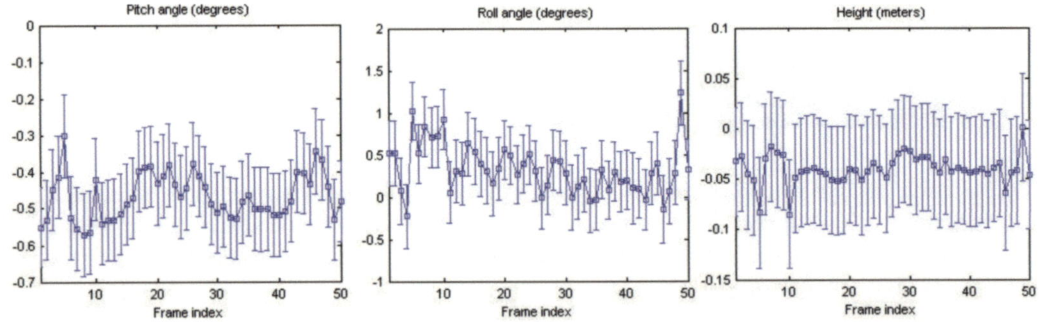

Fig. 15. Uncertainties in road surface model fitting due to uncertainties present in the camera parameters. 100 experiments in each of the 50 successive frames are evaluated. The curves represent mean values and the bars represent the standard deviations. There are outiers in model fitting but the computed sensitivities remain stable over time.

Finally, Figure 16 shows the effects of the computed camera parameter uncertainties on lane geometry reconstruction demonstrated on the frames already shown in Figure 6. Figure 16 demonstrates that the proposed off-line calibration method together with the discussed stereo lane reconstruction method gives acceptable lane reconstruction accuracy, but in the meantime, the derived errors are not insignificant, and thus, they can not be neglected, even if special care has been taken at calibration.

7. Conlusions

A novel off-line static method has been proposed for calibrating the cameras of a stereo vision-based driver assistance system. We formulated the maximum likelihood cost function for the stereo calibration problem. The resulting method involves the optimization of the 3D marker locations and covariance information of the measurements. Therefore, the method is only applicable, when an appropriate preliminary estimation of the uncertainties of the calibration measurements can be given. Moreover, a method for estimating the sensitivity of the parameters has been presented. It has been shown on real data, that when measurement uncertainties are available, our approach co-minimizing errors in the image together with errors in 3-space gives significantly better results than one can achieve by using the common reprojection error minimization. Thus, we have put extra effort in estimating measurement uncertainties at calibration. A stereo lane reconstruction algorithm has also been presented and by Monte-Carlo simulations of a triangulation method, we have demonstrated how the computed parameter uncertainties affect the precision of 3D reconstruction. The estimated reconstruction errors can be used when defining the safety margins in a decision algorithm that may trigger an actuation in a critical situation (e.g. unexpected lane departure or collision). The study should draw attention to the reconstruction errors arising from the non-ideal nature of camera calibration which is increasingly important in safety-critical systems. Covariance information is also required when using a Kalman-filter for lane tracking.

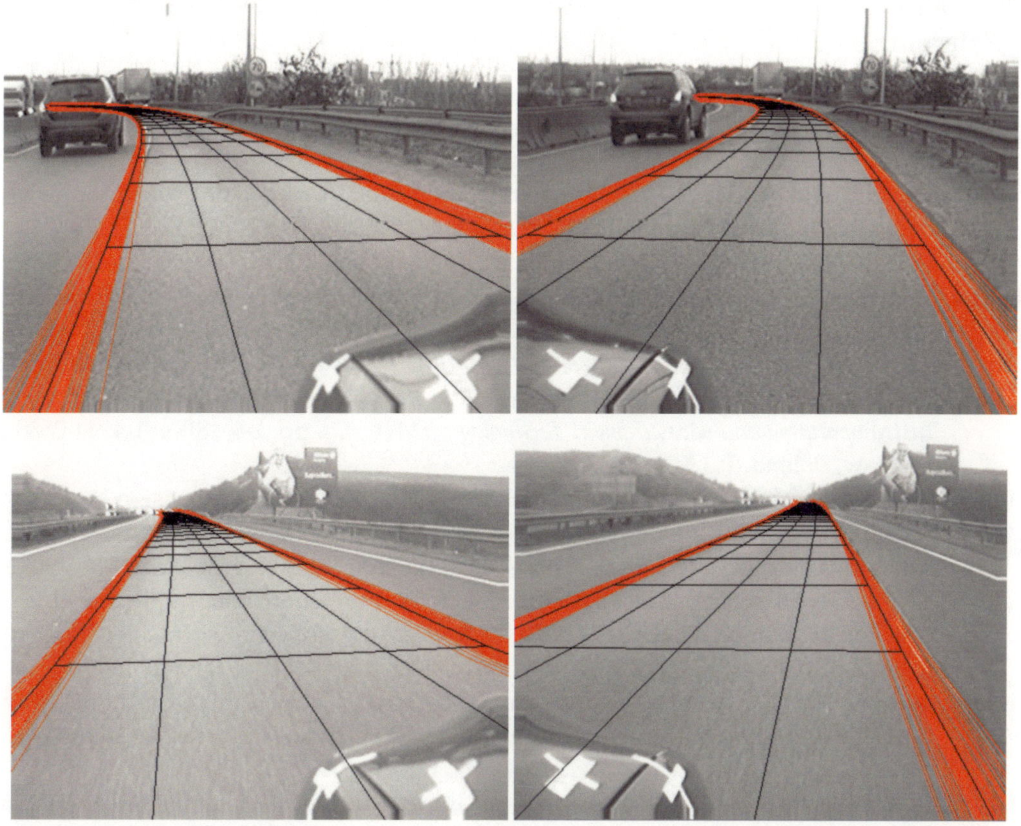

Fig. 16. Uncertainties in lane reconstruction due to uncertainties present in the camera parameters. 100 experiments are overlayed.

The procedure followed in the estimation of 3D point reconstruction uncertainties can be applied to estimate the output quality of a vehicle or obstacle detection algorithm. This is meant without a tracking algorithm that should decrease the errors by involving temporal information. It should be noted that, in general, tracking increases robustness, but the vision algorithm without tracking should still be reliable in itself, as well.
The proposed optimization method and far-range calibration arrangement of "ground thruth" control points is relatively elaborate compared to markerless on-line methods, while it is indispensable to integrate some kind of on-line parameter estimation - or at least parameter checking – in such systems. This is critical because the cameras are subject to shocks and vibrations and the parameters (mostly the extrinsic parameters) may change over time. Thus, the presented methods and results will serve as a reference to evaluate some on-line calibration methods that are presently developed.

8. Acknowledgements

This work was supported by the Hungarian National Office for Research and Technology through the project "Advanced Vehicles and Vehicle Control Knowledge Center" (OMFB-01418/2004) which is gratefully acknowledged.

9. References

Bellino, M.; Meneses, Y. L.; Kolski, S. & Jacot J. (2005). Calibration of an embedded camera for driver assistant systems, *Proceedings of the IEEE Intelligent Transportation Systems*, pp. 354-359, ISBN: 0-7803-9215-9, Vienna, Austria, 13-15 Sep 2005

Bertozzi, M. & Broggi, A. (1998). GOLD: A parallel real-time stereo vision system for generic obstacle and lane detection. *IEEE Transactions on Image Processing*, Vol. 7, No. 1 (Jan. 1998), 62-81, ISSN: 1057-7149

Bertozzi, M.; Broggi, A.; Cellario, M.; Fascioli, A.; Lombardi, P. & Porta, M. (2002). Articial vision in road vehicles. *Proc. of the IEEE*, Vol. 90, No. 7, 1258-1271, ISSN: 0018-9219

Bodis-Szomoru, A.; Daboczi, T.; Fazekas, Z. (2007). A far-range off-line camera calibration method for stereo lane detection systems, *Proceedings of the IEEE Conference on Instrumentation and Measurement Technology (IMTC'07)*, pp. 1-6, ISBN: 1-4244-0588-2, Warsaw, Poland, 1-3 May 2007

Broggi, A.; Bertozzi, M. & Fascioli, A. (2001). Self-calibration of a stereo vision system for automotive applications, *Proceedings of the IEEE International Conference on Robotics and Automation*, pp. 3698-3703, ISBN: 0-7803-6578-X, Seoul, Korea, 21-26 May, 2001

Eidehall, A. & Gustafsson, F. (2004). Combined road prediction and target tracking in collision avoidance, *Proceedings of the IEEE Intelligent Vehicles Symposium (IV'04)*, pp. 619-624, ISBN: 0-7803-8310-9, Parma, Italy, 14-17 June 2004

Fischler, A. & Bolles, R. C. (1981). Random sample consensus: A paradigm for model fitting with applications to image analysis and automated cartography, *Communications of the ACM*, vol. 24, no. 6 (June 1981), 381-395, ISSN: 0001-0782

Hartley, R. & Zisserman, A. (2006). *Multiple View Geometry in Computer Vision, Second Edition*, Cambridge University Press, ISBN: 0521-54051-8, Cambridge, United Kingdom

Heikkila, J. & Silvén O. (1997). A Four-step camera calibration procedure with implicit image correction, *Proceedings of the Conference on Computer Vision and Patter Recognition (CVPR'97)*, pp. 1106-1112, ISBN: 0-8186-7822-4, San Juan, Puerto Rico, 17-19 June 1997

Kastrinaki, V.; Zervakis, M. & Kalaitzakis, K. (2003). A survey of video processing techniques for traffic applications, *Image and Vision Computing*, Vol. 21, No. 4 (April 2003), 359-381, DOI: 10.1016/S0262-8856(03)00004-0

Kumar, R. & Hanson, A. R. (2004). Robust methods for estimating pose and a sensitivity analysis, *Computer Vision Graphics and Image Processing: Image Understanding*, Vol. 60, No. 3 (Nov. 1994), 313-342, ISSN: 1049-9660

Lepetit, V. & Fua, P. (2005). Monocular model-based 3D tracking of rigid objects: A survey, *Foundations and Trends in Computer Graphics and Vision*, Vol. 1, No. 1, 1-89, ISSN: 1572-2740

Malm, H. & Heyden, A. (2001). Stereo head calibration from a planar object, *Proceedings of the Conference on IEEE Computer Society*, pp. 657-662, ISBN: 0-7695-1272-0, Kauai, Hawaii, 8-14 December 2001

Malm, H. & Heyden, A. (2003). Simplified intrinsic camera calibration and hand-eye calibration for robot vision, *Proceedings of the Conference on IEEE Intelligent Robots and Systems*, pp. 1037-1043, ISBN: 0-7803-7860-0, Las Vegas, Nevada, USA, Oct. 2003

Marita, T.; Oniga, F.; Nedevschi, S.; Graf, T. & Schmidt, R. (2006). Camera calibration method for far range stereovision sensors used in vehicles, *Proceedings of the IEEE Intelligent Vehicles Symposium (IV'06)*, pp. 356-363, ISBN: 4-901122-86-X, Tokyo, Japan, 13-15 June 2006

Nedevschi, S.; Danescu, R.; Marita, T.; Oniga, F.; Pocol, C.; Sobol, S.; Graf, T. & Schmidt, R. (2005). Driving environment perception using stereovision, *Proceedings of the IEEE Intelligent Vehicles Symposium (IV'05)*, pp. 331-336, ISBN: 0-7803-8961-1, Las Vegas, Nevada, USA, 6-8 June 2005

Nedevschi, S.; Oniga, F.; Danescu, R.; Graf, T. & Schmidt, R. (2006). Increased accuracy stereo approach for 3D lane detection, *Proceedings of the IEEE Intelligent Vehicles Symposium (IV'06)*, pp. 42-49, ISBN: 4-901122-86-X, Tokyo, Japan, 13-15 June 2006

Sun, W. & Cooperstock, J.R. (2005). Requirements for camera calibration: Must accuracy come with a high price?, *Proceedings of the Seventh IEEE Workshop on Application of Computer Vision*, pp. 356-361, Breckeneidge, Colorado, USA, 5-7 Jan. 2005

Sturm, P. F. & Maybank, S. J. (1999). On plane-based camera calibration: A general algorithm, singularities, applications, *Proceedings of the Conference on Computer Vision and Patter Recognition (CVPR'99)*, pp. 1432-1437, ISBN: 0-7695-0149-4, Fort Collins, Colorado, USA, 23-25 June 1999

Tsai, R. Y. (1987). A versatile camera calibration technique for high-accuracy 3D machine vision metrology using off-the-shelf TV cameras and lenses, *IEEE Journal of Robotics and Automation*, Vol. 3, No. 4 (Aug. 1987), 323-344, ISSN: 0882-4967

Weber, J.; Koller, D.; Luong, Q.-T. & Malik, J. (1995). New results in stereo-based automatic vehicle guidance, *Proceedings of the Intelligent Vehicles Symposium (IV'95)*, pp. 530-535, ISBN: 0-7803-2983-X. 386, Detroit, Michigan, USA, 25-26 Sep. 1995

Zhang, Z. (2000). A flexible new technique for camera calibration, *IEEE Transactions on Pattern Analysis and Machine Intelligence*, Vol. 22, No. 11 (Nov. 2000), 1330–1334, ISSN: 0162-8828.

Zollner, H. & Sablatnig, R. (2004). Comparision of methods for geometric camera calibration using planar calibration targets, *Proceedings of the 28th Workshop of the Austrian Association for Pattern Recognition*, pp. 237-244, Hagenberg, Austria, 2004

2

Stereo Correspondence Estimation based on Wavelets and Multiwavelets Analysis

Asim Bhatti and Saeid Nahavandi
Intelligent Systems Research Lab., Deakin University
Australia

1. Introduction

Stereo vision has long been studied and a number of techniques have been proposed to estimate the disparities and 3D depth of the objects in the scene. This is done by estimating the pixel correspondences of similar features in the stereo perspective views originated from the same 3D scene (O. Faugeras 2001; R. Hartley 2003). However, finding correct corresponding points is subject to a number of potential problems like occlusion, ambiguity, illuminative variations and radial distortions (D. Scharstein 2002). A number of algorithms have been proposed to address some of the aforementioned issues in stereo vision however it is still relatively an open problem.

Current research in stereo vision has attracted a lot of focus on multiresolution techniques, based on wavelets/multiwavelets scale-space representation and analysis, for correspondence estimation (S. Mallat 1993; He-Ping Pan 1996). However, very little work has been reported in this regard. The main advantage of these algorithms is their hierarchical nature that exhibit behaviour similar to iterative optimization algorithms. These algorithms generally operate on an image pyramid where results from coarser levels are used to constrain regional search at finer levels. The coarse-to-fine techniques adopted in these algorithms are considered to be a middle approach unlike other existing algorithms where correspondences are established based either purely on local search (L. Di Stefano 2004; Q`ıngxiong Yang 2006) or as a global cost-function optimization problem (D. Scharstein 1998; M. Lin 2003) with their respective shortcomings.

In this work, earlier contributions regarding the application of wavelet/multiwavelets in stereo vision is presented highlighting shortcomings, involved in those techniques such as translation and rotational variance (R. R. Coifman 1995; S. Mallat 1999) inherited in the discrete wavelet/multiwavelet transformation. Furthermore, correspondence estimation, directly using wavelet coefficients for aggregation costs, is very sensitive to noise and illuminative variation that generally exists in between the stereo perspective views.

In addition, a new novel and robust algorithm is presented (Asim Bhatti 2008) using the hierarchical correspondence estimation approach where wavelets/multiwavelets transform modulus maxima (*WTMM*) are considered as corresponding features, addressing the issue of translation invariance. *WTMM* defines translation invariant features with phases pointing normal to the surface. The proposed algorithm introduced a new comprehensive selection criterion called *strength of the candidate* (*SC*) unlike many existing algorithms where selection is solely based on different aggregation costs. The *SC* involves the contribution of probabilistic weighted normalized correlation, symbolic tagging and geometric topological refinement. Probabilistic weighting involves the contribution of more than one search spaces especially in the case of multi-wavelet based multi-resolution analysis. Symbolic tagging procedure helps to keep the track of different candidates to be an optimal candidate. Furthermore, geometric topological refinement addresses the problem of ambiguity due to geometric transformations and distortions that could exist between the perspective views. The geometric features used in the geometric refinement procedure are carefully chosen to be invariant through many geometric transformations such as affine, metric and projective (M. Pollefeys 2000).

The developed vision system based on the proposed algorithm is very simple and cost effective as it consists of only a stereo cameras pair and a simple fluorescent light. The developed system is capable of estimating surface depths within the tolerance of 0.5 mm. Moreover, the system is invariant to illuminative variations and orientation of the objects in the input image space, which makes the developed system highly robust. Due to its hardware simplicity and robustness, it can be implemented in different factory environments without a significant change in the setup.

2. Wavelets analysis in stereo vision: a background

Wavelet theory has been explored very little up to now in the context of stereo vision. Some work has been reported in applying wavelet theory for addressing correspondence problem. To the best of author's knowledge, Mallat (S. Mallat 1991; S. Mallat 1993) was the first who used wavelet theory concept for image matching by using the *zero-crossings* of the wavelet transform to seek correspondence in image pairs. In (S. Mallat 1993) Mallat also explored the signal decomposition into linear waveforms and signal energy distribution in time-frequency plane. Afterwards, Unser (M. Unser 1993) used the concept of multi-resolution (*coarse to fine*) for image pattern registration using orthogonal wavelet pyramids with *spline bases*. Olive-Deubler-Boulin (J. C. Olive 1994) introduced a block matching method using orthogonal wavelet transform whereas (X. Zhou 1994) performed image matching using orthogonal Haar wavelet bases. Haar wavelet bases are one of the first and simplest wavelet basis and posses very basic properties in terms of smoothness, approximation order (A. Haar 1910), therefore are not well adapted for most of the imaging applications, especially correspondence estimation problem.

A more comprehensive use of wavelet theory based multi-resolution analysis for image matching was done by He-Pan in 1996 (He-Ping Pan 1996; He-Ping Pan 1996). He took the application of wavelet theory a bit further by introducing a complete stereo image matching algorithm using complex wavelet basis. In (He-Ping Pan 1996) he explored many different properties of wavelet basis that can be well suited and adaptive to the stereo matching problem. A number of real and complex wavelet bases were used and transform is performed using wavelet pyramid, commonly known by the name *Mallat's dyadic wavelet filter tree* (*MDWFT*) (S. Mallat 1999). The common problem with *MDWFT* is the lack of translation and rotation variance (R. R. Coifman 1995; I. Cohen 1998) especially in real valued wavelet basis. Furthermore similarity measures were applied on individual wavelet coefficients which are very sensitive to noise. Similarly, Magarey (J. Magarey 1998; J. Margary 1998) introduced algorithms for motion estimation and image matching, respectively, using complex discrete *Gabor-like* quadrature mirror filters. Afterwards, Shi

(Fangmin Shi 2001) applied *sum of squared difference* (*SSD*) technique on wavelet coefficients. He uses translation invariant wavelet transformation for matching purposes, which is a step forward in the context of stereo vision and applications of wavelet.
More to the wavelet theory, multi-wavelet theory evolved (G. Plonka 1998) in early 1990s from wavelet theory and enhanced for more than a decade. Their success, over scalar wavelet bases, stems from the fact that they can simultaneously posses the good properties of orthogonality, symmetry, high approximation order and short support, which is not possible in the scalar case (Özkaramanli H. 2001; A. Bhatti 2002). Being a new theoretical evolution, multi-wavelets are still new and are not yet applied in many applications. In this work we will devise a new and generalized correspondence estimation technique based wavelets and multiwavelets analysis to provide a framework for further research in this particular context.

3. Wavelet and multiwavelets fundamentals

Before proceeding to the correspondence estimation algorithms it seems wise to provide brief background of wavelets and multiwavelets theory as the algorithm presented is based heavily on this theory. Furthermore this background will assist the user to understand the features that are used to establish correspondences between the stereo pair of images.
Classical wavelet theory is based on the dilation equations as

$$\phi(t) = \sum_h c_h \, \phi(Mt - h) \tag{1}$$

$$\psi(t) = \sum_h w_h \, \phi(Mt - h) \tag{2}$$

where c_h and w_h represents the scaling and wavelet coefficients whereas M represents the band of filter bank (A. Bhatti 2002). In addition, multiresolution can be generated not just in the scalar context, i.e. with just one scaling function and one wavelet, but also in the vector case where there is more than one scaling function and wavelet are involved. The latter case leads to the notion of multiwavelets. Multiwavelets bases are characterized by r scaling functions and r wavelets in contrast with one scaling function and one wavelet, i.e. $r = 1$. Here r denotes the multiplicity in the vector setting with $r > 1$.
In the case of multiwavelets, scaling functions satisfy the matrix dilation equation as

$$\Phi(t) = \sum_h C_h \Phi(Mt - h) \tag{3}$$

Similarly for the multiwavelets the matrix dilation equation can be expressed as

$$\Psi(t) = \sum_h W_h \Phi(Mt - h) \tag{4}$$

In equations (3) and (4), c_h and w_h are real matrices of multi-filter coefficients whereas $\Phi(t)$ and $\Psi(t)$ can be expresses in terms of r scaling functions and r wavelets as

$$\mathbf{\Phi}(t) = \begin{bmatrix} \phi_0(t) \\ \phi_1(t) \\ \vdots \\ \phi_{r-1}(t) \end{bmatrix} \tag{5}$$

And

$$\psi(t) = \begin{bmatrix} \psi_0(t) \\ \psi_1(t) \\ \vdots \\ \psi_{r-1}(t) \end{bmatrix} \quad (6)$$

Generally only two band multiwavelets, i.e. $M = 2$, defining equal number of wavelets as scaling functions are used for simplicity. For further information about the generation and applications of multiwavelets, with desired approximation order and orthogonality, interested readers are referred to (S. Mallat 1999; A. Bhatti 2002).

3.1 Wavelet filter banks

Wavelet transformation produces scale-space representation of the input signal by generating scaled version of the approximation space and the detail space possessing the property

$$A_{s-1} = A_s \bigoplus D_s \quad (7)$$

where A_s and D_s represents approximation and detail space at lower resolution/scale and by direct sum constitutes the higher scale space A_{s-1}. In other words A_s and D_s are the subspaces of A_{s-1}. Expression (7) can be better visualized by Figure 1.

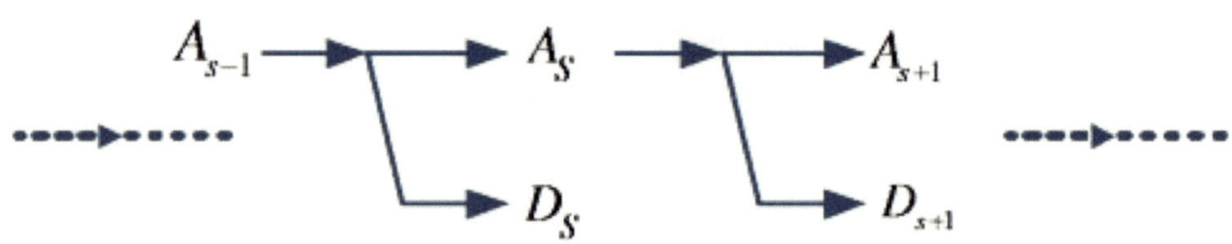

Fig. 1. wavelet theory based Multiresolution analysis

The use of Mallat's dyadic filter-bank (S. Mallat 1999) results in three different detail space components that are the horizontal, vertical and diagonal. Figure 2 can best visualize the graphical representation of the used filter-bank, where and W represents the low-pass and high-pass filters consisting of the scaling functions and wavelets coefficients, respectively.

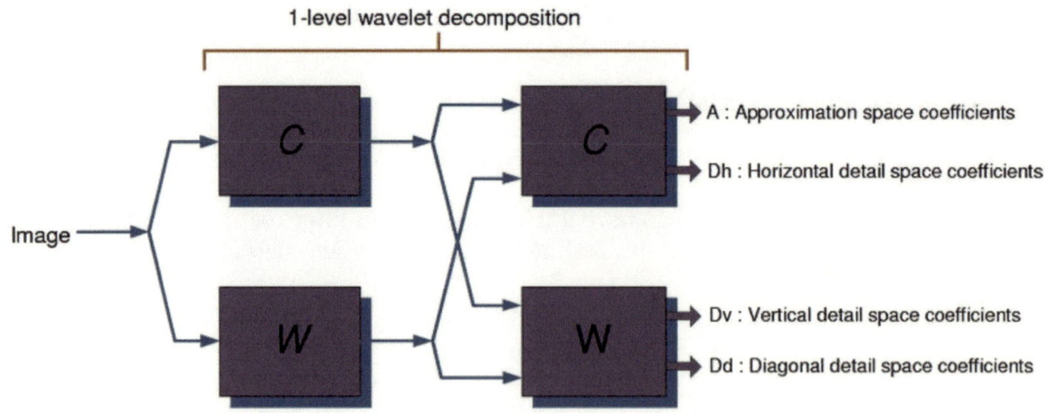

Fig. 2. Mallat's dyadic wavelet filter bank

3.2 Wavelet transform modulus

The *Wavelet Transform Modulus* (*WTM*), in general vector representation, can be expressed as

$$WTM_{s,k} = W_{s,k} \angle \Theta_{W_{s,k}} \tag{8}$$

Where $W_{s,k}$ is

$$W_{s,k} = \sqrt{|D_{h,s,k}|^2 + |D_{v,s,k}|^2} \tag{9}$$

where $D_{h,s,k}$ and $D_{v,s,k}$ are the kth horizontal and vertical detail components at scale s. Furthermore $\Theta_{W_{s,k}}$ can be expressed as

$$\Theta_{W,s,k} = \begin{cases} \alpha(s,k) & if \quad D_{h,s,k} > 0 \\ \pi - \alpha(s,k) & if \quad D_{h,s,k} < 0 \end{cases} \tag{10}$$

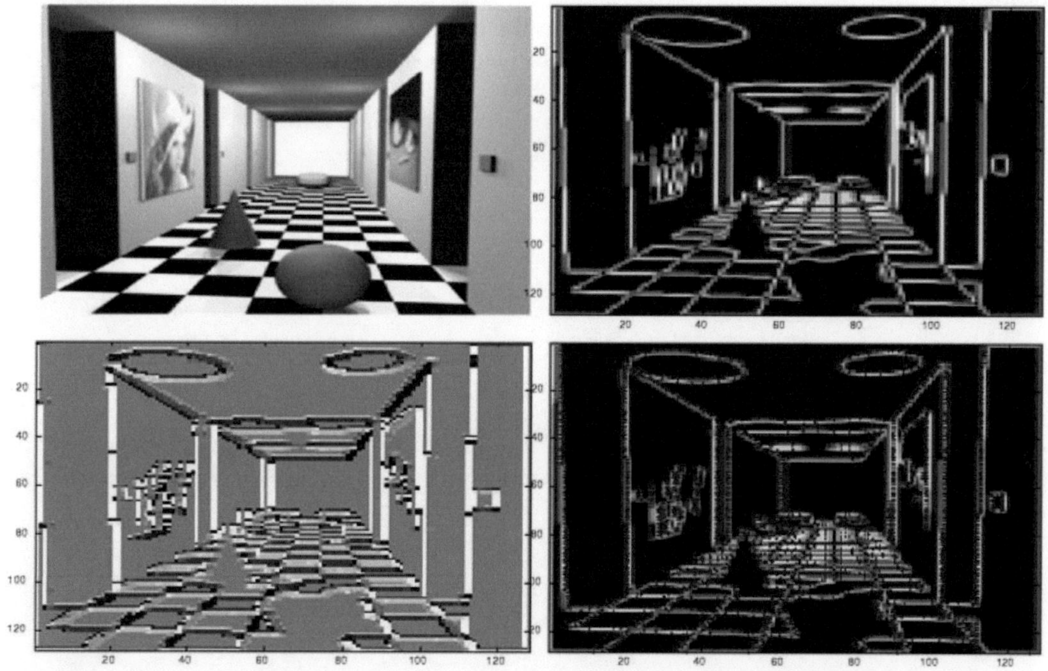

Fig. 3. Top Left: Original image, Top Right: Wavelet Transform Modulus, Bottom Left: wavelet transform modulus phase, Bottom Right: Wavelet Transform Modulus Maxima with Phase vectors

Where

$$\alpha(s,k) = tan^{-1}\left({D_{v,s,k}} / {D_{h,s,k}} \right) \tag{11}$$

The vector $\vec{n}(k)$ points to the direction normal to the edge surface as

$$\vec{n}(s,k) = \left[cos(\Theta_{W,s,k}), sin(\Theta_{W,s,k})\right] \quad (12)$$

An edge point is the point p at some scale s such that $WT_{s,k}$ is locally maxima at $k = p$ and $k = p + \varepsilon\vec{n}(k)$ for ε small enough. These points are known as *wavelet transform modulus maxima* (*WTMM*), and are shift invariant through the wavelet transform. For further details in reference to wavelet modulus maxima and its translation invariance, reader is kindly referred to (S. Mallat 1999).

4. Correspondence estimation

The matching process of the proposed algorithm is categorized into two major parts. The first part of the algorithm defines the correspondence estimation process only at the coarsest scale level, whereas the second part defines the iterative matching process from finer up to the finest scale level. Correspondence estimation at the coarsest scale is the most important part of the proposed algorithm as the algorithm uses the hierarchical approach for correspondence estimation. Therefore, the part of the algorithm related to the correspondence estimation at finer scale levels is very much dependent on the outcomes of coarsest level matching. Finer level matching involves the local search at the locations where any predecessor candidates have already been selected, in the coarsest level. A block diagram, as shown in Figure 4, presents a detailed visual representation of the correspondence estimation algorithm.

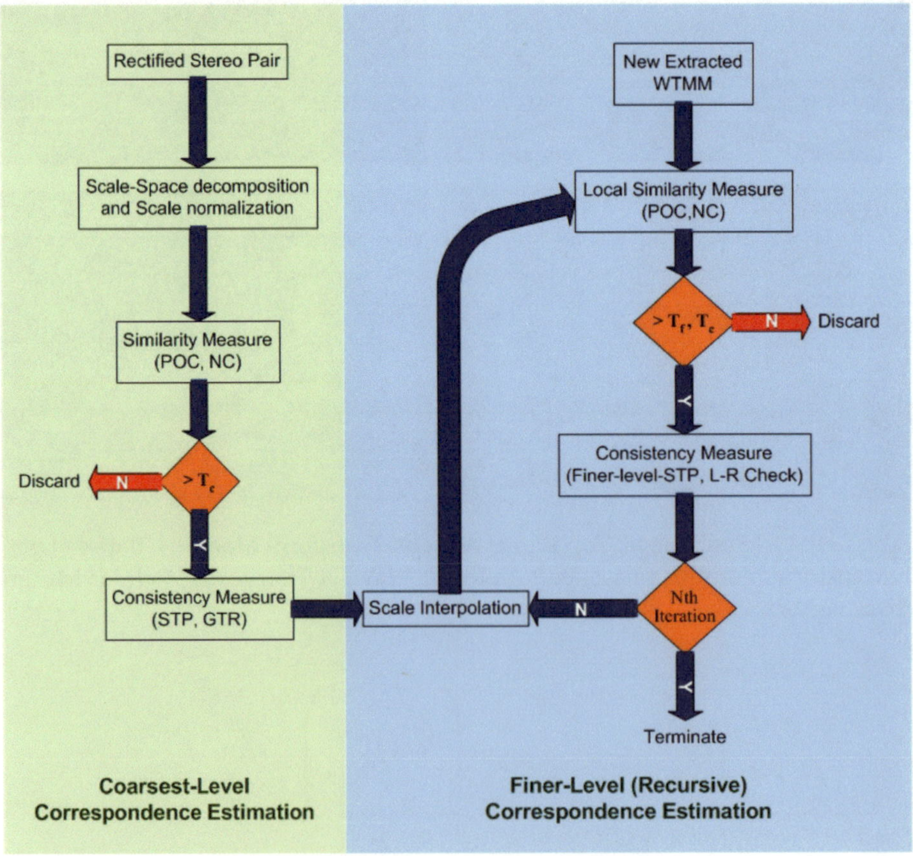

Fig. 4. Block diagram of the correspondence estimation algorithm

4.1 Coarsest-level correspondence estimation

Coarsest level matching (*CLM*) is very important and crucial step of the whole matching process as correspondence estimation at finer levels are very much dependent on the outcome of *CLM*. All matching candidates at finer levels are arranged according to the matched locations found at the coarsest level. Considering the significance of (*CLM*) in the overall matching process there is a great need of keeping this process error free as much as possible. For this purpose, a comprehensive check is performed to exploit the likelihood of each candidate to be a credible match, before accepting or discarding it.

The matching process starts with wavelet decomposition up to level *N*, usually taken within the range of [4-5] depending on the size of the image. Before proceeding to the similarity measure block, wavelet scale normalization (*WSN*) is performed along with normalized correlation measure that helps to minimize the effect of illuminative variation that could exist in between the perspective views. The reason this comprehensiveness normalization is the nature of the application that we are trying to address in this work. The objects that we are concerned with, for the depth estimation, are aluminium die-castings with highly shiny and reflective surfaces. Therefore, there is a great need for the illuminative variation compensation before proceeding to the main correspondence estimation block.

The *WSN* is performed on each level of wavelet transform decomposition. It is done by dividing the details space with the approximation space and can be defined as

$$NW_{s,k} = {D_{dc,s,k}}/{|A_{s,k}|} \qquad \forall\, dc \in \{h, v, d\} \tag{13}$$

Where $\{h,v,d\}$ represents horizontal, vertical and diagonal detail components, respectively, s represents the scale of decomposition and A represents the approximation space.

4.1.1 Similarity measure

After the extraction of *wavelet transform modulus maxima* (*WTMM*) correlation based similarity measure is performed to obtain an initial estimation of the disparity map. A multiwindow approach (A. Fusiello 2000) is used to enhance the performance of correlation based similarity measure. The details about the improvements from single window to multiwindow approach can be found in (A. Fusiello 2000). The multi-window correlation score can be defined as

$$NC(s,k) = \overline{NC_{s,k,W_0} + \sum_{j=1}^{\frac{n_W}{2}} NC_{s,k,W_j}} \tag{14}$$

where NC_{s,k,W_0} represents the *normalized correlation* (*NC*) with respect to the central window whereas NC_{s,k,W_j} defines the *NC* respect to jth surrounding windows with n_W number of surrounding windows. In (14), the second term represents the summation of the best ${n_w}/{2}$ windows out of n_W. An average of the correlation scores from these windows is considered to keep the score normalized within the range of [*0-1*].

4.1.2 Probabilistic weighting

To make the correlation based selection criteria more comprehensive, a probabilistic weighting for the correlation measure in (14), is introduced. It is the probability of selection of a point, let say ith point, from within each search space, as a candidate C_i and out of r^2 search spaces as

$$P_c(C_i) = {}^{n_{C_i}}/_{r^2} \qquad where \qquad 1 \leq n_{C_i} \leq r^2 \tag{15}$$

where n_{C_i} is the number of times a candidate C_i is selected and r is the multiplicity of multifilter coefficients. As all matching candidates have equal probability of being selected, the probability of occurrence of any candidate through one search space is ${}^1/_{r^2}$. It is obvious from expression (15) that the $P_c(C_i)$ lies between the range of $\left[{}^1/_{r^2} - 1\right]$. We would like to call that probability term, *probability of occurrence* (POC) as it is the probability of any candidates C_i to appear n_{C_i} times in the selection out of r^2 search spaces. More specifically, if jth candidate C_j is selected ${}^{r^2}/_2$ times out of r^2 search spaces then POC of jth candidate is 0.5, i.e. $P_c(C_j)$=1/2. The correlation score in expression (15) is then weighted with POC as

$$CS_{C_i} = P_c(C_i) \sum_{n_{C_i}} NC_{C_i}(x,d) \qquad \forall_{n_{C_i}} \in \mathbb{Z}: n_{C_i} \leq r^2 \tag{16}$$

The probabilistic weighted correlation score CS_{C_i}, in (16), can be defined as *candidate strength* (*CS*). It represents the potential of the candidate to be considered for further processing and the involvement of the specific candidate in the selection of other potential candidates.

4.1.3 Symbolic tagging

Filtration of candidates, based on the *CS*, is followed by symbolic tagging procedure, which divides the candidates into three different pools based on three thresholds T_c, T_{c1} and T_{c2} possessing the criterion $T_{c2} > T_{c1} > T_c$. The threshold T_c acts as a rejection filter which filters out any candidate possessing lower *CS* than T_c. The rest of the candidates are divided into three pools as

$$\begin{array}{llll} NC_{s,k} \geq T_{c_1}, & and & P_c(C_i) = 1, & \Rightarrow Op \\ NC_{s,k} \geq T_{c_1}, & and & 0.5 \leq P_c(C_i) < 1, & \Rightarrow Cd \\ NC_{s,k} \geq T_{c_2}, & and & {}^2/_{r^2} \leq P_c(C_i) < 0.5, & \Rightarrow Cr \end{array} \tag{17}$$

It can be seen from the first expression in (17), there is no ambiguity for the matches with tag *Op* as the *POC* is 1, whereas ambiguity does exist for the matches with tags *Cd* and *Cr*. Ambiguity is the phenomenon where there exists more than one correspondences for a single point in the reference image (R. Hartley 2003).

4.1.4 Geometric refinement

To address the issue of ambiguity, a simple geometric topological refinement is introduced in order to extract the optimal candidate matches out of the pool of ambiguous candidate matches. For that purpose, the geometric orientation of the ambiguous points with reference to Op from (17) is checked and the pairs having the closest geometric topology with respect to the Op are selected as optimal candidates. Three geometric features that are *relative distance difference* (*RDD*), *absolute distance difference* (*ADD*) and *relative slope difference* (*RSD*), are calculated to check the geometric orientation similarity. These geometric features are invariant through many geometric transformations, such as Projective, Affine, Metric and

Euclidean (M. Pollefeys 2000). The geometric measurement is then weighted with the *CS* of the candidates to keep the previous achievements of the candidates in consideration.
In order to calculate the geometric statistics a number of candidate pairs with tag *Op* are randomly selected. Let say n_r is the number of randomly selected pairs from n_{Op} candidate pairs possessing tag *Op*. Before proceeding to the calculation of *ADD* between the ambiguous pair of points we calculate average absolute distance *AAD* between selected pairs as

$$d_{Op_{n,i}} = \left\|Op_{1_i} - Op_{2_i}\right\|_{n_r \,:\, n_r \leq n_{Op}} \tag{18}$$

Where ||-|| defines the Euclidean distance between the pair of points with tags *Op* referring to image 1 and 2, respectively. The process in (18) is repeated *n* times to obtain *n* values of *AAD* in order to minimize the involvement of any wrong candidate pair that could have assigned the tag *Op*. Similarly for ambiguous candidate pairs with tag *Cd* the absolute distance can be calculated as

$$d_{Cd_j} = \left\|C_{Cd,1} - C_{Cd,2_j}\right\|_{m:\ j=1\cdots m} \tag{19}$$

Where m is the number of candidates $C_{Cd,2_i}$ selected from second image with potential to make a pair with $C_{Cd,1}$ in the first image. From (18) and (19) we can define ADD as

$$d_{AC_i} = \left|\frac{d_{Cd_i} - d_{Op_{n,i}}}{d_{Cd_i} + d_{Op_{n,i}}}\right|_n \tag{20}$$

Where d_{AC_i} is the *ADD* for *i*th candidate in the second image related to $C_{Cd,1}$ in the first image. Obviously we are interested in the candidate with minimum *ADD*. It is worth mentioning that absolute distances are invariant through Euclidean Transformation (R. Hartley 2003).

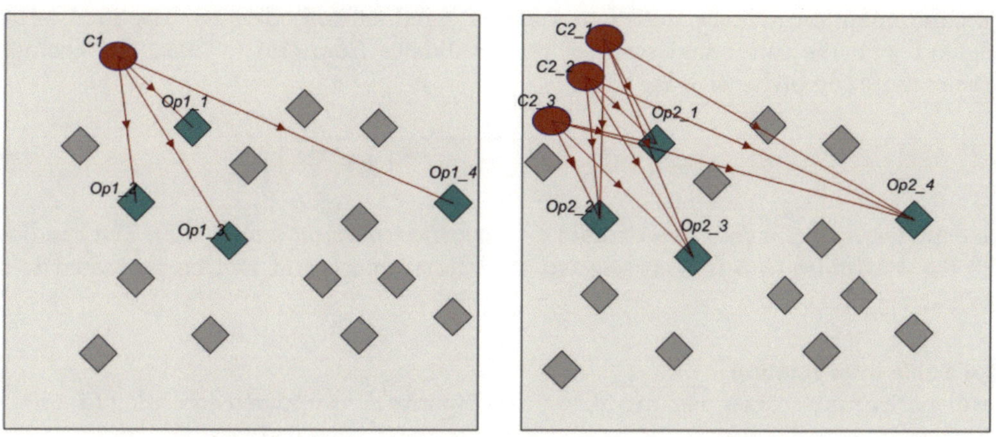

Image 1 Image 2

C1: Reference candidate
Op i _ j: Randomly selected candidates out of Op candidates' pool
C2_i: Potential candidate matches for C1
Pool of candidates with tag Op

Fig. 5. Geometric refinement procedure

Before proceeding to the definition of *RDD* it is worthwhile to visualize the geometri refinement procedure as in Figure 5. There the candidate C_1 in the first image pairs with three potential candidates C_{2i} in the second image. The pairs with tag *Op*, shown by the gray colour, are spread all over the image and act as reference points in addressing the problem of ambiguity. The points with green colour are randomly selected points out of the pool of reference points with tag *Op*. Similarly, *RDD* can be defined by the following expression

$$d_{R_{C_i}} = min_j \left(\frac{d_{R_{C_{1,i}}} - d_{R_{C_{2,i,j}}}}{d_{R_{C_{1,i}}} + d_{R_{C_{2,i,j}}}} \right)_n \quad (21)$$

Where

$$d_{R_{C_{1,i}}} = \left\| C_1 - Op_{1,i} \right\|_{i \in n_{Op}} \quad where \quad i = 1 \cdots n_r \quad (22)$$

And

$$d_{R_{C_{2,i,j}}} = \left\| C_{2,j} - Op_{2,i} \right\|_{i \in n_{Op}, j \in m} \quad where \quad i = 1 \cdots n_r,\ j = 1 \cdots m \quad (23)$$

Similar to *ADD*, *RDD* is also calculated *n* times to minimize the effect of any wrongly chosen point with *Op* tag. Finally to calculate the relative slope difference we need to define relative slope for both images and between candidate points and the reference points. Thus, *RSD* can be defined as

$$d_{S_{C_i}} = min_j \left(\left| \frac{S_{C_{1,i}} - S_{C_{2,i,j}}}{S_{C_{1,i}} + S_{C_{2,i,j}}} \right|_n \right) \quad (24)$$

The term (–)$_n$ defines the average over *n* repetitions, where *n* is usually taken within the range of [3-5]. Using (20), (21) and (24) a general and common term, as a final measure, to select the optimal candidate out of m potential candidates, is defined. The final term is weighted with the correlation score of the candidates from (16) to make the geometric measure more comprehensive as

$$Gc_i = max_i \left(CS_{C_i} \left(e^{-d_{A_{C_i}}} + e^{-d_{R_{C_i}}} + e^{-d_{S_{C_i}}} \right) \right) \quad (25)$$

The expression in (25) could be defined as *geometric refinement score* (*GRS*). The candidate with the maximum *GRS* is then selected as optimal match and will be promoted to the symbolic tag of *Op*.

4.1.4 Scale interpolation

The disparity from coarser disparity d_c to finer disparity d_F is updated according to

$$d_F = d_L + 2d_c \quad (26)$$

Where d_L is the local disparity obtained within the current scale level. This process is repeated until the finest resolution is achieved which is the resolution of input image. An example of the outcome of coarsest level correspondence estimation can be seen in Figure 6.

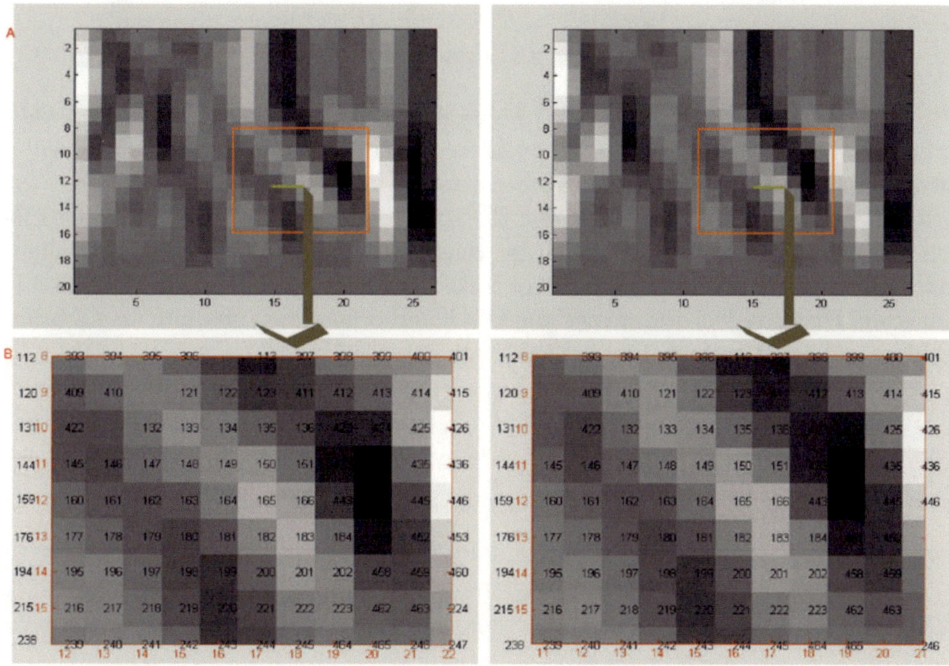

Fig. 6. An outcome of coarsest level matching

4.2 Finer-level correspondence estimation

Correspondence estimation at the finer level constitutes an iterative local search, based on the information extracted from the coarsest level. Due to the simpler nature of this search no geometric optimization is performed to deal with ambiguity but rather simpler approach is used, generally known as *left-right consistency* (*LRC*) check and can be defined as

$$-d_{s,k,1} = d_{s,k,2}\left((s,k)_x + d_{s,k,1}\left((s,k)_x,(s,k)_y\right),(s,k)_y\right) \tag{27}$$

Where $d_{k,i}$ is the estimated discrete disparity of kth coefficient in ith image, whereas k_x and k_y are the x and y coordinates of the kth coefficient at scale s.

Similar to coarsest level search, interpolated coefficients are assigned candidate strength based on correlation scores and probability of occurrence as in (16), however the search is only defined over the relevant interpolated areas. Before proceeding further to the symbolic tagging procedure and *LRC* check, any coefficient with insignificant *CS* is discarded.

Symbolic tagging procedure is very similar to the one presented in section 4.1.3 however new assignation of tags depends on their ancestors' tags. In other words the coefficients that are interpolated from the coefficient, at the coarsest level, with tag *Op* will be dealt with different conditions than the one with tag *Cd*. The coefficients interpolated from *Op* are assigned tags as

$$\forall\, Op \Rightarrow \begin{cases} Op & if \quad P(C_k) \geq 0.5,\ \ NC(s,k) \geq T_{f_1} \\ Cd & if \quad P(C_k) \geq 0.2,\ \ NC(s,k) \geq T_{f_1} \end{cases} \tag{28}$$

Whereas the coefficients having predecessor with tag *Cd*, we have

$$\forall\, Cd \Rightarrow \begin{cases} Op & if \quad P(C_k) = 1,\ NC(s,k) \geq T_{f_1} \\ Cd & if \quad P(C_k) \geq 0.2,\ NC(s,k) \geq T_{f_1} \end{cases} \tag{29}$$

Where T_{f_1} is usually chosen within the range of [0.4 - 0.5]. Similar to the coarsest level matching, the coefficients with Op are considered as the reference locations that will assist in rearranging the *Cd* coefficients using the expression in (28).

After this step, some gaps still left in the disparity map which is required to be filled to achieve dense depth map. These gaps are due to coefficients that were not taken into account before due to the unavailability of linked ancestors and have just appeared in the current scale. These coefficients are assigned *Cd* if and only if their strength i.e. *CS* from (16), is greater than T_{c1} and perform best in *LRC* check provided in (27).

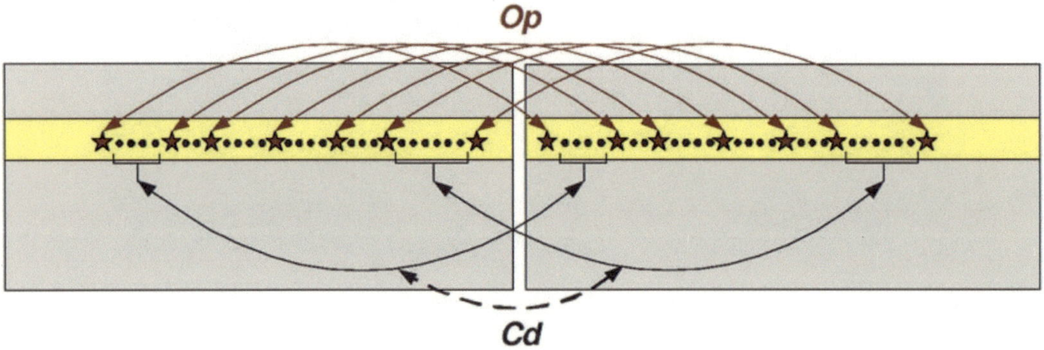

Fig. 7. Local search constellation relation between the images after symbolic tagging

The process of finer level correspondence estimation is repeated until the finest resolution is achieved, i.e. the resolution of the input image. Using a number of thresholds and symbols makes the appearance of the algorithm a little bit complicated and computationally expansive however comparing to existing algorithms it is not much different. Currently no explicit comparative information is extracted to support our claim but is intended for future works. No argument, there are many correlation based algorithms that are very fast due to low computational cost but dose not provide very promising qualitative performance. In addition, most of the algorithms, existing in the literature, perform post processing to coverup the deficiencies occurred during the correspondence estimation process which is itself is very computationally expansive. On the other hand proposed algorithm, due to the comprehensive criteria of selection/rejection, dose not require any post processing.

Furthermore, due to hierarchical nature, the disparity search is only $2^{level-1}$ of the original disparity search required at the input image level, that is, for a required search of 32 the proposed algorithm only required to search 4 disparities with decomposition of level 4.

5. Disparity estimation

The algorithm presented is exploited to its maximum capacity in terms of the stereo correspondence estimation performance. Four popular synthetic images are chosen from the database of the University of Middlebury. The relevant disparity maps are shown in Figure 8 to 9. In addition, error images are also calculated for each of the estimated disparity maps that simply are the absolute difference, defined in (30), between the ground truth and estimated disparity maps in terms of gray scale intensity values, as shown in subfigures 8(D) and 9(D). The absolute error can be expressed as

$$E = |d_G(x,y) - d_E(x,y)|_{\forall\, x \in X,\mathbb{Z},\ y \in Y,\mathbb{Z}} \tag{30}$$

Where $d_G(x,y)$ is the discrete ground truth disparity map, whereas $d_E(x,y)$ is the estimated one.

In order to find the statistical deviation of the estimated disparity maps from the provided ground truth disparity, two statistics are calculated as

$$\boldsymbol{R} = \sqrt{\frac{1}{N} \sum_{x,y} |\boldsymbol{d}_E(\boldsymbol{x},\boldsymbol{y}) - \boldsymbol{d}_G(\boldsymbol{x},\boldsymbol{y})|^2} \tag{31}$$

and

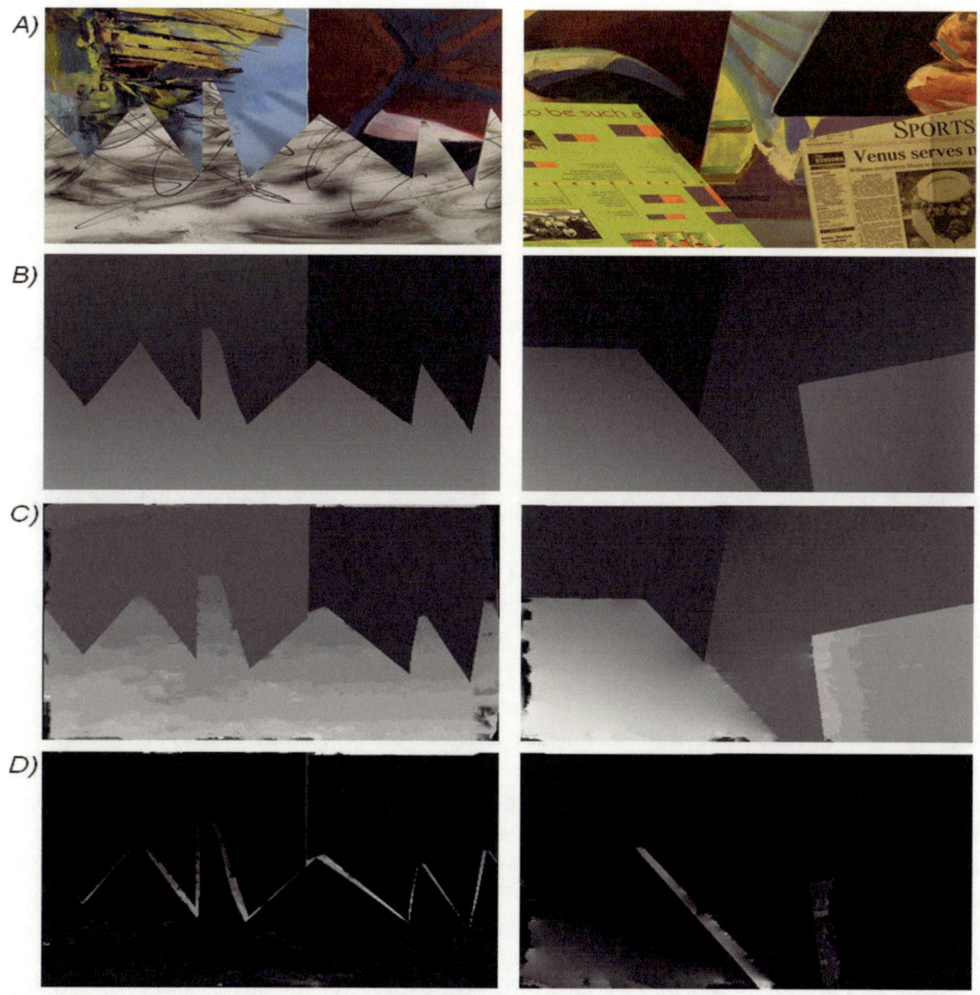

Fig. 8. A) Right images of the Sawtooth (left) and Venus (right) stereo pair, B) Ground truth disparity maps, C) Estimated disparity maps, D) Disparity error

Fig. 9. A) Right images of the *Cones* (left) and *Teddy* (right) stereo pair, B) Ground truth disparity maps, C) Estimated disparity maps, D) Disparity error

$$\boldsymbol{B} = \frac{1}{N}\sum_{x,y}|\boldsymbol{d_E}(\boldsymbol{x},\boldsymbol{y}) - \boldsymbol{d_G}(\boldsymbol{x},\boldsymbol{y})|^2 > \boldsymbol{\xi} \quad (32)$$

Where R and B represent the *Root Mean Squared Error* (*RMSE*) and *Percentage of Bad Disparities* (*PBD*), respectively. N represents the total number of pixels in the input image whereas ξ represents the acceptable deviation of the estimated disparity value from the ground truth and is fixed to 1 in this particular work.

The images are taken into consideration with different complexities, in terms of pixel intensity variation and surface boundaries. First pair of stereo images is shown in Figure 8 with related ground truth disparity maps, estimated disparity maps and the error between the ground truth and estimated disparity maps. As it can be seen in Figure 8 the edges of the discontinuities are extracted to high accuracy and estimated disparity is very much similar to the ground truth disparity, visually. The *RMSE* and *PBD* score for *Sawtooth* and *Venus* are R = 1.9885, B = 0.0262 and R = 0.1099, B = 0.0381, respectively.

Similarly, another pair of disparity maps are shown for images *Cones* and *Teddy* and related *RMSE* and *PBD* scores are $R = 3.3798$, $B = 0.1270$ and $R = 2.7629$, $B = 0.1115$, respectively, as shown in Figure 9.

6. Disparity estimation

To further validate the claims about the performance of the proposed algorithm a comparison is performed between the proposed algorithm and a number of selected algorithms from the literature. Eight algorithms are chosen, known for their performance, within the computer vision research community. These estimated disparity maps are related to the images *Cones, Venus* and *Teddy*. The chosen algorithms for comparison purpose are *Double-bp (Q`ıngxiong Yang 2006), Graph Cuts (D. Scharstein 2002), Infection (G. Olague 2005), Layered (L. Zitnick 2004), Scanline Optimization (D. Scharstein 1998), SSD min. Filter (D. Scharstein 2002)* and *Symmetric-Occlusion (J. Sun 2005).*
The estimated disparity map selected for comparison against the aforementioned algorithms from the literature is generated using MW2 (Özkaramanli H. Bhatti A. and 2002). These calculated statistics, i.e. R and B, for the analysis of comparative performance with respect to the estimated results are shown in Table 1 and Figures 10 to 12. It is obvious from Table 1 and Figures 10 to 12, the proposed algorithm has performed best in the case of *Venus* image. However, in case of *Cones* and *Teddy* images the proposed algorithm has ranked 3rd, though very competitive to the algorithm ranked 1. Specifically in case of B, the proposed algorithm has outperformed all other algorithms. This reflects the true consistency and robustness of the proposed algorithm as number of bad disparity values estimated are lowest in all cases. It also reflects the comprehensiveness of the selection criteria defined by consistency measure from expressions (16) and (25).

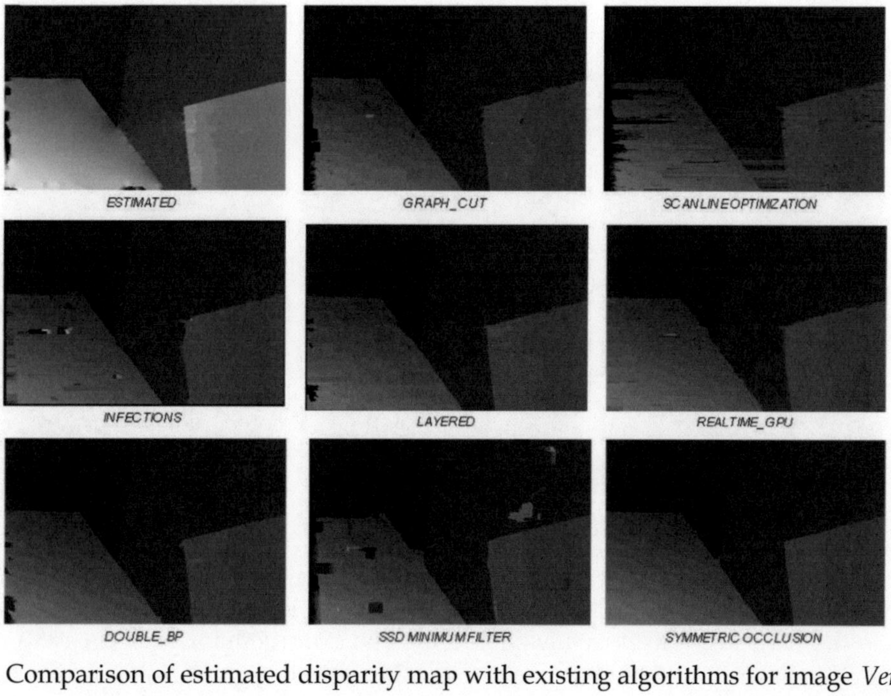

Fig. 10. Comparison of estimated disparity map with existing algorithms for image *Venus*

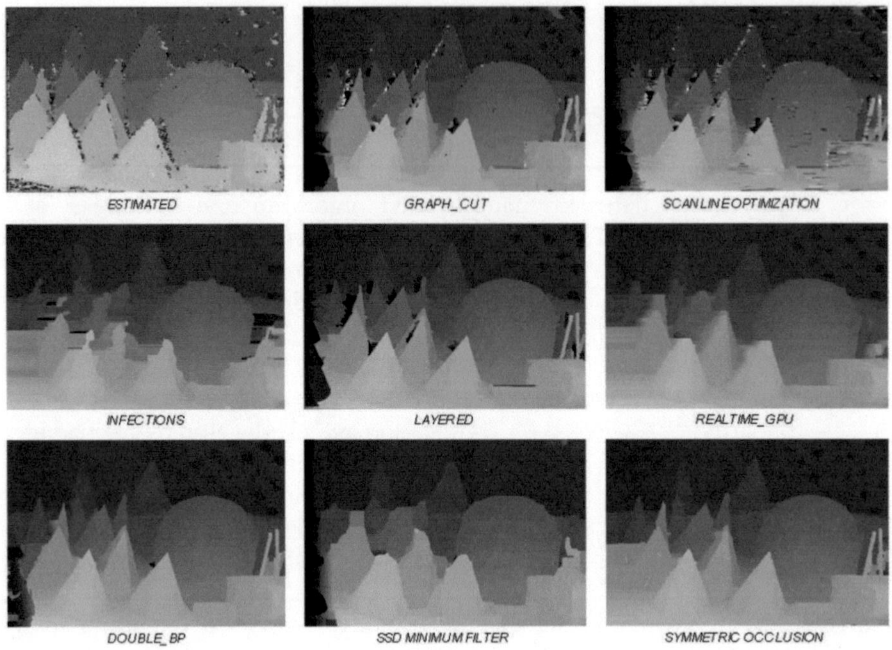

Fig. 11. Comparison of estimated disparity map with existing algorithms for image *Cones*

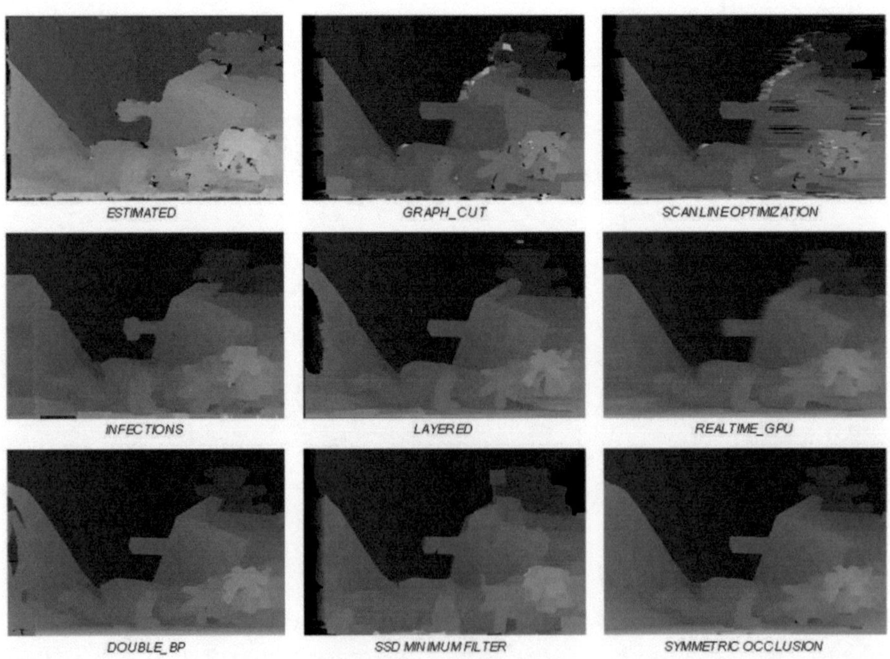

Fig. 12. Comparison of estimated disparity map with existing algorithms for image *Teddy*

Algorithms	*Cones*		*Venus*		*Teddy*	
	R	*B*	*R*	*B*	*R*	*B*
Estimated	3.3798_3	0.1270_1	1.9885_1	0.0262_1	2.7629_3	0.1115_1
Double-Bp	3.4898	0.2329	2.2114_3	0.2860_3	2.9360	0.2842
Graphcut	4.9694	0.2732	3.3977	0.3065	5.6912	0.3314
Infection	4.2949	0.2147_3	4.4952	0.3119	4.5092	0.2439_3
Layered	4.6167	0.2638	3.1955	0.3186	4.3622	0.3096
Realtime-Gpu	3.2784_2	0.2456	2.0780_2	0.2609_2	2.7535_2	0.2815
Scanline Opt.	5.4622	0.2989	4.2491	0.3090	5.7917	0.3538
SSD min. Filter	4.5599	0.2248	3.733	0.2960	5.9532	0.3111
Sym. Occlusion	3.1457_1	0.2145_2	15.7478	1.000	2.6445_1	0.2421_2

Table 1. A comparison of the estimated disparity with a number of existing well known algorithms

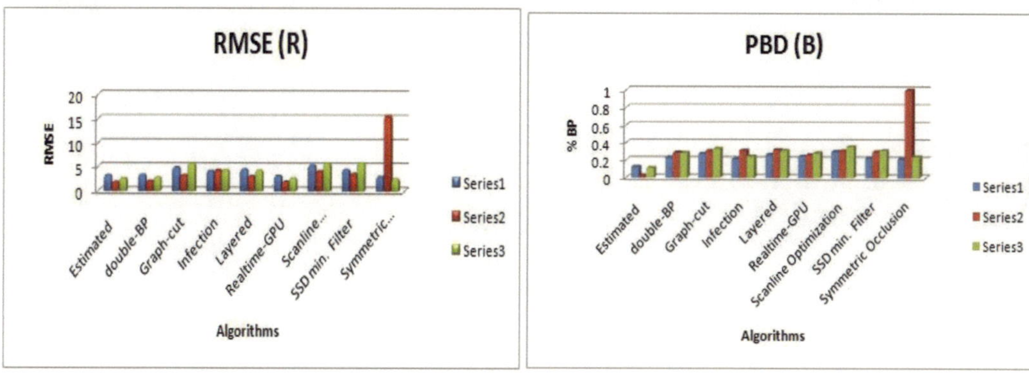

Fig. 13. comparison of the estimated disparity with a number of existing well known algorithms

7. Depth estimation

The main objective of the presented work is to obtain accurate depth estimations for quality inspection of the metallic parts in automotive industry. Consequently, it is important to keep the system simple with minimal hardware involvement, as the cost is an important factor for industry acceptance of the technology. While there are some existing techniques, such as (D. Scharstein 2003), capable of producing accurate disparity maps by using light or laser projectors however are both difficult and expensive to use in a typical production environment. In the presented work, only a basic hardware setup is used namely stereo cameras and circular florescent light. The simplicity of the hardware helps in keeping the implementation costs down and to make the automated fault detection in the automotive manufacturing industry as flexible as possible. Furthermore, hardware simplicity helps in deploying the system in different factory environments without a significant change in the setup.

For performance estimation of the proposed algorithm, two different metallic parts are used. In this section of the work, the main concern is to estimate accurate depth of the object, i.e. z dimension, and not the x, y as these dimensions can be estimated using a single 2D view. For the validation of the estimated depth map, one-dimensional cross-section of the depth difference between the estimated and the real depth maps are shown in Figures 14 and 15. One-dimensional cross-section of the depth difference is also shown for each of the two parts to give an idea about the quality of the estimated depth maps and their accuracy. As can be seen from the estimated depth maps the part defects can be accurately detected.

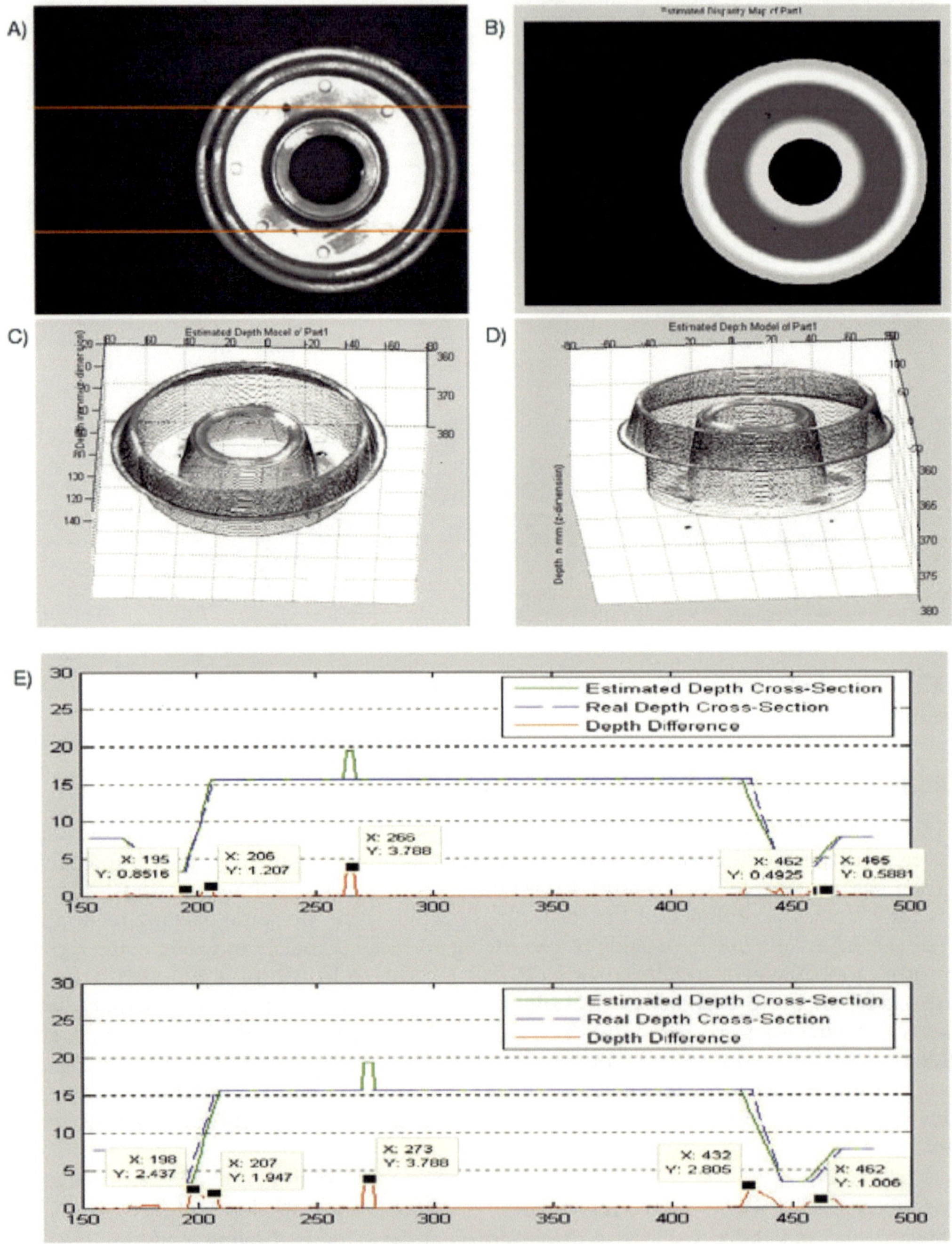

Fig. 14. A: Original image of Part-1 (Bad sample) B: Estimated Disparity Map of Part-1 C: Estimated 3D Depth of Part-1 in (mm) D: Estimated 3D Depth of Part-1 in (mm) (difference view) E: Different view of the Estimated Depth map (mm)

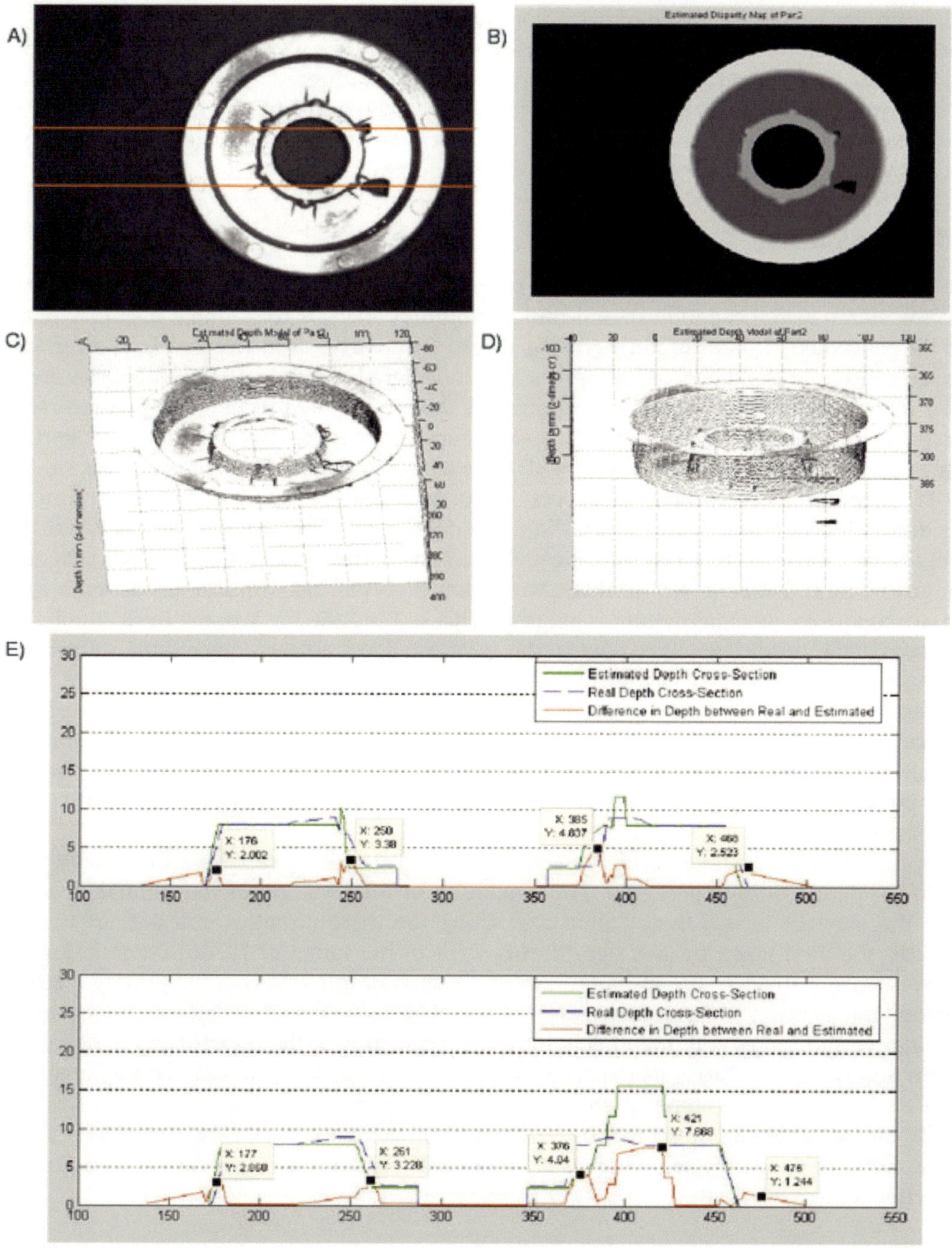

Fig. 15. A: Original image of Part-2 (Bad sample) B: Estimated Disparity Map of Part-2 C: Estimated 3D Depth of Part-2 in (mm) D: Estimated 3D Depth of Part-1 in (mm) (difference view) E: Different view of the Estimated Depth map (mm)

Referring to Figure 14 and 15, the difference between the real and estimated depth is very small across the area with no defect. For Part 1, the difference between the estimated and real depth lies within the range of [0.80-1.49*mm*] whereas for Part 2 the difference is [0.018-1.42 *mm*]. Therefore, the maximum depth difference regarding Part 1 and Part 2 is 1.49mm and 1.42mm, respectively. From the upper value of depth difference, an error tolerance can be set for differentiating between good and defective parts in an inspection system. Moreover, in Figure 14(E) and Figure 15(E) the sharp peaks are in fact due to the difference in x and y dimensions rather than the difference in depth.

8. Conclusion

The developed vision system consists of a novel robust algorithm. The proposed algorithm uses the stereo vision capabilities and multi-resolution analysis to estimate disparity maps and the concerned 3D depths. Furthermore, it uses multiwavelets theory that is a newer way of scale space representation of the signals and considered as fundamental as Fourier and a better alternative. The proposed algorithm uses the well-known technique of *coarse-to-fine* matching to address the problem of stereo correspondence. The translation invariant *wavelets transform modulus maxima* (*WTMM*) are used as corresponding features. To keep the whole correspondence estimation process consistent and resistant to errors, optimized selection criterion strength of the candidate is developed. The strength of the candidate involves the contribution of probabilistic weighted normalized correlation, symbolic tagging and geometric refinement. Probabilistic weighting involves the contribution of more than one search spaces, whereas symbolic tagging helps to keep the track of the most significant and consistent candidates throughout the process. Furthermore, geometric refinement addresses the problem of geometric distortion between the perspective views. The geometric features used in the geometric refinement procedure are carefully chosen to be invariant through many geometric transformations, such as affine, metric, Euclidean and projective. Moreover, beside that comprehensive selection criterion the whole correspondence estimation process is constrained to uniqueness, continuity and smoothness.

A novel and robust stereo vision system is developed that is capable of estimating 3D depths of objects to high accuracy. The maximum error deviation of the estimated depth along the surfaces is less than 0.5mm and along the discontinuities is less than 1.5mm. Similarly the time taken by the algorithm is with in the range of [12-15] seconds for the images of size [640-480]. The proposed system is very simple and consists of only a stereo cameras pair and a simple fluorescent light. The developed system is invariant to illuminative variations, and orientation, location and scaling of the objects, which makes the system highly robust. Due to its hardware simplicity and robustness, it can be implemented in different factory environments with out a significant change in the setup of the system. Due to its accurate depth estimation any physical damage, regarding the object under consideration, can be detected which is a major contribution towards an automated quality inspection system.

9. References

A. Bhatti, H. Özkaramanli (2002). M-Band multiwavelets from spline super functions with approximation order. International Conference on Acoustics Speech and Signal Processing, (ICASSP 2002), IEEE. 4: 4169-4172.

A. Fusiello, V. R., and E. Trucco (2000). "Symmetric stereo with multiple windowing." International Journal of Pattern Recognition and Artificial Intelligence.

A. Haar (1910). "Zur Theorie der orthogonalen Funktionen-Systeme." Math 69: 331-371.

Asim Bhatti, and Saeid Nahavandi, (2008). "Depth estimation using multiwavelet analysis based stereo vision approach." International Journal of Wavelets, Multiresolution and Information Processing 6(3): 481-497.

D. Scharstein, a. R. S. (1998). "Stereo matching with nonlinear diffusion." Int. J. of Computer Vision 28(2): 155-174.

D. Scharstein, a. R. S. (2002). "A Taxonomy and Evaluation of Dense Two-Frame Stereo Correspondence Algorithms." Int. J. of Computer Vision 47: 7-42.

D. Scharstein, R. S. (2003). High-accuracy stereo depth maps using structured light. Computer Society Conference on Computer Vision and Pattern Recognition (CVPR). 1.

Fangmin Shi, Neil Rothwell Hughes, and Geoff Robert (2001). SSD Matching Using Shift-Invariant Wavelet Transform. British Machine Vision Conference: 113-122.

G. Olague, F. Fernández, C. Pérez, and E. Lutton, (2005). "The infection algorithm: an artificial epidemic approach for dense stereo correspondence." Artificial Life.

G. Plonka, and V. Strela, (1998). From Wavelets to Multi-wavelets. 2nd Int. conf. on Math. methods for curves and surfaces, Lillehammer, Norway: 375-400.

He-Ping Pan (1996). "General Stereo Matching Using Symmetric Complex Wavelets." Wavelet Applications in Signal and Image Processing 2825.

He-Ping Pan (1996). Uniform full-information image matching using complex conjugate wavelet pyramids. XVIII ISPRS Congress, International Archives of Photogrammetry and Remote Sensing. 31.

I. Cohen, S. Raz, and D. Malah, (1998). Adaptive time-frequency distributions via the shiftinvariant wavelet packet decomposition. Proc. of the 4th IEEE-SP Int. Symposium on Time-Frequency and Time-Scale Analysis, Pittsburgh, Pennsylvania.

J. C. Olive, J. Deubler and C. Boulin, (1994). Automatic registration of images by a waveletbased multiresolution approach. SPIE. 2569: 234-244.

J. Magarey, and N.G. Kingsbury (1998). "Motion estimation using a complex-valued wavelet transform." IEEE Transections on signal proceessings 46(4): 1069-1084.

J. Margary, and A. dick (1998). Multiresolution stereo image matching using complex wavelets. Proc. 14th Int. Conf. on Pattern Recognition (ICPR). 1: 4-7.

J. Sun, Y. Li, S.B. Kang, and H.-Y. Shum, (2005). Symmetric stereo matching for occlusion handling. Computer Society Conference on Computer Vision and Pattern Recognition (CVPR), IEEE.

L. Di Stefano, M. M., S. Mattoccia, G. Neri (2004). "A Fast Area-Based Stereo Matching Algorithm." Image and Vision Computing 22(12): 938-1005.

L. Zitnick, S.B. Kang, M. Uyttendaele, S. Winder, and R. Szeliski, (2004). High-quality video view interpolation using a layered representation. SIGGRAPH.

M. Lin, and C. Tomasi, (2003). Surfaces with occlusions from layered stereo. CVPR: 710-717.

M. Pollefeys (2000). 3D modelling from images. Conjunction with ECCV2000. Dublin, Ireland.

M. Unser, and A. Aldroubi, (1993). "A multiresolution image registration procedure using spline pyramids." SPIE Mathematical Imaging 2034: 160-170.

O. Faugeras, Q. T. L., T. Papadopoulo, (2001). The Geometry of Multiple Images, MIT Press.

Özkaramanli H., Bhatti A. and Bilgehan B. (2001). Multiwavelets From Spline super functions with approximation order. International Symposium on Circuits and Systems, (ISCAS 2001), IEEE. 2: 525 - 528.

Özkaramanli H. Bhatti A. and, Bilgehan B., (2002). "Multi wavelets from B-Spline Super Functions with Approximation Order." Signal Processing, Elsevier Science: 1029-1046.

Q`ıngxiong Yang, L. W., Ruigang Yang, Henrik Stewenius and David Nister (2006). Stereo matching with color-weighted correlation, hierarchical belief propagation and occlusion handling. CVPR. 2: 2347-2354.

R. Hartley, and A. Zisserman (2003). Multiple View Geometry. Cambridge, UK, Cambridge University Press.

R. R. Coifman, a. D. L. D. (1995). Translation-invariant de-noising. Wavelet and Statistics, Lecture Notes in Statistics, Springer-Verlag: 125-150.

S. Mallat (1991). "Zero-Crossings of a Wavelet Transform,." IEEE Transactions on Information Theory 37: 1019-1033.

S. Mallat (1999). A Wavelet Tour of Signal Processing, Academic Press.

S. Mallat, a. S. Z. (1993). "Matching Pursuits With Time-Frequency Dictionaries." IEEE Transactions on Signal Processing 41(12): 3397-3415.

X. Zhou, and E. Dorrer, (1994). "Automatic image-matching algorithm based on wavelet decomposition." IAPRS 30(1): 951-960.

3

Stereo Vision for Unrestricted Human-Computer Interaction

Ross Eldridge and Heiko Rudolph
RMIT University
Melbourne,
Australia

1. Introduction

Since the advent of the electronic computer there have been constant developments in human-computer interaction. Researchers strive towards creating an input method that is natural, compelling and most of all effective.

The ideal computer interface would interpret natural human interactions in much the same way another human would. As humans we are used to giving instructions with our voice, body language and gestures: i.e. pointing at objects of interest, looking at things we interact with and using our hands to move objects. The ideal visual input device, the "holy grail" of human computer interfaces would be capable of accepting visual cues and gestures much as another human being would. Although this theoretical ideal is not feasible with current technologies it serves to guide research in the field.

Stereo vision has been researched for the purpose of studying human motion for many years (Aggarwal & Cai 1997; Cappozzo et al. 2005). Much effort has gone into achieving full body motion capture with multiple cameras, in order to overcome the need for artificial markers. This allows for unrestricted movement for the user, improves flexibility and overcomes occlusion problems with marker based approaches. However the processor time required to locate and track arbitrary objects in 3D has until recently been prohibitive for use in interactive scenarios.

Faster processors, memory, bus/interface speeds are helping overcome a significant constraint in using stereo vision for human-computer interaction (HCI), that of: producing real-time refresh rates and low latency response.

Now that interactive 3D human motion capture is becoming a reality, it's necessary to take stock of what can be achieved with this approach.

This chapter will not cover the theory behind stereo vision. Nor will we be considering technologies such as motion capture, as these are not intended for everyday computer usage. We are interested in applications that allow unencumbered interactions and operate without requiring the user to attach special devices to their body. We will be looking the current state of the art and where the next could be taken.

2. History

2.1 Computer input devices

The majority of computer input methods are tactile in nature. From the humble switch to the joystick or touch pad, these are all based on the user physically interacting with a device, generally with the hand and fingers. This represents the most straightforward way for a computer to obtain information: The user presses a switch and the signal changes in the circuitry. The advantages of this method of interacting with a computer are:

- Input is essentially digital, i.e. 'unambiguous'.
- Low processing of input data when compared to more complex systems.
- Standard method applies across languages, cultures and computer systems.

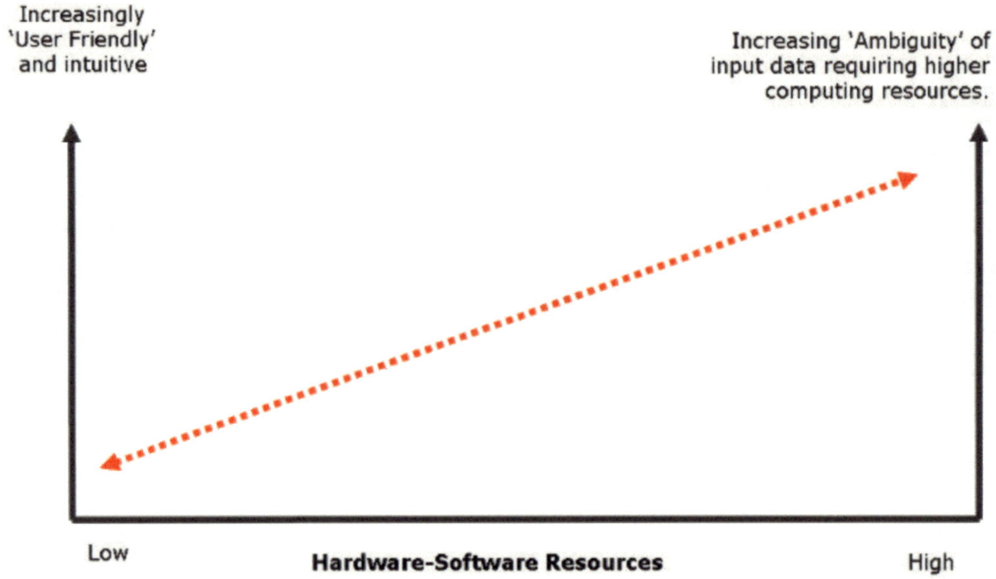

Fig. 1. Conceptual graph illustrating general relationship between 'ambiguity' of input, 'user friendliness' and computing resource requirements. Note that although show linearly, the relationship is likely not linear.

The standard input devices used with modern computers are the keyboard and mouse. Keyboards with physical keys that allow the input of natural language and computer commands, and a mouse to select and manipulate graphical user interface elements. The computer mouse has been available commercially since 1981, and the keyboard can be traced to the mechanical typewriter. Other devices are certainly used for many applications (e.g. touch screens, joysticks, voice input), however the core interaction with a computer is still via these two venerable devices.

There is good reason for the longevity of these interaction methods. The keyboard is one of the simplest ways for a human to enter information into a machine. The keyboard forces the human operator to enter unambiguous information in digital form which a computer can process with minimal effort. The even longer history of keyboard layouts within typewriters has meant that those familiar with traditional typewriters can make the transition easily.

The Graphical User Interface (GUI) and pointing devices were a radical and significant milestone in making personal computers accessible to the general population. The GUI reduced the need for specialized command line input via the keyboard, and reduced the amount of computer knowledge required by average users, while maintaining a clear and unambiguous input structure for the personal computer.

It should however be noted that GUI's require significant more processing power and hardware and software resources than its previous purely text-based interfaces. As a general principle: as the holy grail of human computer interfacing is approached, more and more computing resources are required.

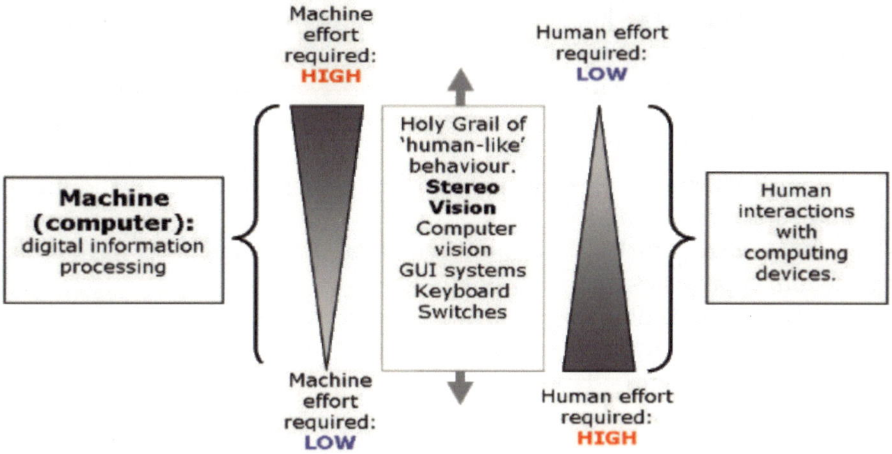

Fig. 2. Placing human computer interfaces into context regarding effort required by both sides.

This can be broken down into two directions:

1. Seen from the direction of human-computer interface, this principle can be understood in terms of degrees of 'ambiguity'. Switches and keyboards are a relative simple input for a machine to process, with GUI's becoming more resource demanding. As we approach computer vision systems the degrees of ambiguity increase and more resources are required to break down the input into information which can be digitally processed unambiguously.
2. Moving in the direction of machine to human communication: the issue is one of relevant feedback for the user. Information from the machine needs to be organized and presented to the user in ways which are easy for the user to process and understand and modify.

Within the past few years several new interaction technologies have gained commercial success, such as touch-screens which can detect multiple inputs simultaneously (Apple iPhone, Microsoft Surface) and wireless controls that use accelerometers to obtain free human movement as input (Nintendo Wii).

2.2 Stereo vision

The applications that have driven the development of computer stereo vision have varied greatly since its inception. The first major use was for mapping the topography of the land by performing calculation disparity in satellite imagery (Barnard & Fischler 1982). Stereo vision later saw applications in human motion capture and allowed a computer to better animate humanoid models by capturing 3D human motion.

A further key area was robotics, especially in autonomous mobile systems. Stereo vision allows the robot to calculate its distance from objects in the environment: enabling it to calculate the dimensions of objects and spaces in the surroundings with greater accuracy.

More recent applications range from cameras in cars to judge distances to obstacles, to tracking people in open areas for surveillance (Cai & Aggarwal 1996).

2.3 What stereo vision brings to HCI

What does stereo vision bring to human computer interactions that can't be achieved using single camera approaches? Stereo vision can:

- Create active/inactive spaces for interaction. Just as we can lift our hands off a mouse to stop interaction, gestures and other interactions can be disabled based on distance from the camera.
- Help distinguish interactive parts of captured image (objects that are valid input) and background parts to be ignored.
- More accurate and reliable 3D position data than single camera (distance approximation) approaches.
- Better face tracking/matching by perceiving a disparity map of a person's face, giving another dimension to matching algorithms.

3. Computer vision for unrestricted human motion tracking

Unlike the previously mentioned types of human input, limb and joint tracking has been researched and used for many years before it became feasible for real-time applications. Early computer based human motion tracking used physical sensors and were primarily for bio-mechanics research. Other researchers began analyzing human movement in prerecorded video footage. Multiple cameras are used to track movements in 3D and to reduce occlusion problems, however this could not be achieved in real-time and required manual post-processing to correct errors (Sturman 1994).

Once real-time analysis became feasible, optical motion capture began seeing extensive use in computer graphics animations. The predominating system for this requires the person to wear a special suit covered in markers, as this improves accuracy and reduces computation requirements.

This required specialized and expensive hardware, not possible on a consumer level computer until more recently (Brown et al. 2003). Goncalves researched estimating 3D arm position without markers using a single camera, to reduce cost and make the system more practical (Goncalves et al. 1995), however computation speed limited the real-time application of such a system.

Within the past decade gesture, fingertip and arm tracking have reached a point where real-time interactions are possible. Various applications have since developed. These range from using 3D arm tracking as input to a robotic arm, achieving an untethered and natural remote interface (Verma et al. 2004), to using stereo vision for rehabilitation and human motion analysis (Cappozzo et al. 2005).

4. Relevant HCI considerations

4.1 Interaction models

A main facet of HCI in this area is how to structure the computer interactions to suit the new input device. Existing modes of interaction may not be best suited to unrestricted HCI, there are new areas where it will excel and different limitations. Will discuss viability of using stereo vision input for:

- Everyday use: Limitations of the standard WIMP (Window, Icon, Menu, Pointing device) model, new interaction models to achieve everyday computing tasks.
- Gaming: Achieve more life-like interactions by having the user physically perform a series of motions to interact with the game.
- Rehabilitation/training: Track limb movement in real-time for analysis and give immediate sense of achievement.
- Alternative input for the disabled: e.g. Gaze direction instead of mouse movement.

4.2 Direct input

Most vision based input devices can still be generalized as cursor devices, as we are interpreting an area of interest or intention, much like mouse input.

A major UI decision in vision based user interaction is the use of direct input. Direct input is any system where the interaction takes place in the same area as the response. For example, a touch pad (used in many portable computers) is an indirect device, as your movements control a virtual cursor on screen. A touch screen however is a direct input devices, as your finger is the cursor itself.

One of the key advantages to a direct input system is its ease of use. We are accustomed to directly interacting with the environment using our hands, so a direct input device is more natural. However, there are some disadvantages. Direct input methods tend to be less precise. A person's finger is larger than average mouse cursor, so interactions with a touchscreen will be less precise than, for instance, mouse input. The user's finger will also obscure part of the displayed image.

Direct input devices usually require a different GUI paradigm to the standard WIMP model, as they are two state devices. A cursor device generally consists of the following three states:

1. Hover
2. Active
3. Active and Moving

For a mouse: State 1 is moving the mouse around. State 2 is a mouse click. State 3 is click and drag. A number of alternate input devices lack one of these states, or have to infer the state from secondary information. A touch screen only has states 2 and 3, as there isn't a way to move the cursor without touching the screen. The only method to determine cursor position (touching) is also the only method to determine click activation.

It is often argued that this isn't relevant in direct input devices such as the touch screen, as your finger is the cursor and hence you don't need visual feedback to see where the cursor is. However there are many GUI conventions that stem from mouse usage which are impossible to achieve with a touch screen alone. (i.e. hovering over an area of interest to see a tool tip describing what the area does).

By tracking movement when the person isn't touching the surface we can create the Hover state in the standard cursor model, adding another degree of freedom. This enables a more direct interpretation of standard mouse actions, given the three state interaction, and hence the user can use existing GUI functionality. It also allows for a virtual cursor when not touching the screen. This gives the user a constant sense of where their interactions will take place within the GUI environment, before they commit to activating UI elements.

5. Stereo vision for interactive surfaces

5.1. Interactive surfaces

An interactive surface generally consists of standard computer screen combined with some form of direct input. The most ubiquitous of these is the touchscreen kiosk, often used in shopping centers or for library catalogs.

Another type of interactive surface is the interactive tabletop environment, also known as an augmented desk.

Traditional interactive surface environments take one of three approaches. The oldest and most common approach is a touchscreen. These devices are usually "resistive" screens. They detect changes in electrical resistance across the screen to determine where the interaction will take place. They therefore require a conductive material (such as the human hand) for interaction, and output a single overall position for the cursor.

Another approach that has seen success recently is "Multi-touch". There are several technologies that can detect multiple touches on a screen surface. Jefferson Han's system uses infrared light that is internally reflected within the surface (Han 2005). When the user touches the surface this breaks the internal reflection and the IR light can be detected by a rear mounted camera.

This and other rear mounted camera approaches greatly increase the space required by such a system when compared to conductive technologies. SmartSkin (Rekimoto 2002) uses a capacitance based approach to achieve multiple input detection in a flat surface. Phillips Entertaible (Hollemans et al. 2006) achieves multiple finger interaction as well as basic shape matching with an LCD panel. It uses a proprietary method based on IR emitters and photo diodes.

The touch screen interface is more traditional. Interaction takes place when the user physically touches the screen. Methods of interaction include clicking and dragging objects, similar to the standard WIMP module. These may be expanded into gesture based interaction, and multiple touch allows for more dynamic gestures: rotation, scaling, etc.

Vision based tracking in these environments tend to not require physical contact. The user does not need to touch the screen as interaction is detected by a camera above or behind the screen, and clicks can be trigger by having the user 'dwell' on a particular spot or using gestures. The University of Tokyo have developed a natural hand tracking augmented desk (Oka et al. 2002) using a far-infrared camera to obtain clear hand silhouette, and user interaction is determined by finger gestures.

The use of stereo vision systems can bring the benefits of a multi-touch system and vision based tracking approaches. In the case of the work by Wilson, A.D (Wilson 2004), a stereo camera rig is placed behind the transparent screen. Distance data is used to determine when the user is interacting with the screen (effectively a multiple-touch screen), but also to morph images of the user for video conferencing. This new dimension can also expand the interactive surface area into an interactive box above the screen, with many possibilities for interaction.

5.2 Depth and its usefulness

The third dimension (e.g. distance from the camera/screen) is generally used to determine whether an object is touching the screen. As such this data is effectively thresholded and binary, touching or not touching.

Researchers at the University of Sydney developed an augmented desk system using stereo cameras (Song & Takatsuka 2005). By thresholding the depth data of their fingertip tracking system just above screen level they achieved a virtual button press, which acts in much the same way as a touchscreen. An earlier system placed the cameras at non-aligned positions (above and to one side) to achieve 3D position tracking of the human arm (Leibe et al. 2000). This effect has also been achieved with a single camera and a touch screen interface. Dohse uses one camera below the table for touch (using infrared reflection), one above for above-screen hand movement (Dohse et al. 2008). Whilst this system uses two cameras, they aren't able to see the same objects and hence can't determine 3D position.

There are many possibilities to enhance user interaction by using depth data in a more granular way. A relevant use of this is seen in (Franco 2004), a gesture based infrared instrument that uses height data to control musical effects.

Another possibility is to track a user's finger and obtain a 3D vector to determine where the user is pointing. Rather than reaching across a large surface to interact, the user need only point with a finger. Researchers have used stereo cameras to extract 3D pointing gesture from a dense disparity map (Jojic et al. 2000; Demirdjian & Darrell 2002).

The 3D information may also be useful in a non-immediate sense. The paths that a user takes when using the interface can be analyzed after the fact to examine how people use the device, in order to improve the system. For example, if it is found that users only ever interact within a subset of the area that the system covers, the system may be altered for increased accuracy or speed.

7. Challenges facing stereo vision in HCI

Over the last few years many new human-computer interaction devices have gained commercial success. (e.g. Nintendo's Wii gaming console, Apple's iPhone) The concepts that we have covered here are making headway in mainstream computing applications, such as multi-touch and gesture based interaction. However the success of stereo vision based approaches has been limited.

There has yet to be a 'killer app' for stereo vision computer interfaces (i.e. an application that would accelerate mass adoption of the technology). It has seen success in niche markets and with more commercial, industrial and military systems. And whilst there are low cost devices that can achieve stereo vision available to the home user, there isn't a major drive for use in the consumer market.

The true advantage of stereo vision over other technologies in this field is the acquisition of real-time and reasonably accurate 3D position data for arbitrary objects in the physical environment. In order to achieve this, the cameras much be able to see the subject in question. This generally means placing the cameras in highly visible and often awkward positions (e.g. pointing down from above a desk). This makes it nearly impossible to integrate into a self-contained device. The system requires setup and some form of calibration.

Another issue is privacy and comfort of the users around active video capture. In order to obtain 3D arm position or determine gaze direction, the computer needs to 'see' the end user. People may have become used to computer's seeing them through a webcam, but this would be an always-on approach: all interactions effectively require video 'surveillance' after a fashion.

Infrared based multi-touch systems (Han 2005) use cameras to detect IR reflections, however these are placed behind the screen: they can't see the end user. Nintendo's Wii system has a camera (within the game controller), however it can only see IR hotspots and is generally pointed away from the user. This is a fundamental issue with the technology, and indeed any technology that wishes to achieve the goal of untethered and unrestricted human movement input.

8. Conclusion

Human computer interfaces have come long way in recent years, but the goal of a computer interpreting unrestricted human movement remains elusive. The use of stereo vision in this field has enabled the development of systems that begin to approach this goal. As computer technology advances we come ever closer to a system that can react to the ambiguities of human movement in real-time.

In the foreseeable future stereo computer vision is not likely to replace the keyboard or mouse. There is at this point no clearly identifiable mass market for stereo vision in the human computer interaction field.

However in this regards stereo vision may be in a similar position to personal computing: in the late 1980's, and early 1990's. In that period personal computers were a somewhat specialized technology with a slowly growing application. The main driving force for personal computers outside research and certain business applications were video games for young people.

Similarly it is still the games industry, and the entertainment industry which drives the adoption of new human computer input systems in general and computer vision systems in particular. Thus it is quite possible that stereo vision is at this point in a similar position personal computers once were.

Certainly the foundations for any expansion into computer vision are increased processing power and software intelligence to drive the systems. Both are proceeding at a rapid pace.

However even as stereo vision advances it is would most likely be used in applications where its strengths are required. Stereo computer vision is at this point a technology in its pre-teenage years, which given the advances in computing, can be expected to mature rapidly from here on.

10. References

Aggarwal, J.K. & Cai, Q. (1997). Human motion analysis: a review. In *Nonrigid and Articulated Motion Workshop, 1997. Proceedings., IEEE*, (pp. 90-102)

Barnard, S.T. & Fischler, M.A. (1982). Computational Stereo. *ACM Comput. Surv.*, 14, 4, (pp. 553-572)

Brown, M.Z., Burschka, D. & Hager, G.D. (2003). Advances in computational stereo. *IEEE Transactions on Pattern Analysis and Machine Intelligence*, 25, 8, (pp. 993-1008)

Cai, Q. & Aggarwal, J.K. (1996). Tracking human motion using multiple cameras. In *Pattern Recognition, 1996., Proceedings of the 13th International Conference on*, (pp. 68-72 vol.3)

Cappozzo, A. et al. (2005). Human movement analysis using stereophotogrammetry: Part 1: theoretical background. *Gait & Posture*, 21, 2, (pp. 186-196),

Demirdjian, D. & Darrell, T. (2002). 3-D articulated pose tracking for untethered diectic reference. In *Proceedings. Fourth IEEE International Conference on Multimodal Interfaces*, (pp. 267-272)

Dohse, K.C. et al. (2008). Enhancing Multi-user Interaction with Multi-touch Tabletop Displays Using Hand Tracking. In *Advances in Computer-Human Interaction, 2008 First International Conference on*, (p. 297—302)

Franco, I. (2004). The AirStick: a free-gesture controller using infrared sensing. In *NIME '05: Proceedings of the 2005 conference on New interfaces for musical expression,* Singapore, Singapore: National University of Singapore, (pp. 248-249)

Goncalves, L. et al. (1995). Monocular tracking of the human arm in 3D. In *Computer Vision, 1995. Proceedings., Fifth International Conference on*, (pp. 764-770)

Han, J.Y. (2005). Low-cost multi-touch sensing through frustrated total internal reflection. In *UIST '05: Proceedings of the 18th annual ACM symposium on User interface software and technology,* New York, NY, USA: ACM Press, (pp. 115-118)

Hollemans, G. et al. (2006). Entertaible: Multi-user multi-object concurrent input. *Adjunct Proceedings of UIST*, 6, (pp. 55-56)

Jojic, N. et al. (2000). Detection and Estimation of Pointing Gestures in Dense Disparity Maps. *IEEE International Conference on Automatic Face and Gesture Recognition, Grenoble, France*

Leibe, B. et al. (2000). The Perceptive Workbench: toward spontaneous and natural interaction in semi-immersive virtual environments. In *Virtual Reality, 2000. Proceedings. IEEE*, (pp. 13-20)

Oka, K., Sato, Y. & Koike, H. (2002). Real-time fingertip tracking and gesture recognition. *Computer Graphics and Applications, IEEE*, 22, 6, (pp. 64-71)

Rekimoto, J. (2002). SmartSkin: an infrastructure for freehand manipulation on interactive surfaces. *Proceedings of the SIGCHI conference on Human factors in computing systems: Changing our world, changing ourselves*, (pp. 113-120)

Song, L. & Takatsuka, M. (2005). Real-time 3D finger pointing for an augmented desk . In *Proceedings of the Sixth Australasian conference on User interface - Volume 40 ,* Newcastle, Australia : Australian Computer Society, Inc., (pp. 99-108)

Sturman, D.J. (1994). A Brief History of Motion Capture for Computer Character Animation. *SIGGRAPH 94, Character Motion Systems, Course notes*

Verma, S., Kofman, J. & Wu, X. (2004). Application of markerless image-based arm tracking to robot-manipulator teleoperation. In *Computer and Robot Vision, 2004. Proceedings. First Canadian Conference on*, (pp. 201-208)

Wilson, A.D. (2004). TouchLight: an imaging touch screen and display for gesture-based interaction. *Proceedings of the 6th international conference on Multimodal interfaces,* (pp. 69-76)

4

A Systems Engineering Approach to Robotic Bin Picking

Ovidiu Ghita and Paul F. Whelan
Dublin City University
Ireland

1. Introduction

In recent times the presence of vision and robotic systems in industry has become common place, but in spite of many achievements a large range of industrial tasks still remain unsolved due to the lack of flexibility of the vision systems when dealing with highly adaptive manufacturing environments. An important task found across a broad range of modern flexible manufacturing environments is the need to present parts to automated machinery from a supply bin. In order to carry out grasping and manipulation operations safely and efficiently, we need to know the identity, location and spatial orientation of the objects that lie in an unstructured heap in a bin.

Historically, the bin picking problem was tackled using mechanical vibratory feeders where the vision feedback was unavailable. This solution has certain problems with parts jamming and more important they are highly dedicated. In this regard if a change in the manufacturing process is required, the changeover may include an extensive re-tooling and a total revision of the system control strategy (Kelley et al., 1982). Due to these disadvantages modern bin picking systems perform grasping and manipulation operations using vision feedback (Yoshimi & Allen, 1994).

Vision based robotic bin picking has been the subject of research since the introduction of the automated vision controlled processes in industry and a review of existing systems indicates that none of the proposed solutions were able to solve this classic vision problem in its generality. One of the main challenges facing such a bin picking system is its ability to deal with overlapping objects. The object recognition in cluttered scenes is the main objective of these systems and early approaches attempted to perform bin picking operations for similar objects that are jumbled together in an unstructured heap using no knowledge about the pose or geometry of the parts (Birk et al., 1981). While these assumptions may be acceptable for a restricted number of applications, in most practical cases a flexible system must deal with more than one type of object with a wide scale of shapes.

A flexible bin picking system has to address three difficult problems: scene interpretation, object recognition and pose estimation. Initial approaches to these tasks were based on modeling parts using 2D surface representations. Typical 2D representations include invariant shape descriptors (Zisserman et al., 1994), algebraic curves (Tarel & Cooper, 2000), conics (Bolles & Horaud, 1986; Forsyth et al., 1991) and appearance based models (Murase & Nayar, 1995; Ohba & Ikeuchi, 1997). These systems are generally better suited to planar object recognition and they are not able to deal with severe viewpoint distortions or objects with complex shapes/textures. Also the spatial orientation cannot be robustly estimated for objects with free-form contours. To address this limitation most bin picking systems attempt to recognize the scene objects and estimate their spatial orientation using the 3D information (Fan et al., 1989; Faugeras & Hebert, 1986). Notable approaches include the use of 3D local descriptors (Ansar & Daniilidis, 2003; Campbell & Flynn, 2001; Kim & Kak, 1991), polyhedra (Rothwell & Stern, 1996), generalized cylinders (Ponce et al., 1989; Zerroug & Nevatia, 1996), super-quadrics (Blane et al., 2000) and visual learning methods (Johnson & Hebert, 1999; Mittrapiyanuruk et al., 2004). The most difficult problem for 3D bin picking systems that are based on a structural description of the objects (local descriptors or 3D primitives) is the complex procedure required to perform the scene to model feature matching. This procedure is usually based on complex graph-searching techniques and is increasingly more difficult when dealing with object occlusions, a situation when the structural description of the scene objects is incomplete. Visual learning methods based on eigenimage analysis have been proposed as an alternative solution to address the object recognition and pose estimation for objects with complex appearances. In this regard, Johnson and Hebert (Johnson & Hebert, 1999) developed an object recognition scheme that is able to identify multiple 3D objects in scenes affected by clutter and occlusion. They proposed an eigenimage analysis approach that is applied to match surface points using the spin image representation. The main attraction of this approach resides in the use of spin images that are local surface descriptors; hence they can be easily identified in real scenes that contain clutter and occlusions. This approach returns accurate results but the pose estimation cannot be inferred, as the spin images are local descriptors and they are not robust to capture the object orientation. In general the pose sampling for visual learning methods is a problem difficult to solve as the numbers of views required to sample the full 6 degree of freedom for object pose is prohibitive. This issue was addressed in the paper by Edwards (Edwards, 1996) when he applied eigenimage analysis to a one-object scene and his approach was able to estimate the pose only in cases where the tilt angle was limited to 30 degrees with respect to the optical axis of the sensor.

In this chapter we describe the implementation of a vision sensor for robotic bin picking where we attempt to eliminate the main problem faced by the visual learning methods, namely the pose sampling problem. This chapter is organized as follows. Section 2 outlines the overall system. Section 3 describes the implementation of the range sensor while Section 4 details the edge-based segmentation algorithm. Section 5 presents the viewpoint correction algorithm that is applied to align the detected object surfaces perpendicular to the optical axis of the sensor. Section 6 describes the object recognition algorithm. This is followed in Section 7 by an outline of the pose estimation algorithm. Section 8 presents a number of experimental results illustrating the benefits of the approach outlined in this chapter.

2. System overview

The operation of the system described in this chapter can be summarized as follows (see Fig. 1). The range sensor determines the depth structure using two images captured with different focal settings. This is followed by the image segmentation process that decomposes the input image into disjoint meaningful regions. The resulting scene regions from the image segmentation process are subjected to an orthographic projection that aligns them to be perpendicular to the optical axis of the sensor. This operation will determine 2 degrees of freedom (DOF) for each object (rotations about x and y axes). The recognition framework consists of matching the geometrical primitives derived from the segmented regions with those contained in a model database. The object that gives the best approximation with respect to the matching criteria is then referred to the pose estimation algorithm which constrains the object rotation around the optical axis of the range sensor (z axis) using a Principal Components Analysis (PCA) approach. Once the object pose is estimated, the grasping coordinates of the identified object are passed to the bin picking robot.

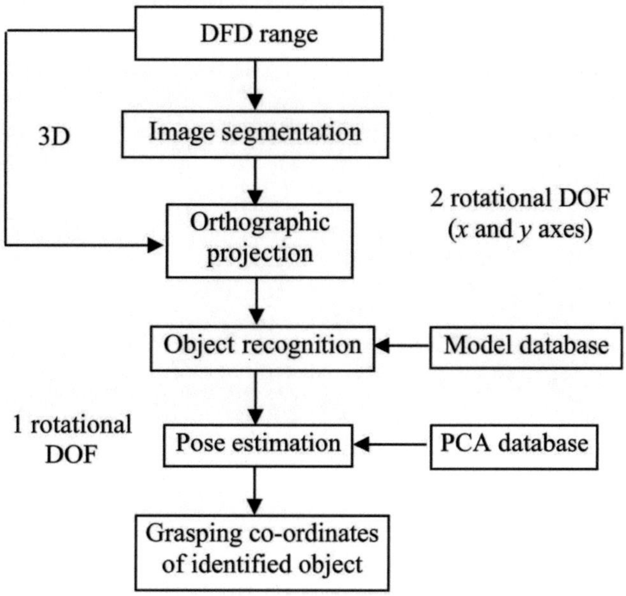

Fig. 1. Overall system architecture (Ghita & Whelan, 2003).

3. Range sensor

The range sensor employed by this application is based on active depth from defocus (DFD). This ranging technique has been initially developed as a passive ranging strategy by Pentland (Pentland, 1987). The principle behind DFD range sensing extends from the fact that the scene objects are imaged in relation to their position in space. In this fashion, the objects that are placed on the focal plane are sharply imaged on the sensing element of the camera, while the points situated on the surface of the objects shifted from the focal plane are refracted by the lens into a patch whose size is in direct relationship with the distance from the focal plane to the imaged object. It has been demonstrated in (Subbarao, 1988; Nayar et al., 1995) that the diameter of the defocus (blur) patch is dependent on the object distance u, lens aperture D, sensor distance s and focal length f. While one image is not sufficient to solve the uncertainty whether the scene object is placed in front or behind the focal plane, the depth can be uniquely estimated by measuring the blurring differences from two images captured with different focal settings. In our implementation the defocused images are captured by changing the sensor distance s (Ghita et al., 2005).
Since the level of blurriness in the image can be thought of as a convolution with a low pass filter (that is implemented by the point spread function (PSF)), to estimate the level of

blurriness in the image we need to convolve the image with a focus operator that extracts the high frequency information derived from the scene objects (Pentland, 1987). Nonetheless this approach returns accurate results only if the scene objects are highly textured. When dealing with weakly and non-textured scene objects this approach returns imprecise depth estimation. To address this issue, a solution is to project a structured light onto the scene that forces an artificial texture on all visible surfaces of the scene. While the artificial texture has a known pattern, the focus operator is designed to respond strongly to the dominant frequency in the image that is associated with the illumination pattern (Girod & Scherock, 1989; Nayar et al., 1995; Ghita et al., 2005).

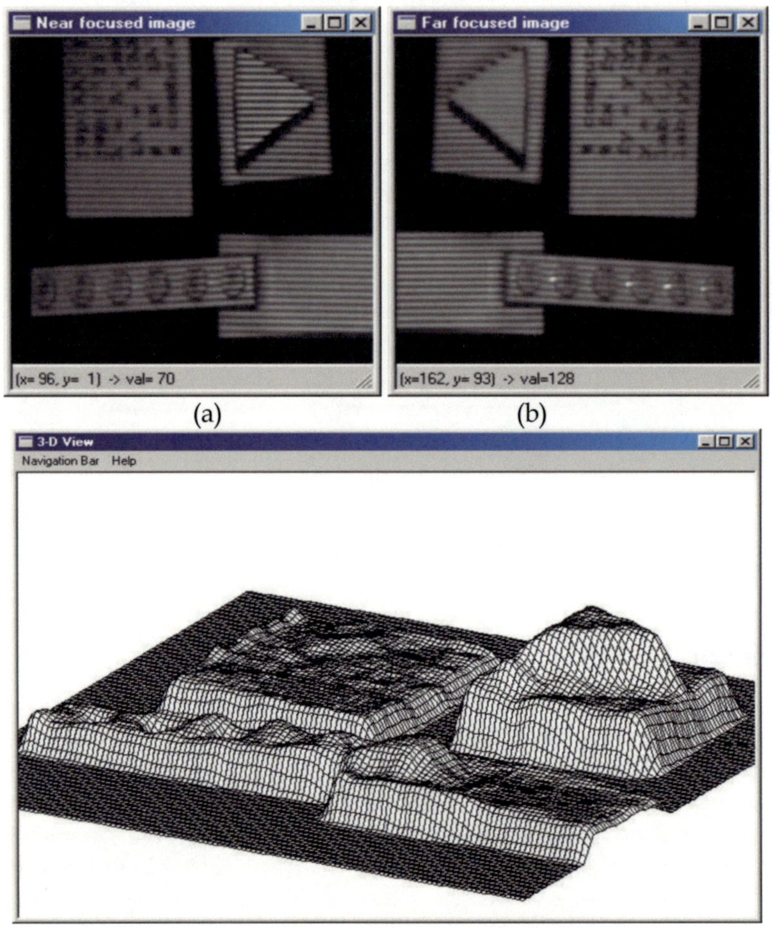

(c)

Fig. 2. Depth estimation for a scene defined by textureless, textured and mildly specular objects. (a) Near focused image. (b) Far focused image. (c) Depth estimation.

In our implementation we used an illumination pattern defined by evenly spaced opaque and transparent stripes and the focus operator is implemented by a tuned Gabor filter (full details about the implementation of our range sensor are provided in Ghita et al. 2005). Fig. 2 depicts the depth map obtained when the range sensor was applied to estimate the depth of a complex scene containing textureless, textured and specular objects.

4. Scene segmentation process

An important decision in developing robotic systems is to decide which sensorial information is better suited for a particular application. Henderson (Henderson, 1983) suggested to approach the scene segmentation using the information about the objects that define the scene. In this regard, if the scene objects are highly textured and depth discontinuities are significant best results will be achieved if range data is analysed. Conversely, if the scene is defined by small textureless objects better results may be obtained if the segmentation process is applied on intensity images (Ghita & Whelan, 2003).

While our application deals with the recognition of a set of textureless polyhedral objects, we developed an edge-based segmentation scheme to identify the visible surfaces of the scene objects. Edges are associated with sharp transitions in pixel intensity distribution and they are extracted by calculating the partial derivatives in the input data. Edge detection is one of the most investigated topics in computer vision and to date there is no edge detector that is able to adapt to problems caused by image noise and low contrast between meaningful regions in the input data. Thus, the edge structure returned by the edge detector is either incomplete, gaps are caused by the low variation in the distribution of the input data, or contains false edges that are caused by image noise, shadows, etc. Thus after the application of edge detection, additional post-processing is applied to eliminate the spurious edge responses and bridge the gaps in the edge structure (this operation is referred to as edge linking). Approaches that have been used to bridge the gaps in the edge structure include morphological methods (Hajjar & Chen, 1999), Hough transform (Davies, 1992), probabilistic relaxation techniques (Hancock & Kittler, 1990), multi-scale edge detection methods (Farag & Delp, 1995) and the inclusion of additional information such as colour (Saber et al., 1997). From these techniques the most common are the morphological and multi-scale edge linking strategies. In general, the morphological edge linking techniques use the local information around edge terminators while multi-scale approaches attempt to bridge the gaps in the edge structure by aggregating the information contained in a stack of images with differing spatial resolutions (Ghita & Whelan, 2002). The main disadvantage associated with multi-scale approaches resides in the high computational cost required to calculate the image stack and in our implementation we developed a morphological edge linking scheme that evaluates the direction of edge terminators in identifying the optimal linking decisions.

4.1 Edge linking

To extract the surfaces of the imaged scene objects we have developed a multi-step edge linking scheme that is used in conjunction with an edge detector that extracts the partial derivatives using the ISEF (Infinite Symetrical Exponential Filter) functions (Shen & Castan, 1992). The reason to use the ISEF-based edge detector was motivated by the fact that its performance in detecting true edges matches that achieved by the more ubiquitous Canny edge detector (Canny, 1986), but the computation of the ISEF edge detector entails a lower computational cost than that associated with the Canny edge detector. In our implementation we have set the scale parameter to 0.45 and the threshold parameters required by the hysteretic threshold are selected using a scheme that minimise the incidence of small edge segments that are usually generated by image noise.

As mentioned earlier, the edge structure returned by the ISEF detector will be further post-processed using a multi-step morphological edge linking strategy. The first step of the edge linking algorithm (Ghita & Whelan, 2002) involves the extraction of the edge terminators (endpoints). The edgepoint extraction requires a simple morphological analysis where the edge structure is convolved with a set of 3×3 masks (Vernon, 1991). The second step of the algorithm determines the direction of the edge terminators by evaluating the linked edge points that generate the edge terminators. The application of the edge linking process for two iterations is illustrated in Fig. 3.

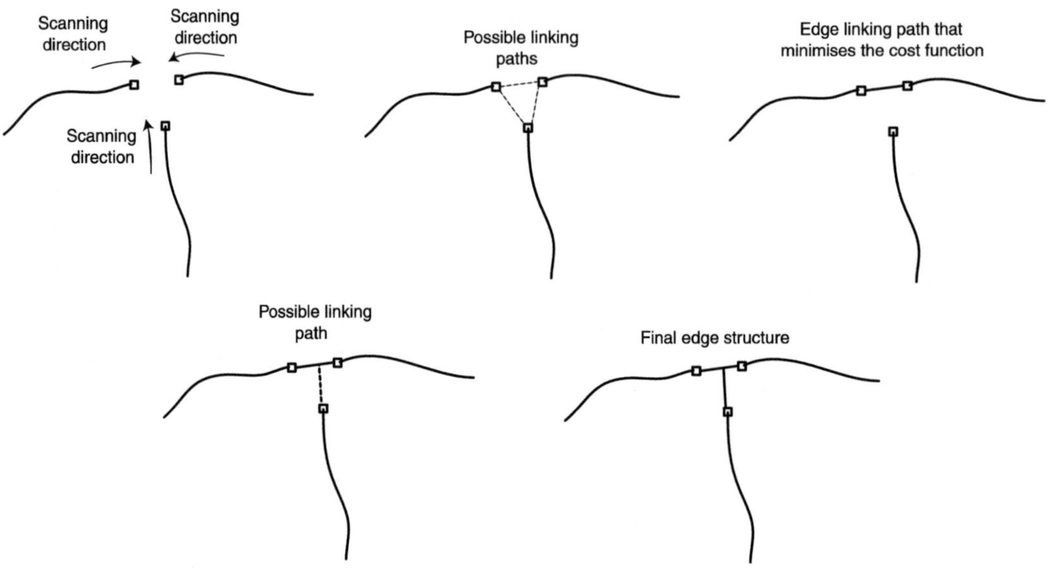

Fig. 3. The edge linking process. The algorithm evaluates all linking decisions around each edge terminator and the optimal linking path minimises the cost function depicted in equation (1). In this diagram the edge pixels are marked in black and the edge terminators are marked with black squares.

The third step of the edge linking scheme attempts to find the possible paths to bridge the gaps in the edge structure by analysing the edge pixels at the side indicated by the endpoint direction in an 11×11 neighbourhood. In this way, for each edge point situated in the endpoint's neighbourhood a linking factor is calculated using the following cost function,

$$Cost(ep) = k_d dist(et, ep) + k_e + k_{dir} \tag{1}$$

where *et* and *ep* are the co-ordinates of the endpoint and the edge pixel under analysis and *dist* defines the Euclidean distance. In equation (1) k_d, k_{dir} and k_e are some pre-defined parameters (a detailed description of these parameters and a discussion in regard to their optimal selection is provided in Ghita & Whelan, 2002). The cost function is calculated for each edge pixel situated in the neighbourhood indicated by the endpoint direction and the minimal value determines the optimal linking path. The gap in the edge structure between the edge terminator and the edge pixel that returns the minimum linking factor is bridged using the Bresenham algorithm (Bresenham, 1965). Fig. 4 illustrates the performance of the edge linking algorithm when applied to an image detailing a cluttered scene.

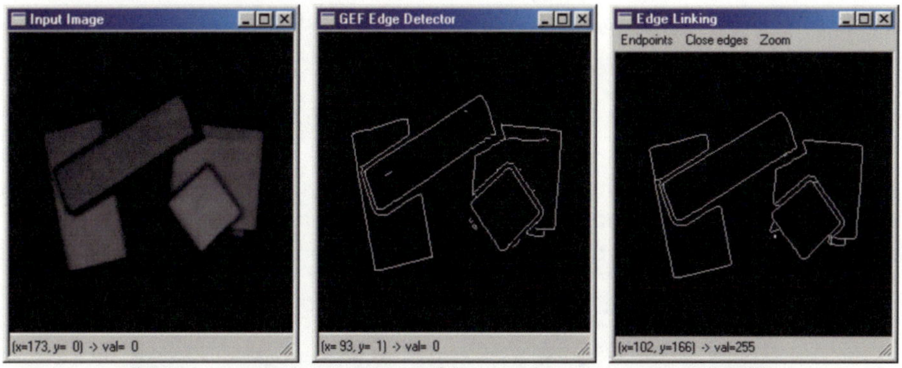

(a) (b) (c)

Fig. 4. The results of the scene segmentation process. (a) Input image. (b) Edge information. (c) Edge linking results. Note the removal of unconnected edge segments.

5. Data formatting

Our application implements a vision sensor able to determine the information required by a bin picking robot to perform object manipulation. Since the objects of interest are polyhedral, a convenient representation is to describe them in terms of their surfaces that are identified by the scene segmentation algorithm detailed in the previous section. Thus, the object recognition task can be formulated in terms of matching the objects' visible surfaces with those stored in a model database. Although conceptually simple, this approach is quite difficult to be applied in practice since the geometrical characteristics of the object's surfaces are viewpoint dependent. To address this problem, we need to align all visible surfaces resulting from the scene segmentation process to a planar that is perpendicular to the optical axis of the range sensor. In this fashion, we attempt to constrain two degrees of freedom (rotations about x and y axes) using the 3D information returned by the range sensor.

The first operation of the data formatting procedure involves the calculation of the normal vector for each surface resulting after the application of the scene segmentation procedure. Since the object surfaces are planar, the normal vector can be calculated using the knowledge that elevation (z co-ordinate) is functionally dependent on the x and y co-ordinates. Then given a set of n points from range data that belong to the segmented surface, the normal vector can be statistically computed by a planar fitting of the 3D points as follows,

$$Err(\hat{a}) = \sum_{i=1}^{n} (\hat{a}_1 x_i + \hat{a}_2 y_i + \hat{a}_3 - z_i)^2 \tag{2}$$

where $\hat{a} = [\hat{a}_1, \hat{a}_2, \hat{a}_3]$ are the estimated values. Equation (2) generates a simultaneous system where the unknown values are $\hat{a}$. The normal vector associated with the surface under analysis is represented in homogenous form as $\overline{N} = [n_x, n_y, n_z, 1]^T = [\hat{a}_1, \hat{a}_2, -1, 1]^T$ (Ghita et al., 2007). As mentioned previously, our aim is to calculate the rotations about x and y axes. The rotation angle about x axis (A_x) is calculated using the following expression: $A_x = \tan2^{-1}(n_y, n_z)$. The rotation angle about y axis (A_y) is computed using the transform $N_{Rx} = R_x N = [n_{rx}, n_{ry}, n_{rz}, 1]^T$, $A_y = -\tan2^{-1}(n_{rx}, n_{rz})$, where $tan2^{-1}$ is the four quadrant inverse tangent. Once the angles A_x and A_y are estimated, the required transformation that is applied to align the surface under analysis to the planar perpedicular to the axis of the range sensor can be formulated as follows,

$$H = T_o^{-1} R_y R_x T_o \tag{3}$$

where T_0 is the transformation that translates the 3D points that define the surface about the origin and R_x and R_y are the rotation matrices about x and y axes. Fig. 5 illustrates the results obtained after the application of the orthographic projection.

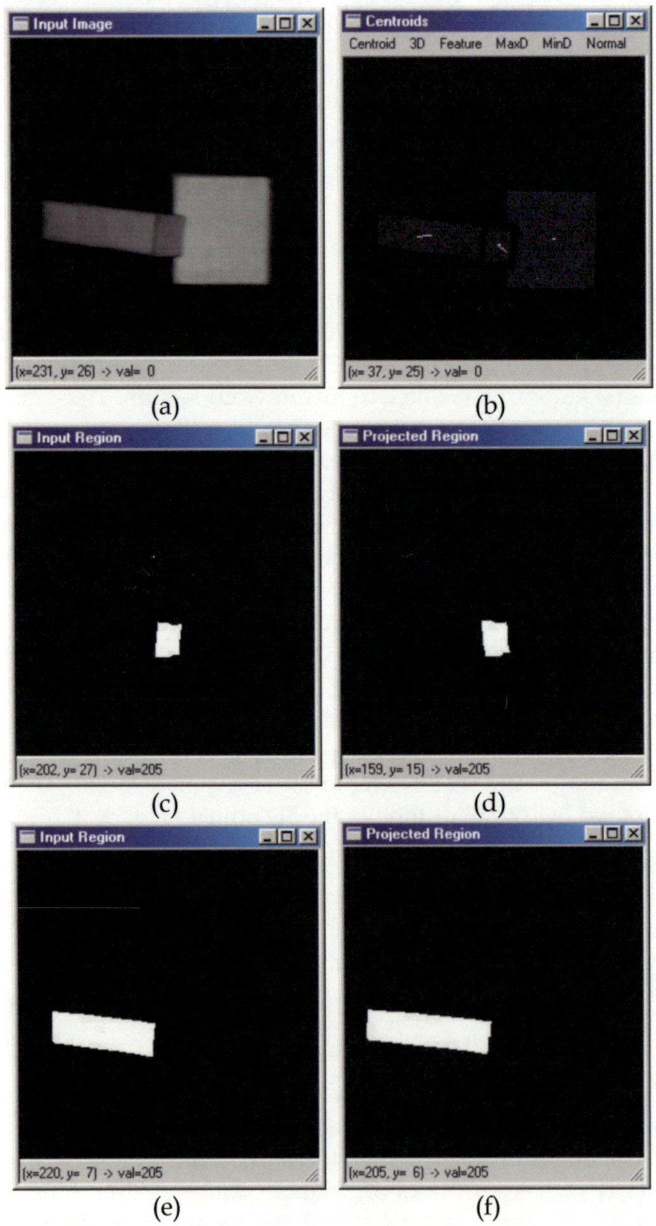

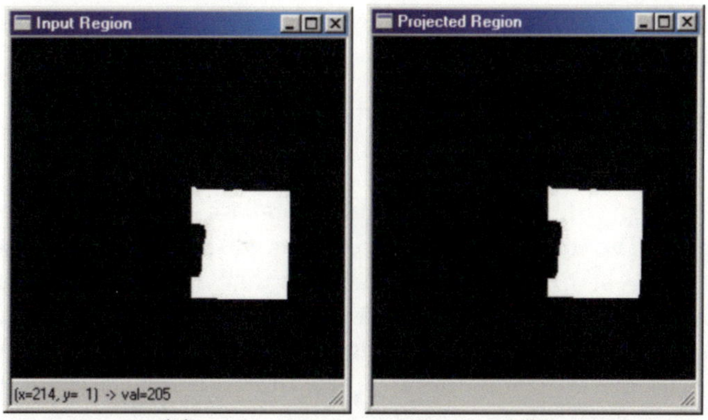

(g) (h)

Fig. 5. Orthographic projection of the segmented scene regions. (a-b) Input image and scene regions resulting from the segmentation process (normal vectors relative to the range sensor position). (c-d) Orthographic projection of the first region (A_x= 26.79^0, A_y=-18.61^0). (e-f) Orthographic projection of the second region (A_x= -36.19^0, A_y=-4.76^0). (g-h) Orthographic projection of the third region (A_x= 6.08^0, A_y=2.68^0).

6. Object recognition

As indicated in the previous section, the recognition of the scene objects is formulated as the recognition of their visible surfaces resulting after the application of the scene segmentation process using an approach that calculates features that sample the geometrical properties of the object surfaces. While the geometric characteristics of the object surfaces are dependent on their orientation in space, in order to eliminate the viewpoint distortions the segmented surfaces were subjected to a 3D data formatting procedure that aligns them to a planar whose normal vector is aligned to the optical axis of the range sensor (z axis). The next step of the algorithm deals with the extraction of geometrical primitives that are used to perform the scene to model recognition process. Approaches that have been used include the extraction of local features such as junctions, lines and partial contours (Bolles & Horaud, 1986; Lowe, 2004) and macro features such as area, perimeter and statistical features (Ghita & Whelan, 2003). Local features may appear better suited when dealing with scenes affected by clutter and occlusions than macro features. But it is useful to note that approaches based on local features rely on a detailed structural description of the objects of interest and when dealing with complex scenes a large number of hypothesis are generated, a fact that requires the development of complex scene to model matching procedures. While our goal is the recognition of a set of polyhedral objects, macro features represent a better option since the segmented surfaces are planar and they can be easily indexed to describe the object structure. To this end, we have adopted features such as area, perimeter, shape factor and radii (maximum and minimum) distances calculated from the surface's centroid to the surface border (Ghita & Whelan, 2003).

The developed object recognition algorithm consists of two main stages. The training stage consists of building the database by extracting the aforementioned features for each surface of the object. Since the features involved have different ranges, to compensate for this issue we have applied a feature normalisation procedure where each feature is normalised to zero mean and unit variance (Duda et al., 2001). The matching stage consists of computing the

Euclidean distance between the normalised features calculated for scene surfaces and object surfaces contained in the model database.

$$dist_j = \sqrt{\sum_{i=1}^{n}(X_j[i]-Y[i])^2} \qquad for \quad i=1,..,n \tag{4}$$

where X_j is the jth pattern contained in the model database and Y defines the pattern derived from an input region. The input scene surface is contained in the database if the minimum distance that gives the best approximation is smaller than a predefined threshold value.

One problem with this approach is the fact that most scene surfaces are affected by occlusions. As the object recognition algorithm is included in the development of a robotic application, we focus the attentions only on the topmost objects since they can be easily manipulated and their surfaces are not affected by severe occlusions. The selection of the topmost object is achieved by eliminating the surfaces that are affected by occlusions based on the 3D information supplied by the range sensor. The scene to model verification procedure is applied only for surfaces that pass the 3D selection criteria (for additional details refer to Ghita & Whelan, 2003).

7. 3 DOF pose estimation

The orthographic transformation illustrated in equation (3) can constrain only two degrees of freedom (DOF), the rotations about x and y axes. The surfaces subjected to this orthographic transformation are perpendicular to the axis of the range sensor and the estimation of the surface rotation about z axis can be carried out using Principal Components Analysis (PCA). This procedure involves the calculation of an eigenspace representation from a set of training images that are generated by rotating the object surfaces in small increments. To estimate the rotation about z axis, all recognized scene surfaces are projected onto the eigenspace and their projections are compared to those stored in the model database (whose rotations about the z axis are known). The minimal distance between the projection of the input surface and those contained in the model database gives the best match.

8. Experiments and results

The vision sensor detailed in this chapter consists of four main components, range sensing, scene segmentation, object recognition and pose estimation. Our implementation employs an active DFD range sensor whose implementation has been outlined in Section 3. To test the performance of the developed range sensor we have applied it to recover the depth information from scenes defined by textured and textureless objects. The relative accuracy was estimated for successive measurements and was formulated as the maximum error between the real and estimated depth values. During the operation the range sensor was placed at a distance of 86cm above the baseline of the workspace. The relative accuracy attained by the developed sensor when applied to scenes containing non-specular objects with bright surfaces is 3.4% normalised in agreement with the distance from the sensor.

The developed bin picking system has been applied to 5 different polyhedral objects that are used to create various cluttered scenes. The edge-based segmentation algorithm detailed in Section 4 is applied to identify the object surfaces. The surfaces resulting after the application of the scene segmentation algorithm are subjected to data formatting in order to constrain 2 rotational DOF (rotations about x and y axes). Since data formatting involves 3D analysis, the precision of this procedure is influenced by the accuracy of the depth estimation. The performance of the data formatting procedure is illustrated in Fig. 6.

The third major component of the algorithm addresses the object recognition task. The algorithm was able to identify the topmost objects in all situations and is able to identify correctly the scene objects if the occlusion cover less than 20% of the object's total surface. The last component of the algorithm is applied to identify the rotation about z axis. In our implementation we have created a PCA model database for each object of interest and the object rotation has been sampled uniformly by acquiring 24 training images with the object lying flat on a dark worktable. This generates 24 PCA projections that are able to sample the object rotation with a resolution of 15 degrees. To increase the resolution of the PCA projections we have applied a linear interpolation procedure that generate 30 interpolated projections between any adjacent projections generated by the 24 images contained in the training set (Ghita & Whelan, 2003; Ghita et al., 2007). The performance of the pose estimation is affected by the accuracy of the data formatting procedure and the experiments indicate that the pose is more precise for low values of the tilt angles (rotations about x and y axes). This is motivated by the relative low resolution of the range sensor in sampling depth discontinuities. In our experiments the rotation about z axis was measured with an error of 2.1 degree under the condition that the rotations about x and y axes are smaller than 25 degrees.

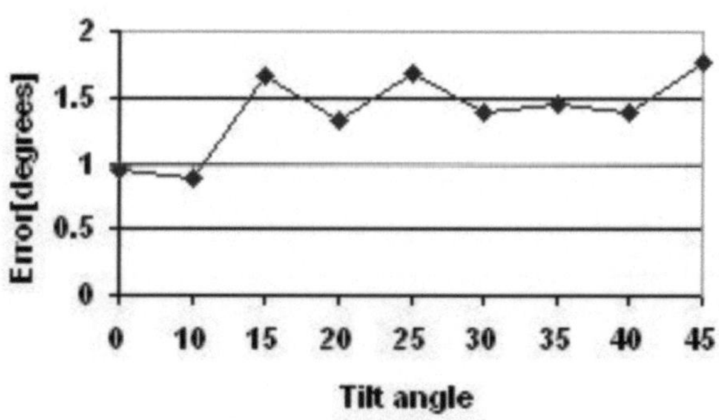

Fig. 6. Data formatting estimation accuracy (rotation about x axis).

9. Conclusions

This chapter describes the development of a fully integrated vision sensor for robotic bin picking. The developed vision sensor is able to provide the information to a bin picking robot to perform scene understanding and object grasping/manipulation operations. Our implementation employs a range sensor based on active depth from defocus that is used in conjunction with a multi-stage scene understanding algorithm that is able to identify and estimate the 3D attitude of the scene objects. In this regard, the scene segmentation scheme attempts to separate the scene regions that are associated with object surfaces using an edge based implementation. The novel part of this scheme is the edge linking procedure that is able to return quality connected edge structures. The object recognition scheme performs scene to model verification using the global attributes extracted from the segmented scene surfaces. As these features are vulnerable to viewpoint distortions we have devised a data formatting scheme that re-format the orientation of the scene surfaces on a planar perpendicular to the optical axis of the sensor. This transformation eliminates the viewpoint distortions and allows the application of standard PCA to sample the rotation about z axis. The experimental results indicate that the approach detailed in this chapter is particularly useful in the development of bin picking systems that are applied to the manipulation of polyhedral objects.

10. References

Ansar, A. & Daniilidis, K. (2003). Linear pose estimation from points or lines, *IEEE Trans. Pattern Anal. Mach. Intell.*, vol. 25, no. 5, pp. 578-589.

Birk, J.R.; Kelley, B. & Martins, H. (1981). An orienting robot for feeding workpieces stored in bins, *IEEE Trans. Syst. Man Cybern.*, vol. 11, no. 2, pp. 151-160.

Blane, M.; Lei, Z.; Civi, H. & Cooper, D.B. (2000). The 3L algorithm for fitting implicit polynomial curves and surfaces to data, *IEEE Trans. Pattern Anal. Mach. Intell.*, vol. 22, no.3, pp. 298-313.

Bolles, R.C. & Horaud P. (1986). 3DPO: A three dimensional part orientation system, *Intl. J. Robotics Res.*, vol. 5, no. 3, pp. 3-26.

Bresenham, J.E. (1965). Algorithm for computer control of a digital plotter, *IBM Systems Journal*, vol. 4, no. 1, pp. 25-30.

Campbell, R. & Flynn, P. (2001). A survey of free-form object representation and recognition techniques, *Computer Vision and Image Understanding*, vol. 81, no. 2, pp. 166-210.

Canny, J. (1986). A computational approach to edge detection, *IEEE Trans. Pattern Anal. Mach. Intell.*, vol. 8, no. 6, pp. 698-700.

Davies, E.R. (1992). Locating objects from their point features using an optimised Hough-like accumulation technique, *Pattern Recognition Letters*, vol. 13, no. 2, pp. 113-121.

Duda, R.O.; Hart, P.E. & Stork, D.G. (2001). *Pattern classification*, (2nd edition), Wiley, ISBN 0-471-05669-3, USA.

Edwards, J. (1996). An active, appearance-based approach to the pose estimation of complex objects, *Proc. of the IEEE Intelligent Robots and Systems Conference*, Osaka, Japan, pp. 1458-1465.

Fan, T.; Medioni, G. & Nevatia, A. (1989). Recognizing 3-D objects using surface description, *IEEE Trans. Pattern Anal. Mach. Intell.*, vol. 11, no. 11, pp. 1140-1157.

Farag, A. & Delp, E.J. (1995). Edge linking by sequential search, *Pattern Recognition*, vol. 28, no. 5, pp. 611-633.

Faugeras, O.D. & Hebert, M. (1986). The representation, recognition and locating of 3-D objects, *Intl. J. Robotics Res.*, vol. 5, no. 3, pp. 27-52.

Forsyth, D.; Mundy, J.L.; Zisserman, A.; Coelho, C.; Heller, A. & Rothwell C. (1991) Invariant descriptors for 3-D object recognition and pose, *IEEE Trans. Pattern Anal. Mach. Intell.*, vol. 13, no. 10, pp. 971-991.

Ghita, O. & Whelan, P.F. (2002). Computational approach for edge linking, *Journal of Electronic Imaging*, vol. 11, no. 4, 2002, pp. 479-485.

Ghita O. & Whelan, P.F. (2003). A bin picking system based on depth from defocus, *Machine Vision and Applications*, vol. 13, no. 4, pp. 234-244.

Ghita, O; Whelan, P.F. & Mallon, J. (2005). Computational approach for depth from defocus, *Journal of Electronic Imaging*, vol. 14, no. 2, 023021.

Ghita, O.; Whelan, P.F.; Vernon D. & Mallon J. (2007). Pose estimation for objects with planar surfaces using eigenimage and range data analysis, *Machine Vision and Applications*, vol. 18, no. 6, pp. 355-365.

Girod, B. & Scherock, S. (1989). Depth from defocus and structured light, *Optics, Illumination and Image Sensing for Machine Vision IV, Proc. of the Soc. of Photo-Opt. Instrum.*, vol. 1194, pp. 209-215.

Hajjar, A. & Chen, T. (1999). A VLSI architecture for real-time edge linking, *IEEE Trans. Pattern Anal. Mach. Intell.*, vol. 21, no. 1, pp. 89-94.

Hancock, E.R. & Kittler, J. (1990). Edge labelling using dictionary-based relaxation, *IEEE Trans. Pattern Anal. Mach. Intell.*, vol. 12, no. 2, pp.165-181.

Henderson, C. (1983). Efficient 3-D object representation for industrial vision systems, *IEEE Trans. Pattern Anal. Mach. Intell.*, vol. 5, no. 6, pp. 609-617.

Johnson, A. & Hebert, M. (1999). Using spin images for efficient object recognition in cluttered 3D scenes, *IEEE Trans. Pattern Anal. Mach. Intell.*, vol. 21, no. 5, pp. 433–449.

Kelley, B.; Birk, J.R.; Martins, H. & Tella R. (1982). A robot system which acquires cylindrical workpieces from bins, *IEEE Trans. Syst. Man Cybern.*, vol. 12, no. 2, pp. 204-213.

Kim, W. & Kak, A. (1991). 3-D object recognition using biopartite matching embedded in discrete relaxation, *IEEE Trans. Pattern Anal. Mach. Intell.*, vol. 13, no. 3, pp. 224-251.

Lowe, D.G. (2004). Distinctive image features from scale-invariant keypoints, *Intl. Journal of Computer Vision*, vol. 60, no. 2, pp. 91-110.

Mittrapiyanuruk, P.; DeSouza, G.N. & Kak, A. (2004). Calculating the 3D-pose of rigid objects using active appearance models, *Intl. Conference in Robotics and Automation*, New Orleans, USA.

Murase, H. & Nayar, S.K. (1995). Visual learning and recognition of 3-D objects from appearance, *Intl. Journal of Computer Vision*, vol. 14, pp. 5-24.

Nayar, S.K.; Watanabe, M. & Noguchi, M. (1995). Real-time focus range sensor, *Proc. of the Intl. Conf. on Computer Vision (ICCV)*, pp. 995-1001.

Ohba, K. & Ikeuchi, K. (1997). Detectability, uniqueness and reliability of eigen window for stable verification of partially occluded objects, *IEEE Trans. Pattern Anal. Mach. Intell.*, vol. 19, no. 9, pp. 1043-1048.

Pentland, A. (1987). A new sense for depth of field, *IEEE Trans. Pattern Anal. Mach. Intell.*, vol. 9, pp. 523-531.

Ponce, J.; Chelberg, D. & Mann, W. (1989). Invariant properties of straight homogenous generalized cylinders and their contours, *IEEE Trans. Pattern Anal. Mach. Intell.*, vol. 11, no. 9, pp. 951-966.

Rothwell, C. & Stern, J. (1996). Understanding the shape properties of trihedral polyhedra, *Proc. of European Conference on Computer Vision*, 1996, pp. 175-185.

Saber, E.; Tekalp, A.M. & Bozdagi, G. (1997). Fusion of color and edge information for improved segmentation and edge linking, *Image and Vision Computing*, vol. 15, no. 10, pp. 769-780.

Shen, J. & Castan, S. (1992). An optimal operator for step edge detection, *CVGIP: Graphical Models and Image Processing*, vol. 54, no. 2, pp. 112-133.

Subbarao, M. (1988). Parallel depth recovery by changing camera parameters, *Proc. of the IEEE Conf. on Computer Vision*, pp. 149-155.

Tarel, J.P. & Cooper, D.B. (2000). The complex representation of algebraic curves and its simple exploitation for pose estimation and invariant recognition, *IEEE Trans. Pattern Anal. Mach. Intell.*, vol. 22, no. 7, pp. 663-674.

Vernon, D. (1991). *Machine vision: Automated visual inspection and robot vision*, Prentice-Hall International, ISBN 0-13-543398-3, UK.

Yoshimi, B.H. & Allen, P. (1994) . Visual control of grasping and manipulation tasks, *Proc. of the IEEE International Conference on Multisensor Fusion and Integration for Intelligent Systems*, Las Vegas, USA.

Zerroug, M. & Nevatia, R. (1996). 3-D description based on the analysis of the invariant and cvasi-invariant properties of some curved-axis generalized cylinders, *IEEE Trans. Pattern Anal. Mach. Intell.*, vol. 18, no. 3, pp. 237-253.

Zisserman, A.; Forsyth, D.; Mundy, J.; Rothwell, C.; Liu, J. & Pillow, N. (1994). 3D object recognition using invariance, Technical report, Robotics Research Group, University of Oxford, UK.

5

Stereo Vision in Smart Camera Networks

Stephan Hengstler
Department of Electrical Engineering, Stanford University
United States of America

1. Introduction

Stereo vision inherently comes with high computational complexity, which previously limited its deployment to high-performance, centralized imaging systems. But recent advances in embedded systems and algorithm design have paved the way for its adoption for and migration into smart camera networks, which notoriously suffer from limited processing and energy resources. Such networks consist of a collection of relatively low-cost smart camera motes, which—in their simplest form—integrate a microcontroller, an image sensor, and a radio into a single, embedded unit capable of sensing, computation, and wireless communication. When deployed in an in- or outdoor environment, they form ad-hoc or mesh networks that can perform a wide range of applications. Their application areas range from ambient intelligence, building automation, elderly care, autonomous surveillance, and traffic control to smart homes. The underlying network tasks include object localization, target tracking, occupancy sensing, object detection and classification. Stereo vision can bring increased performance and robustness to several of these tasks possibly even at overall reductions in energy consumption and prolonged network lifetime.

This chapter will provide a brief description of the building blocks, characteristics, limitations and applications of smart camera networks. We will then present a discussion of their requirements and constraints with respect to stereo vision paying special attention to differences to conventional stereo vision systems. The main issue arises from the high data rate, which image sensors particularly in a stereoscopic configuration generate. Conventional centralized computing systems can easily handle such rates. But for smart camera networks, their resource constraints pose a serious challenge to effective acquisition and processing of this high-rate data.

Two possible approaches to address this problem have emerged in recent publications: one suggests the design of a custom image processor whereas the other solution proposes utilization of off-the-shelf, general-purpose microprocessors in conjunction with resolution-scaled image sensors. Our discussion focuses on these two state-of-the-art stereo architectures. NXP's WiCa mote is the primary example deploying a dedicated image processor, while Stanford's MeshEye mote pioneered the idea of resolution-scaled stereo vision. More specifically, the WiCa mote deploys an application-specific image processor based on a vector single-instruction, multiple-data architecture, which is able to process the data streams of two VGA camera modules. In contrast, Stanford's MeshEye mote deploys a low-resolution stereo vision system requiring only a general-purpose 32-bit ARM7 processor. Additionally, it hosts a VGA camera module for more detailed image acquisition.

The low-resolution stereo imagers can guide the focus of attention of the higher-resolution camera. Hence, we refer to this novel combination of hybrid-resolution image sensors as a hybrid vision system, which is capable of approaching the performance of a VGA stereo system in the context of smart camera networks.
The primary focus of this chapter is to describe Stanford's MeshEye mote in detail. The description will cover its processing algorithms, hardware architecture, and power model. Special attention will be given to the image processing algorithms underlying the hybrid vision system. Their advantages, limitations, and performance will be discussed and compared against those of the WiCa architecture. Finally, the chapter will outline how smart camera motes equipped with stereo or hybrid vision can collaborate to accomplish target tracking tasks in a distributed fashion.
The remainder of this chapter is organized as follows. Section 2. provides background information and outlines applications for stereo vision in smart camera networks. Section 3. identifies requirements for embedded stereo architectures and summarizes the vision architectures of the WiCa and MeshEye motes. In Section 4., we discuss MeshEye's underlying hybrid vision algorithms that perform object detection, localization and acquisition as building blocks for higher-level tracking algorithms. In Section 5., we describe the implementation of the MeshEye mote in more detail, i.e., its hardware architecture and a power model that facilitates lifetime predictions for battery-powered operation. Section 6. presents an experimental deployment of four MeshEye motes and reports on performance results for indoor target tracking. Section 7. concludes with a summary of the key challenges and solutions presented for stereo vision in smart camera networks. It also identifies areas in need of further work and projects promising directions for future research.

2. Background and applications

Smart camera networks have received increased focus in the research community over the past few years. The notion of spatially distributed smart camera motes, which combine image sensing with embedded computation and are interconnected through radio links, opens up a new realm of intelligent vision-enabled applications (Liu & Das, 2006). Real-time image processing and distributed reasoning made possible by smart cameras can not only enhance existing applications but also motivate new ones. Potential application areas (Hengstler & Aghajan, 2006a; Rahimi et al., 2005; Hampapur et al., 2005; Maleki-Tabar et al., 2006; Qureshi & Terzopoulos, 2007) range from assisted living, smart environments, traffic monitoring, and habitat observation to security and surveillance in public spaces or corporate buildings. Critical issues deciding upon the success of smart camera deployments for such applications include reliable and robust operation with as little maintenance as possible.
In comparison to scalar sensors, such as temperature, pressure, humidity, velocity, and acceleration sensors, vision sensors generate much higher bandwidth data due to the two-dimensional nature of their pixel array. The sheer amount of raw data generated precludes it from human analysis in most applications. Hence distributed information fusion algorithms (Nakamura et al., 2007) supported by in-node image processing are required to successfully operate scalable networks of smart cameras.
The majority of smart camera applications target the presence (or absence) of objects in the network's observation area, their spatio-temporal movement, or—in case of humans or animals—their gestures. Hence, the underlying tasks performed by the vision system reduce

to object detection, localization, tracking, and classification (Zhao & Guibas, 2004). The image processing common to all these tasks consists of background subtraction and foreground segmentation followed by additional foreground processing steps. While commercial deployments and most research have focused on networks of monocular camera motes, there has been increasing interest in exploiting stereoscopic camera motes. A stereo vision mote can accomplish the same underlying tasks at increased accuracy or reduced camera resolution in comparison to a monocular vision mote. With respect to the entire network, this can result in fewer required motes or prolonged network lifetime at the same level of accuracy. The key challenge however lies in performing these vision tasks efficiently, i.e., with a minimum of required complexity, processing resources and energy consumption, such that it fits into low-cost, battery-operated embedded systems.

3. Embedded stereo architectures

Requirements and constraints for stereo vision in smart camera networks differ considerably from conventional stereo vision systems. This section discusses how these differences lead to different designs of embedded stereo architectures for smart camera motes.

Conventional, high-performance stereo vision systems (Bramberger et al., 2006) commonly consist of a pair of high-resolution cameras and at least one computationally powerful processor to extract stereoscopic information. Processing power and energy resources can readily be chosen to meet application requirements. They combine general-purpose processors with digital signal processors (DSPs) or field-programmable gate arrays (FPGAs) for demanding image processing tasks. Moreover, scalability and form factor are typically not of concern in such standalone, centralized stereo vision systems. Their stereo vision tasks range from multi-target tracking, depth map generation, to visual hull reconstruction, which generally need to be performed in real-time, i.e., at frame rates of 15 or 30 frames per second (fps). The information generated oftentimes consists of raw or preprocessed stereo image representations intended for human analysis and interpretation.

Smart camera networks, on the other hand, almost exclusively utilize stereo vision for object localization and tracking as discussed in Section 2.. The information reported to the human operator may consist of the number of objects present, their location, trajectory, and object class or may even be condensed into higher levels of representation. The smart camera network needs to generate this information autonomously using machine vision and information fusion techniques. It is these application requirements that drive the design of stereo vision systems in smart camera networks. To cover varying extent of deployment areas, the network needs to be easily deployable and scalable. This translates into smart camera motes that have a small form factor and are cheap in volume production. Moreover, outdoor deployments may require battery operation in contrast to indoor deployments, where wired power grids are commonly available. Delay and accuracy requirements are less stringent in target tracking applications for smart camera networks, in which distances between cameras and objects are relatively large and multiple cameras track the same objects. Hence, the required frame rates are more in the order of about 0.5 to 5 fps; localization accuracy (Rahimi et al., 2006) of the embedded vision system can be well below those of high-performance stereo vision systems. We deem these the two single most important differences between conventional stereo vision and stereo vision in smart camera networks. For instance, it is unnecessary and prohibitive to use multiple processors with high power consumption in the interest of frame rate and energy dissipation, respectively.

Going forward, we can identify the following list of design guidelines for embedded stereo architectures:

- Usage of low-complexity algorithms: The single most important guideline is to use algorithms of low complexity. This is especially true for vision processing, which deals with large image arrays directly. The complexity of such algorithms drives hardware requirements as well as energy consumption.
- Avoidance of high processing clock frequencies: Power consumption grows quadratically with clock frequency but processing duration only reduces linearly with frequency. Hence, it is advantageous to perform vision processing at clock frequencies slow enough to still meet frame rate requirements.
- Utilization of intermediate memory: It is more energy costly to recompute intermediate results than to compute them once and store them in memory for later use. Hence, enough memory should be available to hold all intermediate results that are used more than once.
- Reduction of component count: The number of components, especially processing elements, should be minimized. This improves both mote cost and energy consumption. Parallel computations can be transformed into sequential computations at the expense of frame rate but at the benefit of fewer processing elements.
- Usage of commercial off-the-shelf components: Compared to custom-designed (semi-conductor) components, commercial off-the-shelf (COTS) components benefit from mass production and pricing and hence the overall mote cost is minimized.
- Selection of low-power components: Low-power components should be chosen wherever applicable. As these may be more expensive, an appropriate trade-off between mote cost and power consumption needs to be established.
- Implementation of power management: To reduce power consumption further, the mote should make use of power saving modes, which only turn on the components required to carry out its current operation. For example, the cameras should only be active during image acquisition and be put into sleep or power-down otherwise. This also includes clock scaling according to the current processing load.

Two stereo vision architectures for smart camera motes have been proposed recently, which strive to satisfy these design guidelines in different ways. The two vision architectures are embedded in NXP's WiCa and Stanford's MeshEye smart camera motes.

3.1 Parallel processing architecture

In 2006, NXP (formerly part of Philips Electronics) introduced WiCa (Kleihorst et al., 2006): a smart camera mote with a high-performance vision system. Its vision system consists of two VGA camera modules (640×480 pixel, 24-bit color), which feed video to IC3D, a remarkable dedicated parallel processor based on a vector single-instruction multiple-data (SIMD) architecture. IC3D alternately acquires and processes frames from the left and right VGA camera. It exchanges image processing results with an 8051-based Atmel AT89C51 host processor, which runs at 24 MHz, through a 128 Kbytes dual-port RAM.

IC3D's parallel architecture running at 80 MHz (scalable) is capable of processing an entire image row in only a few clock cycles at frame rates up to 30 fps. Processing operations include, for example, two-dimensional filtering (with kernel sizes up to 64×64 pixel), edge detection, histogram generation, and template or stereo correlations. Higher level processing steps like object localization and tracking need to be carried out by the host processor.

The key advantage of WiCa's vision system (Kleihorst et al., 2007) lies in its ability to perform complex image processing in real time (in a conventional stereo vision sense) at moderate clock rates. Its reported low power consumption makes it attractive for smart camera networks although its performance exceeds their typical requirements. Its main disadvantage leading to increased mote cost is the use of a custom-designed parallel processor in addition to a general-purpose host processor. Furthermore, dual-port RAM devices have a rather large power consumption, typically in the order of 300 mW.

3.2 Hybrid vision architecture

In Stanford's hybrid vision architecture, introduced in 2006 (Hengstler & Aghajan, 2006b), two low-resolution imagers (30×30 pixel, 6-bit grayscale) in stereo configuration guide the focus of attention of a higher-resolution VGA camera module (640×480 pixel, 24-bit color). Thus, hybrid vision can utilize its low-resolution stereo vision to localize foreground objects, which makes it well suited for tracking tasks in smart camera networks. Most of all, hybrid vision enables foreground segmentation without having to process the background in high resolution. As we will discuss in the following, this leads to significantly simplified vision processing, which can be handled by a general-purpose, sequential microcontroller at moderate clock frequency. The main drawback of the hybrid vision architecture is that its low-resolution stereo system has reduced bearing and ranging resolution compared to the parallel processing architecture.

The configuration of the hybrid vision system is illustrated in Fig. 1. All three pixel arrays are parallel facing the same direction. The camera module is centered between the two low-resolution imagers, which—owing to their total pixel count—are referred to as kilopixel imagers. The three image sensors are focused to infinity and their field of view (FoV) angles should be approximately the same although ideally the kilopixel imagers should have a slightly larger FoV angle. Hence the three imagers have an overlapping FoV only offset by their baseline, that is, the distance separating them.

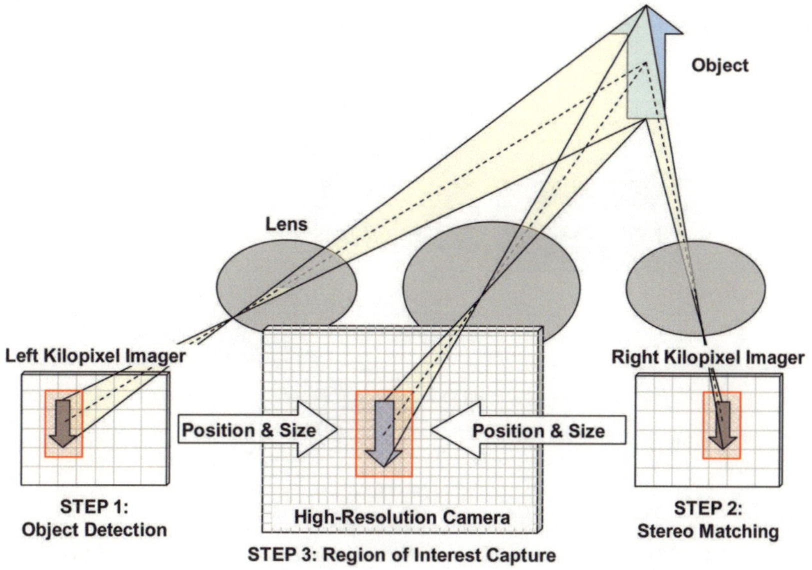

Fig. 1. Hybrid vision system.

In the simplest operation of the hybrid vision system, one of the kilopixel imagers is used to continuously poll for moving objects entering its FoV. Once one or possibly more objects have been detected, position and size within a kilopixel image can be determined for each object. Stereo vision of the two kilopixel imagers yields bearing and distance to the object. This information allows us to calculate the region of interest (RoI) containing the object within the VGA camera's image plane. Subsequently, the microcontroller triggers the VGA camera module to capture a high-resolution grayscale or color RoI including only the detected object. After optional additional low-level processing, the object's RoI will then be handed over to intermediate-level processing functions.

4. Hybrid vision algorithms

The image processing algorithms behind the hybrid vision system, which were first described in (Hengstler et al., 2007), are designed to detect and localize objects entering its FoV. Since a generic 32-bit RISC architecture without dedicated DSP engines needs to execute these algorithms, they are intentionally kept at low computational complexity.
The overall vision processing flow is shown in the flowchart of Fig. 2. Both low-resolution (LR) imagers continue updating their background image and estimate of temporal pixel noise when no objects are present. Upon detection of a moving object in the left kilopixel image, the vision system determines the bounding box and stores the object's RoI. This RoI serves as a template to locate the object's position within the FoV of the right kilopixel imager. If this stereo matching cannot establish a positive match to the template, the left LR imager will continue polling for moving objects. This case occurs for example when the object lies outside the overlapping FoV of the two kilopixel imagers. Knowing the object's position within both LR image arrays and its size, the vision system triggers the VGA camera module to acquire a high-resolution (HR) snapshot of the object. The left kilopixel imager can continue tracking the object until it leaves its FoV and initiate additional HR RoI acquisitions of the object as required by the application. Note that the following discussion is limited to one object inside the FoV, but it can be easily extended to multiple foreground objects.

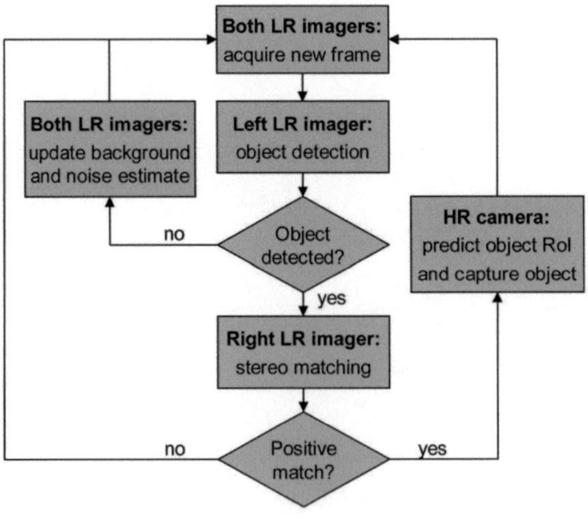

Fig. 2. Flowchart of the vision processing algorithm.

To model the low-resolution stereo system observing a moving object at location **x**, we apply the planar pinhole camera observation model shown in Fig. 3. The intrinsic camera parameters are pixel array half-width D, camera resolution R, focal length f, half-width FoV angle Ψ, and baseline B. These parameters relate the quantities of interest, object bearing θ and range r, to the measurements of the vision system, bearing projection d_{left} and disparity $d = |d_{left} - d_{right}|$ through

$$\tan\theta = \frac{d_{left}}{D}\tan\Psi, \tag{1}$$

and

$$r = \frac{Bf}{d}, \tag{2}$$

respectively.

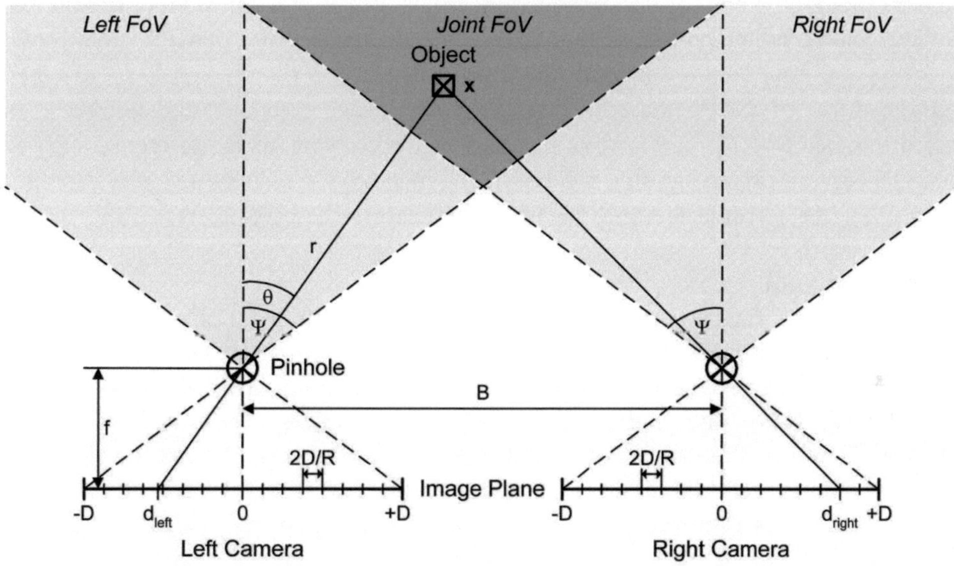

Fig. 3. Planar pinhole camera observation model.

4.1 Object detection

Prior to any further processing, the raw image arrays from both kilopixel imagers are normalized to each frame's average pixel value. This mitigates changes in brightness and exposure time. We found this normalization especially effective in coping with oscillations of the digital shutter control loop inside the kilopixel imagers.

The kilopixel imager performs object detection through background subtraction (Radke et al., 2005) on a frame-by-frame basis. That is, it calculates the frame difference between the current frame at time t, $I^t_{left}(x,y)$, $x,y = 1,2,\ldots,30$, and the latest background $B^t_{left}(x,y)$ as

$$D^t_{left}(x,y) = \left|I^t_{left}(x,y) - B^t_{left}(x,y)\right|. \tag{3}$$

All pixels, whose frame difference $D^t_{left}(x,y)$ exceed a preset multiple k of the temporal noise standard deviation σ_n, are set in a binary motion mask $M^t_{left}(x,y)$ as potential candidates of motion,

$$M^t_{left}(x,y) = \begin{cases} 1 & , D^t_{left}(x,y) \geq k\sigma_n \\ 0 & , \text{otherwise.} \end{cases} \tag{4}$$

In MeshEye's vision system, σ_n is estimated on background frames and it typically ranges around 1.75; k is set to around 6. To eliminate objects smaller than 2×2 pixel, we low-pass filter the binary motion mask according to

$$F^t_{left}(x,y) = \text{conv}_2\left(\frac{1}{4}\begin{bmatrix} 1 & 1 \\ 1 & 1 \end{bmatrix}, M^t_{left}(x,y)\right), \tag{5}$$

where conv_2 denotes two-dimensional convolution.

In the final processing step of object detection, a blob search algorithm identifies all regions as moving objects, which consist of four-connected groups of unity pixels within the binary mask $M^t_{left}(x,y)$ that contain at least one unity pixel of the filtered mask $F^t_{left}(x,y)$. It is then straightforward to determine the bounding box of each object through extraction of the object's difference RoI $D^t_{object}(x,y)$ as a subset of the current frame difference $D^t_{left}(x,y)$. The object's bearing projection d^t_{left} follows as the mean of the object's intensity distribution, which is $D^t_{object}(x,y)$ projected onto the horizontal axis of the pixel array.

4.2 Stereo matching

The objective of the stereo matching algorithm lies in locating the object's RoI within the right kilopixel image array. Since the pixel arrays of both LR imagers are aligned in parallel, the object will appear as a shifted version in the right kilopixel array if it is located within their joint FoV. This satisfies the requirements for template matching based on cross-correlation. Therefore, the vision system computes the cross-correlation of the object's array $D^t_{object}(x,y)$ along the epipolar lines (for an introduction to epipolar geometry, refer to (Foresti et al., 2005) for example) of the right kilopixel difference array $D^t_{right}(x,y)$,

$$C^t_{right}(u,v) = \text{xcor}_2\left(D^t_{right}(x,y), D^t_{object}(x,y)\right), \tag{6}$$

where xcor_2 denotes two-dimensional unbiased cross-correlation without boundary padding such that $u, v = 1, 2, \ldots, 30$.

Lastly, the (u, v) coordinate with the largest cross-correlation value—or the average (u, v) coordinate in case of multiple largest cross-correlation values—is assumed as the object's position within the right LR image array. However, to qualify a positive match and remove objects outside the joint FoV, the maximum cross-correlation value has to be sufficiently close to the center autocorrelation value of the object's RoI, which may be expressed as

$$\max_{u,v} C^t_{right}(u,v) \geq 0.75 A^t_{object}(0,0). \tag{7}$$

$A^t_{object}(u,v)$ denotes the unbiased autocorrelation of the object's difference array

$$A^t_{object}(u,v) = \text{xcor}_2\left(D^t_{object}(x,y), D^t_{object}(x,y)\right). \quad (8)$$

If this condition is not met, no positive match can be established and the object is discarded. Otherwise, the object's bearing projection d'_{right} is computed as the horizontal value of the mean of the object's cross-correlation distribution $C^t_{right}(u,v)$.

4.3 Object acquisition

The hybrid vision system can finally determine the object's position within the HR pixel array using the object's extent and positions d^t_{left} and d^t_{right} within the LR arrays. With this information, a high-resolution snapshot containing only the moving object can be efficiently acquired by using the camera's built-in windowing function. As mentioned before, the HR background is not analyzed and not even acquired with the hybrid vision system. The HR snapshot of the object is stored for further processing or exchange with neighboring smart cameras. Such processing can include object classification or even distributed, multi-camera view identification based on its shape, orientation, aspect ratio, or color histogram for instance. Two examples captured with Stanford's MeshEye smart camera mote are shown in Fig. 4. It contains an indoor (Fig. 4a) and an outdoor (Fig. 4b) snapshot of a moving person. Notice that the LR views have a non-ideal smaller FoV than the HR view.

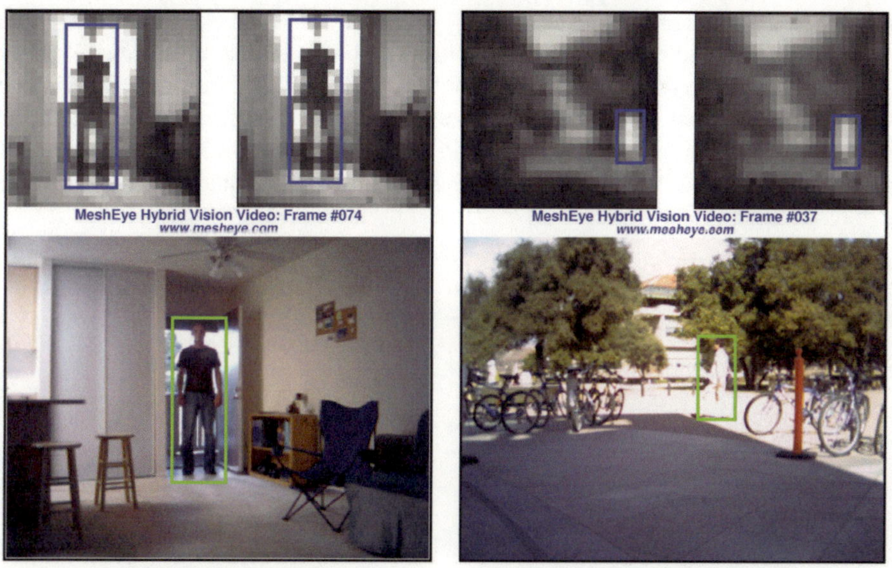

(a) Indoor Snapshot (b) Outdoor Snapshot

Fig. 4. Hybrid vision system performing person localization: indoor (a) and outdoor (b) snapshots. Bounding boxes mark the person's locations in the LR and HR views.

4.4 Computational efficiency

To conclude this section, let us consider the computational savings that the hybrid vision system achieves over the WiCa smart camera mote. WiCa's dual high-resolution camera system combined with the IC3D SIMD dedicated image processor establishes a fair basis of comparison among published embedded stereo vision systems to carry out moving object localization and RoI extraction.

For simplicity, this consideration of computational efficiency is limited to one moving object within the vision system's FoV. The MeshEye vision system carries out the algorithms described in the previous subsections. For the dual-camera SIMD system, our calculation assumes that it performs the same computations on its two HR cameras rather than two LR imagers, which result in a moving object's pixel array and estimated bearing and range. Whenever possible, the computations utilize IC3D's line-parallel processor architecture, which executes an instruction across an entire line of pixel data within one instruction cycle. Of course, an alternative approach would be to downsize incoming frames from VGA to kilopixel resolution prior to object detection and extraction. This however would underutilize the line-parallel processor and not take full advantage of its 640-pixel wide line buffers.

The computational efficiency of MeshEye's vision system relative to the WiCA vision system is shown in Fig. 5. More specifically, it graphs the ratio of number of computations of the dual-camera SIMD system over the hybrid vision system as a function of object size in high resolution when both systems perform object detection, localization and RoI extraction. Hybrid vision achieves a fivefold reduction in the number of computations for small object sizes, but the gains in efficiency diminish down to 0.7 as object sizes increase. Smaller objects require fewer computations and hence hybrid-resolution processing outperforms the parallel processing system. Large objects cause a heavier processing load and hence the SIMD processor excels over hybrid vision.

For objects sized around 132,800 high-resolution pixel^2, which amounts to rather large objects of for example 364×364 pixel—or about 40% of the VGA frame, both vision systems have equal computational efficiency. Note that it is for these significant reductions in the number of computations for moderately sized objects that makes hybrid vision well suited for smart camera embedded vision without the need for a dedicated, high-performance DSP engine.

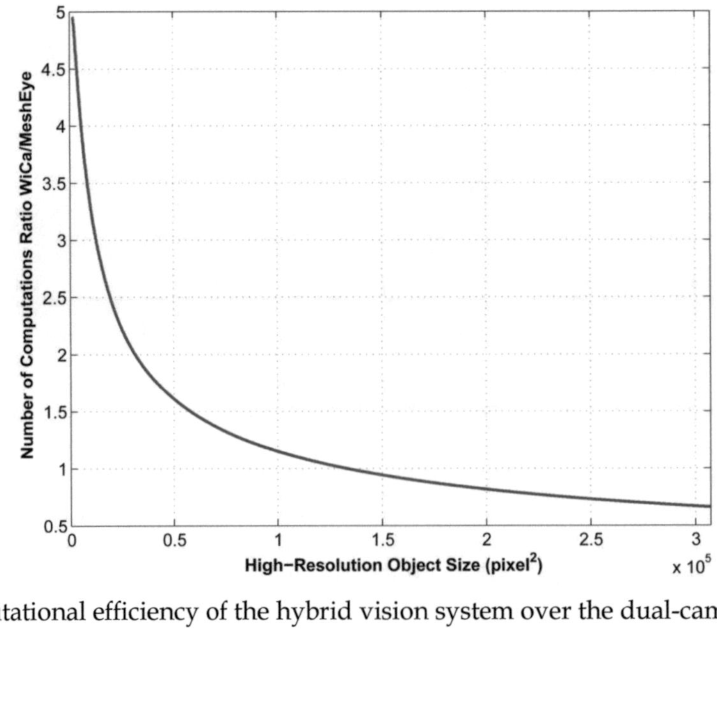

Fig. 5. Computational efficiency of the hybrid vision system over the dual-camera SIMD system.

5. Hybrid vision implementation

This section briefly describes the architecture of Stanford's MeshEye™ mote as a hardware implementation of the hybrid vision system. Its design targets the provision of sufficient processing power for hybrid vision while minimizing component count and power consumption.

5.1 Hardware architecture

The block-level architecture of the MeshEye smart camera mote is shown in Fig. 6. The architecture is centered around an Atmel AT91SAM7S family microcontroller (Atmel, 2006), which contains up to 256 KBytes of flash memory and 64 KBytes of SRAM. Its leading power-efficient ARM7TDMI architecture can be clocked up to 55 MHz. The mote features a USB 2.0 full-speed port and a serial interface for wired connection. Furthermore, the mote can host up to eight kilopixel imagers and one VGA camera module, for which we chose Agilent Technologies' ADNS-3060 high-performance optical mouse sensor (Agilent, 2004) (30×30 pixel, 6-bit grayscale) and Agilent Technologies' ADCM-2700 landscape VGA resolution CMOS camera module (Agilent, 2005) (640×480 pixel programmable, grayscale or 24-bit color), respectively. An MMC/SD flash memory card provides sufficient and scalable non-volatile memory for temporary frame buffering or even image archival. Wireless connection to other motes in the network can be established through a Texas Instruments CC2420 2.4 GHz IEEE 802.15.4/ZigBee-ready RF transceiver (Texas, 2006), which supports data transmission at 250 Kbits per second according to the IEEE 802.15.4 standard with a maximum transmit power of 1 mW in the unlicensed 2.4 GHz ISM band. The mote can either be powered by a stationary power supply if available or battery-operated for mobile applications or ease of deployment.

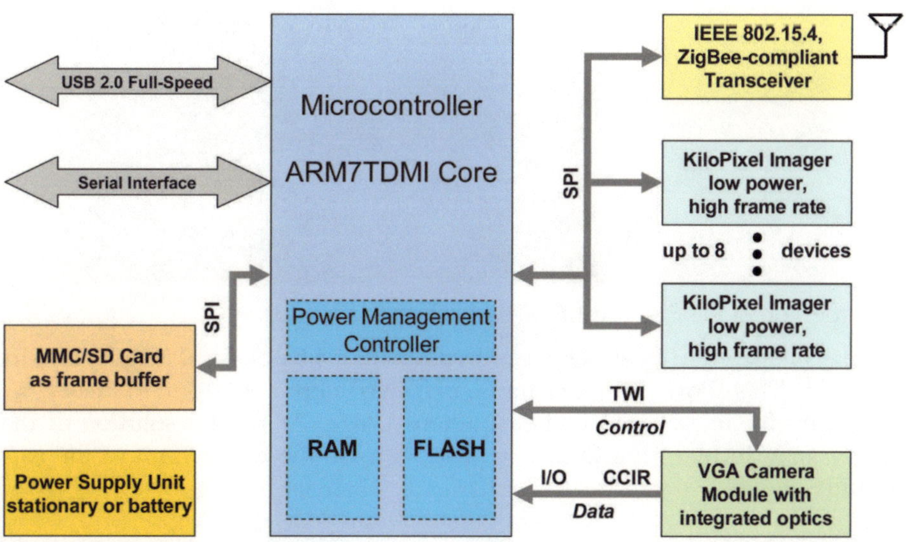

Fig. 6. Block diagram of the MeshEye™ architecture.

The objectives guiding the electrical design of the MeshEye architecture have been the integration of low-power, COTS components, use of standard interfaces, and most of all minimization of total component count. The main motivation in keeping the component

count small lies in reduced power consumption and mote cost. The use of standard interfaces manifests itself in that a single SPI interface connects flash memory card, kilopixel imagers, and radio to the microcontroller. A TWI interface controls the camera module while its CCIR video is read out through general-purpose I/O pins. Note that this CCIR read-out method is not common but avoids additional interface components at the expense of a reduced frame rate of about 3 frames per second. Most other solutions interface CCIR image sensors to microcontrollers through a combination of an FPGA or CPLD and static memory for frame buffering. While such solutions may enable video streaming, they oftentimes add significantly to mote cost and the power budget.
Fig. 7 pictures the first prototype of the MeshEye smart camera mote. It consists of a base board and a sensor board. The base board hosts the voltage regulators, microcontroller, radio, MMC/SD flash card (not visible) and external interface connectors. Power can be supplied through an external source, the Mini-USB port, or pairs of AA batteries. The sensor board, which sits on top of the base board, contains two kilopixel imagers, the VGA camera module, and two white LEDs for short-range illumination.

Fig. 7. Photograph of the first MeshEye™ smart camera mote.

The kilopixel imagers use plastic aspheric lenses with a 4.6 mm focal length *f* and have a 57 mm baseline *B*. This allows for a maximum perceived depth of 8.74 m in theory, which we deem adequate for indoor and limited outdoor usage. At fixed resolution of the image sensors, one may increase the focal length to increase the depth limit at the expense of narrower FoV angles.

5.2 Power model

An important performance metric for battery-operated motes is their lifetime during deployment. For this reason, generating a power model for the MeshEye smart camera mote is a crucial step in analyzing its energy consumption and predicting its lifetime during object localization operation.

The power model assumes that the MeshEye mote is powered by two non-rechargeable AA batteries (capacity 2850 mAh) at a conversion efficiency of 90%. It accounts for current consumption of the following main mote components: Atmel AT91SAM7S64 microcontroller running at a processor clock of 47.92 MHz, PQI 256 MB MultiMediaCard flash memory, two Agilent ADNS-3060 kilopixel imagers, Agilent ADCM-2700 VGA camera module, and Texas Instruments CC2420 IEEE 802.15.4 transceiver. Table 1 summarizes the components' typical current and runtime values. Estimated runtimes and current draw values quoted in each component's datasheet are shown in regular font style. Italicized values have actually been measured on the mote prototype. For the most part, the datasheet estimates are in good agreement with the measurements although the estimated active currents for the flash card and the image sensors turn out to be rather conservative. The power model uses measured values whenever possible. The two runtime states *Poll* and *Event* occur when an object is absent or present, respectively, in the mote's FoV. In the *Event* state, the mote localizes the object and captures its HR region of interest. Subsequently, the object's location is wirelessly exchanged with neighboring motes for localization refinement and object tracking purposes.

Component	**Current (mA)**		**Runtime (ms)**	
	Active	**Sleep**	**Poll**	**Event**[a]
Microcontroller	29.40	0.034	200	2000
	29.72		*125*	
MMC Flash Card	34.00	0.050	off	2000
	14.03	*0.116*		
Kilopixel Imager	30.60	0.005	95	off
	14.89	*<0.010*	*24*	
Camera Module	48.00	0.005	off	1000
	9.65	*0.015*		*600*
Radio Transceiver	18.10	0.426	off	500
	19.81	*0.419*		

a) Event assumes an average object region of interest size of 64×64 pixel.

Table 1. Estimated (regular) and *measured (italicized)* component current and runtime values used in the power model.

For the basic object localization and tracking application considered here, the power model predicts an asymptotic, i.e., no events occur, lifetime of 22 days at a moderately fast poll period of 1 second. At a 0.5 second poll period, the asymptotic lifetime shortens to 11 days. Of course, the lifetime gradually reduces with the frequency of events, that is, more frequent appearance of moving objects in the mote's FoV.

This power model also enables us to compare the hybrid vision system with the parallel processing system in terms of energy consumption. In particular, the MeshEye mote consumes 121 mJ, while the WiCa mote only spends 106 mJ to localize and acquire an object of 64×64 pixel in size. Hence, compared to dual-camera parallel processing, hybrid vision is not as efficient with respect to energy consumption as it is in terms of computational efficiency. Equal energy consumption occurs at a number of computations ratio of 4.3.

6. Experimental deployment

To evaluate functionality and performance of the stereo system within hybrid vision, an indoor network of four MeshEye smart camera motes was put in place. The topology of the network is shown in Fig. 8: the four motes are located on the corners of a rectangle with their camera axes oriented towards the rectangle's center. Each mote records the kilopixel images to an MMC flash memory card at 10 fps over a one minute duration. During this time, a person follows the ground truth path drawn in Fig. 8, which was marked on the room's floor as an inner and outer rectangle. More specifically, the target first walks twice along the inner and then twice along the outer rectangle.

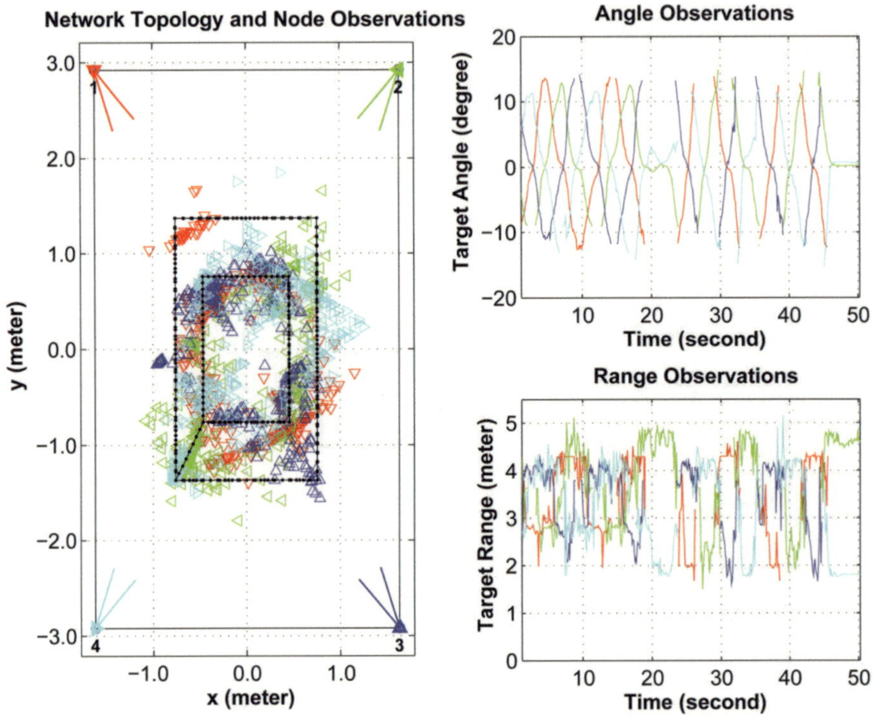

Fig. 8. Topology and observations of experimental network deployment.

The recorded image sequences were then post-processed with the hybrid vision algorithms to determine each mote's target localization over time. These location observations are overlaid onto the network topology in Fig. 8 using different markers. The associated measurements of target bearing angle and range are shown in the two plots to the right. In the final step, a sequential Bayesian tracking algorithm (Hengstler & Aghajan, 2007) fuses the individual location observations to reconstruct the target's track. Fig. 9 shows the resulting track when the motes apply either monocular or stereo vision. Under mono vision, the smart camera motes utilize only the left of the two kilopixel imagers. Note that the target's track is drawn separately for the inner and outer rectangle.

For evaluation purposes, we computed the variance of the tracking error, which is the difference between estimated and ground truth location, as a metric of tracking performance. For the inner rectangle, mono vision achieves an error variance of 0.16 m^2 and stereo vision a variance of 0.11 m^2—a 30% improvement. Generally speaking, mono vision exhibits high localization uncertainty when the axes of the observing cameras are close to parallel. Obviously, this occurs primarily in proximity to the corners of the rectangular track. The outer rectangle shows this ill-posed estimation problem even more frequently. The stereo vision's ranging ability mitigates this problem and we would expect stereo vision to perform even better for the outer rectangle. Indeed, stereo vision outperforms mono vision by 60%; the error variances are $0.28m^2$ and $0.7m^2$, respectively. This difference in tracking performance is not as obvious from the figure when considering error variances. When comparing the bias of the tracking error, the improvement is not as considerable: 0.59 m with mono vision versus 0.44 m under stereo vision. Finally, the estimated track under low-resolution stereo vision reveals its limited ranging resolution as it is not able to reliably dinstiguish between locations on the inner and outer rectangle.

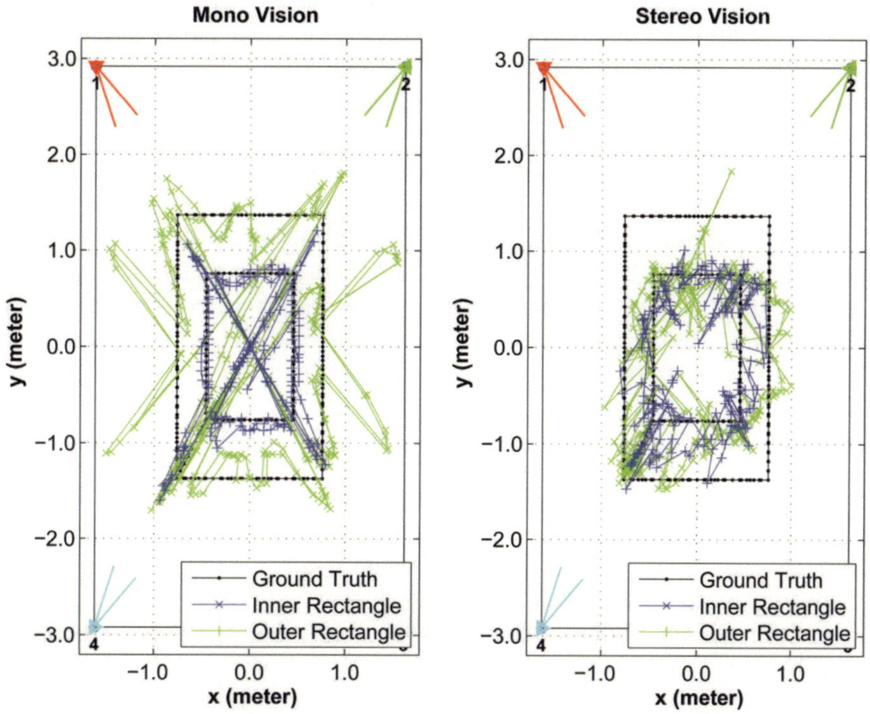

Fig. 9. Tracking performance of experimental network deployment: smart camera motes applying mono (left) and stereo vision (right).

7. Conclusion and future work

This chapter identified requirements and presented solutions for the adoption of stereo vision into resource-constraint smart camera networks. Stereo vision's main benefit lies in improving accuracy in target tracking applications. The two state-of-the-art embedded stereo vision architectures were discussed. Both architectures solve the problem of acquisition and processing of the high-rate image data at moderate clock frequencies. NXP's WiCa mote combines a custom parallel image processor with a dual camera system. Stanford's MeshEye mote requires only a common sequential 32-bit microcontroller to process the data from its hybrid vision system. Both vision systems are able to detect, localize, and capture high-resolution snapshots of foreground objects. Hybrid vision is computationally more efficient and consumes less energy for smaller objects within its field of view. This makes it well suited for a variety of applications in smart camera networks despite its low-resolution ranging capability. The experimental network deployment of four MeshEye motes resulted in a 30% reduction in target tracking error over smart camera motes utilizing solely monocular vision.

The availability of embedded stereo vision motes is the necessary step in their adoption for applications in smart camera networks. While this chapter covered their vision processing and hardware implementation, the challenge of vision calibration has not been addressed. Variations in the optical system and in alignment of the image sensors cause systematic errors in object localization. Calibration techniques, which can be easily and cheaply accomplished in volume production, need to be developed to measure and compensate these variations.

In conclusion, we expect to see more research contributions in the near future that analyze the performance of stereo vision in distributed camera networks with more rigor. This includes studying their merit over mono vision for different network tasks with respect to tracking accuracy, network lifetime, cost of deployment, and number of required camera motes. (Hengstler & Aghajan, 2007), for example, presents an early study of performance trade-offs for target tracking between mono and stereo vision in smart camera networks. Its simulation results encouragingly indicate that (i) stereo vision outperforms mono vision by factors of 2 to 5 in tracking accuracy and (ii) doubling the camera resolution can result in one third the tracking error variance.

8. References

Agilent Technologies (2004). ADNS-3060 High-Performance Optical Mouse Sensor, Datasheet, Oct. 2004

Agilent Technologies (2005). ADCM-2700-0000 Landscape VGA Resolution CMOS Camera Module, Datasheet, Jan. 2005

Atmel Corporation (2006). AT91SAM7Sxxx AT91 ARM Thumb-based Microcontrollers, Datasheet, Apr. 2006

Bramberger, M.; Doblander, A.; Maier, A.; Rinner, B. & Schwabach, H. (2006). Distributed embedded smart cameras for surveillance applications, *IEEE Computer Processing Magazine*, Vol. 39, No. 2, Feb. 2006, pp. 68-75

Foresti, G.; Micheloni, C.; Snidaro, L.; Remagnino, P. & Ellis, T. (2005). Active video-based surveillance system: the low-level image and video processing techniques needed for implementation, *IEEE Signal Processing Magazine*, Vol. 22, No. 2, Mar. 2005, pp. 25-37

Hampapur, A.; Brown, L.; Connell, J.; Ekin, A.; Haas, N.; Lu, M.; Merkl, H. & Pankanti, S. (2005). Smart video surveillance: exploring the concept of multiscale spatiotemporal tracking, *IEEE Signal Processing Magazine*, Vol. 22, No. 2, Mar. 2005, pp. 38-51

Hengstler, S. & Aghajan, H. (2006a). Application Development in Vision-Enabled Wireless Sensor Networks, *Proceedings of the International Conference on Systems and Networks Communications (ICSNC 2006)*, pp. 30-36, Oct. 2006, IEEE Computer Society, Washington, DC, USA

Hengstler, S. & Aghajan, H. (2006b). A Smart Camera Mote Architecture for Distributed Intelligent Surveillance, *ACM SenSys 2006 International Workshop on Distributed Smart Cameras (DSC 2006)*, pp. 6-10, Oct. 2006, ACM, New York, NY, USA

Hengstler, S. & Aghajan, H. (2007). Application-Oriented Design of Smart Camera Networks, *Proceedings of the First ACM/IEEE International Conference on Distributed Smart Cameras (ICDSC '07)*, pp. 12-19, Sept. 2007, ACM Press, New York, NY, USA

Hengstler, S.; Prashanth, D.; Fong, S. & Aghajan, H. (2007). MeshEye: a hybrid-resolution smart camera mote for applications in distributed intelligent surveillance, *Proceedings of the 6th International Conference on Information Processing in Sensor Networks (IPSN '07)*, pp. 360-369, Apr. 2007, ACM Press, New York, NY, USA

Kleihorst, R.; Abbo, A.; Choudhary, V. & Schueler, B. (2006). Design Challenges for Power Consumption in Mobile Smart Cameras, *Proceedings of COGnitive systems with Interactive Sensors (COGIS 2006)*, pp. 9B-3, Mar. 2006, S.E.E., Paris, France

Kleihorst, R.; Schueler, B. & Danilin, A. (2007). Architecture and Applications of Wireless Smart Cameras (Networks), *Proceedings of the 2007 IEEE International Conference on Acoustics, Speech, and Signal Processing (ICASSP '07)*, Vol. 4, pp. 1373-1376, Apr. 2007, IEEE Signal Processing Society, Piscataway, New Jersey, USA

Liu, Y. & Das, S. (2006). Information-intensive wireless sensor networks: potential and challenges, *IEEE Communications Magazine*, Vol. 44, No. 11, Nov. 2006, pp. 142-147

Maleki-Tabar, A.; Keshavarz, A. & Aghajan, H. (2006). Smart Home Care Network using Sensor Fusion and Distributed Vision-Based Reasoning, *Proceedings of the 4th ACM International Workshop on Video Surveillance and Sensor Networks (VSSN 2006)*, pp. 145-154, Oct. 2006, ACM, New York, NY, USA

Nakamura, E.; Loureiro, A. & Frery, A. (2007). Information fusion for wireless sensor networks: Methods, models, and classifications, *ACM Computing Surveys (CSUR)*, Vol. 39, No. 3, Aug. 2007, pp. 9

Qureshi, F. & Terzopoulos, D. (2007). Smart Camera Networks in Virtual Reality, *Proceedings of the First ACM/IEEE International Conference on Distributed Smart Cameras (ICDSC '07)*, pp. 87-94, Sept. 2007, ACM Press, New York, NY, USA

Radke, R.; Andra, S.; Al-Kofahi, O. & Roysam, B. (2005). Image change detection algorithms: a systematic survey, *IEEE Transactions on Image Processing*, Vol. 14, No. 3, Mar. 2005, pp. 294-307

Rahimi, M.; Ahmadian, S.; Zats, D.; Laufer, R. & Estrin, D. (2006). Magic of Numbers in Networks of Wireless Image Sensors, *ACM SenSys 2006 International Workshop on Distributed Smart Cameras (DSC 2006)*, Vol. 1, pp. 77-81, Oct. 2006, ACM, New York, NY, USA

Rahimi, M.; Baer, R.; Iroezi, O.; Garcia, J.; Warrior, J.; Estrin, D. & Srivastava, M. (2005). Cyclops: in situ image sensing and interpretation in wireless sensor networks, *Proceedings of the 3rd International Conference on Embedded Networked Sensor Systems (SenSys '05)*, pp. 192-204, Nov. 2005, ACM, New York, NY, USA

Texas Instruments (2006). Chipcon CC2420 2.4 GHz IEEE 802.15.4/ZigBee-ready RF Transceiver, Datasheet, 2006

Zhao, F. & Guibas, L. (2004). *Wireless Sensor Networks: An Information Processing Approach,* Morgan Kaufmann Publishers Inc., San Francisco, CA, USA

6

Grayscale Correlation based 3D Model Fitting for Occupant Head Detection and Tracking

Zhencheng Hu, Tetsuya Kawamura and Keiichi Uchimura
Kumamoto University
Japan

1. Introduction

With the development of collision safety technology in recent years, delicate control of air bag deployment which adaptively deploys the airbag depending on occupants' body shape, weight and position, has being intensively studied during past few years. The main purpose of the smart air bag system is to deal with the threat that occupants may be seriously injured by the deployment of an air bag at the time of crash if the occupant is too near to the airbag.
The National Highway Traffic Safety Administration (NHTSA) (Federal Motor Vehicle Safety Standards, 2000) specifies different classes for the occupancy including infants in rear facing infant seats, children and small adults, and out-of-position zones for the human occupants, on which the air bag deployment has to be controlled.
Research on detecting the type and position of occupant can be divided into 3 main categories based on different sensing technologies: I) Weight sensors on the seat measure the pressure distribution and classify the occupant into different types(Kennedy 2006, Lasten 2006); II) Electric-magnetic or ultrasound sensors that detect the change in the electric-magnetic field to confirm occupant type and position (Seip 1999); III) Computer vision sensors that directly detect occupant head and body position with 2D or 3D information, and classify the occupants (Trivedi, 2002-2005). Category I and II are the most popular sensors in the market in current stage of air bag control, which requires a reliable classification of adults, children and rear-faced child seats. However they are not adaptable for precisely detection of occupant position and posture, which is vital to the delicate control of air bag deployment.
Vision sensor provides the richest information of occupant position and posture. Depending on the number of cameras used, these studies can be further divided into two categories: monocular camera based methods and stereo vision based methods. Monocular camera always employs edge, contour and other image features to detect ellipse-liked shapes for head detection. By combining with the infrared detector, single camera solution can also obtain satisfied result in some well-controlled environment. However, it suffers from strong shadows, hot weather and insufficient 3D information which is necessary for functions such as the *out-of-position detection*. Stereo vision based methods use two co-planar cameras to calculate the disparity data and detect occupant head position and posture. Many algorithms employ the general 3D model fitting method to detect the ellipsoid-like 3D shape from a range image obtained from the stereo rig. M. Trivedi (Trivedi 2002-2005) uses shape and size constraints to eliminate search regions for less computation purpose, which may

have serious side-effects that the head region can be also eliminated when it appears relative smaller than other ellipsoid-like shapes such as waving arms and shoulders. B.Alefs (Alefs, 2004) uses depth data to recovery the occupant body surface and edge data to generate head candidate. Head recognition was carried out with a large trained dataset.

To achieve real-time performance while keeping high accuracy of occupant head detection, this paper presents a fast 3D parametric model fitting algorithm based on grayscale correlation of range data. Comparing with the traditional 3D parametric model fitting algorithms, this method simplifies the problem of searching 3D model from depth image into 2D grayscale correlation problem, which simultaneously determine all parameters with the best fitting model. By applying the proposed algorithm into occupant head detection application, this paper also proposes a body centerline segmentation method as well as a multi-resolution disparity generation algorithm in order to deal with body occlusion and extra-near disparity calculation problems.

In the remainder of this chapter we will present a brief overview of traditional 3D parametric model fitting and our new approach based on grayscale correlation of range image (Section 2), a detail implementation of our approach (Section 3), and experimental results in the purview of an occupant head detection system (Section 4).

2. 3D parametric model fitting algorithm

2.1 Problem description

Given an image frame (e.g. range image or edge image), the 3D parametric model fitting problem is to find the 3D parameters (e.g. 3D position and orientation, scale factor, intrinsic parameters, etc.) of the model. Figure 1 shows an example of finding an ellipsoid in a range image. The total number of ellipsoid 3D parameters is 9 including 3 rotation and 3 translation parameters, as well as 3 scaling factors along X, Y and Z-axis.

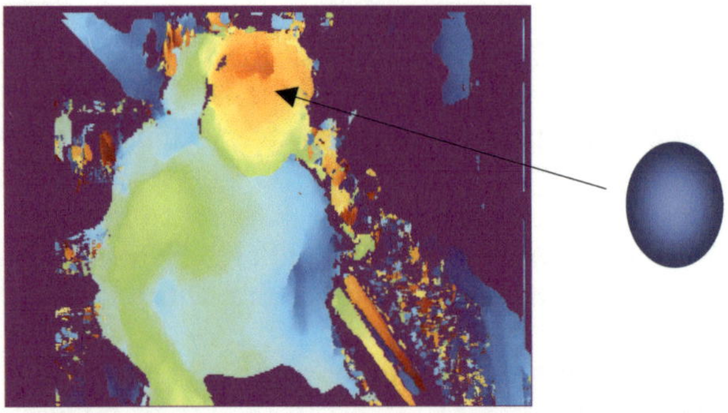

Figure 1. Searching a 3D model in a range image

Research on 3D model fitting leveraged earlier work done in (Lowe, 1991) for generic 3D parametric model fitting. Image formation is modeled as a mapping of a 3D model into the image. Although the inverse mapping is non-linear due to the trigonometric functions of perspective projection, the resulting image changes smoothly as the parameters are changed. Therefore, local linearity can be assumed and several iterative methods can be employed for solving non-linear equations (e.g. Newton's method). Upon finding the solution for one frame, the parameters are used as the initial values for the next frame and the fitting procedure is repeated. The traditional approach can be extremely time consuming and is not adaptive to the real-time required occupant head detection application.

2.2 Our algorithm

With the assumption of local linearity, we can prepare a lookup table of possible combination of all parameters except 3D position (X, Y, Z), which will be determined by the later process of grayscale correlation. Rotation and scale parameters are used to generate the LUT in the case of ellipsoid detection. To simplify the process, only certain combination of rotation and scale parameters are adopted by the constraint of occupant physical position and posture. Here, 3 rotation angles {0, +45°, -45°} along X and Z-axis are combined with 3 different ellipsoid shapes. Scale factors are defined by the possible movement range of the head.

Equation of 3D ellipsoid is shown as follows:

$$Z = Z_0 + c_i \sqrt{1 - \frac{(X - X_0)^2}{a_i^2} - \frac{(Y - Y_0)^2}{b_i^2}} \tag{1}$$

where a_i, b_i, c_i are scale parameters and X_0, Y_0, Z_0 are the 3D world coordinates of ellipsoid center.

Perspective projection equation (2) is adapted to project 3D ellipsoid surface points to the 2D image coordinates.

$$\begin{bmatrix} x \\ y \\ 1 \end{bmatrix} = \begin{bmatrix} f_x & 0 & u_0 \\ 0 & f_y & v_0 \\ 0 & 0 & 1 \end{bmatrix} \begin{bmatrix} R_{11} & R_{12} & R_{13} & 0 \\ R_{21} & R_{22} & R_{23} & 0 \\ R_{31} & R_{32} & R_{33} & 1 \end{bmatrix} \begin{bmatrix} X \\ Y \\ Z \\ 1 \end{bmatrix} \tag{2}$$

where camera's intrinsic parameters like lens focal (f_x, f_y) and optical center coordinates (u_0, v_0) are obtained from some preprocessing steps like camera calibration. Rotation matrix elements $R_{11} \sim R_{33}$ are retrieved from the parameter LUT.

To match with the disparity image, we use the following normalization equation to convert range data into intensity value.

$$I(x, y) = \frac{Z_{\min}}{Z} \times (2^N - 1) \tag{3}$$

where $Z_{\min}$ is the minimum distance from ellipsoid surface points to camera. N is the bit-value of intensity image.

Figure 2 shows some examples of models generated from the parameter LUT.

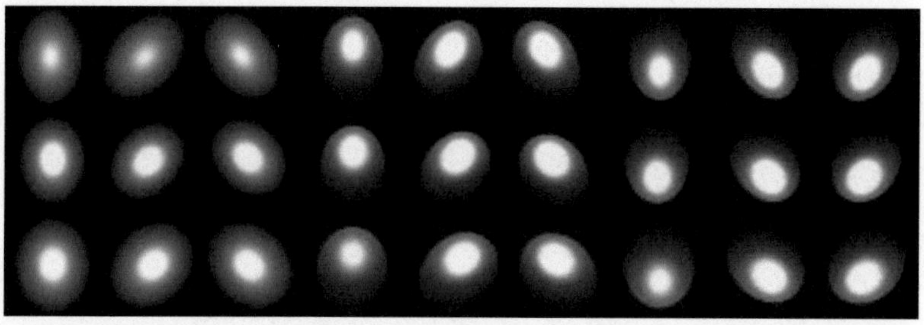

Fig. 2. Parametric models of range data

2.3 3D model fitting

Once the 3D parametric models are generated, we can simply adapt the traditional grayscale correlation algorithm to find a match between models and target range image.
2D grayscale correlation algorithms are well studied for decades and many acceleration techniques like multi-level and pyramid sub-sampling technologies have been proposed. To add tolerance to intensity change, we use the normalized grayscale correlation (NGC) equation to find the best matching from multiple models.

$$r = \frac{S\sum I(x,y)d'(x,y) - (\sum I(x,y))\sum d'(x,y)}{\sqrt{\{S\sum I^2(x,y) - [\sum I(x,y)]^2\}\{S\sum d'^2(x,y) - [\sum d'(x,y)]^2\}}} \tag{4}$$

where $I(x,y)$ and $d'(x,y)$ are intensity values of model image and normalized target image respectively. S is the effective pixel number. Matching score $r = 1$ refers to the perfect match and $r = 0$ means not match at all.
The normalization process on the target image is a general histogram smoothing as described in equation (5).

$$d'(x,y) = (2^N - 1)\frac{d(x,y) - d_{\min}}{d_{\max} - d_{\min}} \tag{5}$$

where $d(x,y)$ is the original disparity value on pixel (x, y), $d_{\min}, d_{\max}$ are the minimum and the maximum disparity value in the region. N is the bit-value of disparity map.
Result of our grayscale correlation algorithm presents not only the position but also the best model, which indicates the rotation and scale parameters simultaneously.

3. System implementation details

The system is designed as a co-planar stereo camera with constructive infrared illumination light source. The stereo rig is mounted on the center roof console near the back mirror. Generally it should have few centimeters baseline and wide-angle lens that can overview the whole passenger's cabinet.

3.1 Constructive illumination lighting system

A fast stereo algorithm (Konolige, 1997) is adapted to generate disparity map with two synchronized video source input at 30 frames per second. To overcome the uneven illumination and shadow problem for real outdoor environment, an infrared pulsed illumination lighting system is installed, combining with band-pass filtered lens to cutoff all un- necessary wavelength light.
A disadvantage of block matching based dense disparity algorithm is the aperture problem. The aperture problem arises as a consequence of the ambiguity of one-dimensional intensity on left and right image through out the horizontal Epipolar line. No disparity data can be derived for an even intensity region like dark or over lighted regions.
We tested different kinds of light patterns, and the cross pattern of light-dark-light with an angle of ±45 degree showed the best performance. Figure 3 shows an example of disparity map result without/with constructive light.

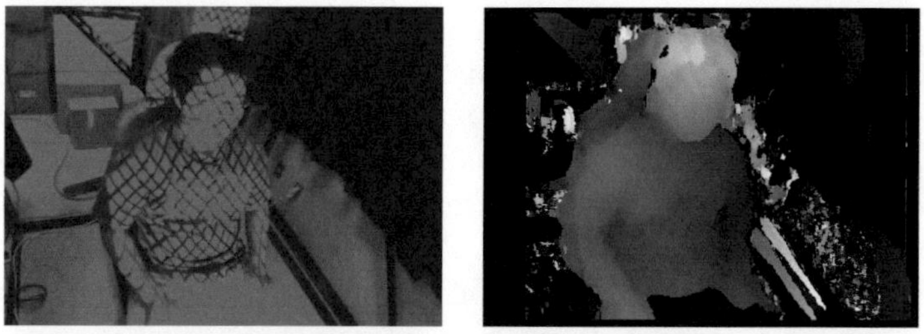

(a) Without constructive light

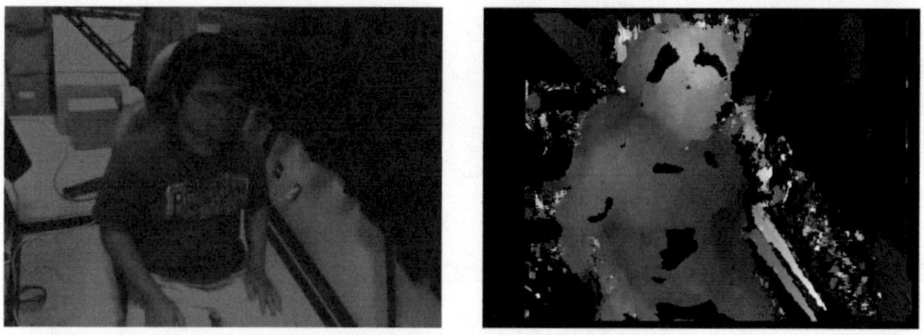

(b) With constructive light

Fig. 3. Disparity map result without/with constructive light

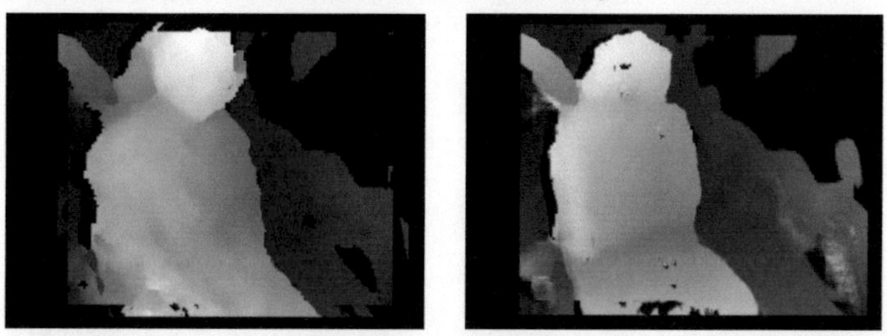

(a)Original disparity map (b)Background disparity map

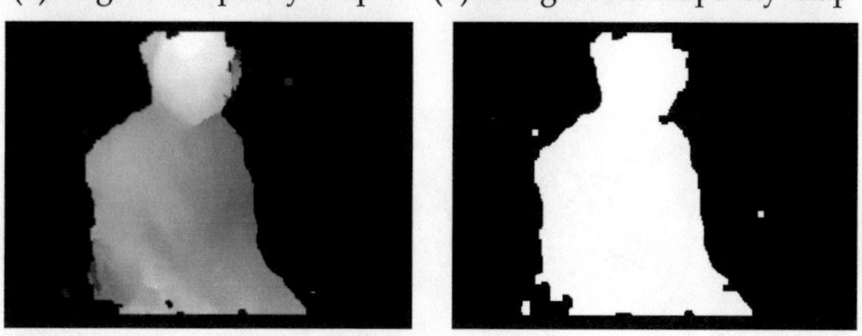

(c)Binarized subtraction result (d) Extracted ROI

Fig.4. Background subtraction result

3.2 Background subtraction

To eliminate passenger's seat, door and other interior regions from the range image, background subtraction is carried out for every new frame. Background range image were generated as an average of 30 frames' range image for the empty seat. Automatic background generation will be further implemented according to the sensors' output of seat lateral position and reclining angle.

Post-processing includes binarizing, morphological process, and blob analysis. The biggest blob that satisfies the position and area constraints will be extracted as occupant body's candidate region. Figure 4 shows the background subtraction results.

3.3 Composition of multi-resolution disparity maps for near distance disparity

Fast stereo processing algorithms (Konolige, 1997) always use a fixed maximum disparity value to accelerate the matching process. For example, a maximum disparity of 32 pixels leads to the maximum searching distance of 32 pixels. Disparities over the maximum disparity will be omitted.

According to the basic equation of stereo disparity shown in (6), the maximum distance leads to the minimum detection distance, as the baseline b and lens focal f is unchanged.

$$d = x_l - x_r = \frac{fb}{z} \tag{6}$$

Figure 5 shows an example of extra-near distance target that cannot obtain disparity data. To enlarge the disparity range for extra-near target detection, we propose a composition algorithm of multiple resolution disparity maps. A lower resolution stereo image pair will generate a wider detection range disparity map since its pixel size is bigger than the general resolution image pair. Figure 6 shows the composition result of disparity maps generated from 160x120 and 320x240 stereo images.

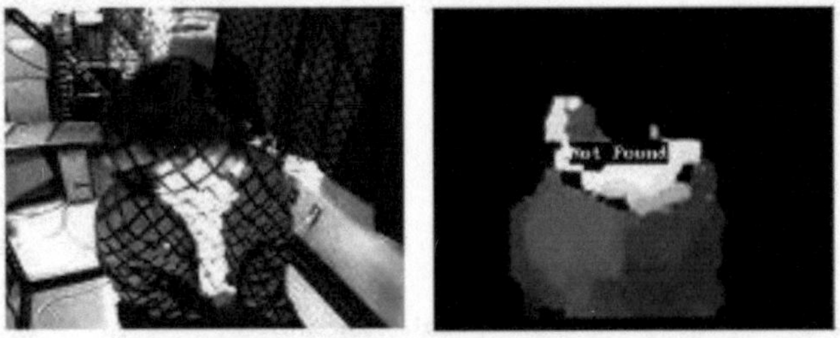

Fig. 5. Near distance target

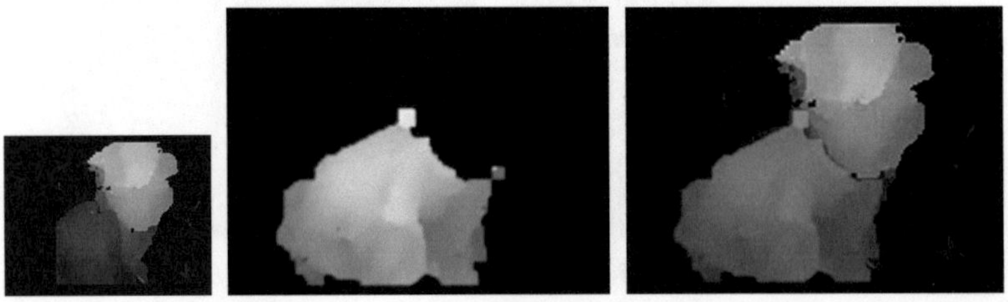

Fig. 6. (a)disparity map generated from 160x120 stereo images, (b) from 320x240 stereo images, (c)Composition of above disparity maps

3.4 Foreground segmentation with body center line

Extracted occupant body ROI may include multiple ellipsoid-liked regions which has similar size with the head, such as shoulders, waving arms, and other objects. Examples are shown in Figure 7. In this paper, we extended Russakoff's concept (Russakoff, 2002) of body center line to 3D region segmentation to eliminate the ambiguities.

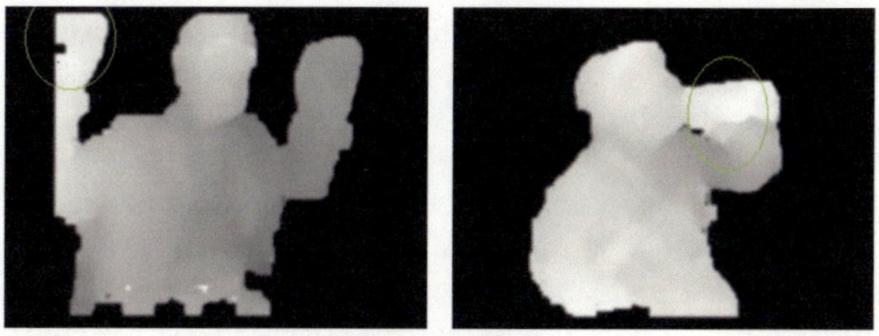

Fig. 7. Examples of ellipsoid-liked objects extracted from body ROI

Assuming passenger is always sitting on the seat, so that the lower part of body's ROI is relatively stable and can be used as the reference part to segment the ROI. Detail steps are shown as follows:

Step 1. After the preprocessing steps described in the above sections, calculate the so-called horizontal median points on each row of the binary ROI image based on Russakoff's algorithm.

Step 2. Detect the upper center position C1 and lower center position C2 along the body center line, where C1 and C2 are on the rows of 1/5 and 4/5 of the ROI height respectively as shown in Figure 8(a).

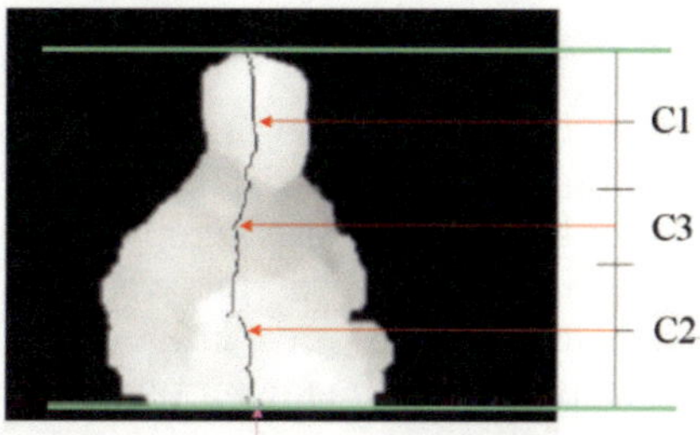

(a) Body Center Line

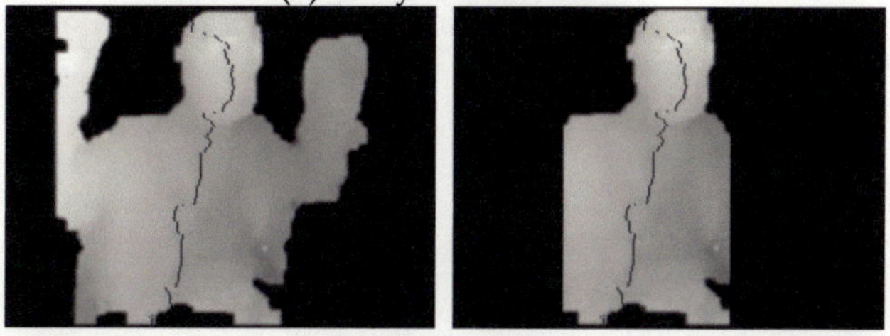

(b) Foreground segmentation result for waving arms

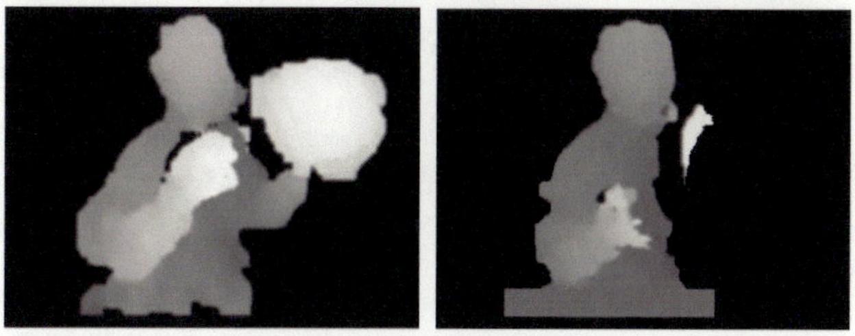

(c) Filtering result by disparity constraint

Fig. 8. Foreground segmentation with body center line

Step 3. If the slope angle of line C1C2 is less than threshold k (the occupant is in the normal seating position), then we can simply vertically cut off the regions that are further than a predefined distance to C1C2's middle point C3. An example is shown in Figure 8(b).

Step 4. If the slope angle is larger than threshold k (the occupant is in the leaning position), the cut-off lines will be parallel to line C1C2, while keeping the predefined distances.

Step 5. Segmented foreground region will be further filtered by the constraint of disparity. The ideal disparity data on each row i can be calculated through the following linear interpolation equation.

$$d_i = d_1 - \frac{y_1 - i}{y_2 - y_1}(d_2 - d_1) \tag{7}$$

This constraint will eliminate most of the outliers and other objects in front of the body ROI. An example is shown in Figure 8(c).

Step 6. The result image will be further normalized for the 3D model fitting process described in Section II.

4. Experimental results

The proposed algorithm was tested under various sizes of passengers and different postures that occupants may behave during the normal driving situations. The stereo vision system was equipped with two gen-locked CCD cameras. Stereo images were captured by a Matrox Meteor2/MC frame grabber board and all processing was done by a Pentium IV 2.66GHz PC. The stereo baseline is 64 mm, and the lens focal is 2.8 mm. 320x240 disparity maps were generated at the speed of 25 ms/frame with the maximum disparity of 32 pixels.

Tester (Posture)	Correct Detected	False Detected	Not Detected	Rate of Correct(%)
1(N)	1500	0	0	100
2(N)	1500	0	0	100
3(N)	1500	0	0	100
4(N)	1500	0	0	100
5(N)	1500	0	0	100
6(N)	1500	0	0	100
7(N)	1500	0	0	100
8(N)	1478	9	13	98.5
9(A)	1500	0	0	100
10(B)	1477	17	6	98.5
11(C)	1416	7	77	94.4
12(D)	1369	2	129	91.3
13(E)	1500	0	0	100
14(F)	1497	0	3	99.8
15(G)	1460	0	40	97.3
16(H)	1484	8	8	98.9
Total	23681	43	276	98.7

Table 1. Test results of different situations

Postures: N=Normal siting position, A=Reading book, B=Playing basketball, C=Moving body in different direction, D=Waving arms around head, E=Reading newspaper, F=Talking with a mobile phone, G=Drinking water, H=Wearing a cap.

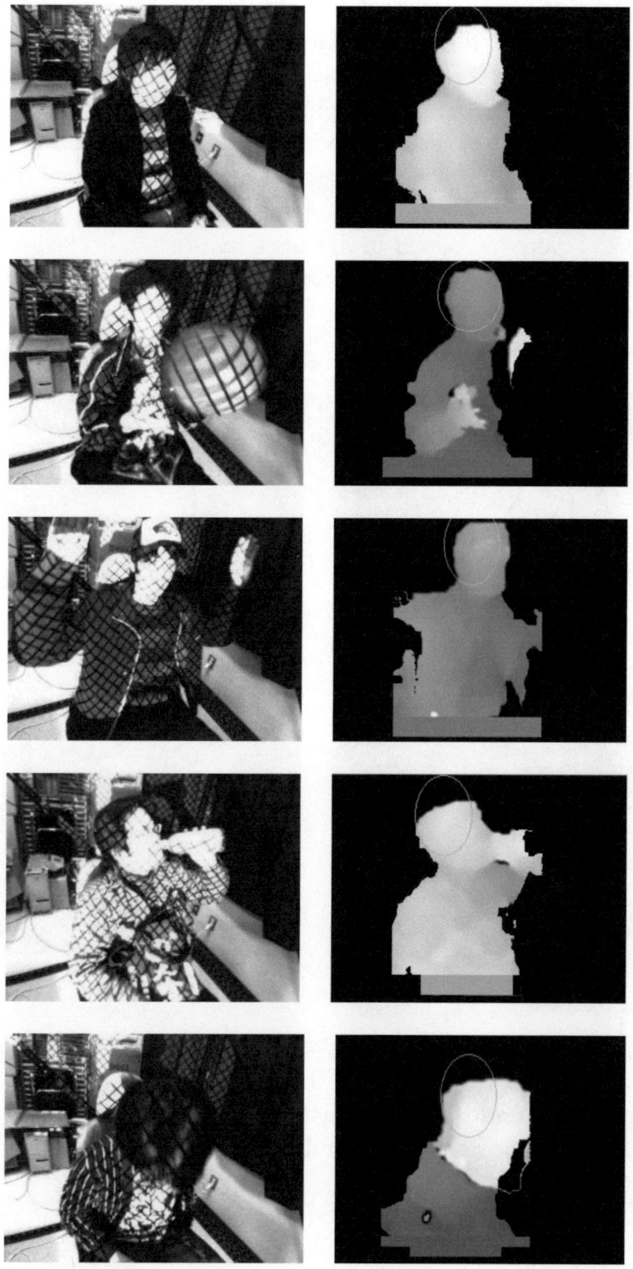

(a)original image (b)detection result

Fig. 9. Occupant Head Detection Results

Totally 16 adults testers including 12 males and 4 females were chosen for the test. With their height distributed from 153cm to 183cm and weight distributed from 50kg to 80kg, the testers were supposed to cover the main range of adult passenger sizes. They were asked to perform all kinds of postures that could be happened during the real driving situations, like readings, waving arms, drinking, etc. Each test was continuously captured for 1500 frames. Table 1 shows the test results of different situations. Tester 1 to 8, who were sitting straightly in the normal position, showed the best performance near 100% correct detection rate. Tester 9 to 16, who were asked to perform different kinds of movement and postures, still showed a very high detection rate about 97.5%. The overall correct detection rate is 98.7%. Some very difficult situations like partially occluded target, extra-near target and multiple ambiguities were also correctly detected. Figure 9 shows some examples.
False detection (<0.2%) were happened under the situations of occlusion and head was not detected (<1.2%) mostly due to the situation that occupant was out of position. Figure 10 shows some false examples.

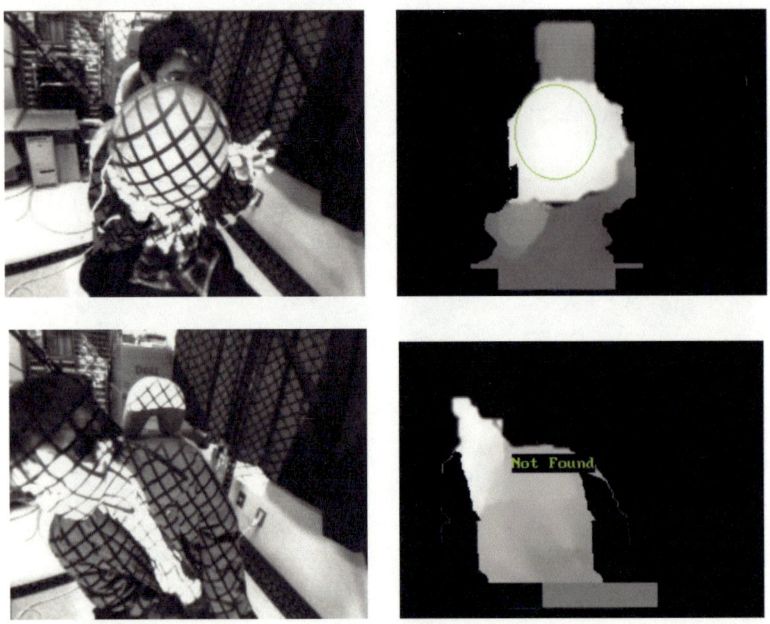

Fig. 10. Falsely Detected Examples

False detection and miss detection generally happen within very short period of time. Tracking of head position in both intensity image and disparity map will largely help to locate the head position even for fully occlusion case. Tracking can also reduce searching area by predicating head position. Some preliminary tests were carried out and showed very satisfied results.

5. Conclusion

Occupant head detection is sensitive to the variation of illumination, occupant posture and body size. To achieve real-time performance while keeping a high accuracy of occupant head detection, this paper presents a fast 3D parametric model fitting algorithm base on grayscale correlation of range data. Evaluation of the method shows over 98% correct head detection. Combining with head tracking algorithm on intensity image and disparity map, the proposed algorithm will perform near 100% correct detection.

6. References

Federal Motor Vehicle Safety Standards (2000). *Occupant crash protection, final rule.* Department of Transportation, Federal Register, vol. 65; no. 93; pp. 30680-30770

Kennedy K.R., Nathan J.F., Shridhar M. (2006). An LVQ-based Automotive Occupant Classification System, *Proceedings of 18th Intl. Conf. on Pattern Recognition,* Vol.2, pp.662-665

Lasten K., *et al.* (2006). iBolt Technology – A Weight Sensing System for Advanced Passenger Safety, *Advanced Micosystems for Automotive Application 2006- Part 2,* VDI-Buch, Jurgen Valldorf and Wolfgang Gessner, pages 171-186

Seip R., Adamczyk B., Rundell D. (1999). Use of Ultrasound in Automotive Interior Occupancy Sensing: Optimum frequency, beam width, and SNR from Empirical Data, *Proceedings of IEEE Ultrasonics Symposium,* Vol.1, Pages. 749-752

Mikic I., Trivedi M. (2002). Vehicle Occupant Posture Analysis Using Voxel Data, *Proceedings of Ninth World Congress on Intelligent Transport Systems*

Krotosky S., Cheng S., and Trivedi M. (2005). Real-Time Stereo-Based Head Detection using Size, Shape and Disparity Constraints, *Proceedings of IEEE Symposium of Intelligent Vehicle 2005*

Trivedi M., Cheng S. Y., Childers E., and Krotosky S. (2004). Occupant posture analysis with stereo and thermal infrared video: Algorithms and experimental evaluation, *IEEE Trans. Veh. Technol.,* vol. 53, no. 6, pp.1968–1712

Krumm J., Kirk G. (1998). Video Occupant Detection for Airbag Deployment, *Proceedings of Fourth IEEE Workshop on Applications of Computer Vision*

Alefs B. *et al.* (2004). Robust occupancy detection from stereo images, *Proceedings of IEEE Intelligent Transportation Systems Conference*

Lowe D.G. (1991). Fitting parameterized three-dimensional models to images, IEEE Transactions on Pattern Analysis and Machine Intelligence, vol. 13, no. 5, pp. 441-450

Konolige K. (1997). Small vision systems: hardware and implementation, *Proceedings of Eighth International Symposium on Robotics Research*

Russakoff D.B., Herman M. (2002). Head tracking using stereo, *Machine Vision and Applications,* Vol. 2002, No.13, pp.164-173

Birchfield S. (1997). An Elliptical Head Tracke, *Proceedings of 31st Asilomar Conference on Signals, System, and Computers*

Yang R. (2002). Model-based Head Pose Tracking With Stereo Vision, *Proceedings of 5th IEEE Intl. Conf. On Automatic Face and Gesture Recognition,* pages 255-260

Hernandez A.M., Devy M. (2000). Application of a Stereovision Sensor for the Occupant Detection and Classification in a Car Cockpit, *Proceedings of 2nd*

International Symposium on Robotics and Automation, LAAS No.00444, pp.491-496

Sun C. (2002). Fast Stereo Matching Using Rectangular Subregioning and 3D Maximum-Surface Techniques, *International Journal of Computer Vision,* vol.47, No.1/2/3, pp.99-117

A Performance Review of 3D TOF Vision Systems in Comparison to Stereo Vision Systems

Stephan Hussmann[1], Thorsten Ringbeck[2] and Bianca Hagebeuker[2]
[1]Westcoast University of Applied Sciences
[2]PMDTechnologies GmbH
Germany

1. Introduction

The most common and well-known principle of 3D image acquisition is stereo vision (SV). This principle of 3D-image acquisition is already known and used for decades in the research community. The advantage of stereo vision to other range measuring devices such as laser scanners, acoustic or radar sensors is that it achieves high resolution and simultaneous acquisition of the entire range image without energy emission or moving parts.

Still, the major disadvantage is the limited Field of View (FOV) and the correspondence problem. To enhance the FOV many techniques are researched such as rotating cameras (Kang et al., 1997; Benosman et al., 1996; Krishnan et al., 1996), increasing the number of the cameras (Kawanishi et al., 1998), and the use of a special optic (Liancheng & Feng, 2005; Lin & Bajcsyg, 2003). Also a combination of 1D-laser scanner and a SV system are proposed to overcome the FOV problem (Cheng et al., 2002). However, these systems are expensive and not easy too synchronize. Due to the correspondence problem algorithms need to be improved to give a lower percentage of false matches as well as better accuracy of depth estimates. Performance of algorithm needs to be evaluated over a broad range of image types in order to test their robustness (Dhond & Aggarwal, 1989).

In the past years the modality of Time-of-Flight (TOF) imaging became more and more attractive to a growing research community (Schwarte_a et al., 1997; Schwarte_b et al., 1997). Because of the enormous progress in TOF-vision systems, nowadays 3D matrix cameras can be manufactured und be used for many application such as robotic, automotive, industrial, medical and multimedia applications. Due to the increasing demand of safety requirements in the automotive industry it can be assumed that the TOF-camera market will grow and the unit price of these systems in the mass production will drop down to ca. 100 € (Hussmann & Hess, 2006).

For all application areas new accurate and fast algorithms for 3D object recognition and classification are needed. As now commercial 3D-TOF cameras are available at a reasonable price the number of research projects is expected to increase significantly. One early example of using a TOF-camera based on the Photonic-Mixer-Devices (PMD)-Technology for 3D object recognition in TOF data sets are presented in (Hess et al., 2003). In this paper the transition from a general model of the system to specific applications such as intelligent

airbag control and robot assistance in surgery are demonstrated. A more current example in using a PMD-TOF camera on a robot system, highlighting the advantages of TOF- compared to SV-vision systems, is reported in (Hussmann & Liepert, 2007).
This chapter is structured as follows. In Section II we expose the issues which motivate a new approach for 3D image acquisition systems. In Section III we derive the equation set needed to design a 3D TOF vision system. In Section IV automotive applications will demonstrate the usability of 3D-TOF vision systems based on the PMD-technology fulfilling the tough target settings in the automotive industry. Concluding remarks will summarize the paper.

2. Comparison of TOF- and SV vision systems

2.1 Working principle

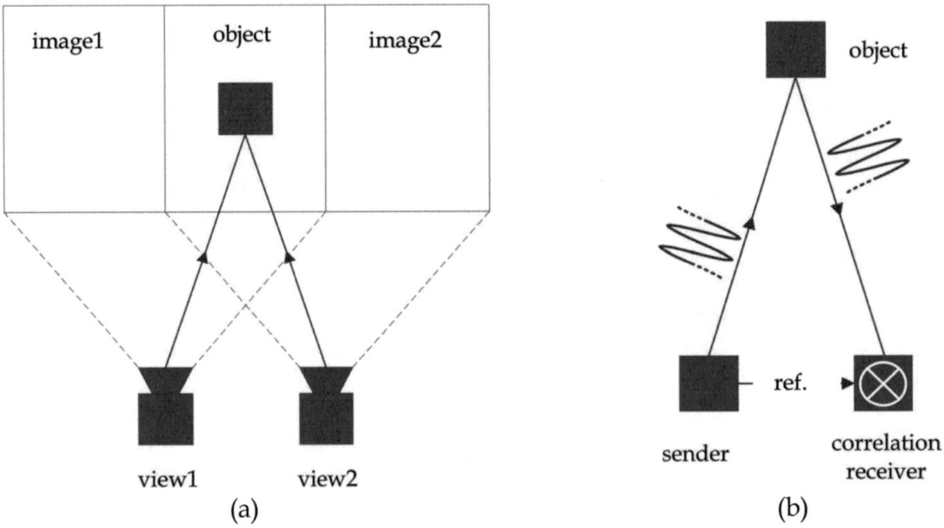

Fig. 1. Working principle of (a) SV vision systems and (b) TOF vision systems (Hussmann & Hess, 2006)

Common stereo vison systems comprise two perspective cameras with limited FOV. As shown in Fig. 1 (a) a physical point is taken up in the observed 3D-space by two perspective cameras. If the corresponding pixel of this point is found in both camera images, the position can be computed with the help of the triangulation principle. The major problem is to detect the corresponding pixels as they are required to estimate the range information correctly (*correspondence problem*). High processing power and time can be consumed if those pixels cannot be find easily. Fig. 1 (b) shows the working principle of TOF vision systems. TOF is an active range system and needs an illumination source. The range information is measured by emitting a modulated near-infrared light signal and computing the phase of the received reflected light signal. Using the PMD-Technology the phase calculation is carried out in each individual pixel of the sensor matrix (Schwarte_a et al., 1997; Schwarte_b et al., 1997). Hence TOF offers a direct depth data acquisition, whereas SV involves a great amount of computational power for the same 3D image. However SV systems can be realized without active illumination.

2.2 Image processing chain

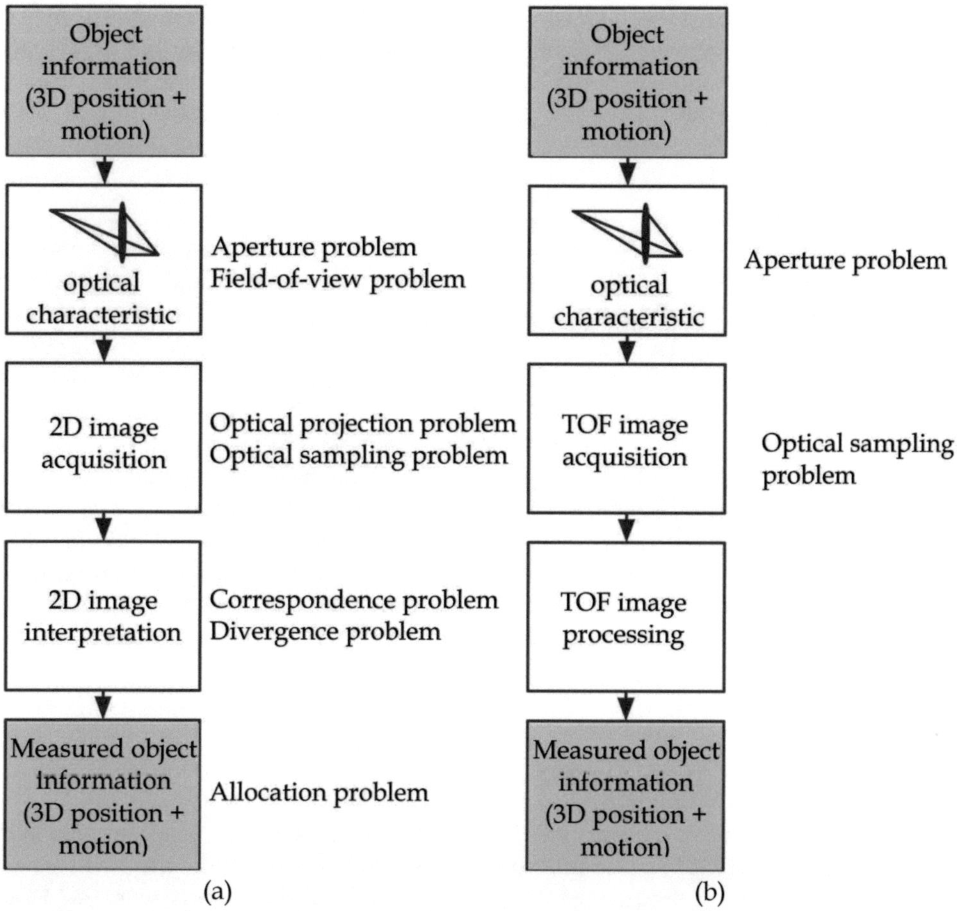

Fig. 2. Functional block diagram of the image processing chain of (a) SV vision systems and (b) TOF vision systems (Hussmann & Hess, 2006)

In Fig. 2 (a) the functional block diagram of the image processing chain of SV systems is shown. Several problems arise during the image processing of the captured object information. Due to the limited aperture angle of the camera optics only a certain part of the environment can be detected (*aperture problem*). The FOV is usually fixed for one application and cannot be changed easily (*FOV problem*). The optical projection of the objects on a planar sensor leads to a complete lost of the 3D-information (*optical projection problem*). Due to the sampling rate of the image sensor the detection of very fast objects is limited (*optical sampling problem*). The *correspondence problem* is already explained in the last section and will be described in more detail in the following section. A change in the gray level values of both cameras is not necessarily caused by an object movement. For example illumination changes could also cause gray level changes. Hence the object motion cannot be described by the gray level values changes (*divergence problem*). At last the *allocation problem* is stated. The estimated range values of the scene are not evenly distributed due to the correspondence problem.

TOF vision systems suffer fewer problems as shown in Fig. 2 (b). The *FOV problem* does not depend on the chosen hardware setup and can be changed by software easily. In the next section more details are provided. The object information is not lost by the optical projection of the objects on a planar sensor. Each pixel of the sensor calculates a range value. A reconstruction of range values based on gray level values does not exist. Hence no *correspondence, divergence* and *allocation problem* is given. Only the *aperture* and *sampling problem* still exists.

2.3 Field-of-view problem

The depth resolution of SV vision systems depends on the chosen optical arrangement of the two cameras which defines the triangle's angles. TOF systems do not depend on geometrical parameters. They are using an active modulated light source. Two modulation techniques are the most common one, pulsed modulation (Moring et al., 1989) or continuous wave (CW) modulation (Beheim & Fritsch, 1986). In the currently available TOF cameras CW-modulation is used as for this mode extremely high rise and fall times are not required and for this reason a larger variety of light sources are available. Mostly square waves are used, but other waveforms such as sinusoidal waves are suitable modulation signals. Using CW-modulation the phase difference between the sent and received optical signal is measured, rather than directly measuring a light pulse's turn-around time. As the modulation frequency f_{mod} is known, the measured phase φ_0 directly corresponds to the time of flight (Lange, 2000). Equation (1) describes how the range of TOF vision systems can be determined. The physical constant for the speed of light (3·10^8 m/s) is given by c and N represents the ambiguity in range estimation.

$$R = \frac{c}{2 \cdot f_{\text{mod}}} \cdot \left(\frac{\varphi_0}{360°} + N \cdot 360° \right) \quad \textit{with } N = 0,1,2,3... \tag{1}$$

A very common modulation frequency of TOF vision systems is f_{mod} = 20MHz. This leads to a non-ambiguity range (*NAR*) of 7.5 m as derived in Equation (2). There are two reasons for choosing this modulation frequency. One reason is that no high-power IR-LEDs are available with a higher modulation frequency at a low price tag and the other reason is that the NAR suits most of the indoor and outdoor applications.

$$NAR = \frac{c}{2 \cdot f_{\text{mod}}} = \frac{3 \cdot 10\text{^}8}{2 \cdot 20 \cdot 10\text{^}6} = 7.5 \text{ m} \tag{2}$$

The FOV of TOF vision systems depend only on the modulation frequency as derived in Equation (1) and (2). A high modulation frequency leads to short measuring range and vice versa.

2.4 Correspondence problem

As mentioned before the correspondence problem is a major problem as the range values can only be estimated in image regions with adequate gray level value changes. As a consequence a SV vision system will always have difficulties acquiring 3D information of objects with equal gray level values such as walls, roadways and so forth. Also objects with the same gray level value at different distances are difficult to distinguish. These problems can be compensated with advanced SV algorithms using a lot of processing power compared to TOF vision systems.

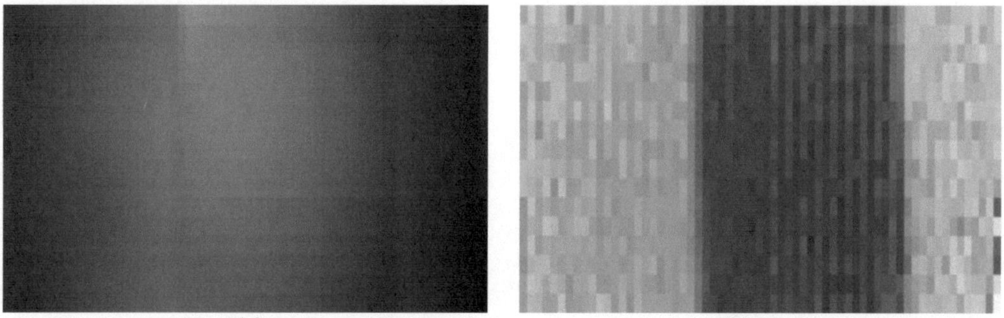

(a) (b)

Fig. 3. (a) Gray level value image and (b) TOF range image of two objects with equal gray level values

The images in Fig. 3 are taken with a PhotonICs® PMD 1k-S (64 x 16 Pixel) sensor chip. Fig. 3 (a) shows the gray level values of an object in front of a wall at different distances. Both objects have almost the same gray level values and are very difficult to distinguish. Hence a SV vision system without active lighting, generating shadows, can not estimate the 3D information of the objects due to the correspondence problem. However, a TOF vision system measures the range to the object for each pixel and therefore delivers evenly distributed 3D information as shown in Fig. 3 (b). Hence even a TOF vision system with fewer pixels than a SV vision system can deliver more 3D information when the SV vision system has too many false range matches. This can happen when the captured scenes have too many areas with equal gray level values. The advantage of TOF over a SV vision system is that the TOF ranging technique does not produce incomplete range data (no shadow effects) because illumination and observation directions can be collinear.

3. Design of PMD TOF vision systems

3.1 Range calculation

A PMD TOF sensor generates a range output voltage which can be described as follows (Schwarte & Heinol, 1999):

$$\Delta U_{ab}(T_L) = K \cdot \int_0^{T_{int}} P_{opt}(t - T_L) \cdot u_m(t) dt \quad (3)$$

The range output voltage $\Delta U_{ab}(T_L)$ is determined by the correlation result of the optical echo $P_{opt}(t\text{-}T_L)$ and the modulation voltage $u_m(t)$ over the integration time T_{int}. K is a system constant. State of the art is to use CW modulation with square waves. Hence the modulation voltages can be easily generated digitally with a high accuracy and stability using programmable logic devices (PLDs) such as microcontrollers, complex programmable logic devices (CPLD) or field programmable gate arrays (FPGA). The low-pass characteristic of the IR-LEDs leads to an attenuation of the square waves' harmonics for larger frequencies. This results in an optical output that gradually looks sinusoidal for frequencies larger than

5-10 MHz. This has to be taken into account if CW modulation with square waves is used. The modulation voltage $u_m(t)$ and the optical echo $P_{opt}(t-T_L)$ is then given by:

$$u_m(t) = \sum_{n=-\infty}^{\infty} rect(\omega t - \pi/2 - n \cdot 2\pi) = \frac{4}{\pi} \sum_{k=1}^{\infty} \frac{\sin((2k-1)\omega t)}{(2k-1)} \tag{4}$$

and

$$P_{opt}(t - T_L) = a \cdot (A + \sin(\omega t - \omega T_L)) \tag{5}$$

Equation (4) shows the square wave signal of the modulation voltage $u_m(t)$. Using Fourier series we can write an ideal square wave as an infinite series of only odd integer harmonics. Equation (5) shows the sinusoidal signal echo $P_{opt}(t-T_L)$ of the IR-LEDs. It is assumed that only the fundamental harmonic is transmitted due to the finite bandwidth of the IR-LEDs. The amplitude a of the optical echo $P_{opt}(t-T_L)$ depends on the reflectivity coefficient. In addition the background light is taken into account by adding a constant A. Furthermore it is assumed that a and A are constant during the capture of one range image. Hence $\Delta U_{ab}(T_L)$ will be different for different amplitudes of the optical echo and will course problems in determining the phase difference between the optical echo $P_{opt}(t-T_L)$ and the modulation signal $u_m(t)$. The amplitude dependency of the output voltage $\Delta U_{ab}(T_L)$ can be avoided by using a phase-shift algorithm. This algorithm is based on the cross correlation of two periodical signals which are shifted in phase to each other. The correlation function $\varphi_{sg}(\tau)$ is defined as follows:

$$\varphi_{sg}(\tau) = s(t) \otimes g(t) = \lim_{T \to \infty} \frac{1}{T} \int_{-\frac{T}{2}}^{\frac{T}{2}} s(t) \cdot g(t+\tau) dt \tag{6}$$

If we deploy equation (6) to our TOF system $s(t)$ and $g(t+\tau)$ can be written as

$$s(t) = P_{opt}(t - T_L) \quad \text{and} \quad g(t+\tau) = u_m(t+\tau) \tag{7}$$

Then the correlation function $\varphi_{sg}(\tau)$ of the TOF system can be represented as follow:

$$\varphi_{sg}(\tau) = \lim_{T \to \infty} \frac{1}{T} \int_{-\frac{T}{2}}^{\frac{T}{2}} a \cdot [A + \sin(\omega t - \varphi_0)] \cdot [\frac{4}{\pi} \sum_{k=1}^{\infty} \frac{\sin((2k-1)\omega(t+\tau))}{(2k-1)}] dt \tag{8}$$

Only the fundamental harmonic of the modulation voltage and the optical echo have the same frequency. Hence the multiplication of these correlated signals results in a DC component and a sinusoidal component with twice the frequency. All other multiplications result in sinusoidal components. After the integration only the DC component remains. Due to this feature of the PMD sensor all other uncorrelated noise sources such as sun light or modulated light sources are suppressed. The DC component is given by:

$$\varphi_{sg}(\tau) = \frac{2 \cdot a}{\pi} \cdot \cos(\varphi_0 + \tau) \tag{9}$$

$\varphi_0 = \omega T_L$ and represents the phase difference between the sent and the received signal. If we now consider equation (9) for example for four different phases τ = 0°, τ = 90°, τ = 180° and τ = 270°, the correlation result is:

$$\varphi_{sg}(0°) = \frac{2 \cdot a}{\pi} \cdot \cos(\varphi_0), \qquad \varphi_{sg}(90°) = -\frac{2 \cdot a}{\pi} \cdot \sin(\varphi_0),$$
$$\varphi_{sg}(180°) = -\frac{2 \cdot a}{\pi} \cdot \cos(\varphi_0) \quad \text{and} \quad \varphi_{sg}(270°) = \frac{2 \cdot a}{\pi} \cdot \sin(\varphi_0) \tag{10}$$

Now we can calculate the phase difference φ_0 without any dependency on the received optical echo's amplitude a.

$$\varphi_0 = \arctan\left(\frac{\varphi_{sg}(270°) - \varphi_{sg}(90°)}{\varphi_{sg}(0°) - \varphi_{sg}(180°)} \right) \tag{11}$$

Equation (8) is equivalent to equation (3), hence equation (11) can be rewritten as:

$$\varphi_0 = \arctan\left(\frac{\Delta U_{ab}(270°) - \Delta U_{ab}(90°)}{\Delta U_{ab}(0°) - \Delta U_{ab}(180°)} \right) \tag{12}$$

The range value R can now be calculated by taken into account the non-ambiguity range (NAR) at a given modulation frequency of f_{mod} = 20MHz:

$$R = NAR \cdot \frac{\varphi_0}{360°} \tag{13}$$

3.2 Range resolution

The performance and hence the range resolution of solid-state imagers is limited by several different noise sources. Three major noise sources exist: (1) photon shot noise, (2) photocharge conversion noise and (3) quantization noise (Theuwissen, 1995). All of these noise sources can be reduced or eliminated by different signal processing techniques or cooling, except photon shot noise. Hence it is the ultimate theoretical limitation of all photo detectors. Since photon shot noise increases with the amount of incoming photons, it finally dominates all other noise sources and hence limits the effective signal-to-noise ratio for higher illumination levels. Shot noise describes the statistical Poisson-distributed nature of the arrival process of photons and the generation process of electron-hole pairs. The standard deviation of shot noise is equal to the square root of the number of photons (optical shot noise) or photogenerated charge carriers (electronic shot noise). In (Lange, 2000) the required number of photoelectrons per sampling point and pixel to achieve a given range resolution using a 4-phase shift algorithm is derived. It is given by:

$$\Delta R = NAR \cdot \frac{\Delta\varphi_0}{360°} = \frac{c}{4 \cdot \sqrt{8} \cdot f_{\text{mod}}} \cdot \sqrt{\frac{B}{M^2}} \tag{14}$$

M is the (de)modulation amplitude, i.e. the number of photoelectrons per pixel and sampling point generated by the modulated light source. M depends on the modulation

depth of the modulated signal and the demodulation contrast of the pixel but also on the optical power of the modulated light source and the target's distance and reflectivity. B is the Offset or acquired optical mean value, i.e. the number of photoelectrons per pixel and sampling point generated by incoming light of the scene's background and the mean value of the received modulated light.
As in practise only the voltage values at the output stages can be measured, equation (14) has to be reformulated. The conversion process of the optically generated charge into an analogous output voltage is characterized using the sensitivity s. The sensitivity is determined by the size of the conversion capacitance C_{int} and the amplification of the output stage A_{os} of the solid-state imager. It is usually specified in terms of volts per electron and is given by Equation (15), where q is the elementary charge (1.6·10^-19 C).

$$s = \frac{A_{OS}}{C_{int}} \cdot q \tag{15}$$

Subsequently the output voltage v_{out} of the solid-state imager can be described by Equation (16), where E is the total number of all electrons accumulated in the conversion capacitance. E is the sum of the acquired optical mean value B and the (de)modulation amplitude M.

$$v_{out} = s \cdot E = s \cdot (M + B) \tag{16}$$

Hence the mean value of v_{out} is related to B and the peak-to-peak value of v_{out} is related to M as given in Equation (17).

$$\overline{v}_{out} = s \cdot B \quad \text{and} \quad v_{outPP} = 2 \cdot s \cdot M \tag{17}$$

Now we can rewrite Equation (14) by taken into account equation (17):

$$\Delta R = \frac{c}{2 \cdot \sqrt{8} \cdot f_{mod}} \cdot \sqrt{\frac{\overline{v}_{out}}{v_{outPP}^{2}} \cdot s} \tag{18}$$

Looking at equation (18) it can be concluded that a large background brightness, which is proportional to $\overline{v}_{out}$, not only restricts the number of available quantization levels but also drastically increases the quantum noise of the system. Background illumination can be reduced by measuring in the dark or by using spectral filters that only transmit the spectrum of the modulated light. Furthermore it can be concluded that an increasing active modulation power density, which is proportional to v_{outPP}, leads to a better range resolution. As the active optical power density increases with decreasing distance to the object, the ranging accuracy also increases for smaller distances. This is an important fact for navigation applications, where a high accuracy is often only needed close to the target.
Equation (18) can be used to determine the resolution of PMD TOF sensors. The intensity output voltage ΣU_{ab} (see (Schwarte & Heinol, 1999)) of these sensors corresponds to the total number of all electrons accumulated in the conversion capacitance. Hence ΣU_{ab} can be expressed as:

$$\Sigma U_{ab} = s \cdot E = s \cdot (M + B) \tag{19}$$

The modulation amplitude M is related to the absolute value of the range output voltage ΔU_{ab} of the PMD TOF sensor:

$$|\Delta\hat{U}_{ab}| = s \cdot M \tag{20}$$

Hence equation (18) can now be reformulated for the use of a PMD TOF sensor:

$$\Delta R = \frac{c}{4 \cdot \sqrt{8} \cdot f_{mod}} \cdot \sqrt{\frac{\Sigma U_{ab} - |\Delta\hat{U}_{ab}|}{\Delta\hat{U}_{ab}^{2}} \cdot s} \tag{21}$$

In (Lange, 2000) it is shown that equation (14) can be expanded to include additional noise scorces such as 1/f-, reset- and thermal noise by adding an additional number of pseudo-background-electrons N_{pseudo} to B. Theses noise sources are not correlated to the modulation signal and thus contribute to B rather than M.

$$\Delta R = NAR \cdot \frac{\Delta\varphi_0}{360°} = \frac{c}{4 \cdot \sqrt{8} \cdot f_{mod}} \cdot \sqrt{\frac{B + N_{pseudo}}{M^2}} \tag{22}$$

The number of pseudo-electrons N_{pseudo} can be obtained by squaring the noise equivalent number of noise electrons N. These can be determined by measuring v_{dark} and dividing it by the sensitivity s of the sensor.

$$N_{pseudo} = N^2 = \left(\frac{v_{dark}}{s}\right)^2 \tag{23}$$

Equation (21) can now be modified to include the additional noise sources. As all electrons generated in the dark should be taken into account the dark voltage is equivalent with the intensity output voltage ΣU_{ab} of the PMF TOF sensor.

$$\Delta R = \frac{c}{4 \cdot \sqrt{8} \cdot f_{mod}} \cdot \sqrt{\frac{(\Sigma U_{ab} - |\Delta\hat{U}_{ab}|) \cdot s + \Sigma U_{ab\,dark}^{2}}{\Delta\hat{U}_{ab}^{2}}} \tag{24}$$

If no background light is present and only the active illumination source is active the highest range resolution can be achieved. The modulation amplitude M is then equal to the optical mean value B. Equation (24) can be simplified to:

$$\Delta R = \frac{c}{4 \cdot \sqrt{8} \cdot f_{mod}} \cdot \sqrt{\frac{s}{|\Delta\hat{U}_{ab}|} + \left(\frac{\Sigma U_{ab\,dark}}{\Delta\hat{U}_{ab}}\right)^2} \tag{25}$$

The range accuracy, which can only be improved by averaging, is the absolute limit of a PMD TOF sensor using a 4-phase shift algorithm. The range resolution also depends on the modulation frequency. The higher the modulation frequency the better is the range resolution. However the measuring range (FOV) will decrease with increasing modulation frequency (see section 2.3). In practice it has shown that increasing the modulation frequency far above 20 MHz will not increase the range resolution because the power of the IR-LEDs will decrease at the same time.

4. Automotive applications using PMD TOF vision systems

4.1 Advantages of PMD TOF vision systems for automotive applications

As mentioned in the introduction section already, TOF vision systems are found in a variety of applications such as industrial sensor systems, automation, robotics, user interfaces, virtual reality and so forth. However the technically most challenging and interesting application is the automotive industry. An appropriate 3D vision system should be able to recognize dangerous situations foresighted to support the driver in the best possible way to avoid accidents. In the case that an accident can not be avoided the system should at least minimize the injury risk for all passengers.

Up to now application-specific sensor units are designed for each particular automotive application. Hence nowadays only pure range measuring systems (long or short range radar, lidar or ultrasonic sensors) or pure opto-electronical 2D camera systems exist in various car assemblies. Increasingly optimized algorithms are developed to fuse the data of different sensor units. This approach is understandable as no system was available to deliver 2D intensity images and range data at the same time. What was missing is a precise, economical, compact, universal sensor technology which can acquire the 3D information (intensity and range information) of a scene in one image capture. Such a technology would deliver the absolute geometrical dimensions of objects without depending on the object surface, - distance, -rotation and -illumination (rotation-, translation- and illumination invariant). SV vision systems were the first to be developed for the automotive industry with the listed disadvantages in section 2. Subsequently the PMD technology was successfully developed which is a TOF vision system with inherent suppression of uncorrelated light signals such as sun light or other modulated light disturbances (Moeller et al., 2005). More advantages of a PMD TOF vision system are the acquisition of the intensity and range data in each pixel without high computational cost and any moving components as well as the monocular setup.

All these advantages lead to a compact and economical design of an automotive 3D TOF vision system. This system can immediately acquire the scene objects parameters without high computational power. The typical frame rate of the system is 100 Hz leading to a high constant 3D data flow. A reliable object plausibility check and calculation of the object motion vectors are possible. Hence a reliable scene interpretation for automotive applications is given. The high frame rate is very important for car safety applications to detect high dynamical changes in the traffic even at a high car speed. With such a system the driver can be best possible supported and the injury risk in a non avoidable collision can be minimized by using active safety measures.

At the moment several automotive manufacturer are using the PMD technology for many different applications such as smart airbag, driver assistant systems, pre crash systems, stop & go systems, pedestrian safety systems, emergency break systems, gesture recognition systems and so force (Ringbeck et al., 2007; Ringbeck & Hagebeuker, 2007; Buxbaum & Hagebeuker, 2005). Fig. 4 and Fig. 5 show some of the realized systems. In the next section a developed PMD TOF vision system usable for pedestrian detection, stop & go and pre crash application is exemplary presented.

Fig. 4. Smart airbag application using a PMD TOF vision systems (Hussmann & Hess, 2006)

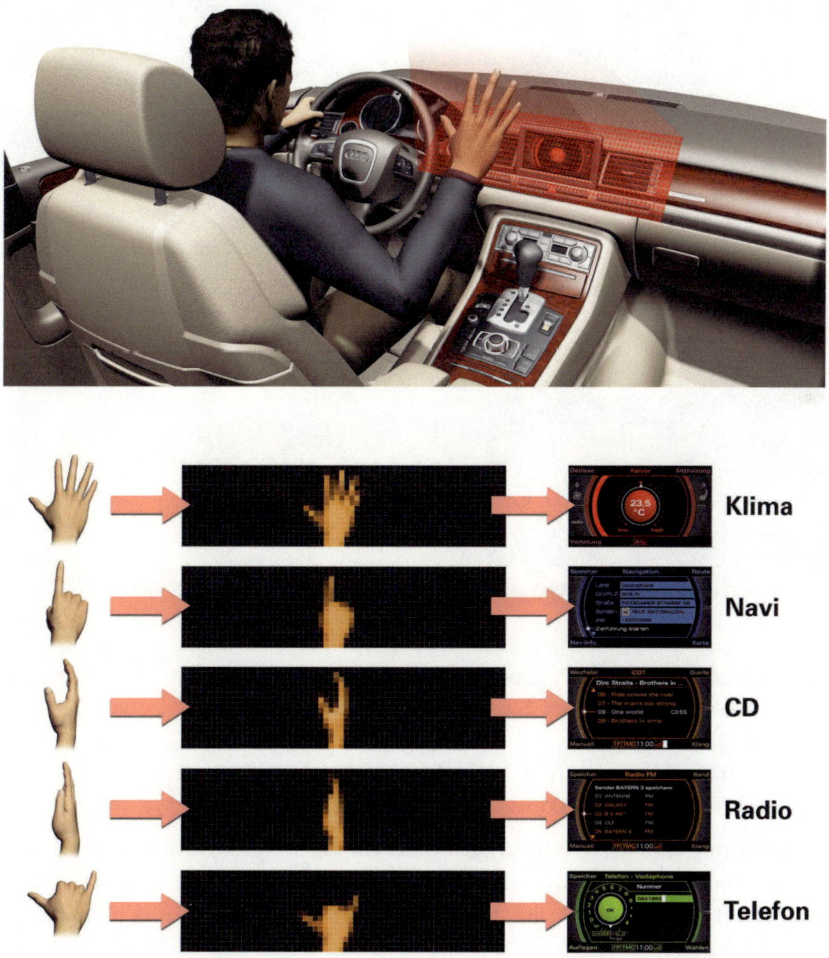

Fig. 5. Gesture recognition using a PMD TOF vision systems (Hussmann & Hess, 2006)

4.2 PMD sensor system for automotive outdoor applications

Fig. 6 shows a PMD sensor system for automotive outdoor applications. This 'A-Muster' camera is designed to demonstrate solutions for several front view applications such as Stop&Go, PreCrash and Pedestrian Safety. The non-ambiguity range (NAR) of the sensor system is 120 m using a modulation frequency of 6,2 and 7,5 MHz. This range is the maximum possible measuring range if enough illumination power is available. To achieve a measuring range as close as possible to the NAR, additional IR-illumination sources can be implemented inside the head lights to increase the transmitted power. For a typical measurement range of 10 m an illumination source with a power rating of 1 W optical power is placed beside the camera system inside the car as shown in Fig. 6 (c). For a measurement range above 35 m a more powerful illumination source has to put into the front of the car as shown in Fig. 6 (b). The range resolution for both measurement ranges is ± 10 cm.

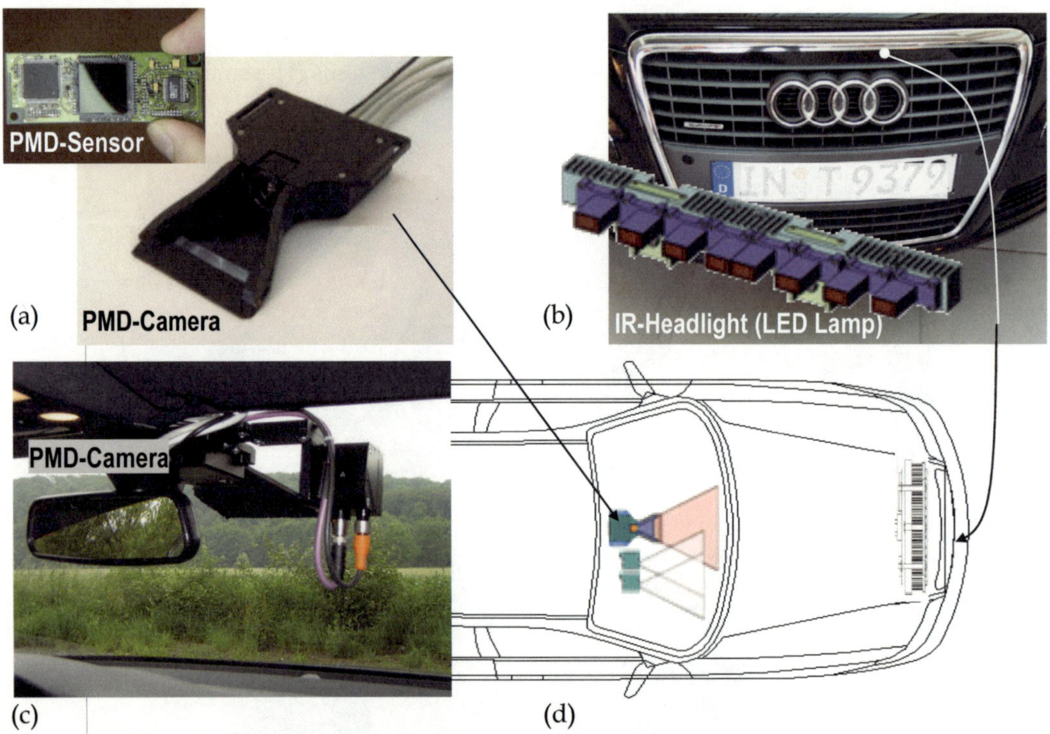

Fig. 6. Illustration of a PMD TOF vision systems for several automotive outdoor applications
(a) 'A-Muster' PMD camera and PCB with the PhotonICs® PMD 1k-S (64 x 16 Pixel)
(b) IR-LED Headlights (8 W)
(c) Internal car setup (PMD camera + 1 W IR-LED illumination source)
(d) Schematic of overall car setup options

The PMD camera delivers an evenly distributed range image of the observed scene in front of the car. The field of view of the observed scene depends on the chosen optic and the active illumination source and has to be adapted to match the requirements for the implemented outdoor application (Stop&Go, PreCrash and Pedestrian Safety).

Fig. 7 shows the processing chain of the PMD TOF vision system. The pre-processing of the camera's row data is simple and does not require a powerful processing unit. Only the range and intensity values are calculated as derived in section 3. Furthermore the verification and selection of the feasible range values and the illumination control are processed in this stage. A clear range image is the result of this stage. In the next stage object segmentation and detection algorithm determined the objects in front of the car. The implemented software varies depending on the chosen application. The output of this stage results in an object list. The dynamical objects in the observed scene are tracked in the next stage and a decision is made based on the dynamical objects' position to activate or not activate the actuators. This decision can be confirmed by collected and fused data of other sensor systems.

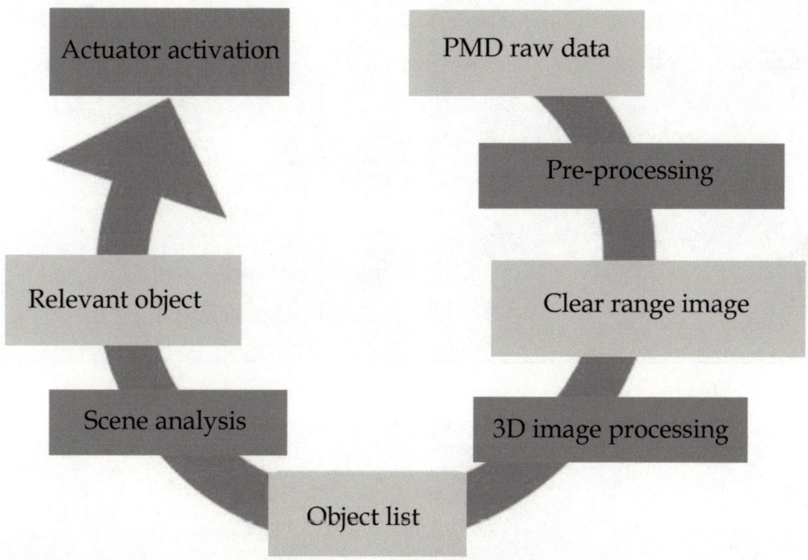

Fig. 7. Processing chain of the PMF TOF vision system

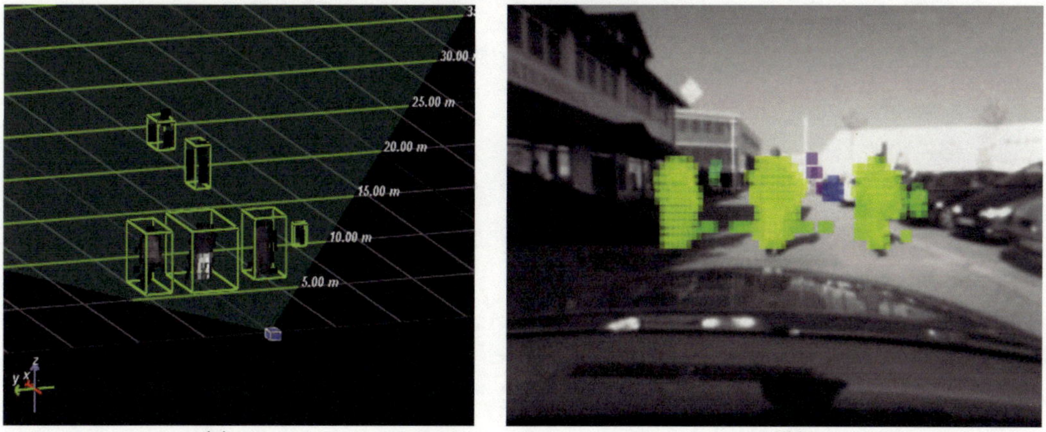

(a) (b)

Fig. 8. Experimental results of an observed traffic scene with the PMD TOF vision system
(a) Range raw data and the resulting object presentations in a virtual 3D space
(b) Conventional video image with chronological superimposed object visualization

In the following figures the raw data and the resulting object representations of the PMD TOF vision system are depicted at different traffic scenarios. The position of the objects is detected in the 3D space. The position changes are used to determine the motion vectors. Now the objects can be tracked and identified in the observed traffic scenes. This process is illustrated in Fig. 8. Three pedestrians are identified as relevant objects and their position are determined, visualized and tracked.

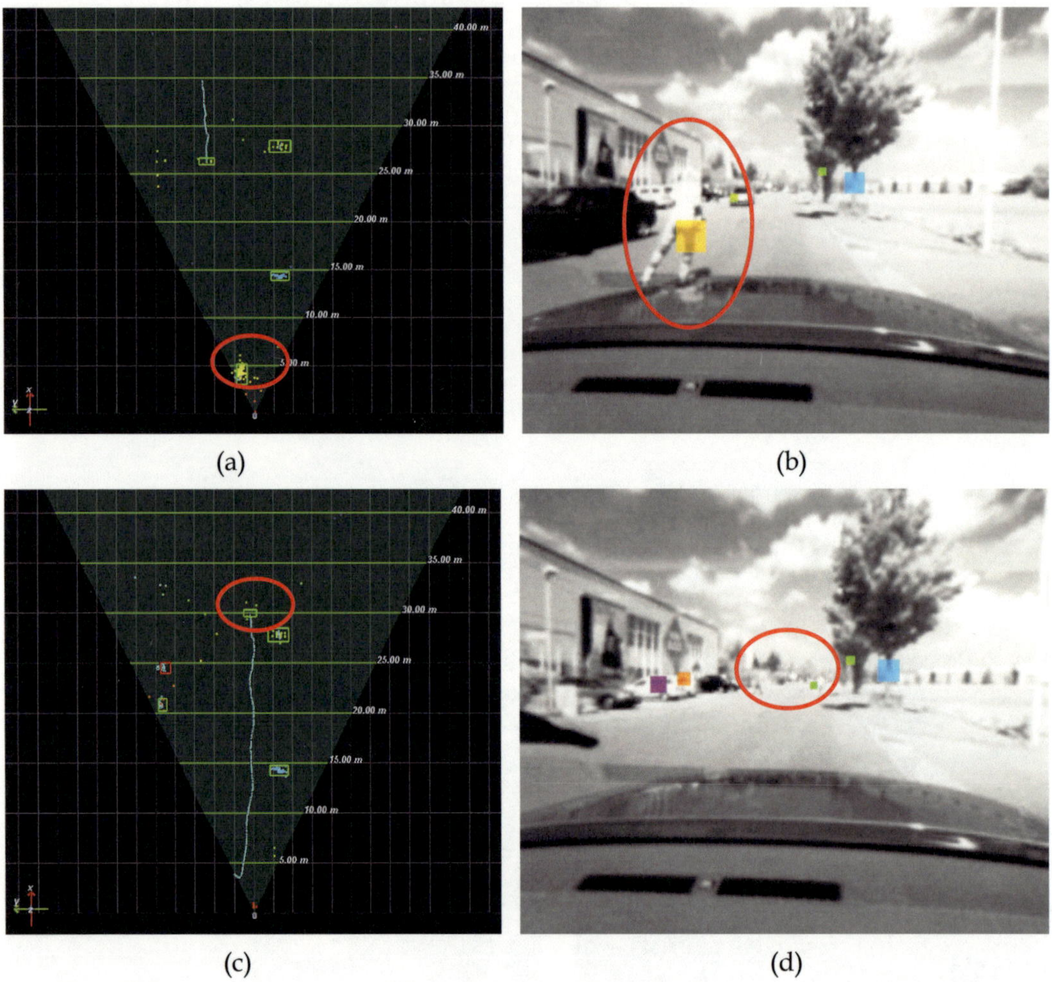

Fig. 9. (a) Range raw data and the resulting object representation of a roller skater (green rectangle in the red circle) after he entered the field of view of the PMD TOF camera
(b) Conventional video image with chronological superimposed object visualization of the roller skater (yellow rectangle in the red circle)
(c) Range raw data and the resulting object representation of the roller skater at 30 m (green rectangle in the red circle) as well as the tracked way (blue line)
(d) Conventional video image with chronological superimposed object visualization of the roller skater (green rectangle in the red circle)

Due to the computational efficient and simple processing chain of the PMD TOF vision system (see Fig. 7), the image processing algorithms are simple and fast. Hence a fast object tracking is possible. Fig. 9 shows a roller skater who is detected and tracked as a relevant object over a range of 30 m. The decision if the roller skater is a relevant object is made straight after he entered the field of view of the PMD TOF camera. This proofs the speed of the processing chain of the PMD system.

The tracking of relevant objects is not only possible at low car speeds but also at high car speeds for example on the motorway. Fig. 10 shows the performance of the object tracking on the motorway of the PMD TOF vision system. It has to be noted that the tracking at 50 m is successful even when the objects are represented by only a few pixels as shown in Fig. 10 (b). This accuracy is only possible because every pixel delivers a range value besides its intensity value. This inherent feature of the PMD TOF vision system distinguishes the PMD technology from typical SV vision systems.

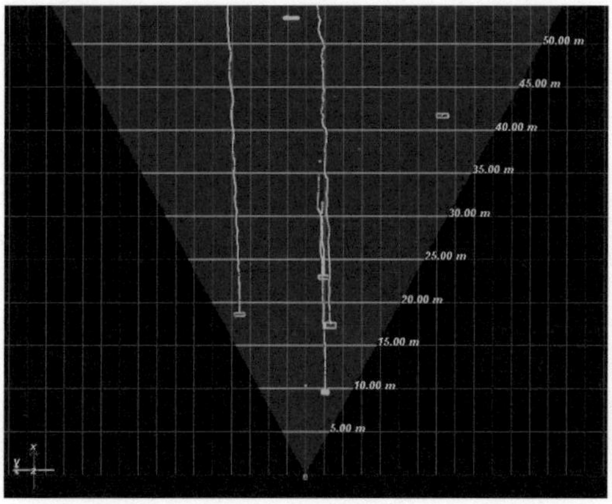

(a)

(b)

Fig. 10. Object tracking on the motorway using the PMD TOF Vision system
(a) Range raw data and the resulting object representation of cars on the motorway at different ranges (green rectangles) as well as their tracked ways (blue lines)
(b) Conventional chronological video image

5. Conclusion

In this chapter we compared conventional stereo vision systems with time-of-flight vision system and highlighted the advantages and disadvantages of both systems. The new PMD technology represents a TOF vision system with many advantages in comparison to conventional stereo vision systems especially in the automotive industry. The equations needed for the design of such a system are derived and demonstrate the simplicity of the extraction of the range information. A PMD camera delivers absolute geometrical dimensions of objects without depending on the object surface, - distance, -rotation and – illumination. Hence PMD TOF vision systems are rotation-, translation- and illumination invariant.

The major advantage of the PMD technology is the delivery of an evenly distributed range and intensity images because each pixel calculates a range and intensity value. Hence the correspondence problem of conventional stereo vision system does not exit. Another advantage of the PMD technology is that the field-of-view can easily extended by varying the modulation frequency without any special optical components. However, the range resolution depends on the chosen modulation frequency. The higher the modulation frequency the better the resolution and the smaller the measuring range. The range resolution depends also on the power rating of the used active illumination source. Furthermore it has been shown that the range data of the 3D-TOF camera is almost independent on the reflection coefficient of the measured objects if a phase-shift algorithm is used. The PMD technology has an inherent suppression of uncorrelated light signals such as sun light or other modulated light disturbances. However if those light sources saturate the sensor, the range information is lost. More advantages of a PMD technology are the acquisition of the intensity and range data in each pixel without high computational cost and any moving components as well as the monocular setup.

All these advantages lead to a compact and economical design of an automotive 3D TOF vision system with a high frame rate. This vision system can be used for many different applications such as smart airbag, driver assistant systems, pre crash systems, stop & go systems, pedestrian safety systems, emergency break systems, gesture recognition systems and so force. In this chapter experimental results of a PMD TOF vision system for automotive outdoor applications such as Stop&Go, PreCrash and Pedestrian Safety are presented and demonstrate that the accuracy of the system is appropriate to fulfil the tough target settings in the automotive industry.

6. References

Beheim, G. & Fritsch, K. (1986). Range finding using frequency-modulated laser diode, *Applied Optics*, 25 (9), pp. 1439-42

Benosman, R., Maniere, T. & Devars, J. (1996). Multidirectional stereovision sensor, calibration and scene reconstruction, *IEEE Int. Conf. on Pattern Recognition, ICPR*, vol.1, pp. 161-5

Buxbaum, B. & Hagebeuker, B. (2005). Dreidimensionale Umfeld-Erfassung, *Trade Journal: "Elektronik automotive"*, WEKA Publisher House, Issue 5, ISSN 1614-0125, pp. 77-81

Cheng, S., Tu, D., Li, G. & Yang, J. (2002). Fusing Range and 2-D images in multisensor for robot vision, *IEEE TENCON*, pp. 565-568

Dhond, U.R. & Aggarwal, J.K. (1989). Structur from stereo – A review, *IEEE Trans. On Systems, Man and Cybernetics*, 19 (6), pp. 1489-1508

Hess, H., Albrecht, M., Grothof, M., Hussmann, S., Oikonomidis, N. & Schwarte, R. (2003). 3D object recognition in TOF data sets, *Proc. SPIE*, vol. 5086, pp. 221-228

Hussmann, S. & Hess, H. (2006). Dreidimensionale Umwelterfassung, *Trade Journal: "Elektronik automotive"*, WEKA Publisher House, Issue 8, ISSN 1614-0125, pp. 55-59

Hussmann, S. & Liepert, T. (2007). Robot Vision System based on a 3D-TOF Camera, *IMTC 2007, IEEE Proc. of the 24th Int. Conf. on Inst. and Meas. Tec.*, pp. 1-5

Kang, S.B. & Szeliski, R. (1997). 3-D scene data recovery using omnidirictional multibaseline stereo, *International Journal of Computer Vision*, 25 (2), pp. 167-83

Kawanishi, T., Yamazawa, K., Iwasa, H., Takemura, H., & Yokoya, N. (1998). Generation of high resolution stereo panoramic images by omnidirectional imaging sensor using hexagonal pyramidal mirrors, *Int. Conf. on Pattern Recognition, ICPR*, pp. 485-489

Krishnan, A. & Ahuja, N. (1996). Range estimation from focus using a non-frontal imaging camera, *International Journal of Computer Vision*, 20 (3), pp. 169-185

Lange, R. (2000). 3D Time-of-flight distance measurement with custom solid-state image sensors in CMOS/CCD-technology, *PhD thesis*, Dep. of Electrical Engineering and Computer Science, University of Siegen, Online publication: http://www.ub.uni-siegen.de/epub/diss/lange.htm

Liancheng, S. & Feng Z. (2005). Design of a novel stereo vision navigation system for mobile robots, *IEEE Int. Conf. on Robotics and Biomimetics (ROBIO)*, pp. 611-614

Lin, S.S. & Bajcsy, R. (2003). High resolution catadioptric omni-directional stereo sensor for robot vision, *Int. Conf. On Robotics & Automation*, pp. 1694-1699

Moeller T., Kraft H., Frey J., Albrecht M. & Lange R. (2005). Robust 3D Measurement with PMD Sensors, *RIM-Day*, ETH Zürich, Online-publication (http://www.pmdtec.com/inhalt/download/documents/RIM2005-PMDTec-Robust3DMeasurements.pdf)

Moring, I., Heikkinen, T., Myllyla, R. & Kilpela, A. (1989). Acquisition of three-dimensional image data by a scanning laser range finder, *Optical Engineering*, 28 (8), pp. 897-902

Ringbeck, T. & Hagebeuker, B. (2007). A 3D Time of flight camera for object detection, *Optical 3-D Measurement Techniques*, ETH Zürich, Online-publication (http://www.pmdtec.com/inhalt/download/documents/070513Paper-PMD.pdf)

Ringbeck, T., Hagebeuker, B., Kraft, H. & Paintner, M. (2007). PMD-basierte 3D-Optosensoren zur Fahrzeugumfelderfassung, In: *Sensoren im Automobil II*, Thomas Tille, pp. 209-222, Expert Verlag GmbH, ISBN: 3816927505

Schwarte_a, R., Xu, Z., Heinol, H., Olk, J. & Buxbaum, B. (1997). New optical four-quadrant phase-detector integrated into a photogate array for small and precise 3D-cameras, *Proc. SPIE*, vol. 3023, pp. 119-128

Schwarte_b, R., Xu, Z., Heinol, H., Olk, J., Klein, R., Buxbaum, B., Fischer, H. & Schulte, J. (1997). New electro-optical mixing and correlating sensor: facilities and applications of the photonic mixer device (PMD), *Proc. SPIE*, vol.3100, pp. 245-53

Schwarte, R. & Heinol, HG. (1999). Optical component combines 3D image detection and mixing, *Elektronik*, Publisher: WEKA-Fachzeitschriften, Germany, 48(12), pp. 80-90

Theuwissen, A. (1995). Solid-State Imaging with Charge-Coupled Devices, *Kluwer Academic Publishers*, ISBN-13: 978-0792334569

8

Construction of an Intelligent Room using Distributed Camera System

Kota Irie, Masaki Wada and Kazunori Umeda
Chuo University / CREST, JST
Japan

1. Introduction

These days, computerization of our living environment is progressing, and the home appliances are becoming more intelligent and function-rich. On the other hand, the increase of their functions makes their operation complicated. For such appliances frequently used in everyday life, intuitive operation is desirable for users. Gestures, which we use frequently and intuitively in our everyday communication, can be one of such human-machine interfaces. Until now, many studies which recognize gestures from a sequence of images have been reported (V. I. Pavlovic et al., 1997). A shortage of many of the gesture recognition studies is that the place where gestures can be recognized is limited and thus they are not suitable for practical applications. There is a trend of studies to make a room or some space itself intelligent (A. Pentland, 2000), (M. Coen, 1999), (Brumitt, B. et al., 2000), (T. Mori & T. Sato, 1999), (J. H. Lee & H. Hashimoto, 2002). This approach is effective for realizing natural human-machine interface. Gesture recognition is one of the important technologies for such an intelligent room or space. We are also constructing an intelligent room in which an operator can control home appliances such as a TV set with intuitive gestures (K. Irie et al., 2004). Characteristics of our intelligent room are that, it focuses on operation of home appliances, it allows users to operate without restriction to the place and without having any instrument, and it uses pan-tilt-zoom cameras only as sensor information.

In this paper, we discuss three dimensional (3D) measurement of the intelligent room using a distributed camera system, especially 3D measurement of hand waving and finger pointing.

2. Outline of intelligent room to operate home appliances

The intelligent room in this paper has some intelligent functions that are intended for a general house or an office. In the room, we can operate home appliances such as a television set and a lighting by gestures. The operator (i.e., a person with an intention to operate an appliance) does not need a special attachment such as a glove or a microphone and can make operations in a natural state. The room specifies the operator autonomously even when two or more persons exist. The operator can make operations wherever in the room without any restriction. Gestures in the room are supposed to be by a hand or fingers.

The room carries CCD cameras with pan, tilt, and zoom functions, and detects an operator and recognizes his/her gestures autonomously with them. Fig.1 shows the conceptual figure of the intelligent room. First, the system performs "detection of waving hands" with two or more cameras to detect the operator and then acquires 3D position information. It carries out pan, tilt and zooming-up of the cameras with the 3D information and restricts the region to process. Then it extracts a hand region using the color information. The skin color of the operator is registered in this stage, which improves the robustness of extracting the hand region to the difference of individual skin colors and the change of lighting environment. Then gestures are recognized for the hand region that is extracted using the color information. The recognized result is presented on a PC monitor and by a speaker, and interaction by the operator is made possible. Based on the recognized operation, a control signal is transmitted to the target appliance by the infrared remote controller connected to PC. For example, turning on/off the TV set, inputting the channel and turning up/down the volume using gestures are possible.

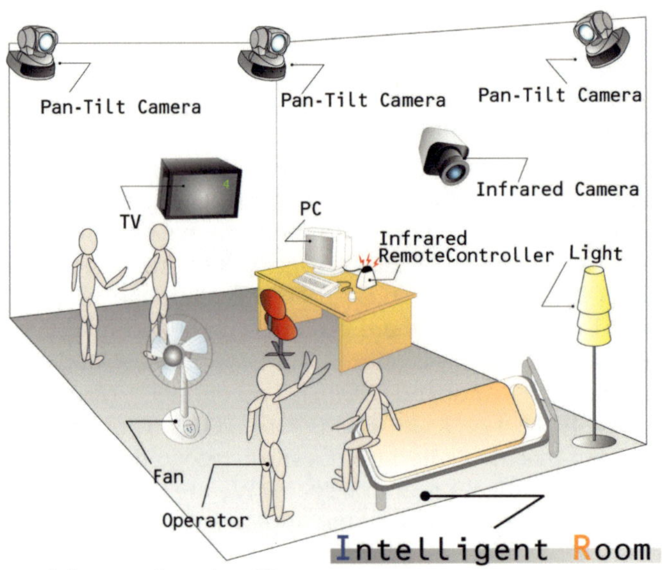

Fig. 1. The conceptual figure of our intelligent room

3. Field of view of Pan-Tilt-Zoom camera

The camera's field of view is an important parameter to use in the intelligent room. When watching the wide space to find an operator, it should be wide. At the same time, too wide field of view is not appropriate for detecting a waving hand. And when focusing on the detected operator to recognize his/her gestures, the field of view should be narrow enough. These days several kinds of pan-tilt-zoom cameras are commercially available. Pan, tilt and zoom of these cameras can be controlled by a PC, and these cameras are suitable for usage in the intelligent room. For example, Sony EVID100, which we adopted in our intelligent room, has 10X optical zoom and its horizontal field of view varies from 65 to 6.6[deg]. Other pan-tilt-zoom cameras have similar zoom specifications. Suppose four cameras are set at the corners of a square room and the field of view is 65[deg]. Then the room is covered by the cameras' field of views as shown in Fig.2. It can be seen that the room is mostly (95.8[%]) covered by two or more cameras, which means that 3D measurement is possible almost everywhere in the room.

And as for detecting a waving hand, suppose the horizontal number of pixels of the low-resolution image is 25, the width of hand waving is 0.3[m], and the observed hand waving should be equal to or larger than 1[pixel] of the low-resolution image. Then the distance to the waving hand can be up to 5.9[m], which is long enough in an ordinary room.

And when the field of view is set to the narrowest 6.6[deg], the corresponding size at the distance 5.9[m] becomes as small as 0.58[m], which means that a hand or other regions can be observed large enough.

Fig.3 shows an example. The two images are captured with the widest and narrowest field of view at 5[m] respectively. We can see this kind of camera is suitable for both watching the whole scene and recognizing each person's gesture.

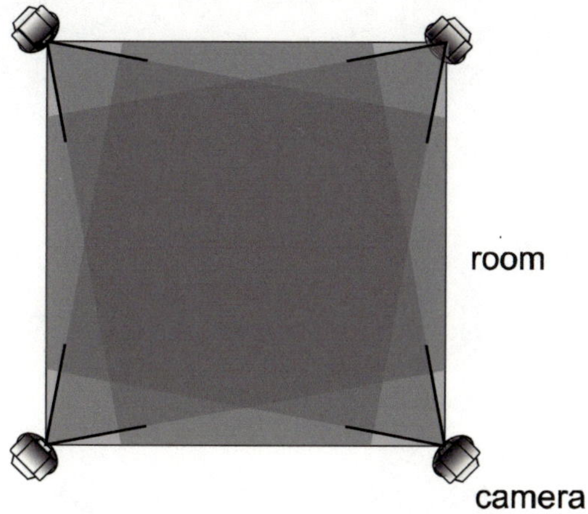

Fig. 2. Covering of a room by the field of view of four cameras

(a) widest field of view

(b) narrowest field of view

Fig. 3. Range of field of view of a pan-tilt-zoom camera EVI-D100 at 5[m]

4. 3D Measurement of waving hands

The 3D position of waving hands is measured by detecting it with two or more cameras.

4.1 Outline of detecting waving hands

We use hand waving to detect an operator in the intelligent room. For human-human interface to communicate one's intention to other person, hand waving is often used and thus it is thought to be appropriate to use in the intelligent room, too. The outline of the method to detect waving hands (K. Irie & K. Umeda, 2002) is as follows (see Fig.4). The images are converted to low-resolution ones, and FFT is applied to each pixel of the low-resolution images. Pixels with high power at the frequencies of hand waving are detected as the pixels of hand waving. The method is robust to lighting condition and individual difference of skin color, because it uses intensities only.

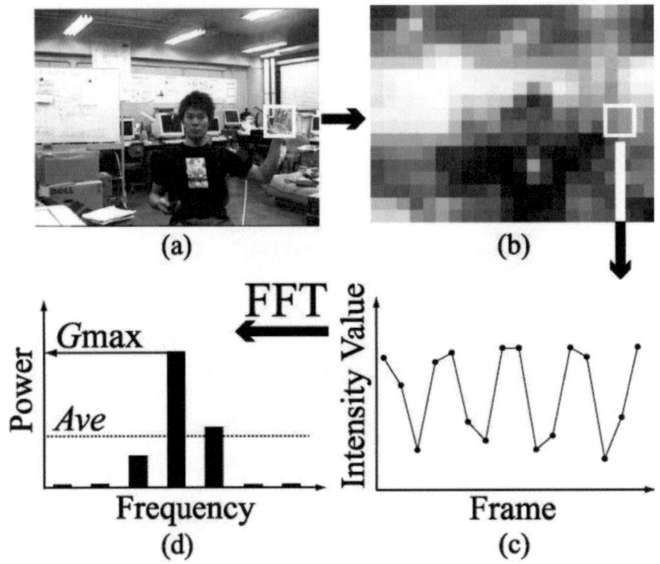

Fig. 4. Detection of hand waving using FFT

4.2 Calibration of the cameras

The intrinsic and extrinsic parameters of the cameras are obtained by the following calibration method that uses a known pattern (O. Faugeras, 1993).

1. Obtain the projection matrix between the 3D points and their projected points on a 2D image.
2. Obtain the intrinsic and extrinsic parameters from the projection matrix. For the point on the image plane $\tilde{\mathbf{m}} = [u \quad v \quad 1]^T$ and the point in 3D space $\tilde{\mathbf{M}} = [X \quad Y \quad Z \quad 1]^T$, the projection equation for the perspective projection is expressed as

$$s\tilde{\mathbf{m}} = \mathbf{P}\tilde{\mathbf{M}} = A[\mathbf{R}, \mathbf{t}]\tilde{\mathbf{M}} \tag{1}$$

where

$$\mathbf{A} = \begin{bmatrix} \alpha_u & -\alpha_u \cot\theta & u_0 \\ 0 & \alpha_v/\sin\theta & v_0 \\ 0 & 0 & 1 \end{bmatrix}, \ \mathbf{R} = \begin{bmatrix} \mathbf{r}_1{}^T \\ \mathbf{r}_2{}^T \\ \mathbf{r}_3{}^T \end{bmatrix}, \ \mathbf{t} = \begin{bmatrix} t_x \\ t_y \\ t_z \end{bmatrix} \tag{2}$$

When **P** is obtained, the matrix **A** that consists of intrinsic parameters, and the rotation matrix **R** and the translation vector t that are extrinsic parameters, are calculated.

4.3 Stereo vision

3D coordinates of the matched point are obtained from the projection matrices of the two cameras. When a 3D point is observed by camera i at $\mathbf{m}_i = [u_i\ v_i]^T$,

$$s_i\tilde{\mathbf{m}}_i = \mathbf{P}_i\tilde{\mathbf{M}} \tag{3}$$

is satisfied. Then for 2 or more cameras,

$$\mathbf{BM} = \mathbf{b} \tag{4}$$

is introduced. The 3D coordinates M of the point is given by least squares method as

$$\mathbf{M} = \mathbf{B}^{+}\mathbf{b} \tag{5}$$

5. 3D measurement of finger pointing

We often point at an object with a finger to select it. Therefore, we use finger pointing in the intelligent room to select an appliance. When it is observed by two cameras, 3D finger pointing can be obtained. Then the object that is closest to the pointing is selected and is operated.

The 3D pointing is obtained as follows. When a finger, hand or arm region is extracted in each image by using skin color extraction, the 2D pointing can be obtained as the principal axis of the extracted region as shown in Fig.5. Here we use the extracted skin color region without distinguishing whether it is a finger, hand or arm. Furthermore, when a face region is also extracted, utilizing the vector from the face region to the hand region as the pointing is possible as shown in Fig.5(b).

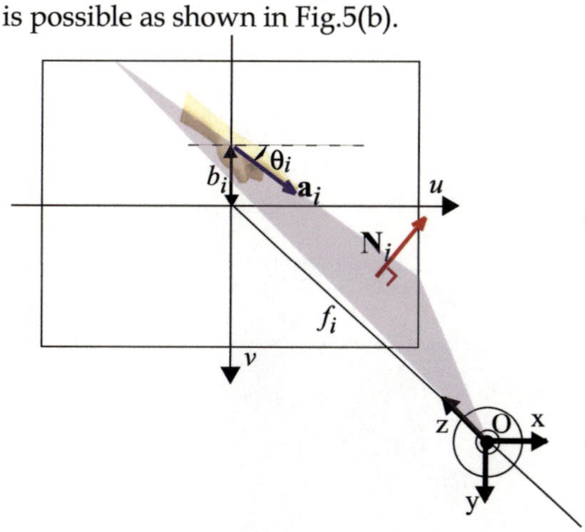

(a) hand only

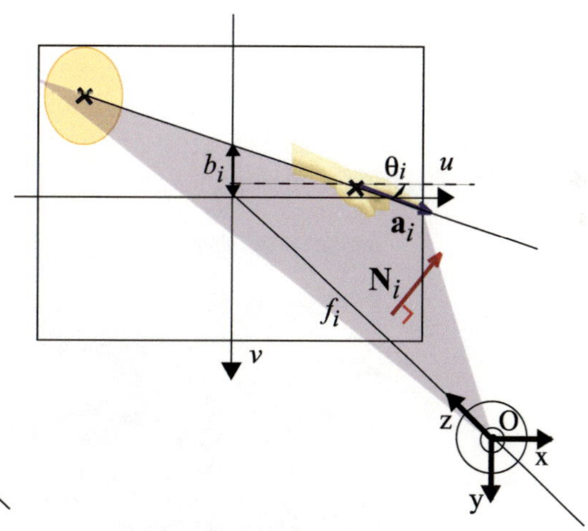

(b) hand and face

Fig. 5. Measurement of 2D pointing

The principal axis and the origin of the camera makes a plane in 3D space. When such a plane is obtained by two cameras, the 3D pointing can be obtained as the intersection of the two planes as shown in Fig.6.

The normal vector of the plane in Fig.5 becomes

$$\mathbf{N} = \begin{bmatrix} f\tan\theta & -f & b \end{bmatrix}^T \tag{6}$$

where f is the focal length of the camera, θ represents the direction in 2D, and b is the v-intercept. By multiplying $\mathbf{R}^T$, the transpose of the rotation matrix $\mathbf{R}$ in eq.(2), the normal vector in the world coordinate system is obtained. Then the 3D pointing direction is obtained as the outer product of the two normal vectors.

Yamamoto et al. (Y. Yamamoto et al., 2004) realizes selection of appliances by arm pointing using multiple stereo cameras. Our method does not use stereo cameras but multiple monocular cameras to realize the 3D measurement of finger pointing.

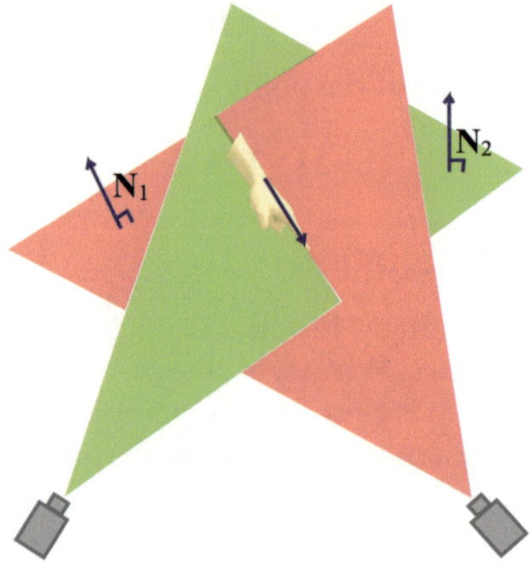

Fig. 6. Measurement of 3D pointing

6. Experiments

The experiments were performed with the constructed intelligent room illustrated in Fig.1. Fig.7 shows the overview of the experiments. MVTec Halcon is used for image processing and other calculations and controls are performed by a DELL PC (Pentium 4 2.2GHz). Three SONY EVI-D100 CCD cameras were set at the corners of a 6.9 × 7.8[m^2] room, at 2.3[m]

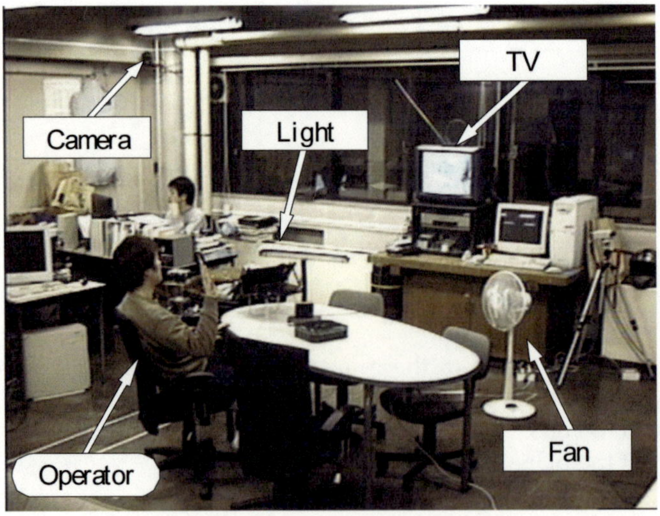

Fig. 7. Overview of experiments for operating home appliances

height. The three images from the three cameras are composed by a Panasonic WJ-MS488 picture division unit and inputted into the PC with a Leutron PicPort Color image capture board (640 × 480[pixel]). A Sugiyama Electron Crossam 2+USB infrared remote-controller is used that can be controlled by the PC to operate home appliances.

6.1 Measurement of 3D position of hand waving

This experiment was carried out with the two cameras of the intelligent room. Fig.8 shows an example of detected hand waving by two cameras. The zoom was set to the widest.
Fig.9 shows the results of error analysis. The hand waving at every 0.5[m] positions for x and y directions was detected. Hand waving was carried out by one subject 5 times for each position with changing the height. The error bars are the standard deviations. The results show that the errors are about 0.1[m], which is small enough for obtaining the 3D position of the operator in the room.

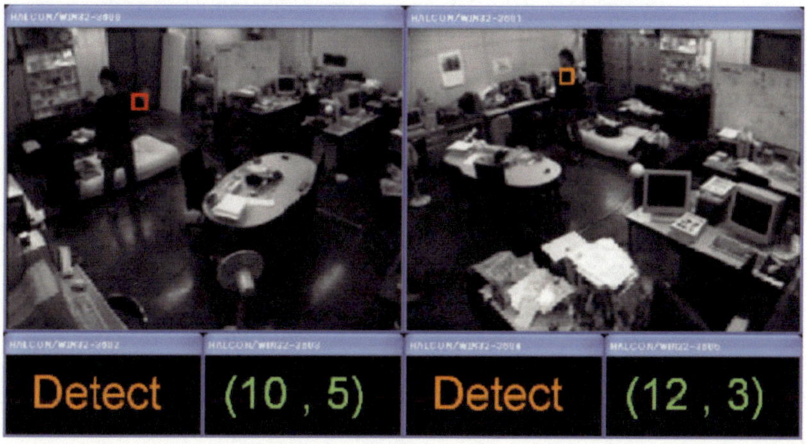

Fig. 8. Example of detected hand waving by two cameras

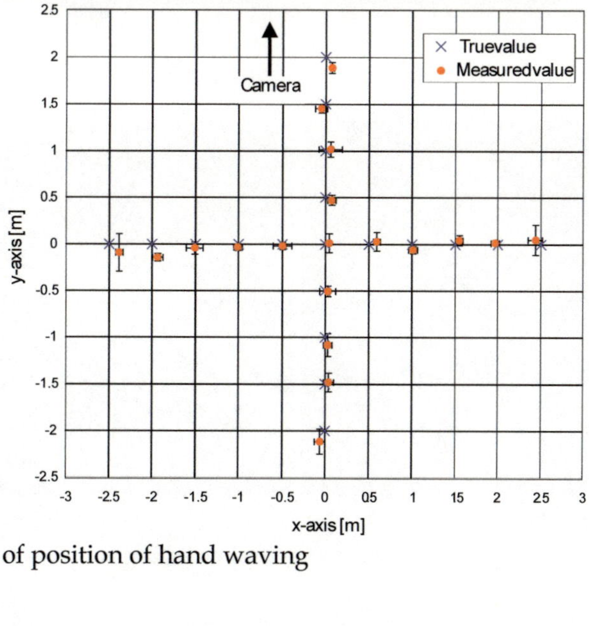

Fig.9. Measurement of position of hand waving

6.2 Measurement of finger pointing

The experiments were performed as shown in Fig.10. The angle between the camera 3 and the pointing direction α was changed every 15[deg] from 0 to 180[deg]. The angle of elevation of the pointing direction β was fixed to 0[deg], i.e. parallel to the floor. Fig.11 shows an example of detected pointing in the images. As can be seen, the vector from the face region to the hand region was used as the pointing. Fig.12 shows the results of the accuracy of detected pointing direction. The best result in the results by three pairs of cameras was selected.

It is shown that the pointing direction was roughly acquired. The reason β has some bias is utilizing the vector from the face region to the hand region as the pointing. However, this was more robust than using the principal axis of the extracted hand region.

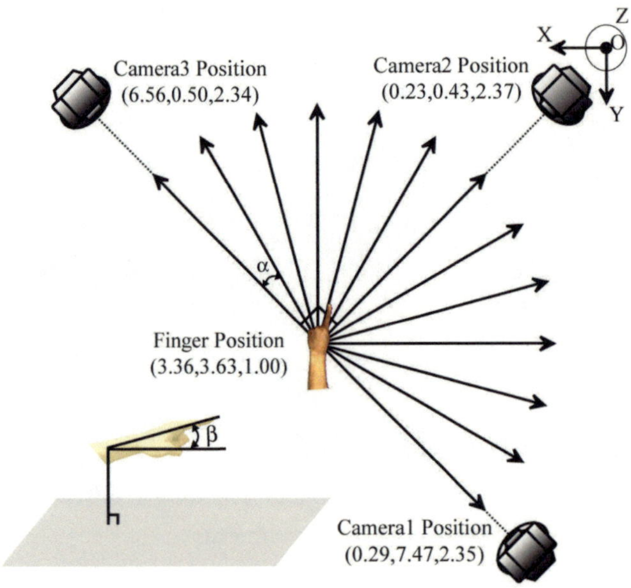

Fig. 10. Experimental condition of measuring pointing direction

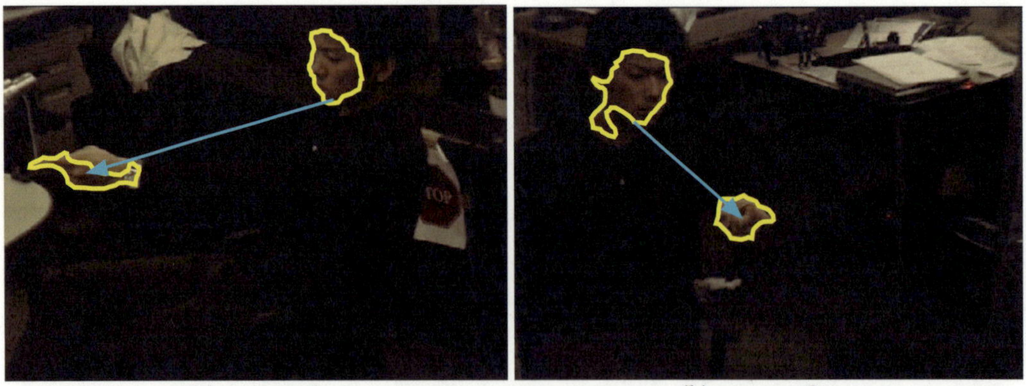

(a) camera 3 (b) camera 2

Fig. 11. Example of detected pointing using face and hand regions

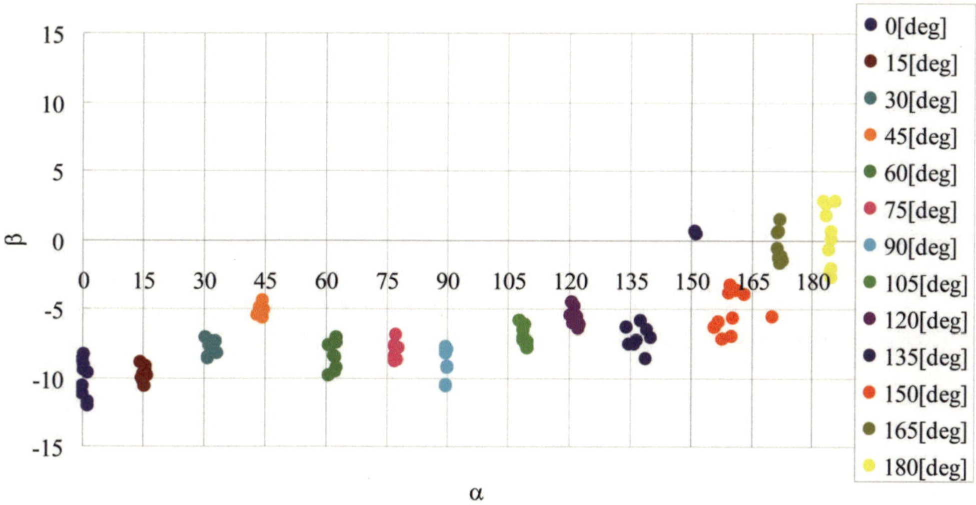

Fig. 12. Experimental results of measuring pointing direction

7. Conclusion

We have discussed three dimensional (3D) measurement of the intelligent room using cameras with pan, tilt and zoom functions. Concretely, 3D measurement of position of operator's waving hand and finger pointing to select an appliance were proposed. Additionally, we showed that the pan-tilt-zoom camera is appropriate to use in the intelligent room. Experiments verified the effectiveness of the proposed 3D measurement methods. Future works include more experimental evaluation, and improvement of the intelligent room.

8. Acknowledgment

We thank Mr. Naohiro Wakamura, who was a master's student at Chuo University, for his support to this study.

9. References

V. I. Pavlovic; R. Sharma & T. S. Huang. (1997). Visual Interpretation of Hand Gestures for Human-Computer Interaction: A Review. *Trans. PAMI*, vol.19, no.7, pp.677-695.

A. Pentland. (2000). Looking at People: Sensing for Ubiquitous and Wearable Computing, *IEEE Trans. on Pattern Analysis and Machine Intelligence*, vol.22, no.1, pp.107-118.

Michael Coen. (1999). The Future Of Human-Computer Interaction or How I learned to stop worrying and love my Intelligent Room. *IEEE Intelligent Systems*, vol.14, no.2, pp.8-19.

Brumitt, B.; Meyers, B.; Krumm, J.; Kern, A. & Shafer, S. (2000). EasyLiving: Technologies for Intelligent Environments. *Proc. Int. Symposium on Handheld and Ubiquitous Computing*, pp.12-27.

T. Mori & T. Sato. (1999), Robotic Room: Its concept and Realization. *Robotics and Autonomous Systems*, Vol.28, No.2, pp.141-144.

Joo-Ho Lee & Hideki Hashimoto. (2002). Intelligent Space - concept and contents -. *Advanced Robotics*, vol. 16, no. 4, pp.265-280.

K. Irie; N. Wakamura & K. Umeda. (2004). Construction of an Intelligent Room Based on Gesture Recognition -Operation of Electric Appliances with Hand Gestures-. *Proc. IEEE/RSJ Int. Conf. on Intelligent Robots and Systems*, pp.193-198.

K. Irie & K. Umeda. (2002). Detection of Waving Hands from Images Using Time Series of Intensity Values. *Proc. 3rd China-Japan Symposium on Mechatronics*, pp.79-83.

O. Faugeras. (1993). *Three-dimensional computer vision: a geometric viewpoint*, MIT Press.

Y. Yamamoto; I. Yoda & K. Sakaue. (2004). Arm-Pointing Gesture Interface Using Surrounded Stereo Cameras System. *Proc. International Conference on Pattern Recognition (ICPR 2004)*, vol.4, pp.965-970.

The Integrated Active Stereoscopic Vision Theory, Integration and Application

Kolar Anthony[1], Romain Olivier[1], Graba Tarik[2], Ea Thomas[3] and Granado Bertrand[4]
[1]*SYEL - University of Pierre et Marie Currie – Paris VI*
[2]*ENST - Paris*
[3]*ISEP – Paris*
[4] *ETIS - CNRS - ENSEA - Univ Cergy Pontoise*
France

1. Introduction

Monolithic integration of stereovision can make a contribution in today's solutions where main limitations are size, power consumption and deployment. Some emergent applications are not equipped with 3D capabilities although the demand is increasing.
One of these applications is 3D-endoscopy which claims good accuracy, small volume, and high autonomy in terms of power consumption and operation. Integrated stereovision used in the medical world and especially for the diagnosis of pathologies of the digestive tract gives an additional advantage for the gastro enterologists in the observation of malignant tumors. In 3D endoscopy, integrated stereovision satisfies four essential needs:
Firstly, precision of diagnosis is required, more exactly the accuracy in the evaluation of the carcinoma size. Without a real 3D representation, estimations are done roughly due to the lack of information (depth parameter) and are defined by the experience of the expert rather than by the system itself. At the present time only an invasive exploration allows an exact location and measurement of tumor size.
Secondly, integrated stereovision meets a growing public need. Improvement in healthcare provides to the population some well-being and people live longer. The diagnosis of early forms of cancer of the digestive tract takes part in the well-being of aged people: the sooner a sick person is treated, the earlier he can recover. In the case of endoscopy, the French Company of Digestive Endoscopy (SFED) stresses in 2006 that the use of endoscopy is increasing: 2.7 million examinations were carried out, 70000 new digestive cancers were discovered and almost 900000 polyps colics were removed. Various types of endoscopy (E.O.G.D, Coloscopy, enteroscopy…) give only planar vision of the digestive tract. The contribution of the 3rd dimension would be an important advantage in the diagnosis of pathology.
Thirdly, integrated stereovision also meets a need for freedom: freedom of movement, freedom to be at home and not in a hospital. The patient swallows the capsule and he can go back to his activities. Optimization of the dimension of the capsule, autonomous operation and long battery life of the sensor are essential constraints.

Finally, integrated stereovision meets economic needs: the number of examinations increases with the growing of old population. To optimize the cost, an early endoscopy would make it possible to improve the forecast and better cure the patient at the beginning of his disease, which reduces the costs of treatment. The sensor has to be low cost to be autonomous.

Another emergent field in which the integrated stereovision can contribute is cartography of dangerous or less accessible areas such as disaster areas, caves, surface of an asteroid or battle fields. This cartography is an aid to navigation of rescuers, robots, drones… Traditional image sensors do not allow us to know the dimension of a scene or to know the distance to the viewed objects. The interest of an integrated stereovision is that it recovers extra information characterizing the scene and its objects, in a 3D world.

For example, drones are useful in reconnaissance operations in confined surroundings, like industrial buildings, caves or houses. In these places, the drones are equipped with stereovision and can carry out photos of its surrounding area and then set up a 3D map, but it is helpful to use the furnished information to help itself for stabilization…

Another way of scoping out unknown areas quickly is to scatter a great number of stereovision sensors and thus to create a wireless network of sensors. These sensors can be easily disseminated on the location of study and supply a maximum of information for mapping danger zones or less accessible fields, like the zones of natural disasters or planets and asteroids.

The ideal 3D integrated sensor has to show a great autonomy, a good accuracy of rebuilding, good performance, a reduced size and low cost. Now, let us look the state of the art in the following chapter.

2. State Of Art: the integrated vision and 3D reconstruction

Fig. 2.1. Wheatstone's stereoscope in 1838

With the time, we had more and more needs to be able to realize a 3D representation of a scene. The advanced in the electronic vision and the image processing domain have allowed the discovery of many important solutions. The first main work in the stereovision was managed by Marr [Marr & Poggio, 1979] in the end of the year 70. These research were fixed on the calculator analyse of the human brain and particularly for the vision.

So today there are many method of rebuilding a scene in three dimensions [Song & Wang, 2006][Szeliski, 2000][Vézien, 1995][Chen & Li, 2003]. The figure 2.2 shows a ranking of the most commonly encountered methods. As we can see, there are two main groups: the reconstruction by triangulation and the reconstruction by temporal delay.

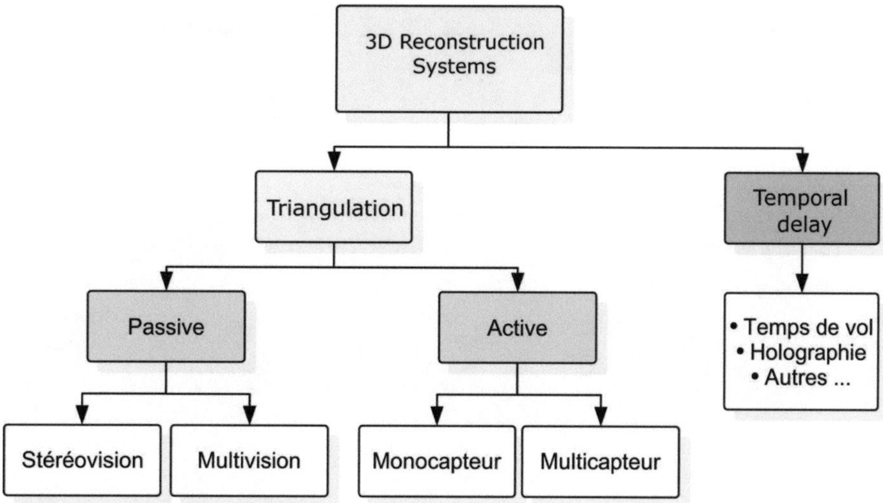

Fig. 2.2. 3D reconstruction methods classification

Since several years, the dynamics of research in this domain is such that many and robust methods to return the relief have been developed with success. For examples, the passive [Belhumeur, 1993][Cochran & Medioni, 1992] and active [Hyun & Gerhardt, 1994][Valkenburg & McIvor, 1998] stereoscopic methods or methods based on time of flight of a light pulse (monochromatic waves)[Gokturk et al., 2004][Gruss et al., 1991][Lange & Seitz, 2001]. Nevertheless, these methods are costly in material and/or temporal resources thus limiting their application in emergent domain such sensor networks for distributed 3D cartography, that could be used for the characterization of an asteroid [Stooke, 1991] and in vivo 3D endoscopy where an important depth and shape information can be added to 2D vision systems as the PillCam by GivenImaging [GivenImaging, 2005]. In fact, the continuous evolutions in the microelectronic domain allow an augmentation of the integration density. This density is such that it's possible to design integrated systems which realize more complex function on a silicon surface in constant decline. These Systems on Chip (SoC) perform the functions of a complete system. In parallel, the Vision Systems on Chip, or electronic retina [Pinna, 2003], were develop. The VSoC regroup all the needed functions for the vision. The processing capabilities and architectures that can be attached to such systems have been studied by A.Moini [Moini, 1999]. Unfortunately, sometime it's impossible have many functions on the same chip which is implement with different technology in order to optimize their performances. For this case it exist another solutions to arrive to the same level of integration. The Multi chip Component (McM) or the System in Package (SiP) [Song et al., 2003] allow to regroup on the same substrate or package many functions which is realize in different technology. A such approach allows the cohabitation of digital processing units in CMOS with telecommunication or vision unit (SiGe or AsGa).

An other form of integration was presented as the ultimate level of integration by Tummala [Tummala, 2005/2006] in order to join the Law of Moore. It is the System on Package, SoP. This approach aims to bring together the various components of technology to smaller dimensions that integrate new functions of communication while being as effective as possible (fig.2.3).

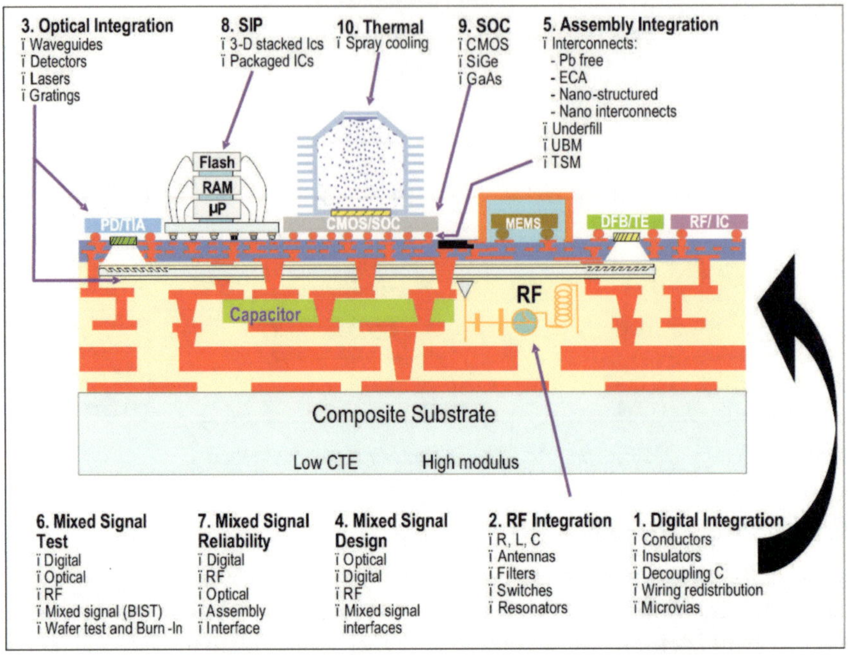

Fig. 2.3. SOP Integration by R.Tummala

Fig. 2.4. PillCam by GivenImaging

A perfect example of VSiP is the endoscopic capsule. GivenImaging developed the project PillCam [GivenImaging, 2005] for the human body exploration. This capsule is designed for examinations that could not be achieved by a classic analysis such the exploration of the small intestine. Once ingested, the PillCam begins to take pictures from inside the digestive system at a rate of two frames per second. For that it combines a CMOS imager, a light source, a unit of telecommunications to transmit the acquire picture and a battery. Unfortunately, the integrated 3D reconstruction into the form of SiP or SoC know a real delay. However, there is a lot of work along these lines we offer systems more and more miniaturized.

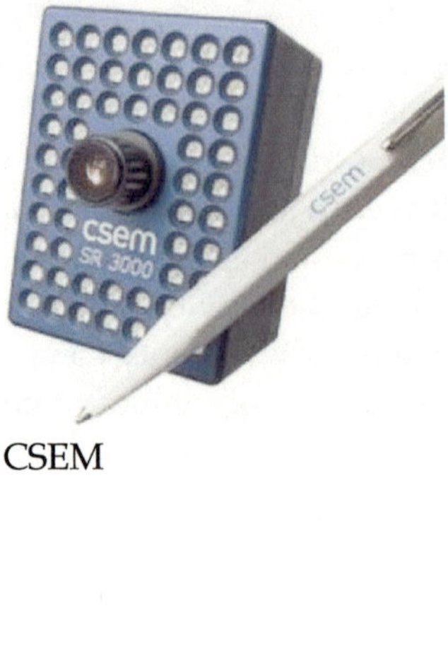

Fig. 2.5. SwissRanger SR3000 by CSEM

CSEM in Switzerland developed a 3D vision sensor based on the phase-measuring time-of-fly principle. The SwissRanger SR3000 [Oggier and al., 2006] looks like an autonomous module (fig2.) and is composed by three blocks: an integrated 3D-TOF image sensor, a 3D-camera electronics block with supplied power and a illumination blocks composed by 55 LEDS with a central wavelength of 850nm.

Fig. 2.6. Photograph of the PCB stack as implemented in the SR3000 camera

A light source emit a wave in the near-infrared which is intensity modulated with a few tens of MHz. When the wave reaches the scene, it is reflected. The result of the acquisition is a depth map on which we can apply the texture of the scene (fig.2.). Images processing are apply in the system in order to determinate in real time the quality of the information and if it is needed to apply a first filter or dedicated algorithms. The transmission protocol is the USB2.0 by default but it can be easily modified in order to by adaptable to owner utilization.

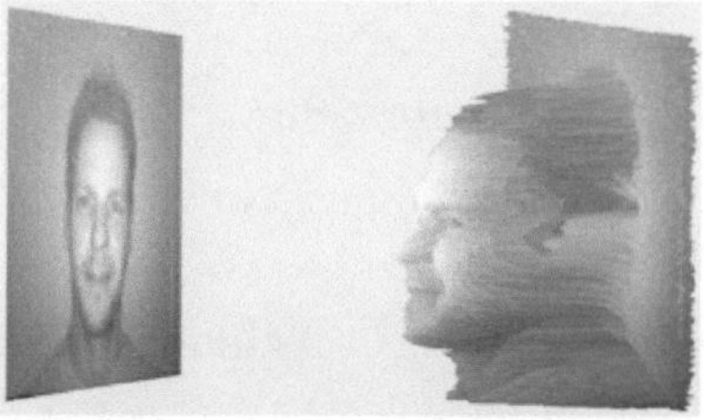

Fig. 2.7. Texture and Depth map of the SR3000 camera

An approach of passive stereoscopy has been presented by K.Konolige [Konolige]. And again it is not really going on the integrated system but on miniaturization. This project, the SRI Small Vision Module (SVM), is based on the principle of passive stereoscopy; it means that the information of distance comes from analysis of the scene from two different points of view. In this case the system embeds an algorithm of disparity on the picture with 64 levels (fig. 9). To realize this, the hardware consists of two CMOS 320x240 grayscale imagers and lenses, low-power A/D converters, a digital signal processor and a small flash memory for program storage. There are also a parallel port for the communication with the host machine and display.

This state of arts proofs that the microelectronic progresses permit to grow considerably an integration density on silicon. So it is possible from now on to make up systems which include multiples complexes functions using in the same time reduced silicon. There exist the numerous applications which can have benefit of this progress such as the endoscopies or sensors networks. Actually the application field of existing systems was too restricted because of their size or consummation but the integration in form of SoC, Sip or SoP gives a new solution opening the way to emergent applications. The systems of integrated vision are increasingly developing for all kind of using like in medical area or even for the drone navigation. Unfortunately, the integrated stereoscopy is rarely today but it stays a challenge. In fact the 3D reconstruction systems are for the moment not more than the miniaturized or embedded systems. This chapter will follow up with the presentation of the adapted methods of reconstruction to the integration and an integrated 3D vision sensor able to answer on the real time restriction using perfectly controlled process of fabrication based on the standard Si technology.

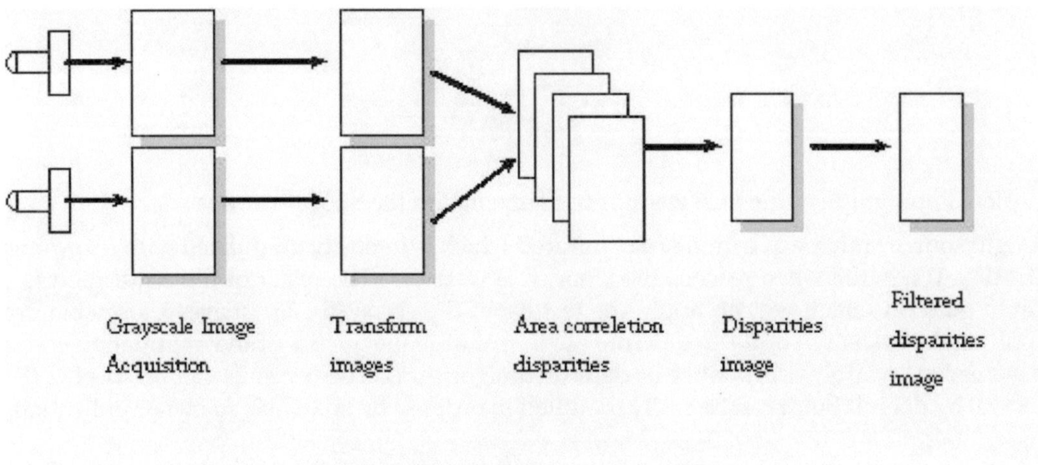

Fig. 2.8. Implemented images processing algorithm

3. Active vision and calibration

In this chapter we present the mathematical background for active vision system modeling.

3.1 System description

The active vision system is composed of an image sensor (a CMOS camera) and a structured pattern projector. The patterns studied here, are simple geometrical pattern obtained with a laser source and an optical diffraction head. This structure generates a repetitive geometrical pattern from a simple geometrical primitive. The primitives used in the pattern generator can be as simple as lines or dots.

The projector has to be simple in order to fit in the constrained size package of the embedded system. Thus complex pattern generators which can use DLP (as the ones used in video projectors) are excluded. The projected pattern considered hereafter is the simplest one, a mesh of regularly distributed points.

A multispectral approach with an IR (Infra-Red) projector and a multispectral (IR and visible images) CMOS sensor allows to grab simultaneously an image of the projected pattern and the scene texture. This is important for real time 3D generation but not only. Indeed, the multispectral approach helps to get easily a filtered image of the IR pattern.

The camera and the projector are considered rigidly fixed together fig.3.1 and the physical dimensions of the system do not change. This is important to allow off-line calibration of the vision system.

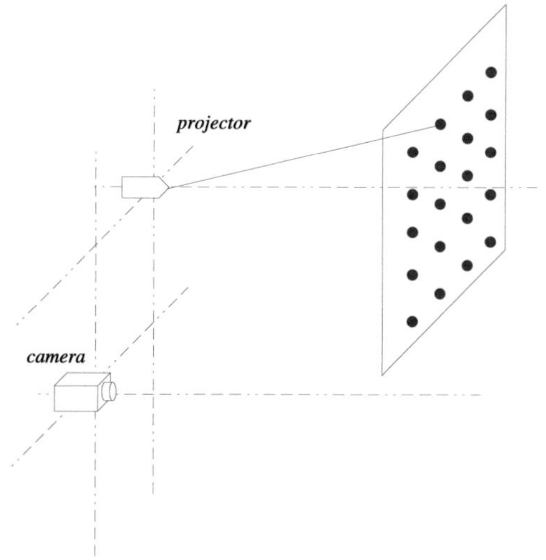

Fig. 3.1. Active stereoscopic system

The 3D reconstruction is achieved through triangulation fig.3.3. Each point of the projected pattern on the scene is the intersection of two lines:

- The line of sight, passing through the pattern point on the scene and its projection in the image plan. This line can be defined with the intrinsic and extrinsic parameters of the camera's pinhole model.
- The laser ray, starting from the projector center and passing through the chosen pattern point. This line can be defined by the projector parameters (position of the center and the angles between the rays).

The "a priori" knowledge of the stereoscopic system characteristics and the projected pattern structure, allows us to recover the distance to a projected spot of the pattern on the scene, from its image coordinate. The pattern simplicity implies correspondence ambiguities [Battle et al., 1998][Salvi et al., 2004][Salvi et al., 1998]. Indeed, we can not distinguish between laser spots in the scene from their shape or color. To solve the correspondence ambiguity problem, we use the epipolar constraint:

A laser impact can only belong, in the image, to the projection of the line supporting the laser ray. This projection represents the epipolar line [Faugeras, 1993]. We can thus simplify the correspondence search, limiting it to this line (one dimension search).

3.2 Calibration methods:

In this part we present two different calibration methods. Both allow us to obtain the correspondence between the position of the laser spots in the image and the distance to the object on which the laser is projected. We also obtain the epipolar lines equations to distinguish between the laser spots.

The first calibration method is described in [Marzani & Voisin, 2002]. Here we take several pictures of the pattern projected on a plan at different distances. As this method uses a big number of experimental points, it gives quite accurate models.

The second is analytic and has the advantage of only needing a unique picture shot to obtain the same results. We use here a calibration chart to calibrate the camera and the whole stereoscopic system and obtain thus a complete model describing the stereoscopic system.

3.2.1 1st method: Macroscopic method

Epipolar model

The pattern is projected on a plan surface parallel to the image plan. The projection plan is moved to different depth and each time a picture is taken. We thus obtain a set of points (u_{jk}^n, v_{jk}^n) where j,k ∈ [1;7] are the position in the pattern mesh and n references the image shot. These coordinates are expressed in pixel in the image. We can then obtain for each laser ray (j ,k), the epipolar line equations by fitting to a linear model

$$v = a \cdot u + b \tag{3.1}$$

so

$$\begin{pmatrix} v_{jk}^1 \\ v_{jk}^2 \\ \vdots \\ v_{jk}^n \\ \vdots \end{pmatrix} = \begin{pmatrix} u_{jk}^1 & 1 \\ u_{jk}^2 & 1 \\ \vdots & \vdots \\ u_{jk}^n & 1 \\ \vdots & \vdots \end{pmatrix} \begin{bmatrix} a_{jk} \\ b_{jk} \end{bmatrix}$$

or

$$V_{jk} = [U_{jk}\ 1] \begin{bmatrix} a_{jk} \\ b_{jk} \end{bmatrix}$$

The parameters *a* and *b* are calculated by a mean square approximation [Marzani & Voisin, 2002], as expressed in equation 3.2.

$$\begin{bmatrix} a_{jk} \\ b_{jk} \end{bmatrix} = \left(\begin{bmatrix} U_{jk}^T \\ 1 \end{bmatrix} [U_{jk}\ 1] \right)^{-1} \begin{bmatrix} U_{jk}^T \\ 1 \end{bmatrix} V_{jk} \tag{3.2}$$

Depth model

In addition to the epipolar lines, we have to express the relation between the position of a laser spot in the image and its distance to the stereoscopic system.

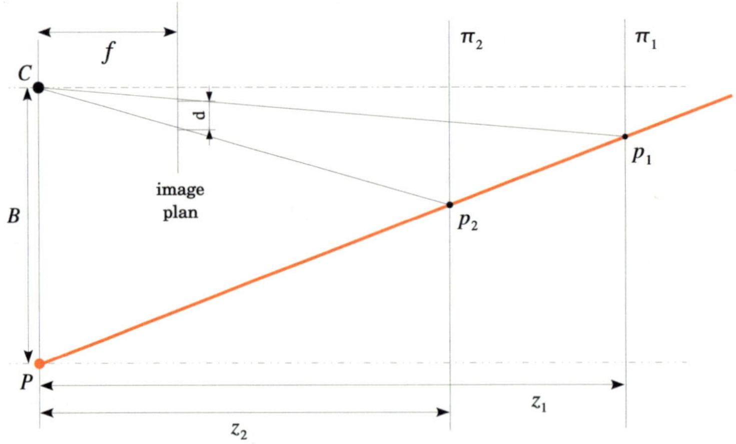

Fig. 3.2. Spot image movement vs. depth

If we consider a laser ray fig.3.2 projected on two different plans π_1 and π_2 located respectively at z_1 and z_1 from the projector/camera plan CP then, the trajectory d of the spot impact in the image (this displacement is constrained to the epipolar line corresponding to the laser ray). Considering the two triangles CPp_1 and CPp_2, we can express d as:

$$d = B\left[(\frac{z_1 - f}{z_1}) - (\frac{z_2 - f}{z_2})\right] = Bf\frac{(z_1 - z_2)}{z_1 z_2} \tag{3.3}$$

Where B is the stereoscopic base and f focal length of the camera and d is the shift between the image positions (u_1, v_1) and (u_2, v_2) on the epipolar line.

$$d = \sqrt{(u_1 - u_2)^2 + (v_1 - v_2)^2}$$

Given the epipolar line equation 3.1 we can have an expression depending only on one coordinate:

$$d = \sqrt{1 + a^2}\,(u_1 - u_2) \tag{3.4}$$

From the lasts equations 3.3, 3.4 and by choosing the appropriate referential origins, we can express the depth as a hyperbolic function of the u image coordinate, equation 3.5.

$$z = \frac{1}{\alpha u + \beta} \tag{3.5}$$

Where the α and β parameters are also calculated using a mean square method.

$$\begin{pmatrix} 1/z^1 \\ 1/z^2 \\ \vdots \\ 1/z^n \\ \vdots \end{pmatrix} = \begin{pmatrix} u^1_{jk} & 1 \\ u^2_{jk} & 1 \\ \vdots & \vdots \\ u^n_{jk} & 1 \\ \vdots & \vdots \end{pmatrix} \begin{bmatrix} \alpha_{jk} \\ \beta_{jk} \end{bmatrix}$$

or

$$\hat{Z} = [U_{jk}\ 1] \begin{bmatrix} \alpha_{jk} \\ \beta_{jk} \end{bmatrix}$$

and so

$$\begin{bmatrix} \alpha_{jk} \\ \beta_{jk} \end{bmatrix} = \left(\begin{bmatrix} U^T_{jk} \\ 1 \end{bmatrix} [U_{jk}\ 1] \right)^{-1} \begin{bmatrix} U^T_{jk} \\ 1 \end{bmatrix} \hat{Z}$$

where $\hat{Z}$ is the inverse depths vector.

3.2.1 2nd method: analytic method

We aim here to obtain a more complete characterization of the whole stereoscopic system from an image in a single snapshot.

Camera calibration

The projective model of the camera (considered ideal), also call pinhole model, links the coordinates of a point in the "world" referential to the image [Faugeras, 1993][Weng et al., 1990][Battle et al., 1998][Horaud & Monga, 1995]. It describes the relative position between the "world" and the image referential and the projection process.

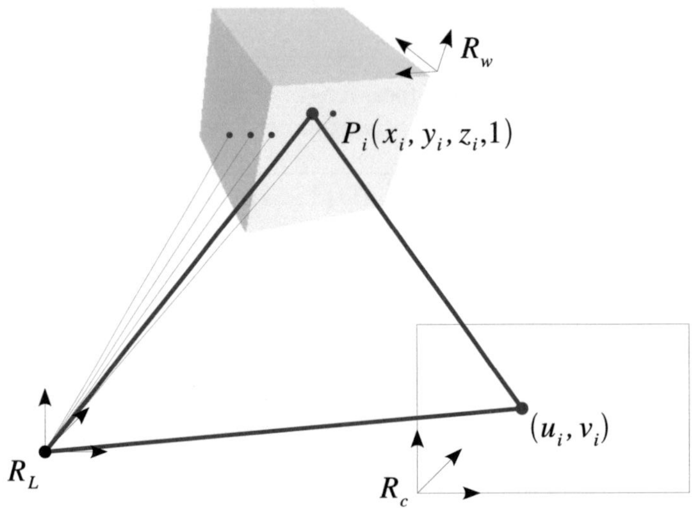

Fig. 3.3. Camera and World referential

For commodity reasons, this model is expressed in homogeneous coordinate. Thus translations and rotations are represented by matrix to vector multiplications.

The model can be split into two parts:

Extrinsic model: Represents the relation between the chosen world referential R_w and the camera referential R_c (see fig.3.3). It expresses all the translations and rotations between the two referential.

$$R_{3\times3} = \begin{bmatrix} r_{11} & r_{12} & r_{13} \\ r_{21} & r_{22} & r_{23} \\ r_{31} & r_{32} & r_{33} \end{bmatrix} \text{ and } T_{1\times3} = \begin{bmatrix} t_1 \\ t_2 \\ t_3 \end{bmatrix}$$

are respectively these rotations and the translation matrix.

The transformation can be expressed in homogeneous coordinates as:

$$\begin{bmatrix} x \\ y \\ z \\ 1 \end{bmatrix}_{R_c} = \begin{bmatrix} R_{3\times3} & T_{3\times1} \\ 0_{1\times3} & 1 \end{bmatrix} \cdot \begin{bmatrix} x \\ y \\ z \\ 1 \end{bmatrix}_{R_w}$$

Intrinsic model: Describes the projective imaging transformation.

It links a point coordinate expressed in the camera referential with the image coordinate expressed in pixels. Figure 3.4 shows the projection process of a point $P_c(x, y, z)^T_{Rc}$ expressed in the camera coordinates. To simplify the representation only x and z axis are represented.

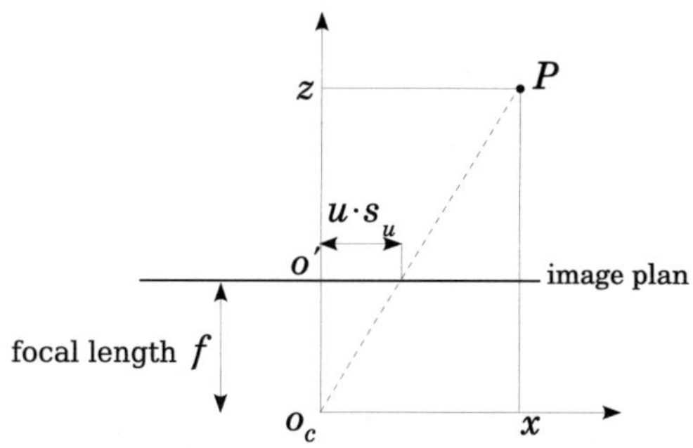

Fig. 3.4. Camera projection

O_c is the camera optical center and O' the projection of this center on the image plan which image coordinates are u_0, v_0. (f) the focal length and ($s_u \times s_v$) the pixel size.
We can express the image coordinate of the point:

$$\begin{cases} z \cdot (u - u_0) = x \cdot \frac{f}{s_u} \\ z \cdot (v - v_0) = y \cdot \frac{f}{s_v} \end{cases}$$

Using homogeneous coordinates we can express this last equation as:

$$\begin{bmatrix} \lambda \cdot u \\ \lambda \cdot v \\ \lambda \end{bmatrix} = I \cdot \begin{bmatrix} x \\ y \\ z \\ 1 \end{bmatrix}_{R_c} = \begin{bmatrix} f/s_u & 0 & u_0 & 0 \\ 0 & f/s_v & v_0 & 0 \\ 0 & 0 & 1 & 0 \end{bmatrix} \cdot \begin{bmatrix} x \\ y \\ z \\ 1 \end{bmatrix}_{R_c}$$

Where I is the intrinsic matrix, and $\lambda = z$ is the projection proportionality reduction factor.
The intrinsic matrix I has a null column and so is irreversible which means that we can not retrieve the depth from the sole image coordinate.
The global camera model is thus a matrix to vector multiplication connecting a point coordinates expressed in the "world referential" to the image coordinates expressed in pixels.

$$\begin{bmatrix} \lambda \cdot u \\ \lambda \cdot v \\ \lambda \end{bmatrix} = M \cdot \begin{bmatrix} x \\ y \\ z \\ 1 \end{bmatrix}_{R_w} = I \cdot \begin{bmatrix} R_{3\times3} & T_{3\times1} \\ 0_{1\times3} & 1 \end{bmatrix} \cdot \begin{bmatrix} x \\ y \\ z \\ 1 \end{bmatrix}_{R_w} \qquad (3.6)$$

Where

$$M = \begin{bmatrix} m_{11} & m_{12} & m_{13} & m_{14} \\ m_{21} & m_{22} & m_{23} & m_{24} \\ m_{31} & m_{32} & m_{33} & m_{34} \end{bmatrix}$$

is the fundamental matrix.

We can rearrange equation 3.6 as:

$$\begin{cases} u = \frac{m_{11}\cdot x + m_{12}\cdot y + m_{13}\cdot z + m_{14}}{m_{31}\cdot x + m_{32}\cdot y + m_{33}\cdot z + m_{34}} \\ v = \frac{m_{21}\cdot x + m_{22}\cdot y + m_{23}\cdot z + m_{24}}{m_{31}\cdot x + m_{32}\cdot y + m_{33}\cdot z + m_{34}} \end{cases} \tag{3.7}$$

To obtain the necessary points for the fundamental matrix parameters estimation, we have design a new calibration 3D pattern fig.3.5 .

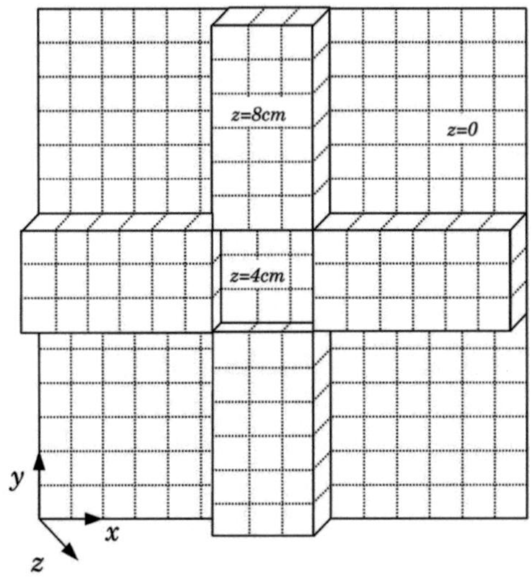

Fig. 3.5. Calibration chart referential

An automatic corner extraction, using Harris filter, from an image of the 3D chart, gives us enough points (184) for a robust estimation [Battle et al., 1998].
Using this model we can obtain the line of sight from any point in the image and especially for the projected pattern points. Thus, for each pattern point i, we determine the line of sight as an intersection between two plans as:

$$\begin{cases} \Gamma_{11}x_i + \Gamma_{12}y_i + \Gamma_{13}z_i + \Gamma_{14} = 0 \\ \Gamma_{21}x_i + \Gamma_{22}y_i + \Gamma_{23}z_i + \Gamma_{24} = 0 \end{cases} \tag{3.8}$$

$$\text{with} \quad \begin{cases} \Gamma_{11}=(m_{11}-u_i\cdot m_{31}) \\ \Gamma_{12}=(m_{12}-u_i\cdot m_{32}) \\ \Gamma_{13}=(m_{13}-u_i\cdot m_{33}) \\ \Gamma_{14}=(m_{14}-u_i\cdot m_{34}) \\ \Gamma_{21}=(m_{21}-v_i\cdot m_{31}) \\ \Gamma_{22}=(m_{22}-v_i\cdot m_{32}) \\ \Gamma_{23}=(m_{23}-v_i\cdot m_{33}) \\ \Gamma_{24}=(m_{24}-v_i\cdot m_{34}) \end{cases}$$

Laser projector characterization:
To validate experimentally the method, we have used a 7x7 laser matrix pattern. Each laser ray can be considered as the intersection of two supporting plans fig.3.6. Thus, we can define the 49 laser rays by the parameters of 7 horizontal plans (H_j) and 7 vertical plans (V_k).

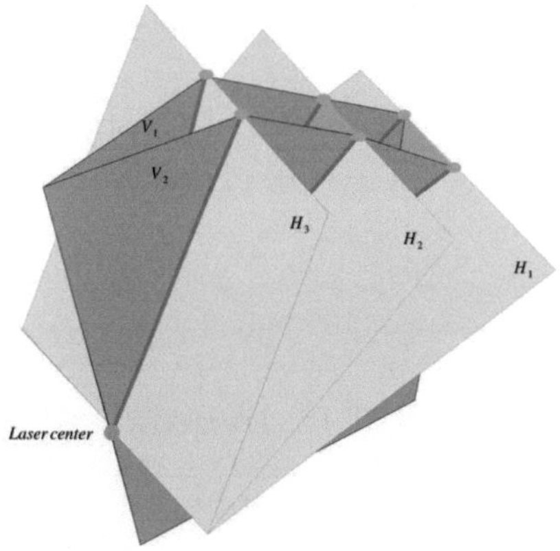

Fig. 3.6. Laser projector model

$$\begin{cases} x_i \cdot h_{1j} + y_i \cdot h_{2j} = z_i + h_{3j} \\ x_i \cdot v_{1k} + y_i \cdot v_{2k} = z_i + v_{3k} \end{cases} \quad with\ j, k \in [1;7] \tag{3.9}$$

To determine the plan parameters, we use an IR image of the laser pattern projected on the 3D calibration chart The chart shape helps us to get easily the z_i coordinate of each spot. The x_i and y_i coordinates are obtained from the image coordinate z_i and the fundamental matrix using (3.8). Arranging the equation, we obtain thus the x_i and y_i world coordinate of each point (3.10).

$$\begin{bmatrix} x_i \\ y_i \end{bmatrix} = M_{uv}^{-1} \cdot M_{uvz} \tag{3.10}$$

where,

$$M_{uv} = \begin{bmatrix} (m_{11} - u_i \cdot m_{31}) & (m_{12} - u_i \cdot m_{32}) \\ (m_{21} - v_i \cdot m_{31}) & (m_{22} - v_i \cdot m_{32}) \end{bmatrix}$$

and

$$M_{uvz} = \begin{bmatrix} z_i \cdot (m_{13} - u_i \cdot m_{33}) + (m_{14} - u_i \cdot m_{34}) \\ z_i \cdot (m_{23} - v_i \cdot m_{33}) + (m_{24} - v_i \cdot m_{34}) \end{bmatrix}$$

Furthermore, we can identify 7 points verifying each plan equation. So we can write for each horizontal plan:

$$\begin{pmatrix} x_{1j} & y_{1j} & 1 \\ \vdots & \vdots & \vdots \\ x_{4j} & y_{4j} & 1 \\ \vdots & \vdots & \vdots \\ x_{7j} & y_{7j} & 1 \end{pmatrix} \cdot \begin{bmatrix} h_{1j} \\ h_{2j} \\ h_{3j} \end{bmatrix} = - \begin{pmatrix} z_{1j} \\ \vdots \\ z_{4j} \\ \vdots \\ z_{7j} \end{pmatrix}$$

or

$$K_h \cdot H_j = -Z_j$$

The horizontal plans parameters can be finally obtained by:

$$H_j = -(K_h^T \cdot K_h^{-1})^{-1} \cdot K_h^T \cdot Z_j$$

The same result can be obtained for the vertical plans:

$$V_k = -(K_v^T \cdot K_v^{-1})^{-1} \cdot K_v^T \cdot Z_k$$

The determination of the plans parameters allows us to determine the projector center in the world referential.
Indeed, this point is the intersection of all the horizontal and vertical plans, and verifies the equation 3.11.

$$\begin{pmatrix} h_{11} & h_{21} & 1 \\ \vdots & \vdots & \vdots \\ h_{17} & h_{27} & 1 \\ v_{11} & v_{21} & 1 \\ \vdots & \vdots & \vdots \\ v_{17} & v_{27} & 1 \end{pmatrix} \cdot \begin{bmatrix} x_{\text{laser}} \\ y_{\text{laser}} \\ z_{\text{laser}} \end{bmatrix} = - \begin{pmatrix} h_{31} \\ \vdots \\ h_{37} \\ v_{31} \\ \vdots \\ v_{37} \end{pmatrix} \tag{3.11}$$

Depth recovery
Once we have characterized the whole stereoscopic system (camera perspective model and the projector parameters), we can recover the coordinate of a laser spot (i) projected on any object from its image coordinate and the laser index (jk).
Starting from the laser ray equation 3.9 we can express the x_i and y_i coordinates as:

$$\begin{bmatrix} x_i \\ y_i \end{bmatrix} = \begin{bmatrix} h_{1j} & h_{2j} \\ v_{1k} & v_{2k} \end{bmatrix}^{-1} \cdot \begin{bmatrix} z_i + h_{3j} \\ z_i + v_{3k} \end{bmatrix} = \begin{bmatrix} \frac{v_{2k}(z_i+h_{3j})-h_{2j}(z_i+v_{3k})}{h_{1j}\cdot v_{2k}-h_{2j}\cdot v_{1k}} \\ \frac{h_{1j}(z_i+v_{3k})-v_{1j}(z_i+h_{3j})}{h_{1j}\cdot v_{2k}-h_{2j}\cdot v_{1k}} \end{bmatrix}$$

$$\text{thus} \quad \begin{bmatrix} x_i \\ y_i \end{bmatrix} = \begin{bmatrix} l_{1jk} \cdot z_i + l_{2jk} \\ l_{3jk} \cdot z_i + l_{4jk} \end{bmatrix} \tag{3.12}$$

with

$$l_{1jk} = \frac{v_{2k} - h_{2j}}{h_{1j} \cdot v_{2k} - h_{2j} \cdot v_{1k}} = \frac{v_{2k} - h_{2j}}{\Delta}$$

$$l_{2jk} = \frac{v_{2k} \cdot h_{3j} - h_{2j} \cdot v_{3j}}{h_{1j} \cdot v_{2k} - h_{2j} \cdot v_{1k}} = \frac{v_{2k} \cdot h_{3j} - h_{2j} \cdot v_{3j}}{\Delta}$$

$$l_{3jk} = \frac{h_{1j} - v_{1k}}{h_{1j} \cdot v_{2k} - h_{2j} \cdot v_{1k}} = \frac{h_{1j} - v_{1j}}{\Delta}$$

$$l_{4jk} = \frac{v_{1k} \cdot h_{3j} - h_{1j} \cdot v_{3k}}{h_{1j} \cdot v_{2k} - h_{2j} \cdot v_{1k}} = \frac{v_{1k} \cdot h_{3j} - h_{1j} \cdot v_{3k}}{\Delta}$$

Where

$$\Delta = h_{1j} \cdot v_{2k} - h_{2j} \cdot v_{1k}.$$

By replacing 3.12 in 3.7 we obtain the image coordinate (u_i, v_i) versus the depth z_i.

$$u_i = \frac{A^1_{jk} \cdot z_i + A^2_{jk}}{A^3_{jk} \cdot z_i + A^4_{jk}}$$

$$v_i = \frac{A^5_{jk} \cdot z_i + A^6_{jk}}{A^3_{jk} \cdot z_i + A^4_{jk}}$$

where

$$\begin{cases} A^1_{jk} = m_{11} \cdot l_{1jk} + m_{12} \cdot l_{3jk} + m_{13} \\ A^2_{jk} = m_{11} \cdot l_{2jk} + m_{12} \cdot l_{4jk} + m_{14} \\ A^3_{jk} = m_{31} \cdot l_{1jk} + m_{32} \cdot l_{3jk} + m_{33} \\ A^4_{jk} = m_{31} \cdot l_{2jk} + m_{32} \cdot l_{4jk} + m_{34} \\ A^5_{jk} = m_{21} \cdot l_{1jk} + m_{22} \cdot l_{3jk} + m_{23} \\ A^6_{jk} = m_{21} \cdot l_{2jk} + m_{22} \cdot l_{4jk} + m_{24} \end{cases}$$

Hence, by eliminating z_i from the last equations we obtain the epipolar line equation:

$$[A^5_{jk}A^4_{jk} - A^3_{jk}A^6_{jk}] \cdot u_i + [A^3_{jk}A^2_{jk} - A^1_{jk}A^4_{jk}] \cdot v_i = [A^2_{jk}A^5_{jk} - A^1_{jk}A^6_{jk}]$$

We identify the epipolar line parameters as in equation 3.1:

$$\begin{cases} a_{jk} = -\frac{[A^5_{jk}A^4_{jk} - A^3_{jk}A^6_{jk}]}{[A^3_{jk}A^2_{jk} - A^1_{jk}A^4_{jk}]} \\ b_{jk} = \frac{[A^2_{jk}A^5_{jk} - A^1_{jk}A^6_{jk}]}{[A^3_{jk}A^2_{jk} - A^1_{jk}A^4_{jk}]} \end{cases}$$

We can also express the z_i coordinate as a function of one of the image coordinate:

$$z_i = \frac{A^4_{jk} \cdot u_i - A^2_{jk}}{A^1_{jk} - A^3_{jk} \cdot u_i}$$

Particular case:

If there is no rotation between the image and the world referential and the Z axis translation is null (as represented in fig.3.2). then, we have:

$$R_{3\times3} = \begin{bmatrix} 1 & 0 & 0 \\ 0 & 1 & 0 \\ 0 & 0 & 1 \end{bmatrix}$$

and

$$T_{1\times 3} = \begin{bmatrix} t_1 \\ t_2 \\ 0 \end{bmatrix}$$

the fundamental matrix will be

$$M = \begin{bmatrix} f/s_u & 0 & u_0 & t_1 f/s_u \\ 0 & f/s_v & v_0 & t_2 f/s_v \\ 0 & 0 & 1 & 0 \end{bmatrix}$$

with m_{31}, m_{32} and m_{34} equal to zero and so $A^4_{jk} = 0$.
Thus, by choosing a good physical configuration a relation, between the world coordinate z and the image coordinate u, equivalent to what expressed in 3.5.

4. Cyclope

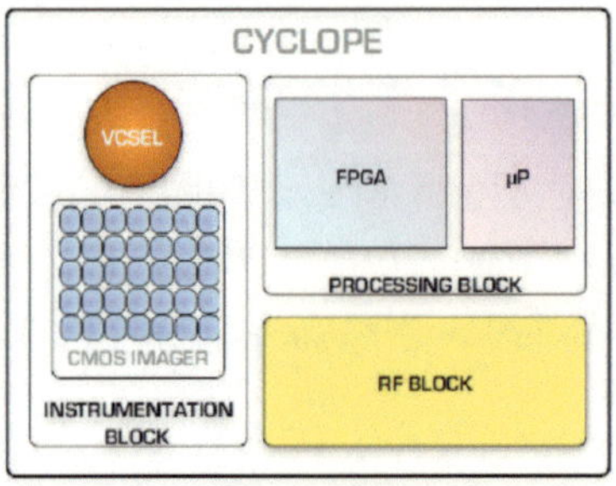

Fig. 4.1. Synoptic of Cyclope

Cyclope is an integrated 3D vision system sensor based on active vision. It is composed of three essential parts:

- An instrumentation block: for image grabbing and the generation of the structured light pattern.
- A processing block: for control and data processing.
- A RF block: for the transmission of the results and the dynamic OTA (Over The Air) reconfiguration.

These parts are realized in different technologies: CMOS for the image sensor and the processing units; GaAs for the pattern projector and RF-CMOS for the communication unit. The monolithic integration of those elements is not feasible, this is why the development of an integrated "SIP" (System In Package) is actually the best solution to overcome the technological constraints and realize a chip scale package. This solution is used in several embedded sensors as The Human++ platform [Gyselinckx et al., 2005] or Smart Dust [Warneke et al., 2001] where digital processing has to cohabit with optical and wireless communication. In Cyclope, we have chosen the realization of an integrated system based on active vision. This choice is the consequence of two facts. The first is the real time processing that we would like to obtain. In active vision, a known pattern is projected on a scene, this projection belong a specific line denoted as epipolar line. This property is very interesting in real time processing, because it could reduce the complex problem of the computation on the z coordinate to a simple problem of line computation. The second fact is that active vision simplifies the matching between a point and is projection, because the only points that we want to match, is the pattern points.

A processing block

As we define in the last part, we have a set of parameters for the epipolar and depth models. Those parameters are used on run time to make the point matching (identify the original position of a pattern point from its image) and calculate the depth using the appropriate parameters.

The architecture is divided into two principal parts:

- Pre-processing unit for low level image processing and feature extraction from the IR image.
- 3D unit for point matching and depth calculation.

Figure 4.2 shows the global processing architecture. A dual-port memory is used for image storage allowing asynchronous image acquisitions. The processing implemented in the Pre-processing unit is thresholding, segmentation and spot center calculation. A FIFO allows the communication between the two units and a final storage FIFO allows communication toward an external UART.

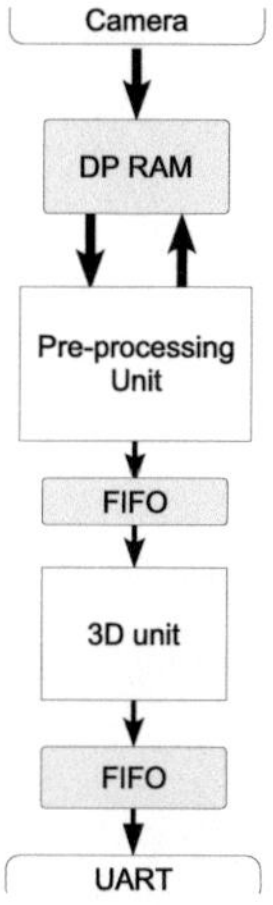

Fig. 4.2. Global processing architecture

4.1.1 The 3D unit

In this part we present the 3D extraction method for on line processing. To achieve this purpose, we have designed a parallel digital processing unit fig.4.3.

After a pre-processing step, where the laser spot center is estimated from the IR image, the coordinate of each point are directed to the processing unit where the (z) coordinate is calculated.

Starting from the point abscissa (u) we calculate its estimated ordinate (~v) if it belongs to an epipolar line. We compare this estimation with the true ordinate (v).

These operations are made for all the epipolar line simultaneously. After thresholding the encoder returns the index of the corresponding epipolar line.

The next step is to calculate the z coordinate from the u coordinate and the appropriate depth model parameters.

These computation blocs are synchronous and pipelined, allowing thus, high processing rates.

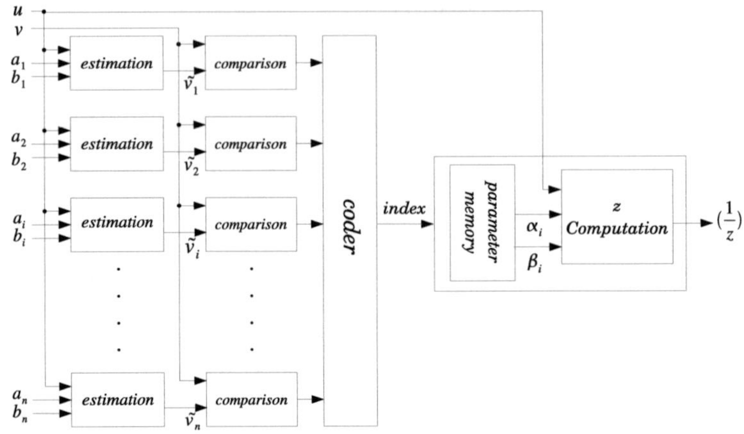

Fig. 4.3. 3D unit

4.1.2 Estimation bloc:

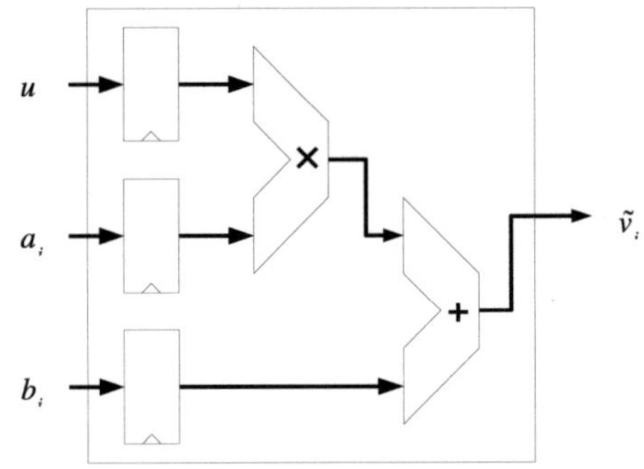

Fig. 4.4. Estimation bloc

In this bloc the estimated ordinate is calculated ~v=a.u+b . The (a,b) parameters are loaded from memory.

4.1.3 Comparison bloc:

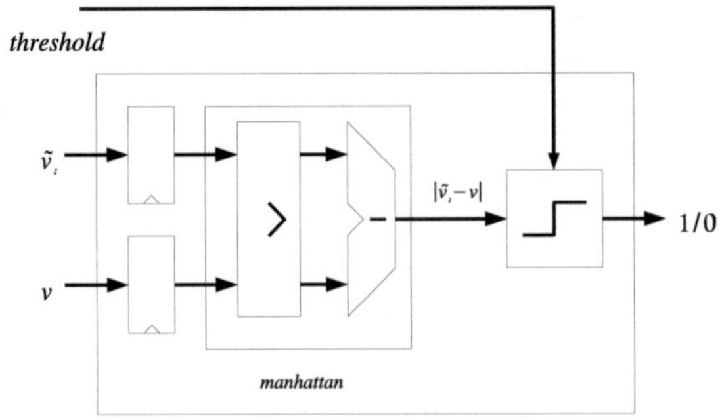

Fig. 4.5. Comparison bloc

In this bloc the absolute value of the difference between the ordinate v and its estimation $\tilde{v}$ is calculated. This difference is then thresholded.
The thresholding avoids a resource consuming sort stage. The threshold was a priori chosen as half the minimum distance between two consecutive epipolar lines. The threshold can be adjusted for each comparison bloc.
This bloc returns a '1' result if the distance is underneath the threshold.

4.1.4 Encoding bloc:

If the comparison blocs returns a unique '1' result, then the encoder returns the corresponding epipolar line index.
If no comparison bloc returns a 'true' result, the point is irrelevant and considered as picture noise.
If more then one comparison bloc returns '1' then we consider that we have a correspondence error and a flag is set. The selected index is then carried to the next stage where the z coordinate is calculated. It allows the selection of the good parameters to the depth model.
We have chosen to calculate (1/z) rather than z to have a simpler computation unit. This computation bloc is then identical to the estimation bloc.

4.2 The instrumentation block

4.2.1 An energetic approach

Having opted for a laser pattern which give out in the near infrared spectrum in order to not misrepresent the scene, many classic solution were possible. We must acquire in the same time an image of the texture therefore in the visible spectrum, and an image of the projected pattern:

- the utilization of a BAYER type optic filter (VI/IR) .
- the conception of a CMOS imager which used a semiconductor technology with a near infrared spectral response [Pei et al., 2003] or by changing the depth and the nature of the junctions [David Starikov & Bensaoula, 2004]
- made the separation of the spectra by electronic or image processing.

Unfortunately none of these options could not completely satisfy us. Indeed, we wanted a solution easily implemented by simplifying the fabrication process and thus stay in a standard Silicon technology without any optic filter, adaptable to already existing systems by providing them with a 3D information without major changes, and finally be able to assure a classic video rate of 25 images per second.
It is in this context that we opted for an original acquisition approach based on the energetic's gap between the scene and the pattern and not based on the wave length.
The idea is to project a pattern much more energetic than the texture could be and to modify the integration time of the imager in order to only acquire the projected pattern. And if is needed, we can also realized an image processing like a binarization.

4.2.2 Image acquisition processing:

To allow real-time acquisition of both pattern and texture, we have developed a multi-spectral 64x64 pixels CMOS imager (fig. 4.6). This sensor has programmable light integration and shutter time to allow dynamic change. The projector pulses periodically on the scene an energetic pattern. An image acquisition with a short integration time allows to grab the image of the pattern without the background texture. A second image acquisition with a longer integration allows to grab the texture when the projector is off. The figure 4.7 shows the sequential scheduling of the images acquisition. To reach a video frame rate of 25 images/s this acquisition sequence must be done in less than 40 ms. The global acquisition time is given in equation 4.1 where T_{rst} is the reset time, T_{rd} is the time needed to read the entire image and T_{intVI} T_{intIR} are respectively the integration time for both visible and IR image.

$$T_{total}=2 \cdot T_{rst}+2 \cdot T_{rd}+T_{intVI}+T_{intIR} \quad (4.1)$$

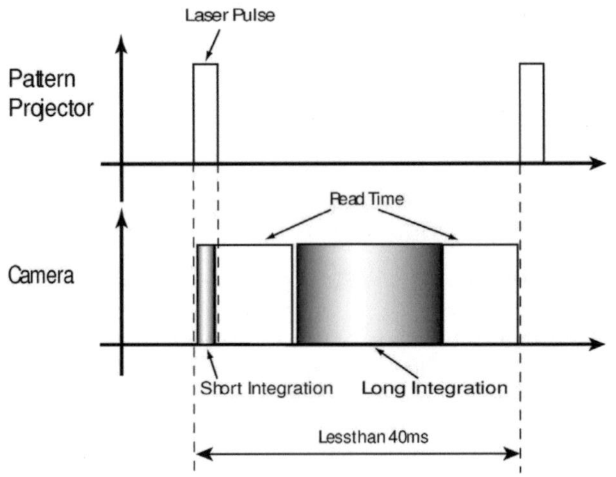

Fig. 4.7. Acquisition sequence

5. Conclusion

In this chapter we saw that the issues of the integrated stereoscopy are really important. Many of emergent domains such as the 3D endoscopy or network sensors for the 3D mapping directly depend on the advance of the integrated 3D vision because the existing solutions is limiting by their application domain, their size or autonomy and there are not suitable to an integrated use. The integrated 3D vision is able to give us a solution where it exist heavy material constraint but we need to develop special architectures and methods in order to have an efficiency solution.

In order to have a complete view of this challenge, we proposed an integrated active stereoscopic vision sensor design for emergent domain and the mathematical model that we developed in order to be used in such applications. Based on this model we have designed a processing architecture allowing a high integration level.

We also proposed an original acquisition method based on the energetic's gap between the texture and the projected pattern allowing an easily spectrum discrimination.

The next step to this work is the chip level integration of both the image sensor and the pattern projector. Evaluate the power consumption of the pulsed laser projector considering the optical efficiency of the diffraction head.

6. References

[Battle et al., 1998] J. Battle, E. Mouaddib, and J. Salvi. Recent progress in codded structurred light as a technique to solve the correspondance problem: a survey. Pattern Recognition, 31(7) :963-982, 1998.

[Belhumeur, 1993] P.N. Belhumeur. A binocular stereo algorithm for reconstructing sloping, creased, and broken surfaces in the presence of half-occlusion. In Fourth International Conference on Computer Vision, 1993.

[Chen and Li, 2003] S.Y. Chen and Y.F. Li. A 3d vision system using unique color encoding. In Internanional conference on robotics, intelligent systems and signal processing, 2003.

[Chichyang and Zheng, 1998] Chen Chichyang and Y.F. Zheng. Passive and active stereo vision for smooth surface detection of deformed plates. IEEE Transactions on Pattern Analysis and Machine Intelligence, 7 :62-81, 1998.

[Cochran and Medioni, 1992] S.D. Cochran and G. Medioni. 3-d surface description from binocular stereo. IEEE Transactions on Pattern Analysis and Machine Intelligence, 14 :981-994, 1992.

[David Starikov and Bensaoula, 2004] Rajeev Pillai David Starikov, Chris Boney and Abdelhak Bensaoula. Dual-band uv/ir optical sensors for fire and flame detection and target recognition. Senson for Industry Conference, 2004.

[Faugeras, 1993] O. Faugeras. Three-Dimensional Computer Vision, a Geometric Viewpoint. MIT Press, 1993.

[GivenImaging, 2005] GivenImaging. Given Diagnostic System, The Platform for PillCam Endoscopy. 2005.

[Gokturk et al., 2004] S.B. Gokturk, H. Yalcin, and C. Bamji. A time-of-flight depth sensor - system description, issues and solutions. In Conference on Computer Vision and Pattern Recognition Workshop, 2004. 33

[Gruss et al., 1991] A. Gruss, L.R. Carley, and T. Kanade. Integrated sensor and range-finding analog signal processor. IEEE Journal of Solid-State Circuits, 26 :184-191, 1991.

[Gyselinckx et al., 2005] B. Gyselinckx, C. Van Hoof, J. Ryckaert, R.F. Yazicioglu, P. Fiorini, and V. Leonov. Human++ : autonomous wireless sensors for body area networks. In IEEE 2005 Custom Integrated Circuits Conference, 2005.

[Horaud and Monga, 1995] R. Horaud and O. Monga. Vision par ordinateur. Edition Hermès, 1995.

[Hyun and Gerhardt, 1994] K. Hyun and L.A. Gerhardt. The use of laser structured light for 3d surface measurement and inspection. In Proceedings of the Fourth International Conference on Computer Integrated Manufac-turing and Automation Technology, 1994.

[Konolige,] K. Konolige. Small vision system : Hardware and implementation.

[Lange and Seitz, 2001] R. Lange and P. Seitz. Solid-state time-of-flight range camera. IEEE Journal ofQuantum Electronics, 37 :390-397, 2001.

[Marr and Poggio, 1979] D. Marr and T. Poggio. A computational theory of human stereo vision. In Proceedings of the Royal Society of London B, volume 204, pages 301-328, 1979.

[Marzani and Voisin, 2002] F. Marzani and Y. Voisin. Calibration of a three-dimentionnal reconstruction system using a structured light source. Optical Engineering, 41 :484-492, 2002.

[Moini, 1999] A. Moini. Vision Chips. Kluwer Academic Publishers, 1999.

[Oggier et al., 2006] T. Oggier, B.Büttgen, F.lustenberger, G.Becker, B. Rüegg, and A.Hodac. Swissranger sr3000 and first experiences based on miniaturized 3d-tof cameras. 2006.

[Pei et al., 2003] Z. Pei, L.S. Lai, H.P. Hwang, Y.T. Tseng, and M.-J. Tsai C.S. Liang. Si1 xgex=si multi-quantum well phototransistor for near-infrared operation. Physical, 2003.

[Pinna, 2003] A. Pinna. Conception d'une rétine connexionniste : du capteur au système de vision sur puce. PhD thesis, Paris 6 - UPMC, 2003.

[Salvi et al., 1998] J. Salvi, J. Batlle, and E. Mouaddib. A robust-coded pattern projection for dynamic 3d scene measurement. Pattern Recognition Letter, 19 :1055-1065, 1998.

[Salvi et al., 2004] J. Salvi, J. Pagès, and J. Batlle. Pattern codification strategies in structured light systems. Pattern Recognition, 37 :827-849, 2004.

[Song and Wang, 2006] LiMei Song and DaNvWang. A novel grating matching method for 3d reconstruction. NDT & E International, 39 :282-288, 2006.

[Song et al., 2003] Yong-Ha Song, S.G. Kim, K.J Rhee, and T.S Kim. A study of considering the reliability issue on asic/memoty integration by sip (system-in-package) technologie. Microelectronics Reliability, 2003.

[Stooke, 1991] J. Stooke. Cartographiy of asteroides and comet nuclei from low resolution data. Asteroids, Comets, Meteors, pages 583-586, 1991.

[Szeliski, 2000] R. Szeliski. Scene reconstruction from multiple cameras. In International Conference on Image Processing, 2000.

[Tummala, 2005] R.R Tummala. Packaging : past, present and future. In International Conference on Electronic Packaging Technology, 2005.

[Valkenburg and McIvor, 1998] R.J. Valkenburg and A.M. McIvor. Accurate 3d measurement using a structured light system. Image and Vision Computing, 16 :99-110, 1998.

[Vézien, 1995] Jean-Marc Vézien. Techniques de reconstruction globale par analyse de paire d'images stéréoscopiques. PhD thesis, Paris 7, 1995.

[Warneke et al., 2001] B. Warneke, M. Last, B. Liebowitz, and K.S.J Pister. Smart dust. communicating with a cubic milimeter computer. Computer, 34 :44 - 51, Jan 2001.

[Weng et al., 1990] J. Weng, P. Cohen, and M. Herniou. Calibration of stereo cameras using a non-linear distortion model. In Pattern Recognition, 1990. Proceedings., 10th International Conference on, volume i 16-21 Jun 1990, pages 246-253 vol.1, 1990.

10

Arrangement of a Multi Stereo Visual Sensor System for a Human Activities Space

Wlodek Kulesza, Jiandan Chen and Siamak Khatibi
Blekinge Institute of Technology
Sweden

1. Introduction

The requirements for autonomous physical services supporting people have become more important in recent years in activities such as taking care of the elderly people, doing the housework, and giving a comfortable living environment. For this reason, our research is aimed towards design and implementation of a high-performance autonomous, distributed vision information system, which would be able to understand human behaviours and living environment, as a temporary substitute for a qualified nurse and housekeeper.

The human-centred computation is proposed in the MIT Oxygen Project, (MIT, 2008). Furthermore, Hashimoto presented the concept of intelligent space: *Intelligent Space can be defined as space with functions that can provide appropriate services for human beings by capturing events in the space and by utilizing the information intelligently with computers and robots,* (Hashimoto, 2003). The Intelligent Space was treated as a platform, which supported people's information and physical needs. It was the interface for both the humans and robots.

The proposed Intelligent Vision Agent System, IVAS, is a high-performance, autonomous, distributed vision and information processing system. Figure 1 illustrates the idea of the IVAS. It consists of multiple sensors and actuators for surveillance of the human activities space involving humans and their surrounding environment including robots and household appliances etc. The system not only gathers information, but also controls these sensors including their deployment and autonomous servo. But first of all it is able to extract required information from images for different applications, especially for three dimensional (3D) reconstruction. The 3D information from a real scene of target objects can be compared with a pattern in order to make decisions. Meanwhile the pattern may also be renewed by the inclusion of a learning phase. These features require the system to dynamically adjust cameras to get optimised 3D information. The intelligent agent consists of a knowledge database, with learning and decision making components that can be used to track, recognize and analyze the objects.

Similar to the human eyes, the stereo vision observes the world from two different points of view. At least two images need to be fused to obtain a depth perception of the world. However due to the digital camera principle, the depth reconstruction accuracy is limited by the sensor pixel resolution. The spatial quantisation is illustrated by iso-disparity maps. The iso-disparity surfaces approach, when calculating the reconstruction uncertainty, has been discussed by Völpel and Theimer, (Völpel & Theimer, 1995). The shape of iso-disparity surfaces for general stereo configurations was studied by Pollefeys & Sinha (Pollefeys &

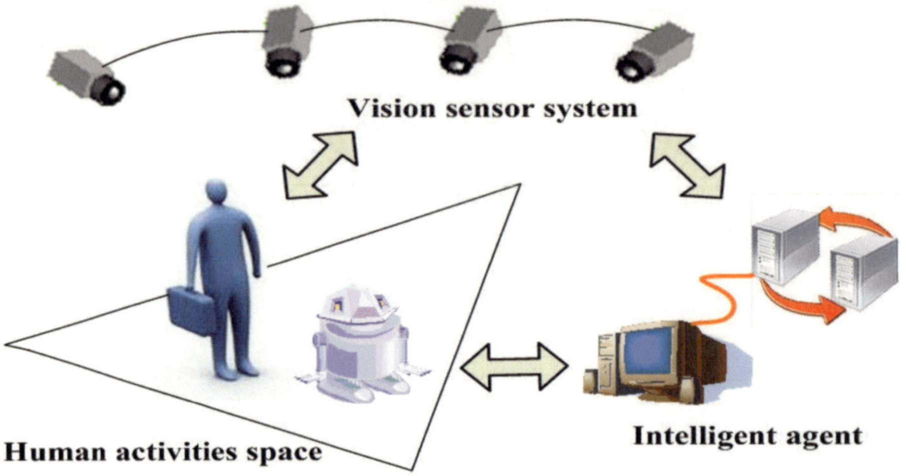

Fig. 1. Overview of an Intelligent Vision Agent System

Sinha , 2004) and Chen et al. (Chen et al., 2007c). The proposed mathematical model of iso-disparity map provides an efficient way of describing the shape of iso-disparity surfaces, and estimating the depth reconstruction uncertainty which is related to the stereo pair baseline length, the target distance to baseline, focal length, convergence angle and the pixel resolution.

The depth spatial quantization uncertainty, caused by a discrete sensor is one of the factors which has the most influence on the depth reconstruction accuracy. This type of uncertainty cannot be decreased by the reduction of pixel size since the signal to noise ratio is also reduced and the sensitivity of the sensor itself is restricted. The selection of an optimal sensor pixel is discussed by Chen et al., (Chen et al., 2000).

The sensor planning by re-annealing software was introduced by Mittal, (Mittal, 2006) and the evaluation of the sensors' configuration by a quality metric was presented in (Chen, 2002). A linear programming method to optimize sensor placement based on binary optimization techniques has been developed, (Chakrabarty et al., 2002; Hörster & Lienhart, 2006; Erdem & Sclaroff, 2006). This is a convenient tool for optimising the visual sensors' configurations when observing target space such as human activities space. Chen at al. described the optimization program for the 3D reconstruction of a human activity space, (Chen et al., 2007a; Chen et al., 2007b). The papers introduce the method of optimizing stereo pair configurations under the required constraints of stereo pair baseline length, visibility, camera movement and depth reconstruction accuracy.

The first part of this chapter introduces a mathematical geometry model which is used to analyze the iso-disparity surface. This model can be used to dynamically adjust the positions, poses and baseline lengths of multiple stereo pairs of cameras in 3D space in order to get sufficient visibility and accuracy for surveillance, tracking and 3D reconstruction. The depth reconstruction accuracy is quantitatively analyzed by the proposed model. The proposed iso-disparity mathematical model presents possibility of reliable control of the iso-disparity curves' shapes and intervals by applying the systems configuration and target properties.

In the second part of this chapter, the key factors affecting the accuracy of 3D reconstruction are analysed. It shows that the convergence angle and target distance influence the depth reconstruction accuracy most significantly. The depth accuracy constraints are implemented in the model to control the stereo pair's baseline length, position and pose. It guarantees a certain accuracy in the 3D reconstruction. The reconstruction accuracy is verified by a cubic reconstruction method. The optimization is implemented by applying the camera, object and stereo pair constraints into the integer linear programming.

2. Quantized depth reconstruction uncertainty

Two images are needed which are fused to obtain a depth perception of the world. Any point in the world scene is captured in these two images as corresponding points which lie on the corresponding epipolar lines. There are two terms related to the depth reconstruction: disparity and quantized depth reconstruction uncertainty. *Disparity* in our approach refers to the displacement of corresponding points along the corresponding epipolar lines for a common scene point, (Pollefeys & Sinha, 2004). In the case where epipolar lines are horizontal, the disparity is measured directly from the difference between the co-ordinates of the corresponding points. The inverse projection of all possible image points with the same disparity will allow reconstruction of the iso-disparity surfaces in 3D space. *Quantized depth reconstruction uncertainty* is defined as the interval between discrete iso-disparity surfaces due to the discrete sensor.

The iso-disparity surfaces of a stereo pair may be simulated by the use of synthetic methods. However for planning real-time multi sensor systems, such simulation is time consuming and a simple mathematical model of the iso-disparity surfaces is needed.

There are two configurations for a stereo pair in common use. The first simple configuration is a parallel stereo pair in which the optical axes of the cameras are parallel. The cameras may have the same focal lengths, or their focal lengths may be different, e.g. to get better reconstruction accuracy of a target placed at the boundaries of the cameras' field of view, FoV. The second common configuration is the convergence stereo pair, where the optical axes cross at a fixation point. The simple mathematical models of the iso-disparity map for these configurations are analyzed in the following subchapters.

2.1 The iso-disparity map of a parallel stereo pair

From the geometry of a parallel stereo pair, two cameras with parallel optical axes and different focal lengths, f_L and f_R for the left and right camera respectively, the iso-disparity plane for disparity $n\Delta D$ can be defined as:

$$z(x,n) = \frac{f_L - f_R}{n\Delta D} x + \frac{B}{2n\Delta D}(f_L + f_R) \tag{1}$$

where B is the baseline length, n is an integer and ΔD is the disparity resolution.

The planes are shown as the thin green lines in Figure 2(a) and Figure 2(c). All the iso-disparity planes intersect with the xy-plane (the stereo pair baseline is a part of the x-axis), and converge on a straight line:

$$x = \frac{B}{2}\frac{f_L + f_R}{f_R - f_L}, \quad z = 0 \qquad (f_L \neq f_R) \tag{2}$$

It is clear from equation (1) that when the focal lengths are equal f_L=f_R=f, z becomes independent of x and the iso-disparity planes are parallel to the xy-plane, see the green lines in Figure 2(b).

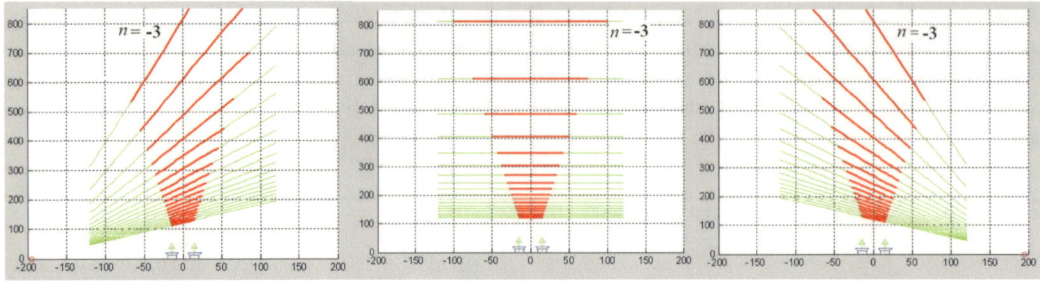

Fig. 2. Iso-disparity planes for parallel stereo pairs from the synthetic simulation (the red lines) and a plot of the mathematical model from equation (1) (the green lines). The lines are plotted with steps of 10 pixels. (a) Cameras with different focal lengths, f_L=3.5 cm, f_R=3.0 cm for left and right camera respectively. The convergence point is (-195 cm, 0) on the *xz*-plane. (b) Cameras with the same focal length of 3.25 cm. (c) Cameras with different focal lengths f_L=3.0 cm, and f_R=3.5 cm for left and right cameras respectively, the convergence point is (195 cm, 0) on the *xz*-plane.

From the inverse projection of the image points and by applying the triangulation method, using the Epipolar Geometry Toolbox, (Mariottini & Prattichizzo, 2005), we can get the synthetic iso-disparity surfaces. Figure 2 shows the synthetic disparity surfaces (the red lines) and the plots from (1) (the green lines). Here the baseline length B is 30 cm and the disparity resolution ΔD=0.04 cm, or ten sensor pixel lengths where p=0.004 cm. Figure 2(a) and Figure 2(c) are plotted for the parallel stereo pair with different focal lengths. The parallel iso-disparity planes for parallel stereo pairs with the same focal lengths are shown in Figure 2(b). The synthetic simulations and the results calculated from (1) give similar results.

2.2 The iso-disparity surface of a convergent stereo pair

Let us consider two cameras with a convergence angle α_c, where $\alpha_{cL0}=\alpha_{cR0}=\alpha_c$ for the left and right camera respectively, with the angles rotated inwards to achieve a fixation point FP_0, as in Figure 3. If the point TP_0 lies on the baseline's axis of symmetry, then the angles, (ψ_{L0}, ψ_{R0}), are the angles between the visual lines and a line perpendicular to the baseline. The zero disparity circle is defined by the fixation point and the left and right camera position points C_L and C_R. This circle is known as the Vieth-Müller circle, and is a projection of the horopter, (Ogle, 1950).

The iso-disparity surface is a cylinder whose cross section on the *xz*-plane is a conic section, which passes through the centers of projection C_L and C_R, and the point M_∞. The point M_∞ is a point imagined at infinity in both images, which can be obtained from the intersection of the normals to the optical axes, going through the projection centres, (Pollefeys & Sinha , 2004). It is possible to prove that for the case when $\alpha_{cL0}=\alpha_{cR0}=\alpha_c$, the conic is an ellipse. To determine the ellipse we need to estimate its five degrees of freedom. Three of these are determined by the points C_L, C_R and M_∞. One of the two remaining degrees is related to the point TP_0 with the disparity $n\Delta D$. The relationship between disparity $n\Delta D$ and focal lengths f_L and f_R for the left and right cameras respectively, is the last required degree of freedom. If the disparity $n\Delta D$ and focal lengths f_L and f_R are known, the unique ellipse can be determined.

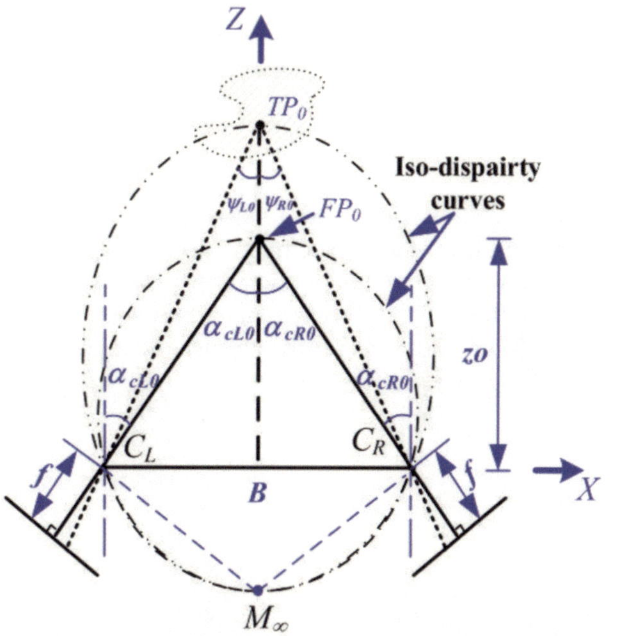

Fig. 3. An example of the iso-disparity curves for the convergence stereo pair in the plane defined by the cameras optical axes. z_0 is the distance from the fixation point to the baseline, f is the focal length.

The iso-disparity surface of quantized disparity $n\Delta D$ for a convergence stereo pair (C_L, C_R) with the same focal length f and same the convergence angles α_{cL0},$=\alpha_{cR0}=\alpha_c$, describes a cylinder, and the ellipses being cross sections of this cylinder on the xz-plane with centres in $0_e(x_{0e}(n), z_{0e}(n))$:

$$\frac{(x - x_{0e}(n))^2}{a^2} + \frac{(z - z_{0e}(n))^2}{b^2} = 1 \tag{3}$$

For the chosen co-ordinates x_{0e}=0 and z_{0e}=b-Btanα_c/2, then:

$$\frac{x^2}{a^2} + \frac{\left(z - \left(b - \frac{B}{2}\tan\alpha_c\right)\right)^2}{b^2} = 1 \tag{4}$$

where B is the baseline length and α_c is the stereo convergence angle.
The ellipse half-axis along the z-axis, b, depends on the discrete disparity $n\Delta D$, baseline length B, focal length f and convergence angle α_c and is described as:

$$b = \frac{\frac{B}{2}}{\sin 2\alpha_c - \frac{n\Delta D}{f}\cos^2\alpha_c} \tag{5}$$

The ellipse half-axis along the x-axis, a, can be found from the relationship:

$$\left(\frac{b}{a}\right)^2 = \tan\alpha_c \frac{1+\dfrac{n\Delta D}{2f}\tan\alpha_c}{\tan\alpha_c - \dfrac{n\Delta D}{2f}} = \frac{\tan\alpha_c}{\tan\psi_c} \tag{6}$$

where $\psi_c = \psi_{L0} = \psi_{R0}$.

Figure 4 shows the iso-disparity surfaces for a stereo pair from the synthetic simulations (bold blue and red lines) and from theoretical model (4) (thin green and light blue lines) in 3D space, in perspective view, Figure 4 (a), and top view, Figure 4 (b). The characteristics of cameras are: a baseline length B=50 cm, convergence angle, α_c=4°, focal length f=2.5 cm and disparity resolution ΔD=0.04 cm. Both results match each other perfectly.

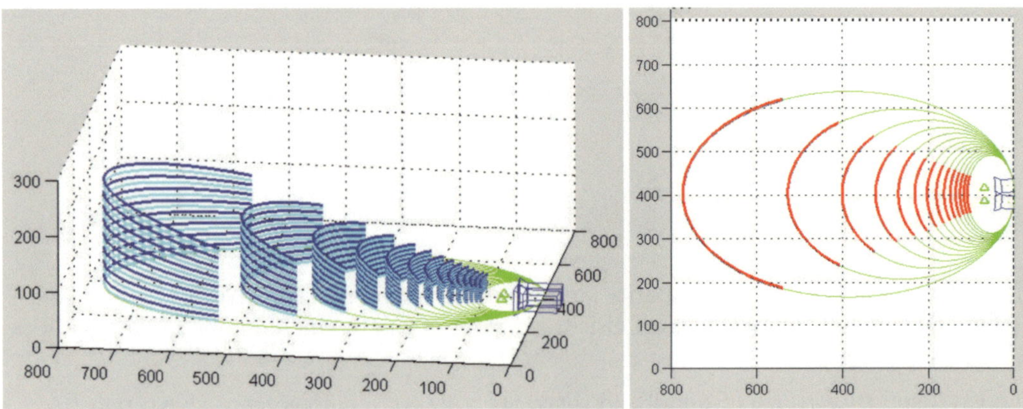

(a) (b)

Fig. 4. The iso-disparity surfaces for a stereo pair from the synthetic model (bold blue and red lines) and from the mathematical model (thin green and light blue lines) with convergence angle, α_c=4°, the baseline length B=40 cm, the focal lengths f=2.5 cm and disparity resolution ΔD=0.04 cm (a) perspective view, and (b) top view.

2.3 Mapping of 2D uncertainty for a stereo pair configuration

Since the gaps between iso-disparity surfaces represent the quantization uncertainty in 3D space, we can generate a 3D depth reconstruction uncertainty map of a particular stereo pair's configuration using the iso-disparity surface geometry, equations (4))-(6). Also, it is possible to generate such a map in 2D on the optical axes plane. This map can be used to optimise the configuration of the stereo setup. Mapping of the 2D uncertainty for a stereo pair configuration can be done in the following three steps:

- Firstly, the plane has to be covered by the stereo pair's FoV, (Chen et al., 2007a). The area is sampled using small grids covered by the stereo pair.
- Secondly, an iso-disparity curve on the optical axes plane should be calculated, passing through each grid point. Knowing that the curve will have a canonical shape then five points are needed. Two of these points can be the grid point and its symmetrical point, with respect to the symmetry axis of the baseline. The three others points are C_L, C_R and M_∞. For a convergent stereo pair, the ellipse axes a and b can be found using the ellipse fitting algorithm, (Halif & Flusser, 1998). Then using equation (5), the two closest

ellipses with discrete disparity values $n\Delta D$ and $(n+1)\Delta D$ respectively, can be found, where the disparity resolution ΔD is one sensor pixel length.

- Finally, the depth reconstruction uncertainty can be calculated as the interval between the iso-disparity surfaces, with the disparity values, $n\Delta D$ and $(n+1)\Delta D$ as the distance between the intersections of these two iso-disparity surfaces, and the line through the grid point and M_∞.

2.3.1 Simulation results

The presented simulations were performed in MATLAB 7.0, and cover a rectangular area of (800 cm × 800 cm). This case study illustrates how depth reconstruction uncertainty in stereo coverage varies with the target distance z for a given stereo baseline length B, focal length f, and sensor pixel length p. The results are presented in Figure 5, where the cameras optical axes are in the xz-plane. The depth reconstruction uncertainty is specified by the positive y-axis of the co-ordinate. However, this uncertainty analysis shows only the area covered by the stereo pair's FoV. To scale the uncertainty on the optical axes plane, a colour map is used. The lowest uncertainty is indicated by the blue colour and the highest uncertainty by the red colour. In order to increase the readability of the iso-disparity curves, the contour is plotted with ten pixel lengths disparity resolution. The map of the iso-disparity curves is

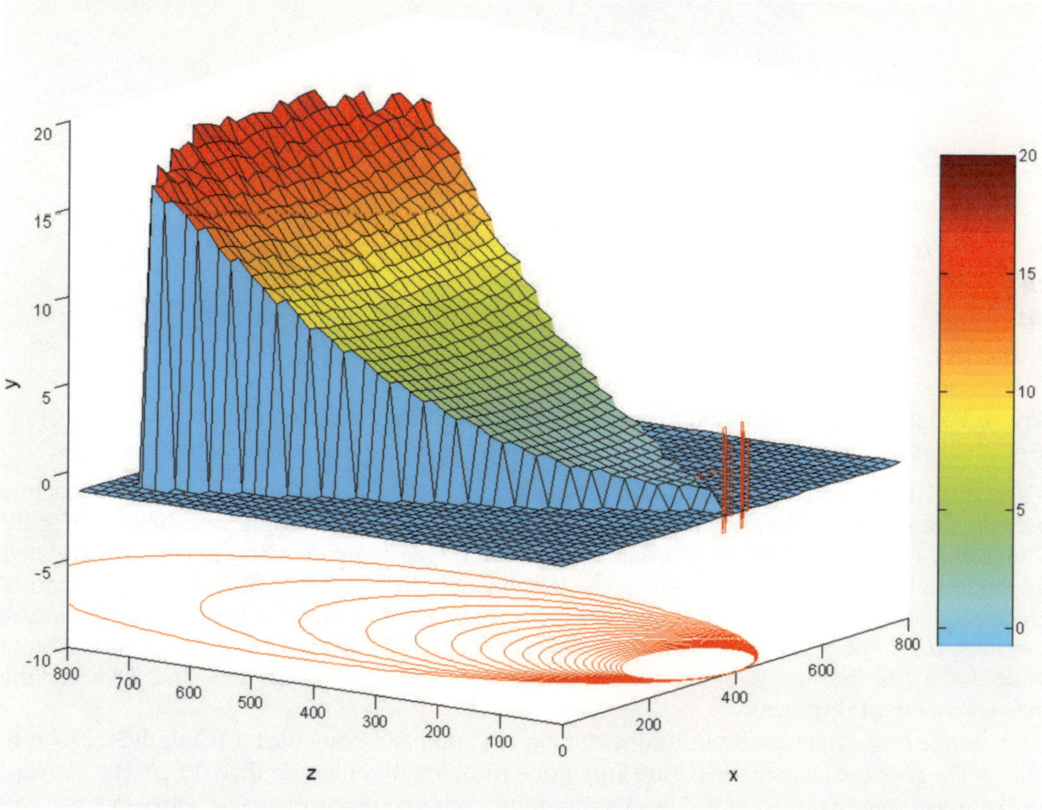

Fig. 5. The depth reconstruction uncertainty map for a stereo pair's FoV, where B=40 cm, f=3.5 cm and p=0.004 cm.

generated with baseline length B=40 cm, focal length f=3.5 cm and pixel length p=0.004 cm, stereo convergence angle, α_c=4° and the FoV is approximately 54°. This case study proves that the quantized depth reconstruction uncertainty increases as the distance to the target increases. To show the quantized properties of depth reconstruction uncertainty, the map of the iso-disparity curves with suitable pixel length is shown in Figure 6. The figure shows only half of the FoV, with a cross section along the ellipses' axes perpendicular to the baseline. The quantization step increases with the target distance.

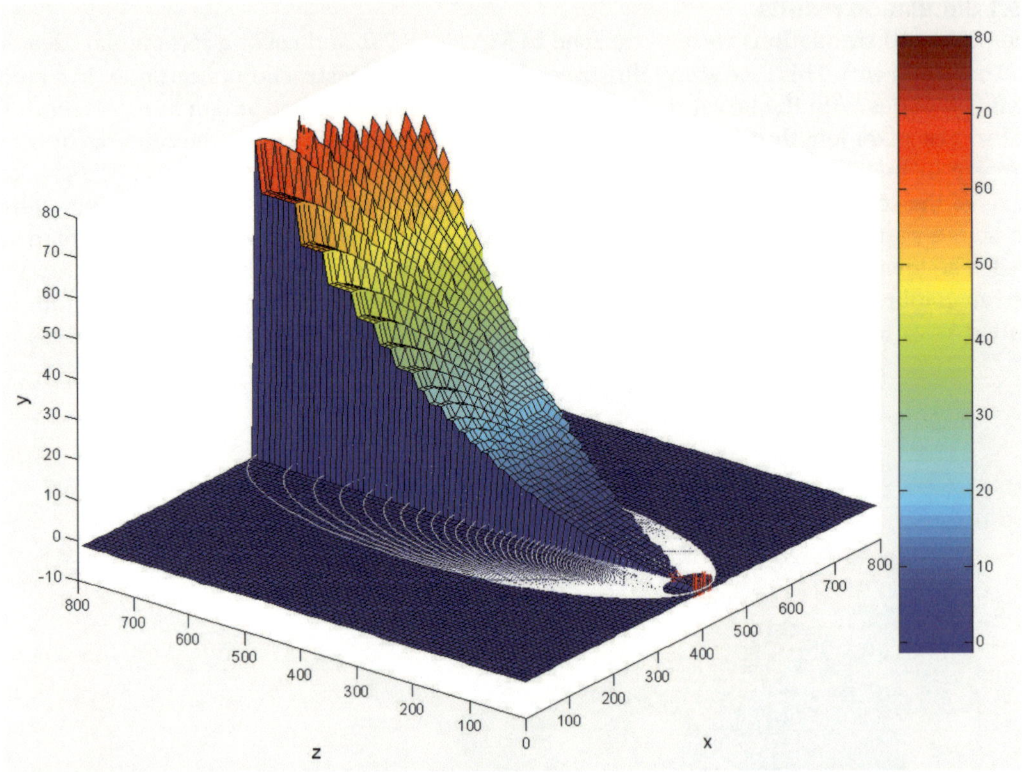

Fig. 6. The depth reconstruction uncertainty map for a stereo pair's half FoV, where B=20 cm, f=3.5 cm and p=0.008 cm.

Exact illustrations of how the depth reconstruction uncertainty varies with the baseline lengths, focal lengths, sensor pixel length and stereo convergence angle, are shown in Figure 7 and Figure 8. Figure 7(a) shows that the relative depth reconstruction uncertainty, related to the target distance, decreases when the baseline length increases. The relative uncertainty decreases slowly for a baseline above about 40 cm. Its minimum value tends to be constantly between 0.5% and 1.5% for target distances of 200 cm and 800 cm respectively. At the same time, for a baseline of about 10 cm, the uncertainty varies between 10% and 2.5% for the respective target distances.

The change of the relative depth reconstruction uncertainty versus the focal length is similar to that of the baseline length; see Figure 7(b). For a focal length of longer than 3.5 cm the increase of the uncertainty is relatively slow. Its minimum tends to be consistently between 1.5% and 0.4% for target distances of 200 cm and 800 cm respectively. Meanwhile, for a focal length of 1 cm, the uncertainty varies between about 9% and 2% for the respective target distances.

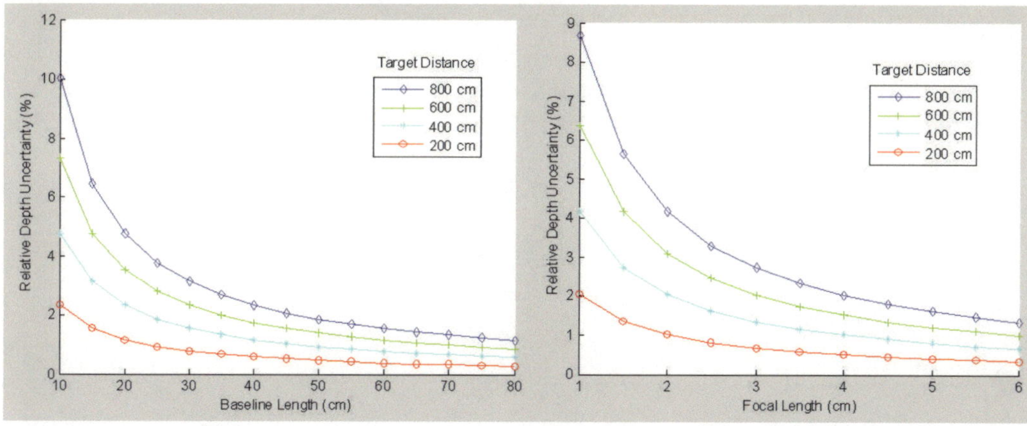

(a) (b)

Fig. 7. The relative quantized depth uncertainty as function of the baseline length (a), focal length (b). The distance from the target to the camera is 800 cm, 600 cm, 400 cm and 200 cm, respectively.

Furthermore, Figure 8(a) illustrates the linear relation of the relative uncertainty and the sensor pixel length. Within the range from 0.001 cm to 0.006 cm, the relative uncertainty varies from 0.2% to 3.5% and also depends on the target distance. Figure 8(b) shows that the stereo convergence angle has a slight influence on the uncertainty but this also depends on the target distance.

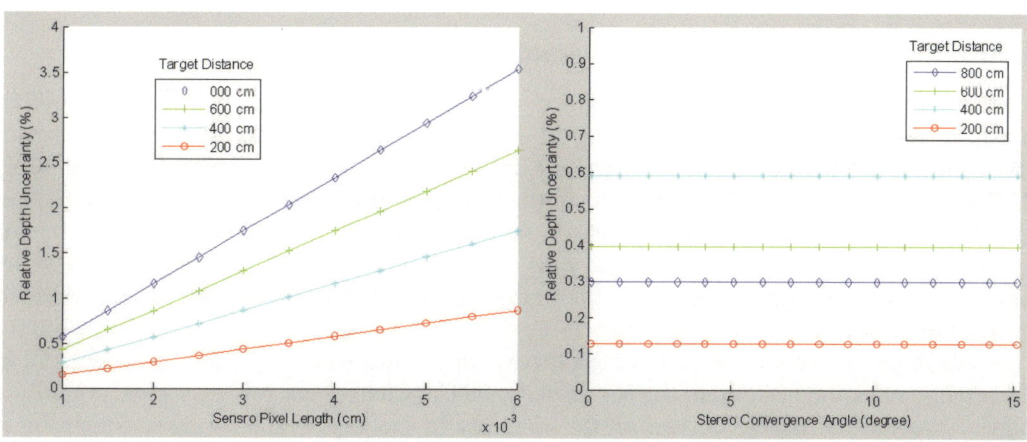

(a) (b)

Fig. 8. The relative quantized depth uncertainty as a function of the sensor pixel length (a), and stereo convergence angle (b). The distance from the target to the camera is 800 cm, 600 cm, 400 cm and 200 cm, respectively.

Figure 9(a) and Figure 9(b) illustrate the variation of the uncertainty when both the focal length and the baseline length are changed for two different target distances, 200 cm and 600 cm, respectively. The uncertainty increases significantly when the baseline length decreases below 40 cm, independent of the location of the target within the FoV. Also, a significant increase in the uncertainty is visible for a focal length below 3.5 cm.

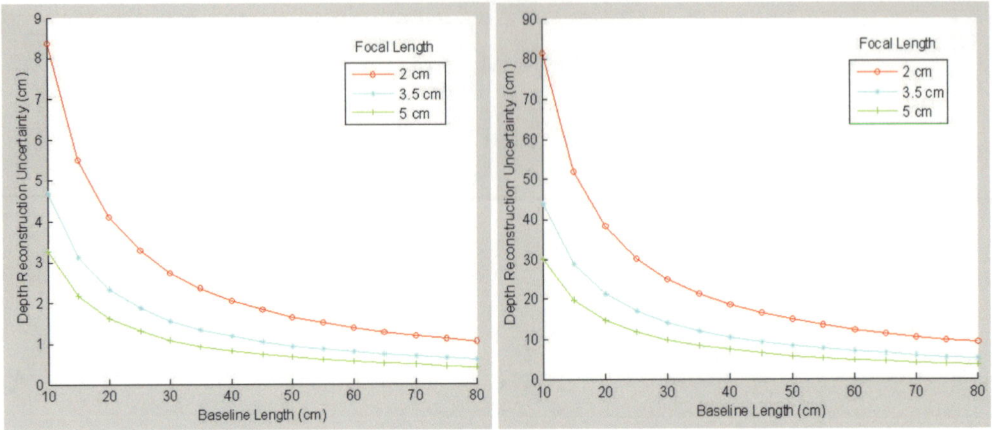

(a) (b)

Fig. 9. The relative quantized depth uncertainty varies with both the focal length and the baseline length. The focal lengths are 2 cm, 3.5 cm and 5 cm, respectively, marked by different type of lines. The target is (a) 200 cm; (b) 600 cm far away from the camera.

The relative accuracy is similar for a target located in different positions, but its absolute value is more significant for a target further from the stereo pair. In order to fulfil the reconstruction accuracy requirement for a distant target, the focal length or baseline has to be adjusted. A longer focal length can be used to compensate for a shorter baseline. And in general, the longer the baseline is, the more difficult the matching becomes.

3. Arrangement of a multi stereo visual sensor system

Human activity space, as a target space for 3D and depth reconstruction, provides constraints to the design and planning of the active stereo camera system. The sensor arrangement can be viewed as an extension to the well-known Art Galley Problem, AGP, (O'Rourke, 1987). The AGP describes a simple polygon, often with holes, and the task is to calculate the minimum number of guards necessary to cover a defined polygon. In our approach, a similar task is required to find the minimum number of stereo pair sensors needed to cover the target space. Here the human activities space as a target space is defined by a tetrahedron.

This subchapter gives an overview of the theory for the multi stereo sensors arrangement in the intelligent vision system. In the (Chen et al., 2007a; Chen et al., 2007b), we suggested the camera constraints, which focused on the visibility to the target. The accuracy constraint is based on the estimation of quantized depth reconstruction accuracy, where a target object has the same target convergence angle. The iso-disparity geometry model gives a deep analysis of depth reconstruction accuracy. It analyzes the whole camera FoV. This can be used to dynamically adjust the position, poses and baselines length of multiple stereo pairs of cameras in order to get the desired accuracy.

The planning algorithm proposed in the (Chen et al., 2007a; Chen et al., 2007b) works in 3D space. The approach dynamically adjusts the stereo pair baseline length according to the accuracy requirement and the target distance as a distance from the target position to stereo pair baseline. The minimal amount of stereo pairs to cover human activity space is solved by means of Integer Linear Programming, ILP, (Chakrabarty et al., 2002; Hörster & Lienhart,

2006; Berkelaar et al., 2005). The 3D reconstruction accuracy, which is ensured by an accuracy constraint, can be verified by a cubic reconstruction.
The constraints of the stereo view optimization model can be defined from the environment, camera properties and human behaviour, which affect the process of identification and reconstruction of the target.

3.1 Constraints for the optimization model

The human activities space is modelled by a tetrahedron as shown in Figure 10. The normal of each tetrahedron's upper triangle gives the orientation of that surface. If the visibility angle, θ, between the triangle normal and a line drawn from the centroid of the triangle to the camera position, increases then the image resolution decreases. In order to get a good image resolution, a visibility angle, θ, of less than the maximum visibility angle, θ_{max}, is required which can be expressed as constrain:

$$\theta \le \theta_{\max} \tag{7}$$

The camera orientation should line up with the centroid of the triangle, thus bringing the target object to the centre of camera FoV and causing less lens distortion. The angle between the camera orientation and the line drawn from the camera position to the centroid of the triangle, φ, of less than the maximum angle φ_{max} is required:

$$\varphi \le \varphi_{\max} \tag{8}$$

In order to follow the movement of the target object, a camera movement distance constraint can be applied. The next-view position for the camera should not be placed too far away from the previous one. This constraint, formulated as the camera maximum movement should be less than the maximum distance of camera movement which the system supported.

$$Dist(StereoPair_{next}, StereoPair_{current}) \le Dis_{\max} \tag{9}$$

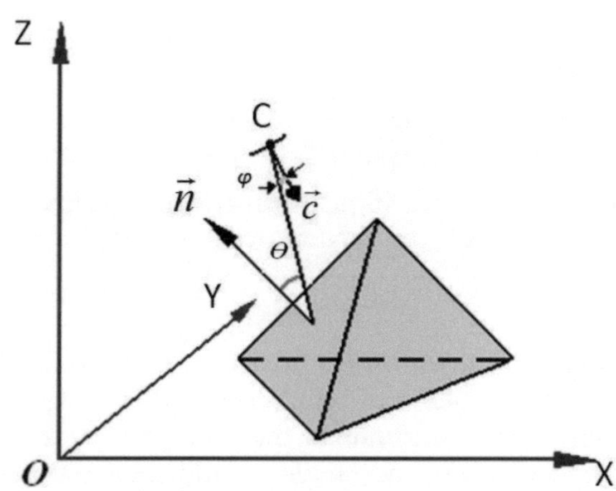

Fig. 10. Illustration of the human space modelled as a tetrahedron; θ - the visibility angle between the triangle normal $\vec{c}$ and a line from the centroid of the triangle to the camera position; φ - the angle between the camera orientation $\vec{c}$ and a line from the camera position to the centroid of the triangle.

A fewer number of potential next-view positions for the cameras restricted by (9)) can simplify computation.
The camera constraints are related to the camera FoV. The camera horizontal and vertical viewable angles, ϕ_h, ϕ_v, and a working distance, r, can be calculated from the camera attributes, see the spherical co-ordinate systems shown in Figure 11. In order to cover the target object feature points by the camera FoV, the following constraints must be fulfilled:

$$\begin{gathered} l \leq r \text{ and} \\ \alpha_c - \phi_h/2 \leq \alpha_o \leq \alpha_c + \phi_h/2 \\ \beta_c - \phi_v/2 \leq \beta_o \leq \beta_c + \phi_v/2 \end{gathered} \tag{10}$$

where: l is the distance between the target position and camera's position; α_o, β_o are respectively the azimuth and elevation of target; α_c, β_c are respectively the azimuth and elevation of the camera's pose.

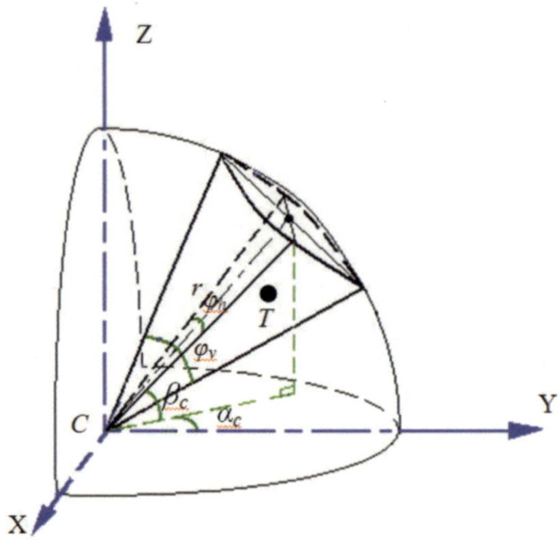

Fig. 11. The spherical co-ordinates system and FoV of a camera where C is camera position and the example target point located at point T.

Since stereo matching becomes more difficult when the baseline distance increases, the baseline length B has to be limited to the maximum stereo baseline length, B_{max}:

$$0 < B \leq B_{\max} \tag{11}$$

The depth reconstruction is one of major focus in this research project. The depth reconstruction accuracy improvement can be adjusted by the baseline length, (Mittal, 2006; Samson et al., 2006). This research introduces the concept of a depth accuracy factor, AF, which is a function of target convergence angle, ψ, and camera pose, α_c. We construct the stereo coverage from the overlapping of two cameras' FoVs. The overlapping FoVs are typically used to extract 3D information. The area of stereo coverage must cover all of the target objects. In the most common case, the cameras form a converging stereo pair. Cameras poses azimuths and baseline are shown in Figure 12. Cameras convergence angles, (α_{cl}, α_{cr}), are the angles of each camera rotated inwards from the parallel to achieve convergence. The target convergence angles, (ψ_l, ψ_r), are the angles between the visual lines of each camera and the baseline perpendicular. From Figure 12, simplifying: $\psi = \psi_l = \psi_r$ and $\alpha_c = \alpha_{cl} = \alpha_{cr}$, we obtain:

$$Z = \frac{B}{2\tan\psi} \tag{12}$$

where B is a baseline length and Z is a target distance.

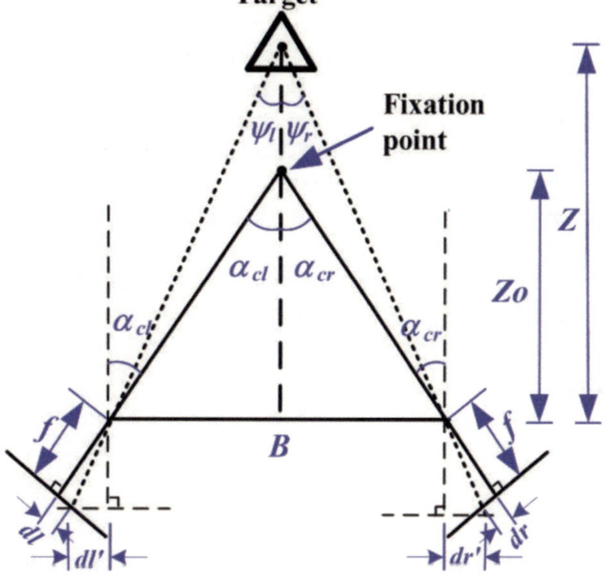

Fig. 12. Model of stereo pair geometry; Z_0 is the depth of the fixation point, $|dl\text{-}dr|$ is stereo disparity, f is focal length.

The equation (12) can be written as:

$$Z = \frac{B}{2}\cdot\left(\frac{1}{\tan\alpha_c - \tan(\alpha_c - \psi)} + \frac{\tan\alpha_c \cdot \tan(\alpha_c - \psi)}{\tan\alpha_c - \tan(\alpha_c - \psi)}\right) \tag{13}$$

In the case of parallel stereo or with the target close to the fixation point, the α_c or (α_c-ψ) varies by a small amount, and the equation (13) can be further simplified. The resolution of the target convergence angle, ψ, is related to a single pixel, p, in the image, thus the relative depth error can be written as:

$$\frac{\Delta Z}{Z} \approx \left|\frac{\cos\alpha_c}{\sin\psi\cos(\alpha_c - \psi)}\right| \cdot \frac{p}{f} = AF \cdot \frac{p}{f} \tag{14}$$

where AF is the depth accuracy factor, f is a focal length, and the depth quantization error is assumed to be one pixel, p.

The depth error is proportional to the depth accuracy factor. In fact, since the depth accuracy factor varies more significantly with respect to the target convergence angle, ψ, than to the camera's pose, α_c, the target convergence angle determines the depth accuracy factor. The accuracy constraint for a given p can be defined as:

$$AF \leq AF_{con} \tag{15}$$

where: AF_{con} is determined from the reconstruction accuracy requirements of the given application.

3.2 Implementation of the camera arrangement

The stereo pair arrangement consists of three stages:

- Firstly, we find potential stereo pairs that satisfy stereo constraints by greedy algorithm from all potential cameras' positions and poses.
- Secondly, the integer linear programming is applied to minimize the total amount of stereo pairs subject to the visibility and baseline length constraints, depth accuracy constraints and camera movement constraints. The objective function minimizes the number of stereo pairs needed to cover all triangles in the target object model, and also ensures that the target object is covered by at least one stereo pair.
- Finally, the 3D reconstruction accuracy can be verified by a cubic reconstruction simulated using a pair of rectified scene images.

3.2.1 Greedy algorithm

This algorithm gives a flexible way of organising cameras into stereo pairs, each potential camera to be included in a stereo pair may be chosen by an algorithm according to the stereo pair constraint. The first step of the algorithm is to sample the potential camera positions $C_n(x_{cn}, y_{cn}, z_{cn})$ and poses $\psi_n(\alpha_{cn}, \beta_{cn})$ of the camera state, $Scamera^k_{C_n,\psi_n}$, where k is the camera state index number. The target object, which must be covered, is modelled as a tetrahedron. In the next step, we calculate all of the potential camera positions and poses needed to cover each upward triangle of this model. Taking this, we combine every two camera states to be a potential stereo pair, $Stereopair_i$, according to the stereo constraint (Chen et al., 2007a; Chen et al., 2007b). The algorithm is sufficiently flexible to add other constraints for stereo pair, e.g. the angle constraint between the cameras' optical axes. Finally the algorithm removes the redundant potential stereo pairs.

3.2.2 Integer linear programming

This model assumes that one type of camera is used throughout, resulting in just one camera FoV being considered. The optimization of the amount of cameras with different FoVs can also be easily extended, by adding one more term for different FoVs. Since the stereo pairs have been found by the greedy algorithm, the integer linear programming can be applied to minimize the total stereo pairs subject to the coverage constraint (Hörster &Lienhart, 2006; Chakrabarty et al., 2002).

A binary variable is calculated and stored in advance. The stereo visibility binary variable table $Stereovis_{j,i}$ is defined by:

$$Stereovis_{j,i} = \begin{cases} 1 & \text{if a } Stereopair_i \text{ covers} \\ & \text{triangle } j \text{ of target object model} \\ 0 & \text{otherwise} \end{cases} \tag{16}$$

which indicates each triangle j as the row j to be covered by the stereo pair i in column i, and $1 \le i \le K_s$, where K_s is the total number of stereo pairs.

This objective function minimizes the number of stereo pairs needed to cover all triangles in the target object model, and also ensures that the target object is covered by at least one stereo pair:

$$\min \sum_{i=1}^{K_s} S_i \tag{17}$$

subject to

$$\sum_{i=1}^{K_s} S_i \times Stereovis_{j,i} \ge 1, \quad \text{for } j = 1,2,3 \tag{18}$$

where the S_i is the binary variable where a "1" indicates the stereo pair to be chosen.
To ensure that only one camera is located at each position and has only one pose, the conflict binary variable table $c_{p,i}$ is also calculated in advance and defined by:

$$c_{p,i} = \begin{cases} 1 & \text{if two pairs } i \text{ and } p \text{ share the same camera} \\ & \text{with different orientations, where } i \ne p \\ 0 & \text{otherwise} \end{cases} \tag{19}$$

for i=1, 2, …, K_s, and p=1, 2, …, K_s.
One more constraint is added into the model:

$$\sum_{i=1}^{K_s} S_i \times c_{p,i} \le 1, \quad \text{for } p = 1, 2, \cdots, K_s \tag{20}$$

The information on the optimal number of stereo pairs, and which pairs to use, are returned as vectors by the ILP model.

3.2.3 Cubic reconstruction

The rectification matrix is calculated directly from the **p**erspective **p**rojection **m**atrix, PPM, (Fusiello et al., 2000), and the rectification algorithm also gives two new PPMs, P_n1 and P_n2. The cubic reconstruction in 3D can be performed with a triangulation method directly from the rectified images, using P_n1, P_n2. The 3D reconstruction error, Δ_{rec}, for a single pixel error along a horizontal direction in the rectified image, has same value as the depth error and is given by:

$$\Delta rec = \sqrt{\frac{1}{8} \sum_{i=1}^{8} \left| \hat{M}_i - M_i \right|^2} \tag{21}$$

where $\hat{M}_i$ gives the co-ordinates of the reconstruction point i in a cube of the rectified images and M_i gives the real coordinates of the target point i in the cube.

3.3 The model validation

The simulations were performed in MATLAB 7.0. The integer linear programs *lpsove package,* (Berkelaar et al., 2005), and the Epipolar Geometry Toolbox, (Mariottini & Prattichizzo, 2005), were used to minimise the number of cameras and transform the object position in 3D. The simulation environment consists of a rectangular room of 8 m × 8 m × 3 m. The modelling of the human activities space as a tetrahedron requires three upward triangles; and each triangle must be visible to at least one pair of cameras. Each model is 2 m high and 1.2 m at the base edges. The cameras' positions are restricted to the ceiling around the room, their potential positions sampled at 0.2 m intervals in the initial phase, and 0.1 m intervals for the next cameras viewing position; Dis_{max} is taken 3 m. The camera's pose is sampled at 12° intervals. The cameras have the same horizontal and vertical viewing angles, ϕ_h, ϕ_v, of 60° and have a working distance, r, of 7 m. The maximum visibility angle, θ_{max}, (7)) and the angle, φ_{max}, (8) are taken to be 70° and 10° respectively. The pixel size of our vision system, p, is 0.02 mm and the focal length, f, is 1.21 cm. The maximum stereo baseline length, $B_{max,}$, is 1.5 m. The cubic centre is located at the centroid of the tetrahedron and each edge is 10 mm.

This case study illustrates how the variable stereo baseline length, camera positions and poses vary according to the accuracy requirement and the target location. In order to illustrate the cameras' positions and poses, the analysis considers the target model at four locations, 1, 2, 3 and 4, see Figure 13. The arrows indicate the optical axes of the cameras.

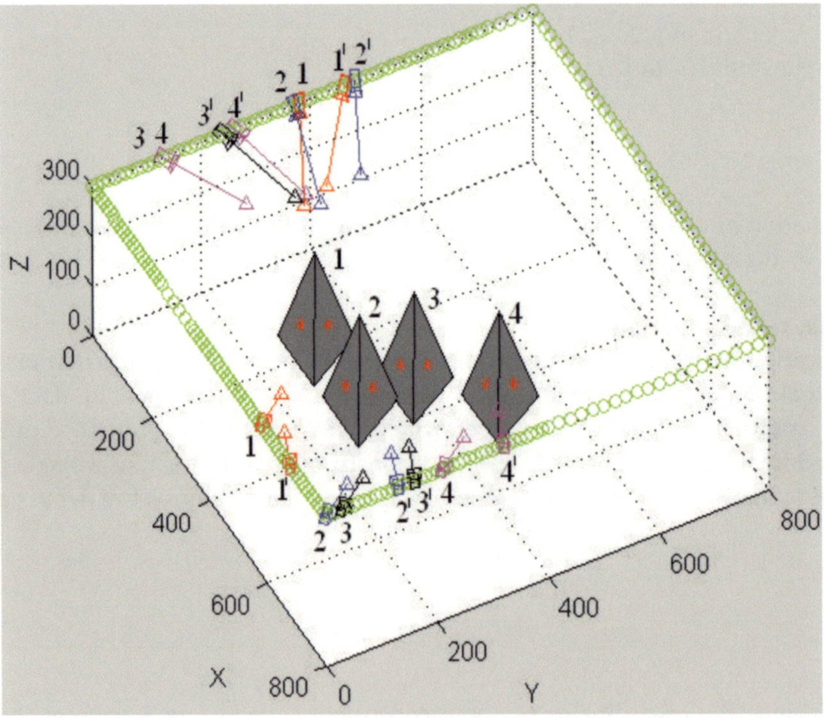

Fig. 13. The stereo pair positions, poses and baselines with the depth accuracy factor AF_{con} = 8, for the moving target.

The index numbers indicate the model locations and corresponding camera positions and poses calculated according to the maximum accuracy factor. The circles are the camera's potential sample positions. The sample positions and intervals are changed according to the camera previous position with the constraint (9). In each position every upward triangular view is visible to at least one stereo pair. The algorithm proves that a set of two pairs is sufficient to cover the three triangle surfaces. The stereo baseline length changes dynamically according to the distance to the target.

Figure 14 illustrates a case of four different values of AF_{con} applied to a target at the same position. The index number indicates the corresponding stereo pair according to AF_{con}. The stereo baseline lengths and reconstruction errors for the different accuracy factors are shown in Tabble 1. It proves that the baseline increases as AF_{con} becomes more restricted, and the reconstruction error is smaller.

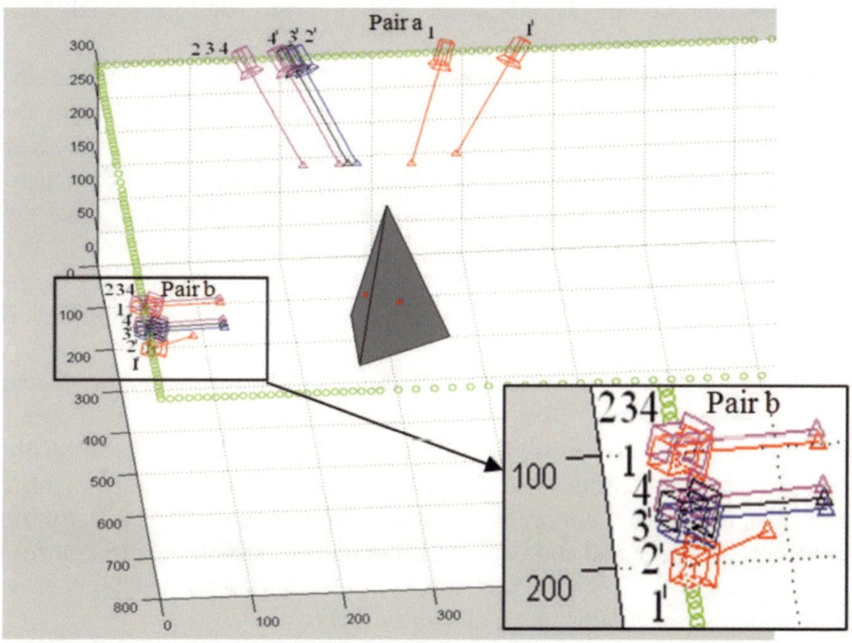

Fig. 14. The stereo pair positions and baseline lengths for the same target location vary according to the different accuracy.

IN	AF_{con}	B_a	B_b	Δ_{rec}	ΔZ_{max}
1	*8*	*100*	*80*	*3.3*	*4.5*
2	*11*	*70*	*60*	*4.5*	*6.2*
3	*14*	*60*	*50*	*4.9*	*7.9*
4	*17*	*50*	*40*	*5.2*	*9.7*

Table 1. The baseline lengths of two pairs and the reconstruction errors for different accuracy factors. *IN*: the index number; AF_{con}: the reconstruction accuracy requirement; B_a, B_b: baseline lengths for each pair, [cm]; Δ_{rec}: the maximum value reconstruction error of pairs, [cm]; ΔZ_{max}: the theoretical maximum depth error, [cm].

4. Conclusion

The chapter focuses on depth reconstruction, the arrangement of multiple stereo sensors for human activities space, which are important issues of IVAS. The research work can be concluded by means of two terms: the uncertainty analysis and human factor handling for vision system.

The analysis presented shows that the quantized depth reconstruction accuracy varies more significantly with respect to the target distance to baseline, baseline length and focal length than to the convergence angle. Small changes in stereo convergence angle do not overly affect the depth accuracy, especially when the target is placed centrally. On the other hand it can have a great impact on the shape of the iso-disparity curves. By using the systems configuration and target properties, we can get reliable control over the shapes and intervals of the iso-disparity curves from the proposed iso-disparity mathematical model.

Modeling the target object as a tetrahedron gives a convenient way to extract the orientation of each surface and guarantee a good observability. Modelling the camera FoV using spherical co-ordinates simplifies the model and constraints, which speeds up computations. Formulating the stereo pairs with the greedy algorithm using stereo constraints is a simple way to get all possible stereo pairs and then minimize the amount of stereo pairs by means of the stereo view ILP model. It is possible to extend this algorithm to dynamic cameras for tracking humans. In order to follow target objects movement, the camera movement distance constraints can be applied (Chen et al., 2007a).

The analysis of key factors which affect the accuracy of 3D reconstruction shows that the convergence angle and target distance are the most significant. The depth accuracy constraint may be sufficient to control the stereo pair's baseline length, position and pose. It is an effective method for system decision making and is easy to implement. From the simulation results, it is readily noticeable that the cubic reconstruction is useful in verifying the reconstruction accuracy and the proposed method of baseline length control has been proven. The two stages sampling of the cameras position has the flexibility to adjust the intervals and position ranges, and speed up computation.

5. References

Berkelaar, M.; Notebaert, P. & Eikland, K. (2005). Lpsolve 5.5: Open Source (mixed-integer) Linear Programming System, Eindhoven Univ. of Technology, http://tech.groups.yahoo.com/group/lp_solve/files/.

Chakrabarty, K.; Iyengar, S; Qi, H. & Cho, E. (2002). Grid Coverage for Surveillance and Target Location in Distributed Sensor Networks. *IEEE Transaction on Computers*, vol. 51, no. 12, pp. 1448-1453, ISSN: 00189340.

Chen, J.; Khatibi, S. & Kulesza, W. (2007a). Planning of a Multi Stereo Visual Sensor System for a Human Activities Space. Proceedings of the 2nd International Conference on Computer Vision Theory and Applications, pp. 480 - 485, ISBN: 9789728865740 Barcelona, Spain, March 2007, INSTICC.

Chen, J.; Khatibi, S. & Kulesza, W. (2007b). Planning of a Multi Stereo Visual Sensor System - Depth Accuracy and Variable Baseline Approach. *Proceedings of IEEE Computer Society 3DTV-Con, the True Vision Capture, Transmission and Display of 3D Video,* ISBN: 9781424407224, Kos, Greece, May 2007, IEEE.

Chen, J.; Khatibi, S.; Wirandi, J. & Kulesza, W. (2007c). Planning of a Multiple Sensor System for Human Activities Space - Aspects of Iso-Disparity Surface. *Proceedings of SPIE on Optics and Photonics in Security and Defence,* vol. 6739, Florence, Italy, September, 2007, SPIE.

Chen, T.; Catrysse, P.; Gamal, A. & Wandell, B. (2000). How Small Should Pixel Size Be? *Proc. of SPIE on Sensors and Camera Systems for Scientific, Industrial, and Digital Photography Applications,* vol. 3965, SPIE.

Chen, X. (2002). Design of Many-Camera Tracking Systems for Scalability and Efficient Resource Allocation. *PhD thesis.* ISBN: 0493628339, Stanford University.

Erdem, U. & Sclaroff, S. (2006). Automated Camera Layout to Satisfy Task-Specific and Floor Plan-Specific Coverage Requirements. *Computer Vision and Image Understanding,* vol. 103, pp. 156-169, ISSN 10773142.

Fusiello A.; Trucco E, and Verri A. (2000). A Compact Algorithm for Rectification of Stereo Pairs, *Machine Vision and Applications,* Vol.12, pp. 16-22, ISSN: 09328092.

Halif R. and Flusser J. (1998). "Numerically stable direct least squares fitting of ellipses," in: *Proc. of the 6th Int. Conf. Computer Graphics and Visualization,* pp. 125 - 132, Czech Republic, February, 1998.

Hashimoto, H. (2003). Intelligent Space: Interaction and Intelligence, *Artificial Life and Robotics,* vol. 7, no. 3, pp.79-85, ISSN: 14335298.

Hörster, E & Lienhart, R. (2006). On the Optimal Placement of Multiple Visual Sensors. *Proc. of ACM International Workshop on Video Surveillance & Sensor Networks.* ISBN: 1595934960, USA, October, 2006, ACM.

Mariottini, G.L. and Prattichizzo D. (2005). EGT for Multiple View Geometry and Visual Servoing Robotics Vision with Pinhole and Panoramic Cameras, *IEEE Robotics & Automation Magazine,* Vol. 12, No. 4, pp. 26-39, ISSN 10709932

MIT Project Oxygen, (2008). http://oxygen.csail.mit.edu/.

Mittal, A. (2006). Generalized Multi-Sensor Planning. *Proc. of 9th European Conference on Computer Vision.* ISBN: 978-3540338321, Austria, May, 2006, Springer.

O'Rourke, J. (1987). *Art Gallery Theorems and Algorithms.* ISBN 0195039653, Oxford University Press.

Ogle, K. (1950). Researches in Binocular Vision. W.B. Saunders Company, Philadelphia & London.

Pollefeys, M. & Sinha, S. (2004). Iso-disparity Surfaces for General Stereo Configurations. *Proc. of the 6th European Conf. on Computer Vision.* ISBN: 9783540219811, Czech Republic, May, 2004, Springer.

Samson, E.; Laurendeau, D.; Parizeau, M.; Comtois, S.; Allan, J. & Gosselin, C. (2006) The Agile Stereo Pair for Active Vision. *Machine Vision and Application,* vol. 17, no. 1, pp. 32-50.

Völpel, B. & Theimer, W. M. (1995). Localization Uncertainty in Area-Based Stereo Algorithms. *IEEE Transactions on Systems, Man, and Cybernetics*, vol. 25, no. 12, pp. 1628 – 1634, ISSN: 00189472.

11

A Stereo Vision Framework for 3-D Underwater Mosaicking

A. Leone, G. Diraco and C. Distante
Institute for Microelectronics and Microsystems, National Research Council
Lecce (Italy)

1. Introduction

Research on automatic mosaic creation for underwater applications has been investigated in the last fifteen years. The reconstruction of complex 3-D structures is useful in several underwater applications; in particular, 3-D mosaicking constitutes an important tool for seabed exploration, improving visualization and navigation in the underwater medium. Moreover, underwater measurement systems have been used extensively in marine research to estimate the size of interesting objects such as organisms and structures. In addition, visual sensing can be an enabling technology for Autonomous Underwater Vehicles (AUVs), which have the critical requirement to maintain an ongoing representation of its relative position with respect to an environmental representation. In these contexts, a better perception of the underwater environment can be achieved by using image processing algorithms for a suitable representation of the seabed. In optic sensing field and underwater environment, shape acquisition concerns with sensing activity including Shape from Stereopsis, Shape from Photometric Stereo, Shape from Motion and Active Stereo. Main aspects about Shape from Stereopsis will be presented focusing the attention on synchronism in the acquisition at hardware/software levels. The usage of triggered expensive equipments in synchronized stereo sequence acquisitions could present some technical limitations, such as low frame rate and poor images resolution, demoting the quality of the 3-D mosaic. For the previous reasons, an ad-hoc algorithmic solution will be detailed to limit asynchronism problems due to time gap (delay) in stereo frames acquisition, when low-cost non-professional (non-triggered) hardware is used. To achieve a metric reconstruction, the presented framework requires a calibration phase whereas a new stereo frame-selection scheme based on the Epipolar Gap Evaluation (EGE) will be discussed to overcome asynchronisms. Normally, dense depth maps obtained by evaluating dense disparity maps allow the construction of high quality 3-D mosaics. A detailed discussion about methodologies for dense stereo matching will be addressed to handle the ill-posed stereo correspondence problem. A joined use of local (respectively global) matching methods and imaging enhancement algorithms will be considered to identify an appropriate amount of right correspondences in severe underwater conditions (backscattering, brightness constancy violation condition). The chapter treats an important aspect of the 3-D mosaic reconstruction as the registration by ego-motion and camera pose estimation. Classical solutions (i.e. Iterative Closest Point algorithm) and new trend in the registration activity (Zhang's Epiflow algorithm) will be discussed emphasizing application limits and how the computational cost grows in size with the amount of considered features. A new registration approach will be presented conjugating Epiflow and ICP, under a simplified motion model.

2. Related works for underwater 3-D reconstruction

In the field of 3-D reconstruction, terrestrial applications have encouraged extensive work over the last three decades; on the other hand a limited amount of underwater applications have been explored primarily for mapping and positioning. The first step of the structure reconstruction is the shape acquisition that can be addressed in different ways, as taxonomically shown in Fig.1. In underwater research the shape acquisition is mainly performed with expensive optical sensing methods (Negahdaripour et al., 2002; Khamene et al., 2001) and non-optical sensing methods, working often in conjunction (these methods are underlined in Fig.1). Shape from X is a generic name for techniques that extract shape from images. Normally, in underwater environment the optical sensing techniques include Shape from Stereopsis (Zhang, 2005), Shape from Photometric Stereo (Negahdaripour et al., 2002), Shape from Motion (Khamene et al., 2001) and Active Stereo (Narasimhan et al., 2005). In the field of non-optic underwater sensing, acoustic cameras are employed for 3-D mosaic reconstruction (Castellani et al., 2004), whereas both acoustic and optic cameras are often used providing scene information that cannot be recovered from each sensor alone. Three-dimensional scene structures captured by a camera may be detected and acquired observing the apparent motion of brightness patterns from images. The primary visual motion cue useful for shape acquisition is the perceived movement of brightness patterns, known as optical flow (Horn, 1986) which is an approximation of the 3-D world motion field. The 3-D reconstruction from differential motion cues requires accurate optical flow computation. In

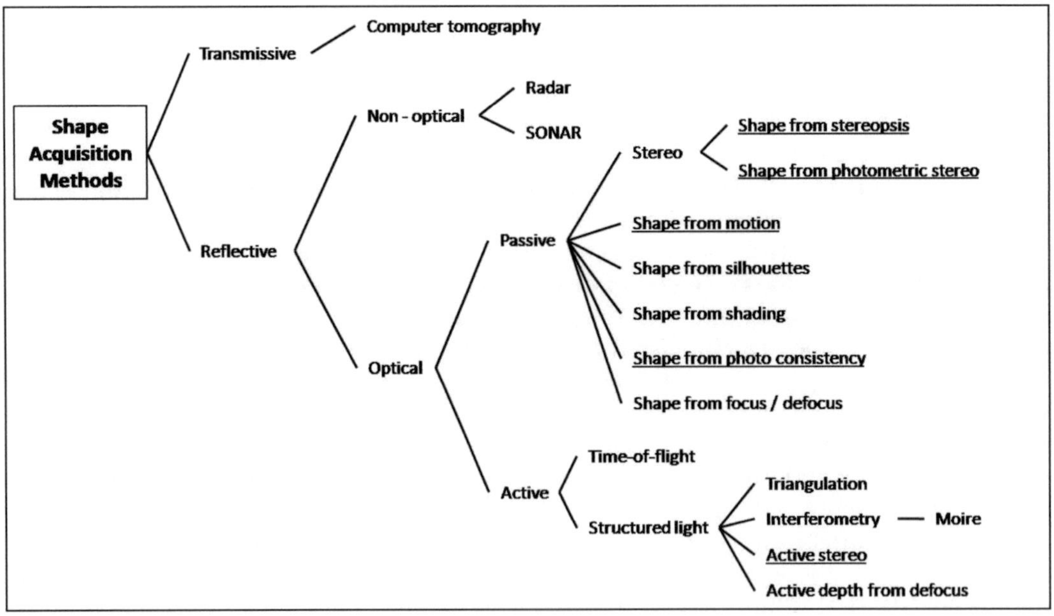

Fig. 1. Short taxonomy of shape acquisition methods.

the last fifteen years, theoretical developments in visual motion studies have established a unified framework for the treatment of the Structure from Motion (SFM) and Structure form Stereo (SFS) problems (Faugeras, 1992), also known as *3-D Reconstruction from Multiple Views*. 3-D reconstruction from multiple views involves extracting target features from one image, matching and tracking these features across two or more images, and using triangulation to determine the position of the 3-D target points relative to the camera. Visual motion methods have been well-studied, requiring densely-sampled image sequences. Instead, the trade-off in stereo vision is between the stereo correspondence problem and a more accurate and robust 3-D reconstruction (Scharstein & Szeliski, 2002). A large amount of works has addressed the correspondence problem, attempting to overcome the various difficulties of the large-displacement correspondence problem: occlusions, large rotations and disparities, photometric and projective distortions (Lucas & Kanade, 1981; Tomasi & Kanade, 1991). The problem of outliers has been solved by the deployment of robust estimation methods (Zhang et al., 1995). The motion and structure may be estimate by using several recursive schemes. Extended Kalman Filter is the most popular approach for jointly estimation of motion and structure with satisfactory results. However, the computation cost of this approach grows cubically with the amount of features, causing a great bottleneck for real-time performance. Stereo matching is one of the most active research areas in computer vision. Stereo matching is a hard problem due to ambiguity in un-textured and occluded areas. Only dominant features, such as points of interest, can be matched reliably. This motivates the development of progressive approaches (Szeliski & Scharstein, 2002). The reduced local disparity in search range makes progressive approaches very efficient in computation and robust. However, the seed initialization remains a computational bottleneck, although, robustness can be improved by enforcing the left-right symmetry constraint. Recently, Graph Cuts and Belief Propagation furnish combinatorial optimization frameworks in which the performance for global stereo algorithms are considerably improved according to an evaluation framework applied to a standard reference stereo data set (Scharstein & Szeliski, 2002). However, the underlying brightness constancy assumption of combinatorial optimization methods severely limits the range of their applications. The earliest attempts for 3-D reconstruction use methods based on volume intersection, as Shape from Silhouette (Laurentini, 1994). Traditional methods, such as stereo, handle large visibility changes between images by solving the correspondence problem between images. The most prominent approaches in 3-D reconstruction are the Voxel Coloring (Slabaugh et al., 2001) and the Space Carving (Kutulakos & Seitz, 1998). These approaches use the color consistency to distinguish surface points from other points in a scene. Cameras with an un-occluded view of a non-surface point see surfaces beyond the point, and hence inconsistent colors, in the direction of the point. Initially, the environment is represented as a discretized set of voxels, and then the algorithm is applied to color the voxels that are part of a surface in the scene. Another promising approach in 3-D reconstruction is the Marching Cubes algorithm for rendering iso-surfaces from volumetric scan data (Lorenson, 1987). The algorithm produces a triangle mesh surface representation by connecting the patches from all cubes on the iso-surface boundary. In underwater scenario, research is directed at exploring the use of vision, potentially in conjunction of other sensors, to automatically control unmanned submersibles, including positioning and navigation by utilizing a photo-mosaic as a two-dimensional visual map. Recent activities combine video imagery taken from multiple views of a scene to derive size and depth measurements and 3-D reconstructions. These activities support (semi-) autonomous or operator-supervised missions pertaining to automatic vision guided station keeping, location finding and navigation, survey and mapping, trajectory following and online reconstruction of a composite image, search and inspection of subsea structures. These tasks require an accurate estimation of camera position, together with fast, accurate correspondence determination, particularly for real-time registration. Common sources of error include non-planar seafloor, moving objects, illumination variations, transect superposition, positioning drift. A number of studies over the last several years have also addressed the 3-D reconstruction for various applications. Khamene and Negahdaripour incorporate cues from stereo, motion and shading flow for 3-D reconstruction in underwater (Khamene & Negahdaripour, 2003); Majidi and Negahdaripour suggest the use of 3-D reconstruction for global alignment of 3-D sensor positions (Madjidi & Negahdaripour, 2005); Nicosevici et al introduce 3-D reconstruction from motion video and representation of the surface topography by piecewise planar surfaces for the construction of orthomosaics (Nicosevici et al., 2005). Hogue et al have developed a stereo vision-inertial sensing device deployed to reconstruct complex 3-D structures in both the aquatic and terrestrial domains (Hogue et al., 2007). The sensor temporally combines 3D information, obtained using stereo vision algorithms with a 3 DOF inertial sensor. The resulting point cloud model is then converted to a volumetric representation and a textured polygonal mesh is extracted using the Marching Cubes algorithm (Lorenson, 1987).

3. Algorithmic framework

3.1 Overview

This section discusses about the framework for 3-D mosaic reconstruction of a seabed based on the optic sensing technique known as *Shape from Stereopsis*. In Shape from Stereopsis two stereo frames are acquired at the same time, so that normally triggered frame grabber (or other techniques that guarantee synchronism in stereo sequence) is required. However, synchronized stereo sequences use expensive equipments and present some technical limitations, such as low frame rate and poor time resolution. For the previous reasons the attention is focused in stereo reconstruction and 3-D structures estimation by using unsynchronized cameras. Svedman proposes to acquire three images sequentially from the left, right, and again from the left camera. A virtual image from the left camera synchronized with the right image is created by interpolating matching interesting points in the two left images (Svedman et al., 2005). Others approaches estimate the time difference between views, synthesizing synchronous image pairs for dense depth information estimation. The depth estimation is based on stereo correspondence evaluation. Depth and ego-motion estimation leads to the ill-posed stereo correspondence problem. Although for egomotion estimation a sparse set of correspondence points is sufficient, depth estimation requires dense correspondences. In underwater environment normalized cross-correlation is generally employed for the robustness to brightness gain. In dense stereo matching, the best performing algorithms use either the Belief Propagation (Sun et al., 2002) or Graph Cuts (Boykov et al., 2001). However, these algorithms are tested on standard data sets under restricted conditions and/or controlled environments (small disparity range, image sequences that satisfy the brightness constancy assumption). The brightness constancy model is often violated, e.g. for most images in the JISCT collection (Bolles, 1993). Zhang proposes a revised data cost to improve global stereo matching algorithms (Graph Cuts,

Belief Propagation) in underwater environment (Zhang, 2005). 3-D mosaic construction requires two-frames registration and then ego-motion estimation. Iterative Closest Point (ICP) algorithm is used for 3-D registration (Besl & McKay, 1992). The epiflow framework proposed by Zhang is based on the integration of motion and stereo epipolar geometries for ego-motion tracking (Zhang, 2005). Summarizing, an inexpensive asynchronous stereo framework for 3-D seabed reconstruction and orthomosaic is presented. In order to achieve a metric reconstruction, a calibration phase is necessary, so that the asynchronism in stereo sequence can be an issue for proper calibration. To overcome this problem a new stereo frame selection scheme based on the Epipolar Gap Evaluation (EGE) is used (Section 5). In order to handle the stereo correspondence problem, local and global matching methods are evaluated in Section 6, with adoption of a suitable similarity measure (ZNCC) and image enhancement technique (CLAHE). Finally, in Section 7 the registration approach is presented conjugating Epiflow and ICP schemes under a simplified motion model.

3.2 Framework architecture

The main building blocks of 3-D seabed mosaic reconstruction system are the following: (a) shape acquisition, (b) depth estimation and (c) mosaic registration and rendering. In Fig. 2 the overall system architecture is shown. The block 1 deals with shape acquisition, asynchronous stereo sequence calibration by using the Epipolar Gap Evaluation (EGE)

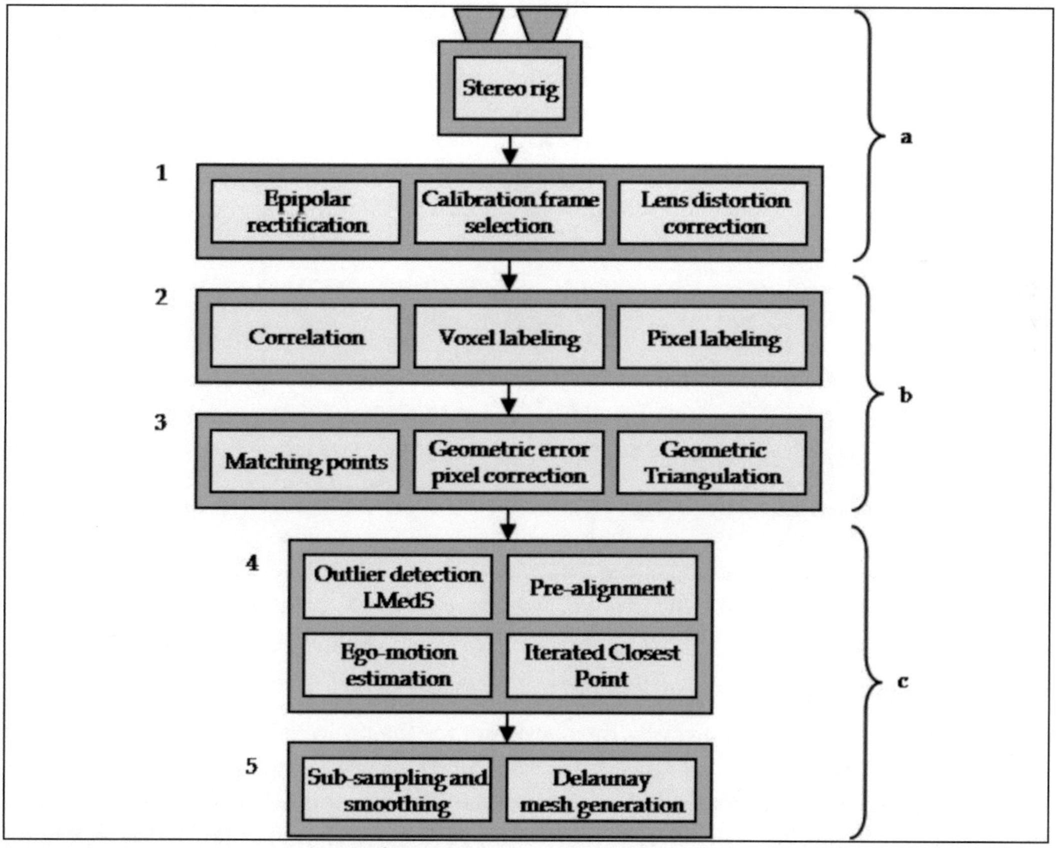

Fig. 2. Overall framework architecture.

scheme, lens distortion correction and histogram equalization. Depth estimation is accomplished in blocks 2 and 3: the former regards the disparity map estimation and the latter computes the subsequent depth map by using geometric triangulation. As described in the following, local and global methods are used in stereo disparity calculation. The block 4 involves fusion and registration of 3-D reconstructions, through system ego-motion estimation, pre-alignment and ICP registration refinement. Finally, triangular Delaunay interpolation and rendering take place in block 5.

4. Structure from stereopsis in underwater environment

4.1 Overview

Starting from two different views of the same scene, it is possible to obtain a 3-D structure reconstruction. For simplicity, the discussion is restricted to scenes consisting of points only; further information about multi-view reconstruction methods can be found in (Hartley & Zisserman, 2003). The input for 3-D reconstruction phase is a set of correspondences in left and right camera images. Let the camera matrices $\mathbf{P}^L$ and $\mathbf{P}^R$, that describe the correspondences $x_i^L \leftrightarrow x_i^R$ in terms of $\mathbf{P}^L X_i = x_i^L$ and $\mathbf{P}^R X_i = x_i^R$, where the point X_i projects to the two given data points x_i^L and x_i^R (see Fig. 3). Unfortunately, neither the projection matrices $\mathbf{P}^L$ and $\mathbf{P}^R$, nor the points X_i and $x_i^L \leftrightarrow x_i^R$ are known *a priori*. The camera matrices estimation is part of the calibration activity (Subsection 5.2), whereas the correspondences estimation $x_i^L \leftrightarrow x_i^R$ deals with the stereo correspondence problem (Subsection 6.2 and 6.3), and the estimation of point X_i is the first step for 3-D reconstruction by using geometric triangulation (Subsection 6.4).

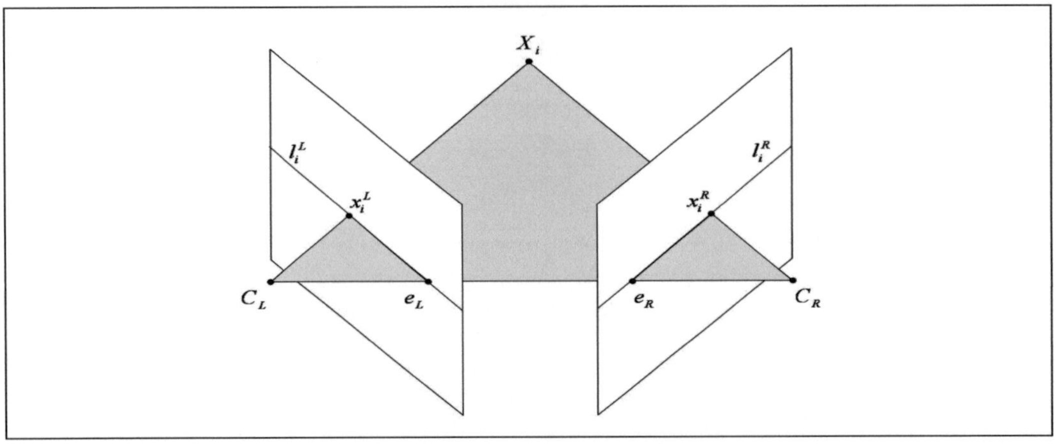

Fig. 3. Epipolar geometry: the optic rays passing through the X_i point and C_L , C_R optical centers intersect the left and right image planes in (x_i^L , x_i^R) image points, respectively.

The main tool in 3-D reconstruction from two views is the *Fundamental Matrix* (Faugeras, 1992), which represents geometrically the constraint for the projected image points (x_i^L , x_i^R) of the same 3-D point X_i. This constraint, also called *Epipolar Constraint*, arises from the coplanarity of the camera centers, the images points (x_i^L , x_i^R) and the space point X_i. Given the Fundamental Matrix **F**, each pair of matching points (x_i^L , x_i^R) must satisfy the following relation:

$$\left(x_i^R\right)^T \mathbf{F} x_i^L = 0 \tag{1}$$

with F a 3×3 matrix of rank 2. Furthermore, Eq.1 can be rewritten as:

$$\left(x_i^R\right)^T l_i^R = 0 \tag{2}$$

where

$$l_i^R = \mathbf{F} x_i^L \tag{3}$$

is the *Epipolar Line* associated with the left point x_i^L. In other words, the Eq.2 states that the point x_i^R belongs to the epipolar line l_i^R. The discussed concepts are summarized in Fig. 3. Points e_L and e_R are the *epipoles*, i.e. the points of intersection of the *baseline* (the line joining the camera centers) with the image planes. A 3-D projective reconstruction can be obtained from the knowledge of the Fundamental Matrix alone, so that the pair of camera matrices can be determined up to a 3-D projective ambiguity and the Fundamental Matrix can be estimated directly from a set of point correspondences. Instead, if the camera matrices are known, by means of calibration activity, a metric reconstruction is possible. Therefore, a reliable calibration activity is needed since a 3-D metric reconstruction of the seabed surface is desired. The calibration activity can be affected by three kind of error: synchronization error, algorithm error and matching error. Algorithm error and marker matching error deal with the specific calibration procedure, whereas synchronization error relates the synchronism in stereo frames. The synchronization error can be minimized by computing calibration matrices for each stereo pair in which calibration pattern is viewed and, then, by evaluating the *Epipolar Gap* (EG). Let a pair of matching points x_i^L and x_i^R, the Epipolar Gap is defined as the Euclidean distance between one of these points (i.e. x_i^R) and the epipolar line associated with the other one, as follow:

$$\mathrm{EG}_i = \mathrm{dist}\left(x_i^R, l_i^R\right) \tag{4}$$

If x_i^L and x_i^R matches exactly, Eq.2 is satisfied and then EG_i=0; otherwise EG_i provides a measure (the Euclidean distance) for the epipolar gap between x_i^L and x_i^R. The calibration activity is realized by a careful selection of a stereo pair from the video sequence in which the calibration pattern is acquired. Since there are many stereo pairs in which the calibration pattern is acquired, the Epipolar Gap Evaluation (EGE) is used as selection criterion. The selection of calibration stereo pairs is motivated by the asynchronism of the acquisition system based on a non-triggered frame grabber. Aim of this step is to minimize the acquisition time distance between the left and the right frame in the stereo pair, selecting a stereo pair affected by minimum time difference. The calibration frame selection requires a large amount of calibration parameters (camera matrices), therefore a calibration pattern (an object with known metric measures) must be employed and detected in the acquired video sequence. In the next section, a calibration object (its marker points imprinted on pattern) detection from video sequence is discussed.

4.2 Metric calibration

4.2.1 Pattern detection

A calibration object can be acquired to provide metric correspondences between points in the image coordinates and points in real world coordinates. Generally, an accurate camera coordinate system-based positioning of the calibration object is hard, especially in underwater environment. Hence, the relationship between target and camera coordinate systems needs to be recovered from reliable correspondence estimation. Several camera calibration methods employing 2-D and 3-D calibration targets were proposed in (Tsai, 1987; Heikkia & Solven, 1996). Calibration algorithms use an object on which a square checker pattern with a known size is printed. In this work a pair of still images of 3-D checkerboard is required, according to the method proposed in (Tsai, 1987). Normally, in underwater applications the checkerboard is arranged on the sea floor and the stereo acquisition system moves around during calibration activity (see Fig. 4.b), in contrast with the ideal calibration conditions in which both checkerboard and stereo acquisition system are more likely to be fixed (see Fig. 4.a). The corners of the checker pattern are used as control points for the camera calibration algorithm (see Fig. 5.a). Bakstein examines several pattern geometries to

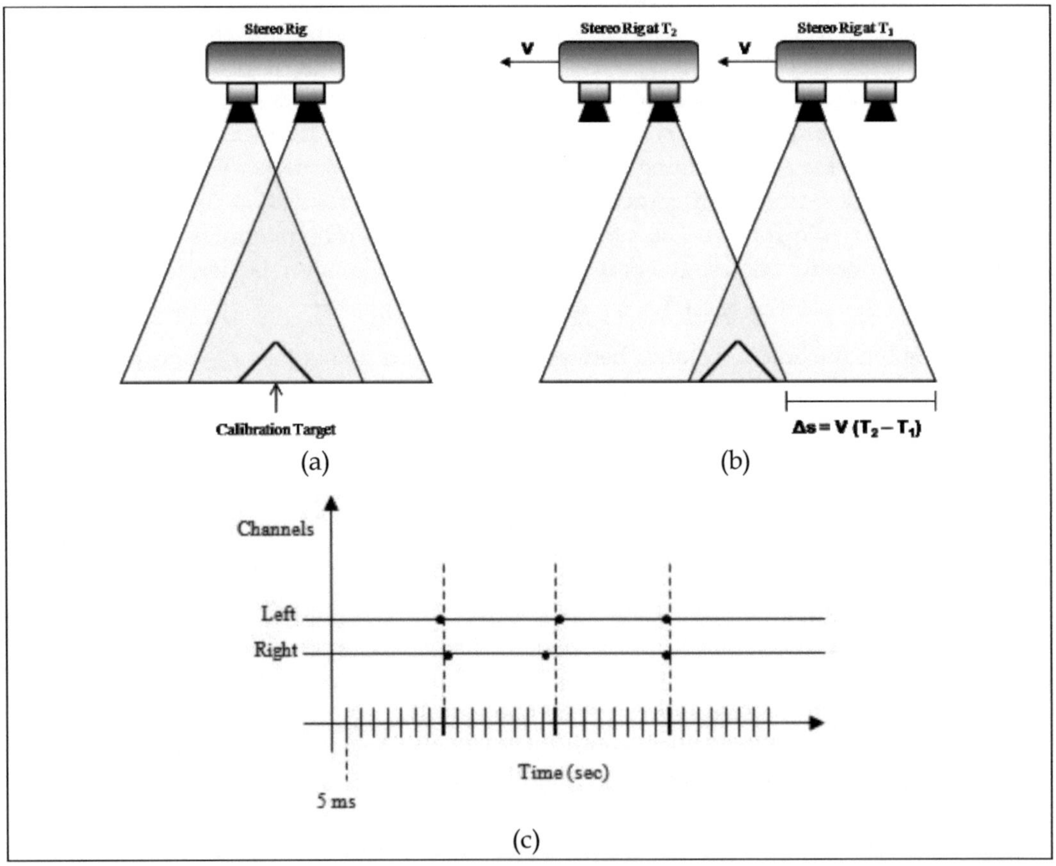

Fig. 4. *a*) Ideal acquisition conditions for proper calibration; *b*) Real calibration conditions: the system moves with velocity **v**; *c*) Stereo acquisition timing at 25 fps. Dashed lines denote ideal acquisition time instants, while dots denote effective left/right acquisition time instants.

derive the control points for camera calibration by using an automated thresholding strategy, so that the checker pattern is chosen for the very low sensitivity by thresholding errors (Bakstein, 1999). Furthermore, a black-and-white checker pattern provides very good contrast and size of the squares could be customized to the object distance for better perception. Asynchronism in stereo acquisition is a problem in the calibration activity, since the epipolar geometry constraint isn't satisfied. This aspect is further explained in Fig.4.c. A 25 fps acquisition system acquires two calibration checkerboard images, but due to asynchronism the effective acquisition time instants (represented by dots in figure) lay around the ideal ones (represented by dashed lines). Unlike others techniques based on view synthesis or time estimation between views, the presented approach tries to solve this problem making only use of calibration parameters encoded in camera matrices. For a generic acquisition system, the calibration error is defined as:

$$\varepsilon = \varepsilon_{syn} + \varepsilon_{cal} + \varepsilon_{match} , \tag{5}$$

(a) (b) (c) (d)

Fig. 5. Calibration object: *a*) Original image; *b*) Segmented image; *c*) Pattern's rectangles; *d*) Centroids.

where ε_{syn} is the synchronization error, ε_{cal} is the calibration algorithm error and ε_{match} is the error affected by calibration pattern points selection. The calibration error minimization takes place by evaluating the epipolar gap for each stereo pairs in which calibration pattern is acquired. The epipolar gap estimation is more accurate as greater amount of calibration pattern images is available. This calibration method detects the calibration object in a rotation independent way and it localizes checkerboard square corners corresponding to calibration target ones by using line-fitting and label-sorting strategies. In order to estimate intrinsic and extrinsic cameras parameters, checkerboard square vertexes must be detected in acquired images and matched with corresponding checkerboard square vertexes in real calibration target. Standard corner detection approaches (Harris & Stephens, 1988) does not give satisfactory results by detecting checkerboard square vertexes, since underwater images present poor quality. To overcome the problem, a projective-geometric approach is investigated, according to the following discussion. Preliminarily, the calibration target is segmented-out (see Fig.5) from undistorted image (Fig.5.a) by using well-known morphological operations (such as connected components for small object removing and thresholding binaryzation). Once the pattern is detected in the image, the checkerboard square vertexes are identified through a line-fitting approach as explained in the following. As well-known, image plane projection preserves the intersection property. Hence in each image the pattern target appears to have squares aligned with the eight non-intersecting lines L_q (see Fig.6.a) and with four non-intersecting lines L_t for each pattern side (see Fig.6.b) due to prospective projection. The previous observation provides a scheme for labeling and

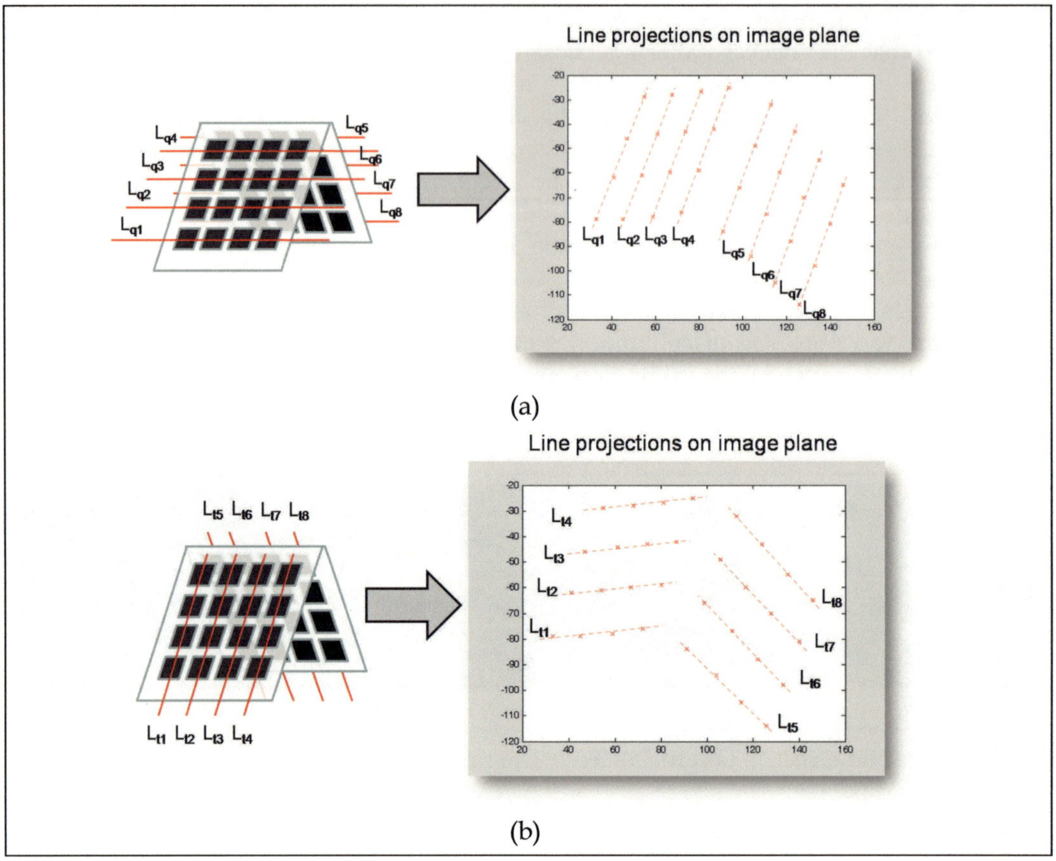

Fig. 6. *a*) projection of 8 non intersecting lines on image plane; *b*) projection of 4 non intersecting lines for each side on image plane.

sorting the centroids as shown in Fig. 7.a. At each iteration the line passing through centroids P_1 and P_i is considered as the reference line *r*, where the P_1 is the outer centroid with the lowest abscissa. Let *L* be a list containing $\left(j, d_j^r, d_j^O\right)$ triples where *j* is the centroid label, d_j^r is the Euclidean distance of P_j from line *r*, and d_j^O is the Euclidean distance of P_j from axes origin *O*. Firstly, centroids are sorted by distance d_j^r producing a new list *L'* in which collinear centroids are grouped. Each group of centroids in *L'* is newly sorted by distance from the axes origin. The loop exit test is based on the evaluation of errors E_θ and E_d. As shown in Fig.7.c, E_θ measures the collinearity along L_q as a maximum shifting from average angular coefficient:

$$E_\theta = \max_{i=1,\dots,32} \left| \frac{\theta_i - \overline{\theta}}{\theta_{\max} - \theta_{\min}} \right|, \tag{6}$$

Experimentally, E_θ <0.7 guarantees a satisfactory collinarity along Lq lines. Instead, E_d measures the maximum distance of middle centroids from line passing through outers centroids. A maximum value of 2 pixel for E_d guarantees a good collinearity along L_t lines.

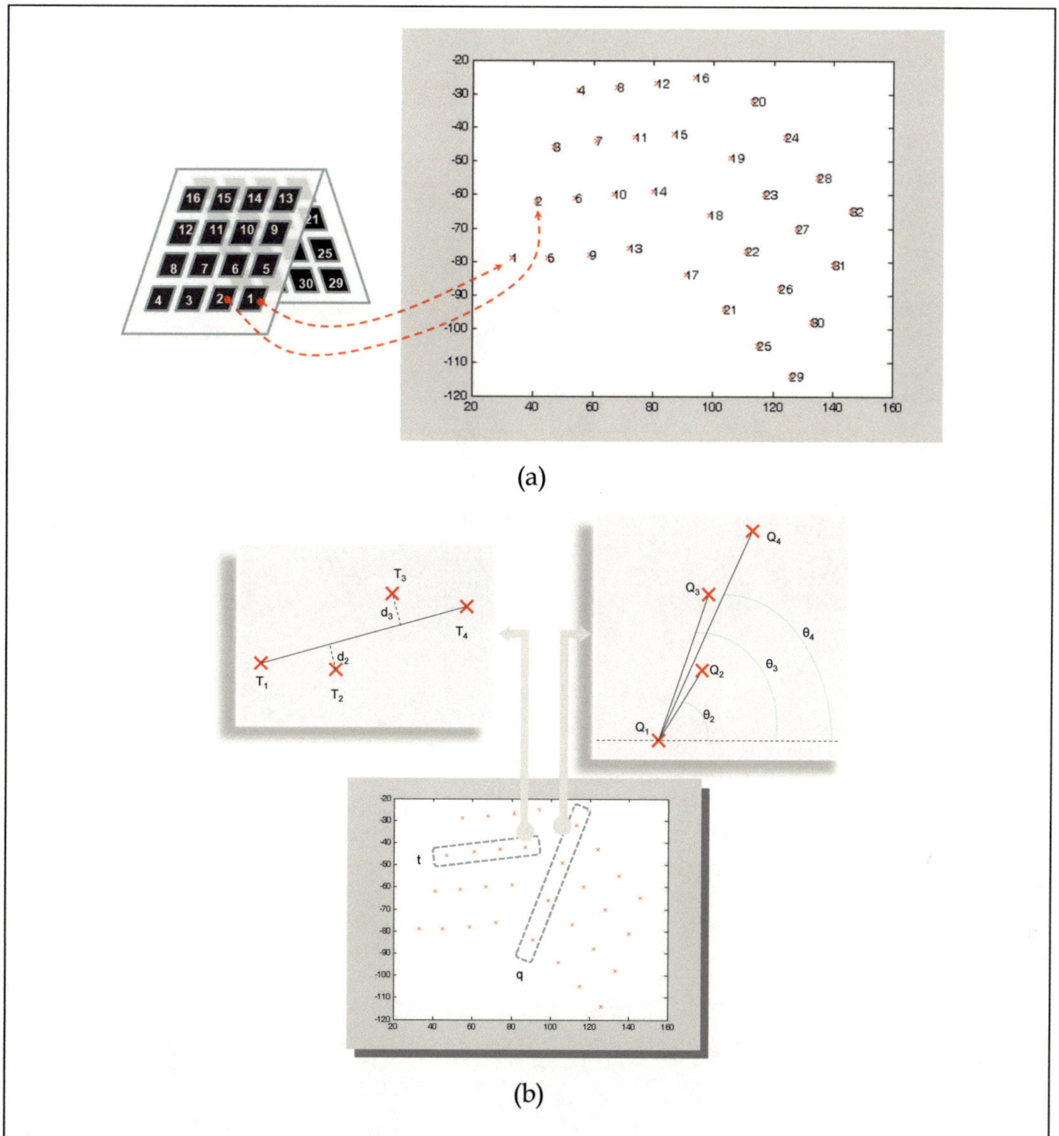

Fig. 7. *a*) Centroid labels mapping and sorting criterion; *b*) Test to evaluate centroid collinearity with L_T and L_Q lines.

Centroid labels are sorted out by evaluating collinearity and distance from the reference line, according to the pseudocode provided in Tab. 1. The goal of the next step is to locate the four vertexes for each labelled and sorted checker square of the calibration pattern. Therefore, for each checkerboard square Northern, Southern, Eastern and Western points are located as intersection of square borders with the two estimated sets of lines L_q and L_t estimated in the previous step (Fig.8.a,b).

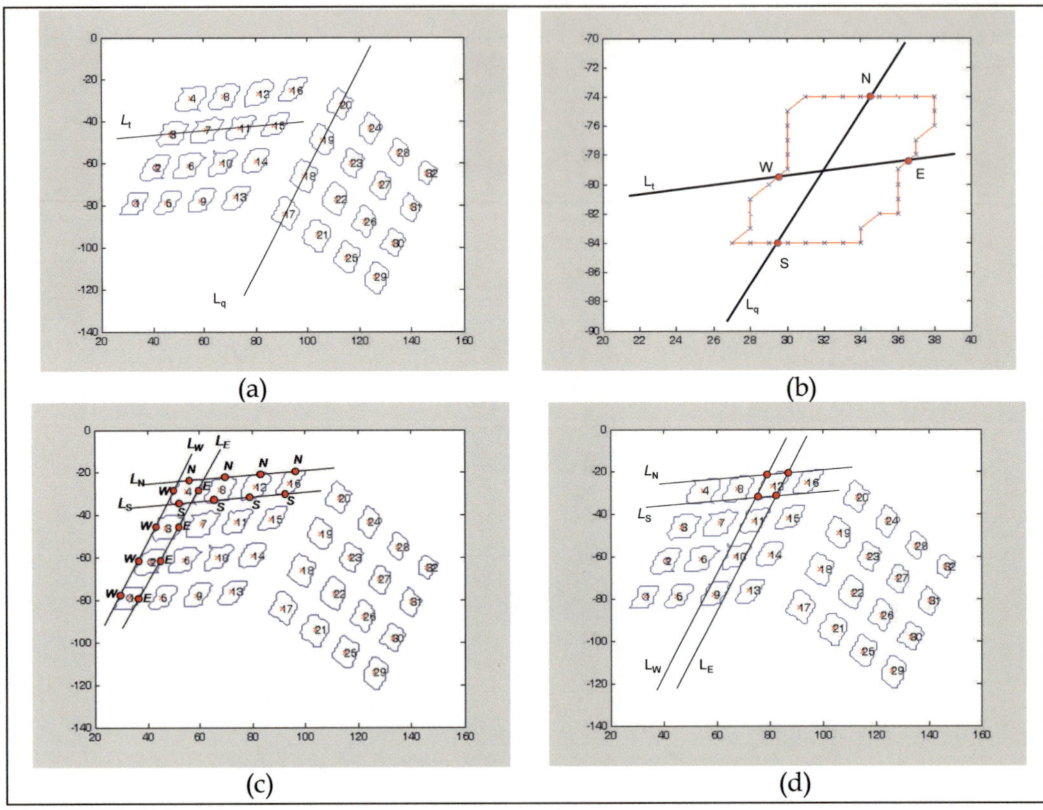

Fig. 8. *a*) Overall viewing of intersections; *b*) Single checker square intersections; *c*) L_N, L_S, L_E, L_W lines are estimated by fitting of N, S, E, W points respectively in each pattern side; *d*) Square vertexes obtained by L_N, L_S, L_E, L_W lines intersection.

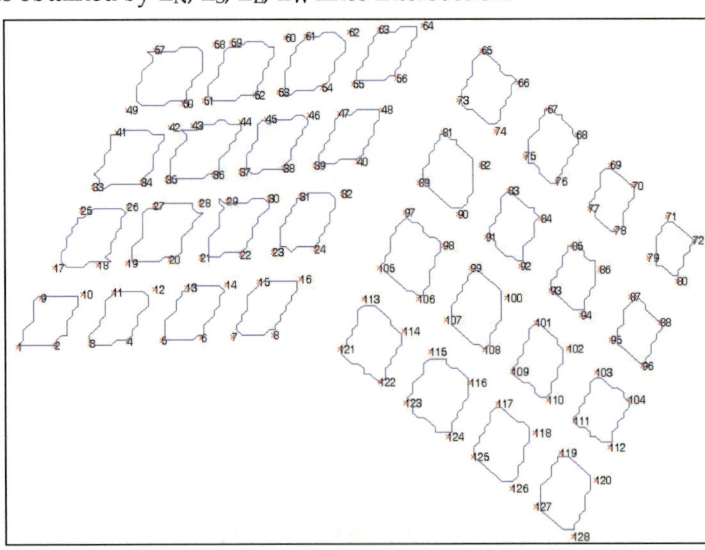

Fig. 9. The result of the calibration object detection algorithm: all square vertexes are labelled and ordered coherently with calibration object marker points.

```
For i = 1 to 32 (amount of checkerboard squares)
    r is the line passing through points P1 and Pi
    ∀j = 1,..., 32 : d_j^r = dist(P_j, r), d_j^O = dist(P_j, O)
    L = {j = 1,..., 32 | (j, d_j^r, d_j^O)}
    L′ = Sort (L) with respect d_j^r
    L″ = Sort (L′) with respect d_j^O
    if (Eθ < 0.7 and Ed < 2) exit loop
next i
```

Table 1. Pseudocode for centroids labeling algorithm.

Once N, E, S, W points are located, the lines L_N, L_E, L_S, L_W can be estimated through line fitting as represented in Fig.8.c. L_N, L_S, L_E, L_W lines are estimated by fitting of N, S, E, W points respectively in each pattern side. Finally, for each checker square the four vertexes are obtained by lines intersection as $L_N \cap L_W$, $L_N \cap L_E$, $L_S \cap L_W$, $L_S \cap L_E$ (see Fig.8.d). The result of the calibration object detection algorithm is shown in Fig.9, where all square vertexes are labelled and ordered in image coherently with the object marker points of the calibration pattern in real world coordinates.

4.2.2 Epipolar gap evaluation

The calibration activity receives a careful selection of a stereo pair from the video sequence in which the calibration pattern is acquired. The selection of calibration stereo pairs is motivated by the asynchronism of the acquisition system based on a non-triggered frame grabber. Since there are many stereo pairs showing the calibration pattern, the *Epipolar Gap Evaluation* (EGE) is used as a selection criterion. Aim of this step is to minimize the acquisition time distance between the left and the right frame in the stereo pair; the purpose is to select a stereo pair affected by minimum time difference (Fig. 4.c). The selection criterion can be schematized with two sequential activities such as: the *Calibration Matrices Estimation* and the *Epipolar Gap Evaluation*. The scheme is depicted in Fig.10.a. The first step receives as input the h-th stereo frame pair, in which the calibration pattern is acquired, providing the corresponding calibration matrices as output, according to Tsai's calibration method. The second block receives as input the calibration matrices from the previous block and the stereo frame pair in which the seabed surface under reconstruction is acquired. The EGE provides the calibration matrices that minimize the Epipolar Gap on the overall reconstructing frames set. In EGE, a sparse set of stereo matches

$$\mathbf{M}_k^{\mathrm{L}} \times \mathbf{M}_k^{\mathrm{R}} = \left\{x_{k,1}^{\mathrm{L}},\ldots,x_{k,n_k}^{\mathrm{L}}\right\} \times \left\{x_{k,1}^{\mathrm{R}},\ldots,x_{k,n_k}^{\mathrm{R}}\right\}, \quad n_k = \left|\mathbf{M}_k^{\mathrm{L}}\right| = \left|\mathbf{M}_k^{\mathrm{R}}\right| \tag{7}$$

is computed for the k-th stereo pair $\left(I_k^L, I_k^R\right)$. Hence, each set $\mathbf{M}_k^{\mathrm{L}} \times \mathbf{M}_k^{\mathrm{R}}$ is compared with the Fundamental Matrix $\mathbf{F}_h$ for frame rejection purpose. The frame rejection is based on the Maximum Epipolar Gap (MEG) evaluated as:

$$\mathrm{MEG}_{k,h} = \max_{j=1,\ldots,n_k} \left\{\mathrm{EG}_{k,j}^h\right\} \tag{8}$$

where $\mathrm{EG}_{k,j}^h$ is the epipolar gap defined in Eq.4 and here generalized as follow:

$$\begin{aligned} EG_{k,j}^{h} &= \operatorname{dist}\left(x_k^R, l_{k,j,h}^R\right) \\ l_{k,j,h}^R &= \mathbf{F}_h x_{k,j}^L, \forall k = 1,\ldots,r, \ \forall j = 1,\ldots,n_k, \ \forall h = 1,\ldots,s. \end{aligned} \quad (9)$$

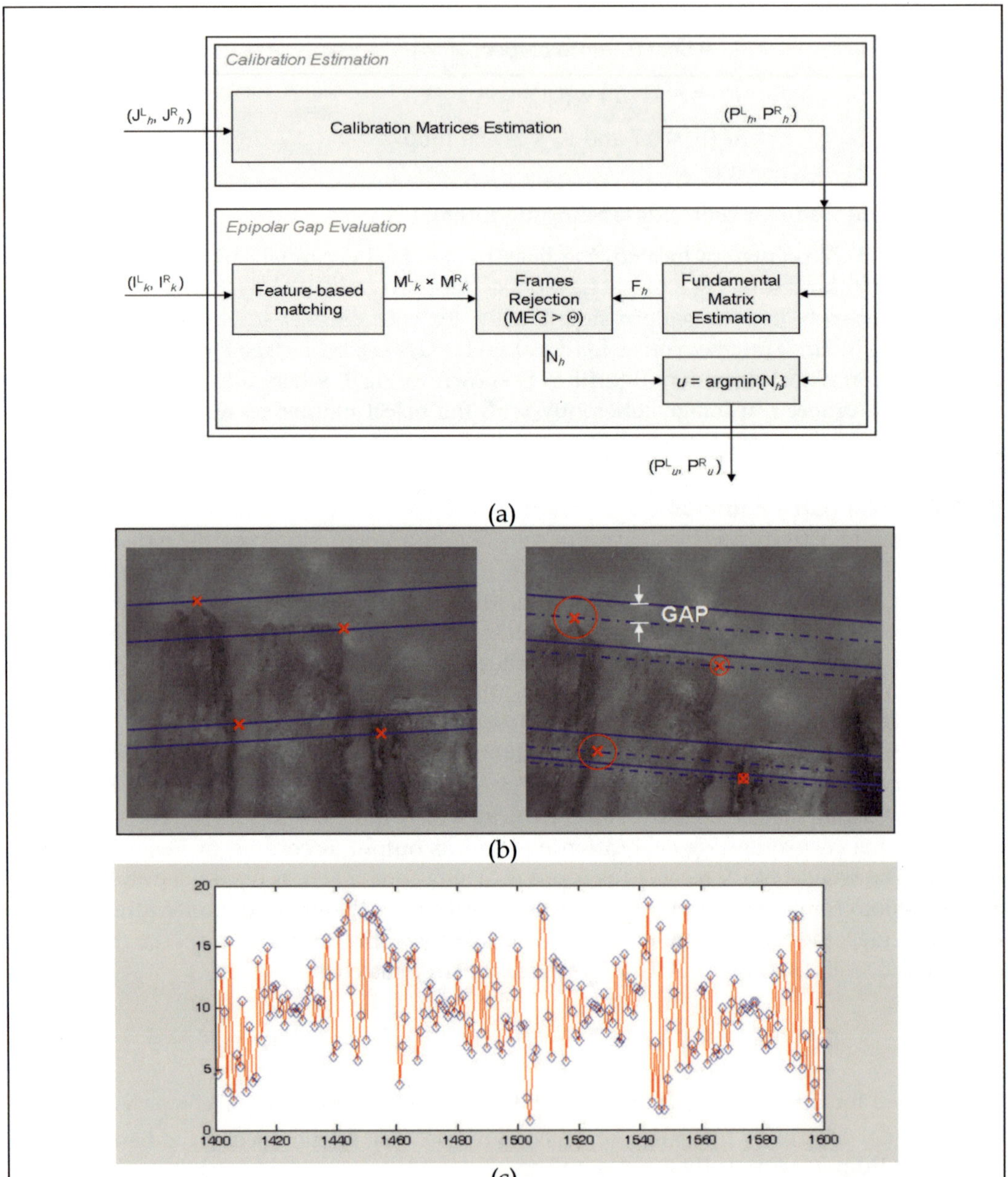

Fig. 10. *a*) Block scheme for the calibration frames selection criterion based on the Epipolar Gap Evaluation. *b*) Epipolar gap evaluated in 4 points ; *c*) Experimental result. The MEG is evaluated within calibration sequence. The optimal value u in Eq.11 is achieved for frame $h = 1504$ with $\theta \approx 1$ pixel.

Thus, MEG is used as a filtering rejection stereo pairs for which the corresponding value exceeds a prefixed threshold. If the amount of rejected couples is given by:

$$N_h = \left|\left\{k = 1,\ldots,r \mid \mathrm{MEG}_{h,k} > \theta\right\}\right| \tag{10}$$

then, the chosen calibration pairs will be $\left(\mathbf{P}_u^{\mathrm{L}}, \mathbf{P}_u^{\mathrm{R}}\right)$ with

$$u = \arg\min_{h=1,\ldots,s} \{N_h\}. \tag{11}$$

5. Seabed reconstruction

5.1 Overview

Stereo reconstruction is based on the epipolar geometry as discussed in Section 5.1 and illustrated in Fig. 3. Let the camera matrices $\mathbf{P}^L$ and $\mathbf{P}^R$ (two 4×4 projection matrices in homogeneous coordinates) and (x_i^L, x_i^R) two corresponding points in the left-right images, the epipolar constraint in Eq.1 must be satisfied. The constraint may be interpreted geometrically in terms of the rays in space corresponding to the two image points x_i^L and x_i^R; in particular each point lies on the epipolar line of the other point according to Eq.2 and Eq.3. Referring to Fig. 3 this means that the two rays $\overrightarrow{X_iC_L}$ and $\overrightarrow{X_iC_R}$ back-projected from image points x_i^L and x_i^R lie in a common *epipolar plane* passing through the two camera centres C_L and C_R. Since the two rays lie in a plane, they will intersect in the 3-D point X_i related to the points x_i^L and x_i^R through cameras projections. X_i can be determined as intersection of the two rays back-projected from corresponding points x_i^L and x_i^R by using a stable triangulation method (Hartley & Zisserman, 2003). Unfortunately, these corresponding points are not known a-priori leading to the *Stereo Correspondence Problem*, so that, given the point x_i^L, the problem is to determine another point x_i^R such as these two points are the projections of the same 3-D point onto the left and the right image plane, respectively. Epipolar constraint allows to simplify the matching algorithmic complexity by reducing the searching area close to the epipolar lines. Further algorithmic simplifications are possible by means of *Epipolar Rectification* process, that is, the calculation of an appropriate projective transformation producing as output a stereo image pair in which the epipolar lines are transformed to lines parallel with the x-axis in the image plane. Rectification allows to perform stereo matching along horizontal scan lines. In this situation, the horizontal offset between corresponding image points $d_i = x_i^L - x_i^R$ is referred as *disparity*. Stereo correspondences are estimated by using matching algorithms that received rectified stereo image pairs in input and provide a dense disparity map as output. The disparity map is a range image in which each pixel describes the disparity value estimated respect to a reference image. Given a left disparity map represented with a 256 grey levels, each right matching point is:

$$x_i^R = x_i^L - \frac{p_i}{256}\Delta - d_{\min} \tag{12}$$

where Δ is the disparity range, $d_{\min}$ is the minimum disparity value, and p_i is the disparity map value corresponding to the x_i^R point (in 256 gray levels). In this framework, for stereo disparity maps calculation local and global methods are explored. In both cases, the major difficulties are related to the underlying assumption of brightness constancy in the stereo algorithms, systematically violated in underwater environment (Bolles et al., 1993).

5.2 Stereo correspondences: local methods

Local approaches are based on a search windows horizontally displaced in one view with respect to the other view for each allowed disparity. Matching measures generally used in local methods are *Sum of Absolute Differences* (SAD), *Sum of Squared Differenced* (SSD), *Normalized Cross Correlation* (NNC) and *Sampling Insensitive Measurement*. In this context the best choice is the *Normalized Zero-Mean Cross-Correlation* (ZNCC), that is quite insensitive at brightness variations typical of underwater environment. ZNCC measure is defined as:

$$\text{ZNCC}\quad(x,y,d)=\frac{\sum\limits_{w_x,w_y\in W}\left(\mathrm{I_L}\left(x+w_x,y+w_y\right)-\bar{\mathrm{I}}_{\mathrm{L}\,x,y}\right)-\left(\mathrm{I_R}\left(x+w_x+d,y+w_y\right)-\bar{\mathrm{I}}_{\mathrm{R}\,x,y}\right)}{\sqrt{\left(\sum\limits_{w_x,w_y\in W}\left(\mathrm{I_L}\left(x+w_x,y+w_y\right)-\bar{\mathrm{I}}_{\mathrm{L}\,x,y}\right)^2\right)\cdot\left(\sum\limits_{w_x,w_y\in W}\left(\mathrm{I_R}\left(x+w_x+d,y+w_y\right)-\bar{\mathrm{I}}_{\mathrm{R}\,x,y}\right)^2\right)}}\tag{13}$$

from which the disparity is calculated as

$$\text{Disparity(x, y)} = \max_{d_{\text{MIN}}\le d\le d_{\text{MAX}}}\ \text{ZNCC}(x,y,d)\,.\tag{14}$$

In Fig.11 is reported en example of ZNCC stereo matching applied on a rectified seabed image.

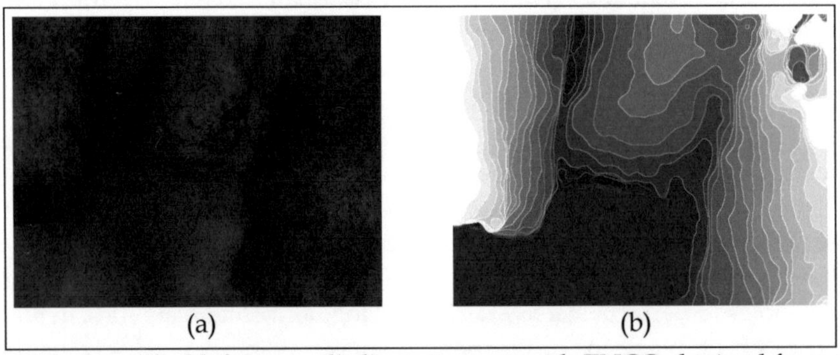

Fig. 11. *a*) original rectified left image; *b*) disparity map with ZNCC obtained from (a).

5.3 Stereo correspondences: global methods

Unlike local approaches, in global approaches the correspondence problem is stated in term of a cost function subject to minimization. The most critical choice is the optimization technique, that essentially falls into two main categories: continuous or discrete optimization. In discrete minimization, the best stereo matching estimation use combinatorial optimization via Belief Propagation (Sun et al., 2002) or Graph Cuts (Boykov et al., 2001). Two Graph Cuts formulation based on the expansion moves scheme are now investigated: Voxel Coloring (Kolmogorov & Zabih, 2002) and Pixel Labeling (Boykov et al., 1998). Graph Cuts based methods address cost function optimization by an energy function that can be represented, in general, as:

$$E(L) = D(L) + V(L), \tag{15}$$

where L is a disparity labeling (i.e. a label set in which each label represents a disparity value), $D(\cdot)$penalizes the variation from observed intensities (data penalties term), and $V(\cdot)$ is the smoothness term that encourages spatial coherence by penalizing discontinuities between neighboring pixels. When cost functions involve convex smoothness term a global minimum is indeed reachable via Graph Cut in polynomial time. Unfortunately, convex smoothness terms do not represent an optimal choice for the stereo problem, in which it is preferable to use a non-convex smoothness function to avoid the over-penalizing of large jumps in disparity.

One simple non-convex function commonly used in stereo vision is the Potts model (Potts, 1952). Despite the simplicity of the Potts model, the resulting problem formulation is proven to be NP-hard (Kolmogorov & Zabih, 2004). However, a strong local optimum can be estimated for non-convex smoothness terms by application of expansion move methods (Boykov et al., 2001). Image enhancement with *Contrast Limited Adaptive Histogram Equalization* (CLAHE) (Pizer et al., 1987) is often employed in underwater imaging application to mitigate brightness variations effects. Both voxel coloring and pixel labeling methods suffer for underwater violation of brightness constancy assumption. However, CLAHE enhancement permits to mitigate brightness variation effects, as explained in Fig.12.

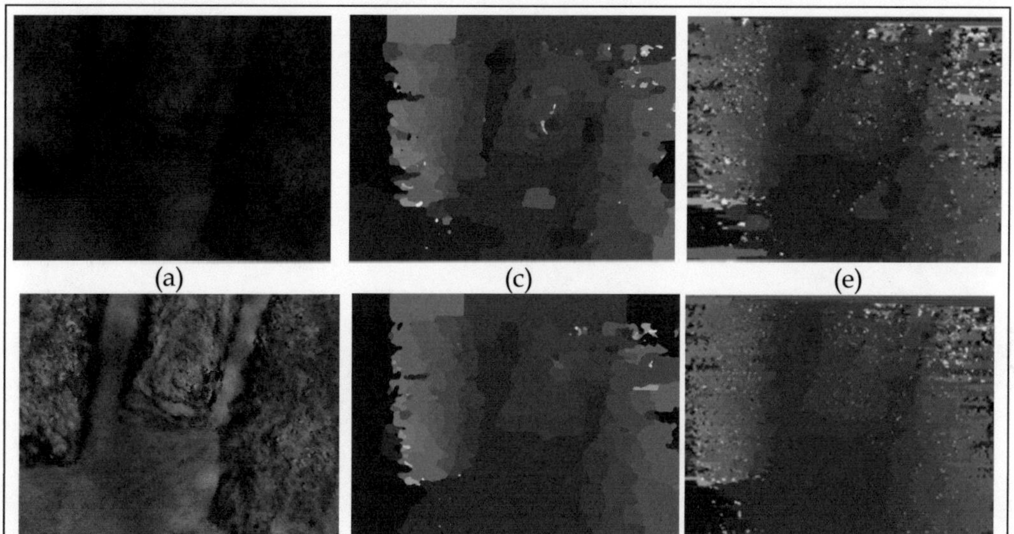

Fig. 12. *a*) original rectified left image; *b*) CLAHE enhanced rectified left image; *c*) disparity map with voxel obtained from (a), *d*) voxel based disparity map obtained from enhanced image (b); *e*) pixel labeling without enhancement; *f*) pixel labeling with CLAHE.

5.4 3-D Reconstruction from stereo correspondences

As prefaced in Section 6.1, given a set of stereo correspondences it is possible to recover the location of the 3-D points by using geometric triangulation. Tsai's calibration method furnishes the camera matrices $\mathbf{P}^L$ and $\mathbf{P}^R$ that relate the image correspondences (x_i^L, x_i^R) with the 3-D point X_i through the relations $\mathbf{P}^L X_i = x_i^L$, $\mathbf{P}^R X_i = x_i^R$, and the epipolar constraint in Eq.1. It follows that x_i^R lies on the epipolar line $l_i^R = \mathbf{F}x_i^L$ and the two rays back-projected from image points x_i^L and x_i^R lie in a common epipolar plane. Since they lie in the same plane, they will intersect at some point. This point is the reconstructed 3-D scene point X_i. While it is possible to recover the 3-D scene point given only the two imaged points (x_i^L, x_i^R), the accuracy is highly dependent upon the exact matching $x_i^L \leftrightarrow x_i^R$. Generally errors are involved, hence a set of points is usually used leading to an over-determined linear system. In particular, camera parameters and image locations are known only approximately. The back-projected rays therefore do not actually intersect in space. It can be shown that intersection equations can be solved in a least squares sense (Kanatani, 1993). Triangulation is addressed in more details in (Hartley & Zisserman, 2003). Fig. 13 shows the 3-D points calculated by triangulation starting from disparity maps obtained by using methods explained in above sections 6.2 and 6.3. The following sections treat how can be obtained a whole mosaic reconstruction starting from single reconstructions like as shown in Fig. 13.

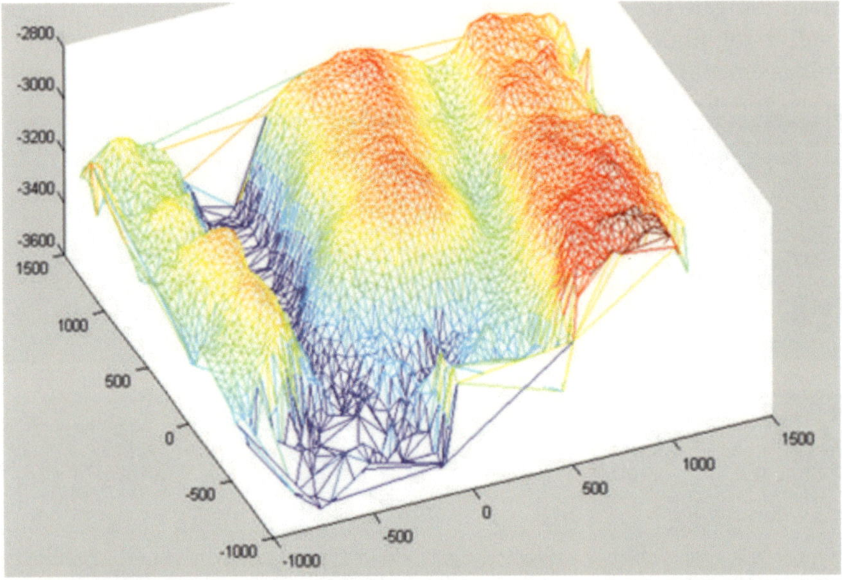

Fig. 13. A reconstruction obtained from a single stereo scan.

6. Seabed mosaic

6.1 Overview

To construct a 3-D mosaic multiple scans must be registered. The registration activity is referred to the alignment of each two-frames reconstruction with the others. A pre-alignment phase can simplify the registration activity, since pitch and roll are not present in motion model (Castellani et al., 2004). For this purpose a pre-alignment activity based on ego-motion estimation through four-frames feature matching (i.e. two following stereo pairs) is used, by evaluating the displacement of left and right matches in two corresponding images. Dimensional effects are normalized by scaling factor according to (Kim & Chung, 2006). Once right correspondences are defined in the four-frame, ego-motion tracking is performed considering displacements through consecutive 3-D point sets.

6.2 Alignment

The correspondences between adjacent frames are estimated with a reliable feature-based matching as explained in the next subsection, while the correspondences between stereo pairs are constrained by the epipolar geometry. Given four frames composed by two following stereo pairs (I_1^L, I_1^R) and (I_2^L, I_2^R), four-frame correspondences are determined according with the scheme shown in Fig. 14.a. Firstly, the correspondences q_1 and q_2

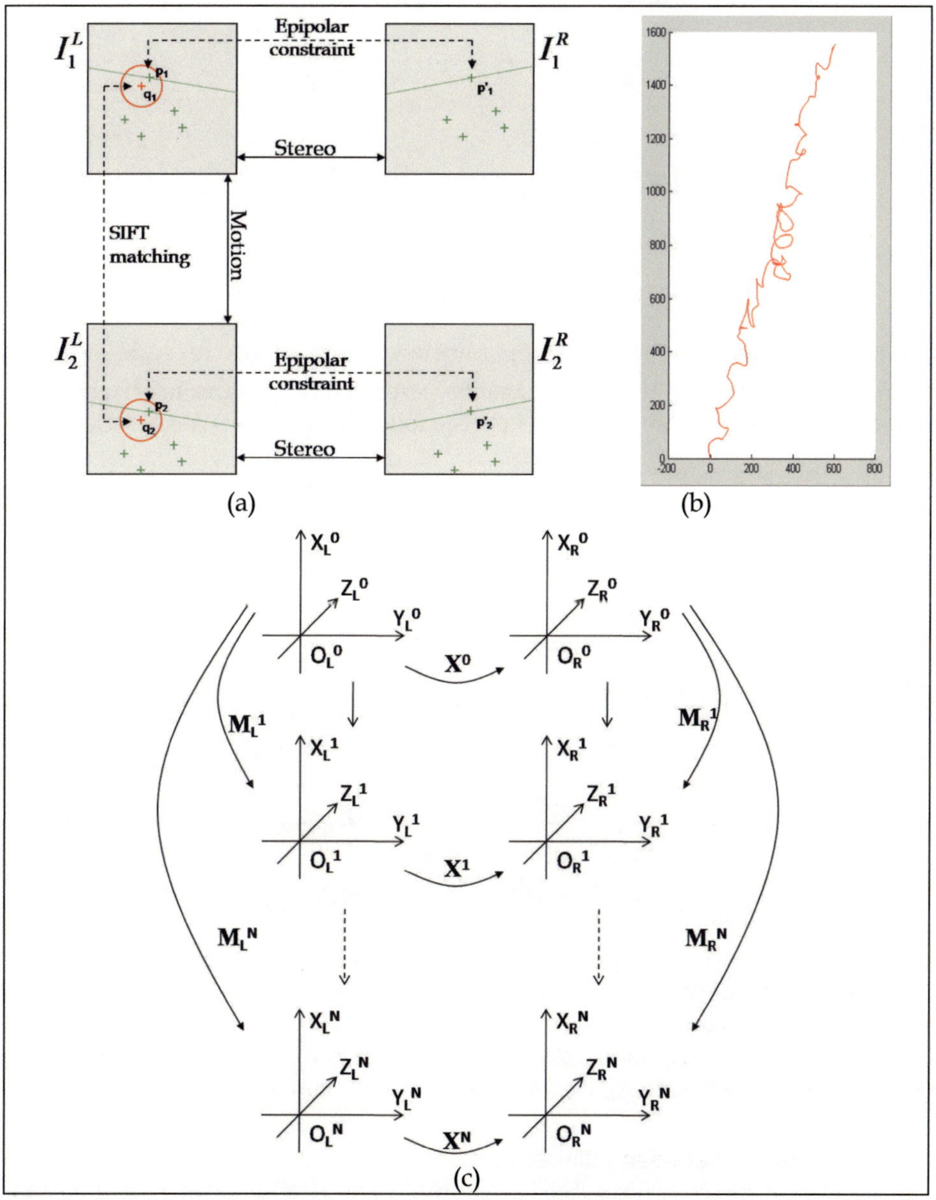

Fig. 14. *a*) Four-frame correspondences estimation scheme; b) Estimated ego-motion trajectory; *c*) The schematic diagram of a moving stereo camera.

between consecutive frames I_1^L and I_2^L are estimated by using the SIFT feature-based matching (see the next subsection). Afterwards, the nearest neighbors p_1 and p_2 to q_1 and q_2 respectively, are selected inside the epipolar constraint. As final result, four matching points p_1, p'_1 and p_2, p'_2 are defined. The scale factor can be determined by using only motion correspondence and the epipolar constrain as explained in the following discussion. Let $\mathbf{R}_x^i$, $\mathbf{R}_L^j$, $\mathbf{R}_R^j$ be rotation matrices and let $\mathbf{t}_x^i$, $\mathbf{t}_L^j$, $\mathbf{t}_R^j$ be translational not zero vectors (for i = 0, …, N and j = 1, …, N). Let be defined the following matrices:

$$\mathbf{X}^i = \begin{pmatrix} \mathbf{R}_x^i & \mathbf{t}_x^i \\ 0 & 1 \end{pmatrix}, \qquad i = 0, \dots, \mathrm{N} \tag{16}$$

$$\mathbf{M}_L^j = \begin{pmatrix} \mathbf{R}_L^j & \mathbf{t}_L^j \\ 0 & 1 \end{pmatrix}, \ \mathbf{M}_R^j = \begin{pmatrix} \mathbf{R}_R^j & \mathbf{t}_R^j \\ 0 & 1 \end{pmatrix}, \qquad j = 1, \dots, \mathrm{N} \tag{17}$$

Let $\bar{\mathbf{t}}_L^j = \mathbf{t}_L^j / \left\| \mathbf{t}_L^j \right\|$ and $\bar{\mathbf{t}}_R^j = \mathbf{t}_R^j / \left\| \mathbf{t}_R^j \right\|$ for j = 1, …, N. Under these assumptions $\mathbf{R}_L^j$, $\bar{\mathbf{t}}_L^j$ and $\mathbf{R}_R^j$, $\bar{\mathbf{t}}_R^j$ can be considered as the motion parameters obtained up to scale using motion correspondence for the left and right cameras respectively. Let the left/right coordinate systems of a moving stereo camera (Fig.14.c). The following relations are satisfied:

$$\mathbf{M}_L^i \mathbf{X}^i = \mathbf{X}^0 \mathbf{M}_R^i, \qquad i = 1, \dots, \mathrm{N} \tag{18}$$

Let $s_L = \left\| \mathbf{t}_L^1 \right\|$ and $s_R = \left\| \mathbf{t}_R^1 \right\|$, the Eq.18 can be rewritten as:

$$\begin{cases} \mathbf{R}_L^i \mathbf{R}_x^i = \mathbf{R}_x^0 \mathbf{R}_R^i \\ \mathbf{R}_L^i \mathbf{t}_x^i + s_L \bar{\mathbf{t}}_L^i = s_R \mathbf{R}_x^0 \bar{\mathbf{t}}_R^i + \mathbf{t}_x^0 \end{cases} \qquad i = 1, \dots, \mathrm{N} \tag{19}$$

The Eq.19 can be rewritten in matricial form as in the following Eq.20:

$$\begin{pmatrix} \bar{\mathbf{t}}_L^1 & -\mathbf{R}_x^0 \bar{\mathbf{t}}_R^1 \\ \vdots & \vdots \\ \bar{\mathbf{t}}_L^{\mathrm{N}} & -\mathbf{R}_x^0 \bar{\mathbf{t}}_R^{\mathrm{N}} \end{pmatrix} \begin{pmatrix} s_L \\ s_R \end{pmatrix} = \begin{pmatrix} \mathbf{t}_x^0 - \mathbf{R}_L^1 \mathbf{t}_x^1 \\ \vdots \\ \mathbf{t}_x^0 - \mathbf{R}_L^{\mathrm{N}} \mathbf{t}_x^{\mathrm{N}} \end{pmatrix} \tag{20}$$

The system in Eq.20 gives a unique solution except in a degenerate motion set as proven in (Kim & Chung, 2006). Given the 4-frame matching points and the scale factors estimated as mentioned above, two corresponding 3-D point sets, X_1 and X_2, are calculated between the two following stereo correspondences. Hence, ego-motion is estimated by evaluating 3-D point displacements from X_1 to X_2. Fig.14.b shows the estimated navigation trajectory.

6.3 Correspondences between adjacent frames

Scale Invariant Feature Transform (SIFT) feature (Lowe, 1999) is used as feature matching method, reducing the effect of outlier by using the Least Median of Square (LMedS) method (Zhang et al., 1995). Generally, feature detection methods, such as the Harris detector

(Harris & Stephens, 1988), are sensitive to the affine distortion of image. Therefore, they are not suitable to build feature sets in image acquired in uncontrolled environments. SIFT feature matching (Lowe, 1999) is widely used because it is invariant to affine transforms. These characteristics are suitable to be employed with imagery obtained at different camera angles by using ROV/AUV. SIFT feature algorithm is based upon finding locations within the scale space of an image. Features are identified by detecting maxima and minima in the *Difference of Gaussian* (DOG) pyramidal space. A subpixel location, scale and orientation are associated with each SIFT feature. In order to achieve high specificity, a local feature is formed by measuring the local image gradients at many orientations in coordinates relative to the location, scale and orientation of the feature. Although the SIFT feature matching algorithm has low bad matching error rate, if the outliers are presented in estimation of transformation matrix, the recovered camera motion is incorrect and it is impossible to register the model correctly. Therefore, the LMedS approach is employed and the amount of sample is estimated by using the following relation:

$$P = 1 - (1 - (1 - \varepsilon)^p)^m, \tag{21}$$

where P is the probability of a good sample for LMedS, and ε, p, m denote the ratio of false matched, the sample size and the number of sample required. Choosing m = 108 with $\varepsilon = 0.5$ and p=4 in Eq.21, the probability of a good sample is 99.9%.

6.4 Registration refinement

After a pre-alignment phase based on ego-motion estimation, Iterated Closest Point (ICP) algorithm (Besl & McKay, 1992) is employed for 3-D registration refinement. The registration activity maps each single scan reconstruction into the same coordinate system, solving the absolute orientation problem when correspondences between each single scan reconstruction and the others are unknown. For each iteration ICP algorithm alternates the following two step: 1) calculation of the closest points between reconstructions and assuming this points as correspondent, 2) overlap of the reconstructions by determining the right transformation using absolute orientation. Given two reconstructed points set X_1 and X_2 subject to registration, the above absolute orientation referred in step (2) is resolved through the following minimization using the Kanatani's method (Kanatani, 1993):

$$\min_{R,t} \sum_{i=1}^{N} \left\| x_{1,i} - \left(R\, x_{2,i} + t \right) \right\|^2 , \tag{22}$$

where (R, t) is the sought transformation (rotation and translation), and (x_{1i}, x_{2i}) is the closest point pair determined in step (1). ICP iterations proceed until the overlapping error doesn't go down a given threshold. The final 3-D mosaic is obtained processing thousands of single scan reconstruction. The final mosaic reported in Fig.15 concerns the reconstruction of a wide underwater area that contains seven columns (only five are visible) in cipolin marble dating to the Roman age.

8. Conclusion and future work

A framework for seabed 3-D mosaic reconstruction has been presented. The three mainly troublesome aspects discussed are asynchronous stereo acquisition, depth estimation and the 3-D mosaic registration. The use of an inexpensive asynchronous stereo sequence is explained, suggesting a new scheme for accurate calibration frames selection. Moreover, disparity map calculation in both points of view, local and global, is considered. Brightness constancy violation is treated adopting cross-correlation and histogram equalization. Results are evaluated by using ground truth data. Moreover, a four-frame features tracking scheme for ego-motion estimation has been suggested, combining epiflow advantages and ICP registration refinements. Current ongoing and future work involve improvements on ICP module, more accurate ground truth results evaluation and the implementation of the Extended Kalman Filter-based ego-motion and structure recovery for off-line 3-D mosaicking.

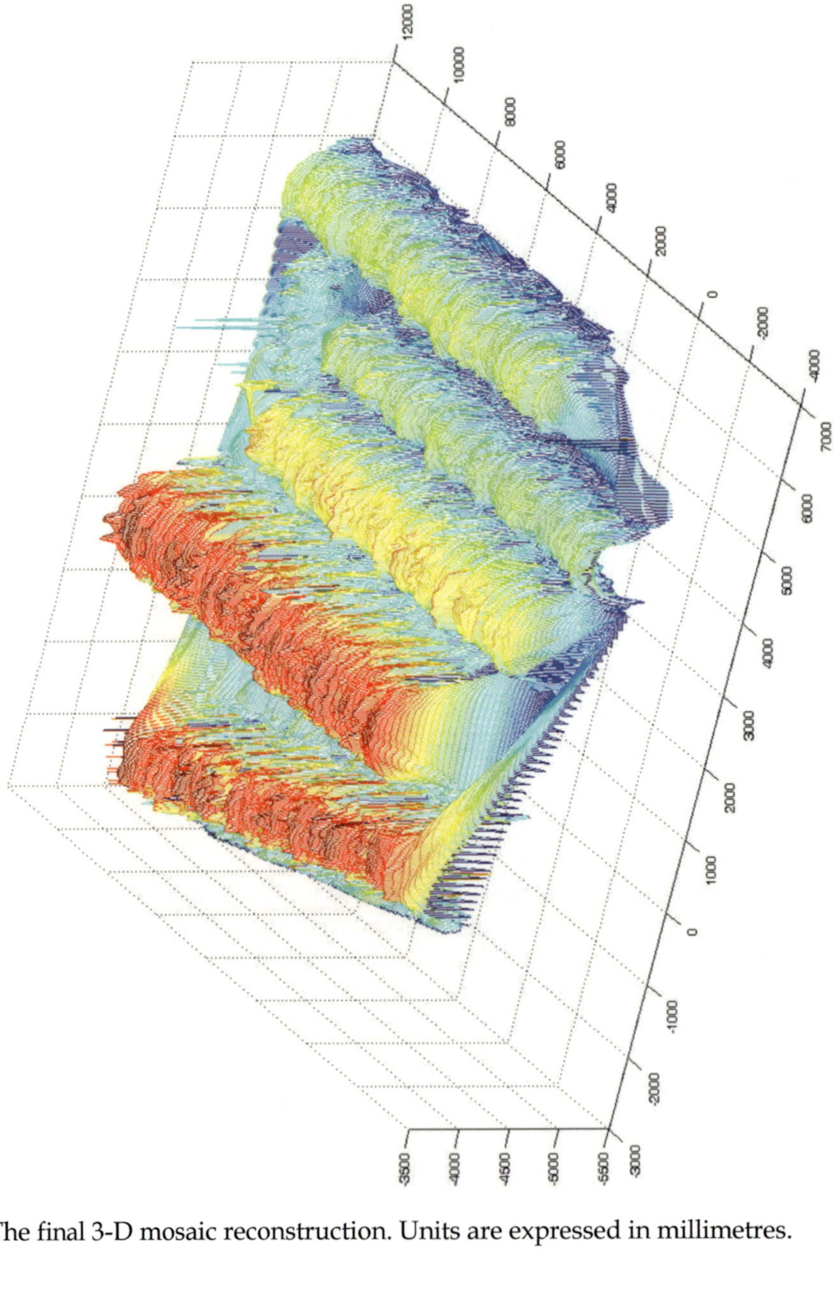

Fig. 15. The final 3-D mosaic reconstruction. Units are expressed in millimetres.

9. References

Bakstein, H. (1999). A Complete DLT-Based Camera Calibration with a Virtual 3D Calibration Object, Diploma Thesis, Charles University, Prague

Besl, P. & McKay, N. (1992). A method for registration of 3D shapes, IEEE Trans. Pattern Analysis and Machine Intelligence, Vol. 14, pp. 239-256, 0162-8828

Bolles, R. ; Baker, H. & Hannah, M. (1993). The JISCT stereo evaluation, In Proc. DARPA Image Understanding Workshop, pp. 263-274, 1993

Boykov, Y.; Veksler, O. & Zabih, R. (1998). Markov Random Fields with Efficient Approximations. IEEE CVPR, p. 648, 1998

Boykov, Y.; Veksler, O. & Zabih, R. (2001). Fast approximate energy minimization via graph cuts. *IEEE Trans. Pattern Analysis and Machine Intelligence*, Vol. 23, pp. 1222-1239

Castellani, U.; A. Fusiello, Murino, V., Papaleo, L., Puppo, E., Repetto, S. & Pittore, M. (2004). Efficient On-line Mosaicing from 3D Acoustical Images. Oceans, MTS/IEEE TECHNO-OCEAN, Vol. 2, pp. 670–677, 9-12 Nov 2004

Faugeras, O. (1992). What can be seen in three dimensions with an uncalibrated stereo rig? In Proc. European Conference on Computer Vision, pp. 563-578, Santa Margherita, Italy, May 1992

Harris, C. & Stephens, M.J. (1988). A combined corner and edge detector, In Alvey Vision Conference, pages 147–152, 1988

Hartley, R. & Zisserman, A. (2003). *Multiple View Geometry In Computer Vision*, Cambridge University Press, Second Edition, 2003

Heikkia, J. & Solven, O. (1996). Calibration Procedure for Short Focal Length Off-the-shelf CCD Cameras, In Proc. Of the 13th International Conference on Pattern Recognition, pp. 166 - 170, Vienna, Austria, 1996

Hogue, A.; German, A. & M. Jenkin. (2007). Underwater environment reconstruction using stereo and inertial data, Systems Man and Cybernetics ISIC IEEE International Conference on, pp. 2372-2377, 7-10 Oct. 2007

Horn, B. (1986). *Robot Vision*, MIT Press, 1986

Kanatani, K. (1993). *Geometric Computation for Machine Vision*. Oxford University Press, 1993

Khamene, A.; Madjidi, H. & Negahdaripour, S. (2001). 3-D Mapping of Sea Floor Scenes by Stereo Imaging, Oceans, MTS/IEEE Conference and Exhibition, Vol. 4, pp. 2576-2583, 5-8 Nov. 2001

Khamene, A. & Negahdaripour, S. (2003). Motion and structure from multiple cues; image motion, shading flow, and stereo disparity, *Computer Vision and Image Understanding*, Vol. 90, pp. 99-127, May 2003

Kim, J.H. & Chung M.J. (2006). Absolute motion and structure from stereo image sequences without stereo correspondence and analysis of degenerate cases, *Pattern Recognition*, Vol. 39, No. 9, pp. 1649-1661, 0031-3203

Kutulakos, K. & Seitz, S. (1998). What do N photographs tell us about 3D shape? Technical Report TR680, Computer Science Department, University of Rochester, January 1998

Kolmogorov, V. & Zabih, R. (2002). Multi-camera scene Reconstruction via Graph Cuts, European Conference on Computer Vision, 2002

Kolmogorov, V. & Zabih, R. (2004). What energy functions can be minimized via graph cuts?, *Transactions on Pattern Analysis and Machine Intelligence*, Vol. 26, pp. 147-159

Laurentini, A. (1994). The visual hull concept for silhouette-based image understanding, *IEEE Trans. Pattern Analysis and Machine Intelligence*, Vol. 16, No. 2, pp. 150-162

Lorenson, W. (1987). Marching Cubes: A High Resolution 3D Surface Construction Algorithm, *Computer Graphics*, Vol. 21, No. 4, pp. 163-169

Lowe, D.G. (1999). Object recognition from local scale-invariant features, IEEE Proc. ICCV, pp. 1150-1157, 1999

Lucas, B. & Kanade, T. (1981). An iterative image registration technique with an application to stereo vision, In Proc. International Joint Conference on Artificial Intelligence, pp. 674-679, Vancouver, Canada, August 1981

Madjidi, H. & Negahdaripour, S. (2005). Global alignment of sensor positions with noisy motion estimates, *IEEE Trans. Robotics and Automation*, Vol. 21, December 2005

Narasimhan, S.G.; Nayar, S.K., Sun, B. & Koppal, S.J. (2005). Structured Light in Scattering Media, Proceedings of the Tenth IEEE International Conference on Computer Vision, Vol. 1, pp. 420-427, 17-21 Oct. 2005

Negahdaripour, S.; Zhang, H. & Han, X. (2002). Investigation of Photometric Stereo Method for 3-D Shape Recovery from Underwater Imagery, Oceans MTS/IEEE, Vol. 2, pp. 1010-1017, 39-31 Oct. 2002

Nicosevici, T.; Negahdaripour, S. & Garcia, R. (2005). Monocular-based 3D seafloor reconstruction and ortho-mosaicing by piecewise planar representation, In Proc. OCEANS MTS/IEEE Conference, Washington, DC, September 2005

Pizer, S.M.; Ambum, E.P., Austin, J.D., Cromartie, R., Geselowitz, A., Greer, T., Romeny, BM.H., Zimmerman, J.B., & Zuiderveld, K. (1987). Adaptive Histogram Equalization and Its Variations, Comp. Vis. Graphics & Im. Proc., pp. 1355-368

Potts, R. (1952). Some generalized order-disorder transformation, In Proceedings of the Cambridge Philosophical Society, Vol. 48, pp. 106-109, 1952

Scharstein, D. & Szeliski, R. (2002). A taxonomy and evaluation of dense two-frame stereo correspondence algorithms, Stereo and Multi-Baseline Vision, IEEE Workshop, pp. 131-140, 9-10 Dec 2002

Slabaugh, G.; Culbertson, W., Malzbender, T. & Schafer, R. (2001). A survey of volumetric scene reconstruction methods from photographs, In Proc. International Workshop on Volume Graphics, pp. 81-100, Stony Brook, NY, June 2001

Sun, J.; Shum, H. & Zheng, N. (2002). Stereo matching using belief propagation, In Proc. European Conference on Computer Vision, pp. 510-524, Copenhagen, Denmark, May/June 2002

Svedman, M.; Goncalves, L., Karlsson, N., Munich, M. & Pirjanian, P. (2005). Structure from Stereo Vision using Unsynchronized Cameras for Simultaneous Localization and Mapping, International Conference on Intelligent Robots and Systems, Intelligent Robots and Systems, IEEE/RSJ, pp. 3069-3074, 2-6 Aug. 2005

Tomasi, C. & Kanade, T. (1991). *Shape and Motion from Image Streams: a Factorization Method - Part 3 Detection and Tracking of Point Features*, Computer Science Department, Carnegie Mellon University, April, 1991

Tsai, R.Y. (1987). A versatile camera calibration technique for high-accuracy 3D machine vision metrology using off-the-shelf TV cameras and lenses, *IEEE Journal of Robotics and Automation RA*, Vol. 3, No. 4, pp. 323 – 344

Zhang, Z.; Deriche, R., Faugeras, O. & Luong, Q. (1995). A robust technique for matching two uncalibrated images through the recovery of the unknown epipolar geometry, *Artificial Intelligence*, Vol. 78, pp. 87-119, 1995

Zhang, H. (2005). *Automatic sensor platform positioning and 3-d target modelling from underwater stereo sequences*, PhD Thesis, Coral Gables, Florida, Dec 2005

12

Moving Target Tracking of Omnidirectional Robot with Stereo Cameras

Jun Ming Kuang, Ming Liu and Xiang Lin
Department of Electrical and Computer Systems Engineering
Monash University, VIC 3800
Australia

1. Introduction

The omnidirectional mobile robots have attracted great attention in last twenty years. Their major advantage superior to the traditional car-like mobile robots, whose motion is subject to the nonholonomic constraints, lies on the feature that their linear and rotational motions can be simultaneously and independently carried out. A typical mechanical structure of omnidirectional robot utilizes three universal driving wheels which can be either driven or slid along any direction [2]. This special motion pattern significantly simplifies the path planning and motion control tasks.

For the motion control of omnidirectional mobile robots, the majority of work so far, however, only consider the kinematic model for motion control. This is equivalent to assuming that the robots are massless bodies and therefore can ideally respond to the input motion control commands, which indeed does not reflect the real situation especially for heavy and fast moving robots. Due to this reason, efforts have been made to develop precise dynamic model to improve robots' performance [1]-[3]. In recent years, the visual servoing which combines control application with vision systems has become a hot topic. However, only a few studies [12], [13] combined the vision system with omnidirectional robot which has excellent maneuverability.

In this study we integrate the motion control with stereo vision for target tracking. Unlike previous work [4], [9], in which control laws were based on image space using the inverse image Jacobian matrix for motion estimation, our algorithm is to directly estimate the relative position and velocity errors in Euclidean space. By using visual feedback, a simple PD controller is proposed. The PD control ensures the asymptotic stability of the tracking error if the target is still which is equivalent to the case of docking, or of the bounded-input-bounded-output (BIBO) stability if the target is moving with varying but bounded rotational and linear velocities.

The paper is organized as follows: Section 2 presents the robot motion equation and tracking task. The dynamic model of the overall system is given in Section 3. The vision based relative motion and tracking error estimation is introduced in Section 4. The PD control law is proposed in Section 5. As a practical justification, the experimental tracking results obtained are given in Section 6. Finally, conclusions and future work are presented in Section 7.

2. Robot dynamics and tracking task definition

2.1 Dynamics of the robot

Considering an omnidirectional mobile robot [2], its configuration space is a smooth 3-manifold and can then be locally embedded in Euclidean space $\mathbf{R}^3$. The robot has three degrees of freedom, i.e. two dimensional linear motion and one dimensional rotational motion. There are three universal wheels mounted along the edge of the robot chassis 120° apart from each other, and each wheel has a set of rollers aligned with its rim, as shown in Figure 1. Because of its special mechanism, the robot is able to simultaneously rotate and translate. Therefore, the path planning can be significantly simplified by directly defining the tracking task with the orientation and position errors obtained by the visual feedback.

On the top of the robot, as indicated in Figure 2, a pair of cameras, C_1 and C_2, are mounted. Let $\mathcal{E} = \{O_e, x_e, y_e\}$ be the earth coordinate frame and $\mathcal{R} = \{O_r, x_r, y_r\}$ be the coordinate frame fixed on the robot body whose origin is located at the mass center of the robot. The orientation of the coordinate frame of the left camera, $\mathcal{C} = \{O_c, x_c, y_c\}$, is assumed to be consistent with that of the robot, i.e. the rotation matrix between those two frames is an identity, but a translation $\mathbf{t}_r^c$ between them exists, as illustrated in Figure 2.

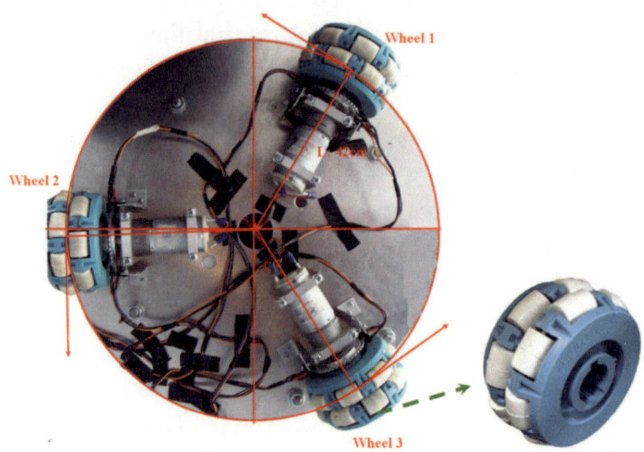

Fig. 1. The configuration of the wheel system

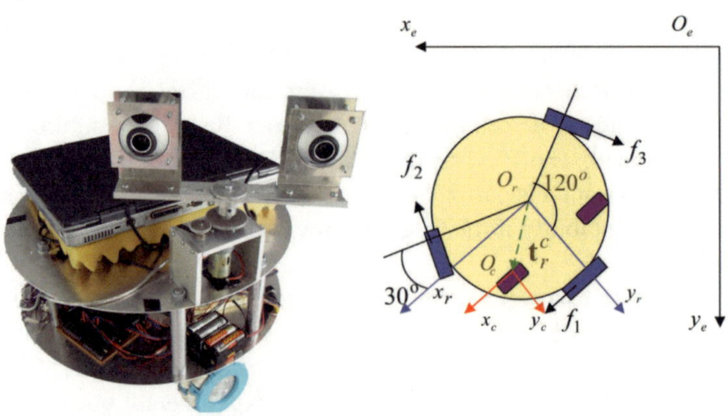

Fig. 2. Top view of omnidirectional mobile robot

Given this geometry layout, the configuration of the robot is given by $(\mathbf{R}_e^r(\theta_e^r), \mathbf{t}_e^r)$ in the earth frame, which can also be locally expressed as follows:

$$\mathbf{q}_r = \begin{bmatrix} \mathbf{t}_e^r \\ \theta_e^r \end{bmatrix} \in \mathcal{E} \tag{1}$$

where $\mathbf{t}_e^r = [x_e^r, y_e^r]^T$ are the position coordinates of the robot in the earth frame and θ_e^r is the relative rotation angle between the robot frame and the earth frame, $\mathcal{E}$.
As shown in [2], the dynamic model of the robot is given by:

$$\mathbf{M}\ddot{\mathbf{q}}_r = \mathbf{F} \tag{2}$$

where $\mathbf{M} = diag\{m, m, I\} > 0$ is the moment matrix in which $m > 0$ and $I > 0$ are the mass and inertia of the robot body, respectively; $\mathbf{F} = [F_x, F_y, \tau_\theta]_T$ is the force vector which includes the driving forces and the rotation torque applied along the directions of the axes, x_e and y_e, in the earth frame and about the mass center of the robot, respectively. Let $\bar{\mathbf{f}} = [f_x, f_y, f_\tau]^T$ be the forces expressed in the coordinate frame of the robot, $\mathbf{F}$ in (2) is revised as follows:

$$\mathbf{F} = \begin{bmatrix} \mathbf{R}_e^r(\theta_e^r) & 0 \\ 0 & 1 \end{bmatrix} \bar{\mathbf{f}} = \bar{\mathbf{R}}\bar{\mathbf{f}}$$

where

$$\mathbf{R}_e^r(\theta_e^r) = \begin{bmatrix} \cos(\theta_e^r) & -\sin(\theta_e^r) \\ \sin(\theta_e^r) & \cos(\theta_e^r) \end{bmatrix}$$

is the rotation matrix of the robot with respect to $\mathcal{E}$. Furthermore, let the forces generated from the three universal wheels be $\mathbf{f} = [f_1, f_2, f_3]^T$, and let the clockwise rotation of the robot viewed from the top of the robot be the positive direction as indicated in Figure 2, the following relation holds

$$\bar{\mathbf{f}} = \mathbf{T}_p\mathbf{f}$$

in which

$$\mathbf{T}_p = \begin{bmatrix} 1 & -1/2 & -1/2 \\ 0 & -\sqrt{3}/2 & \sqrt{3}/2 \\ r & r & r \end{bmatrix}$$

is the force projection matrix, r is the radius of the chassis, and $\sqrt{3}/2 = \cos 30°$ and $1/2 = \sin 30°$. Hence, robot model (2) becomes

$$\mathbf{M}\ddot{\mathbf{q}}_r = \bar{\mathbf{R}}\mathbf{T}_p\mathbf{f}. \tag{3}$$

In this model, the sliding friction of the wheels has been ignored. To handle it, a controller with varying control gains will be applied as those will be shown late.

2.2 Tracking task definition

To define the motion control task, Figure 2 is extended to Figure 3 that includes a moving target T to be tracked by the robot. $\mathcal{T} = \{O_t, x_t, y_t\}$ is denoted as the coordinate frame attached to the target. Let $(\mathbf{R}_e^t(\theta_e^t), \mathbf{t}_e^t)$ be the configuration of the target in the earth frame, which can also be locally expressed by a 3D vector

$$\mathbf{q}_t = \begin{bmatrix} \mathbf{t}_e^t \\ \theta_e^t \end{bmatrix} \in \mathcal{E}. \tag{4}$$

In $\mathcal{T}$ there are n fixed vectors $\mathbf{p}_i = [p_{xi}, p_{yi}]^T$, for $i = 1, 2, \ldots, n$, indicating the locations of a group of feature points on the target as shown in Figure 3.

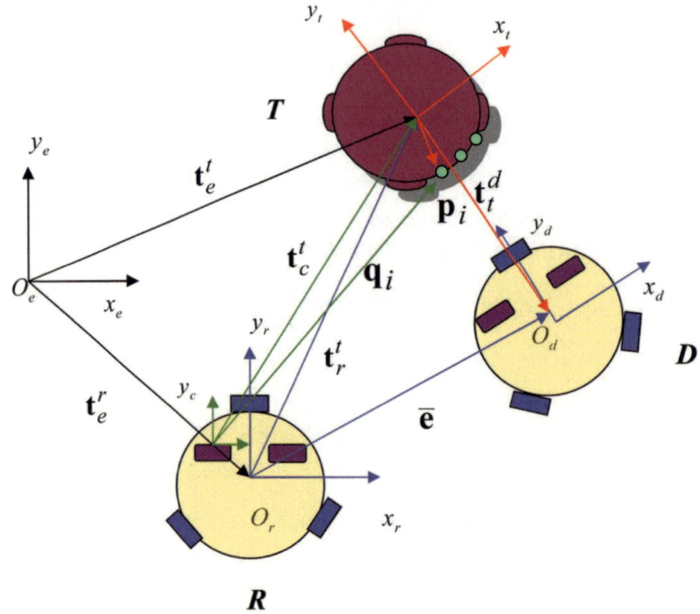

Fig. 3. Geometry layout of omnidirectional mobile robot, the target to be tracked and the destination configuration of the robot

A destination frame also called ideal configuration $\mathcal{D} = \{O_d, x_d, y_d\}$ is further specified and obtained by a constant translation $\mathbf{t}_t^d = [x_t^d, y_t^d]^T$ and a constant rotation angle θ_t^d from the target frame, as shown in Figure 3. The configuration of $\mathcal{D}$ expressed in $\mathcal{E}$ is then given by

$$\mathbf{q}_d = \begin{bmatrix} \mathbf{R}_e^t(\theta_e^t)\mathbf{t}_t^d + \mathbf{t}_e^t \\ \theta_t^d + \theta_e^t \end{bmatrix} \in \mathcal{E}. \tag{5}$$

The control objective is to control the omnidirectional robot, using the feedback from the vision system which grabs the images, recognizes the target and features, and supplies the estimated tracking errors, to track the target in such a way that the motion errors between the robot and configuration $\mathcal{D}$ are driven to be as small as possible.

3. Dynamic models of the overall system

Since the omnidirectional robot is driven by three DC motors, their dynamics are needed to be taken into account for the overall system modeling.

3.1 Motor dynamics and gear transmissions

Based on the mechanical and electrical characteristics of the motors, and denoting $\mathbf{J}'$, $\mathbf{B}'$, $\mathbf{R}'_m$, $\mathbf{L}'$, $\mathbf{K}_m$ and $\mathbf{K}_e$ as the inertia moment matrix, the viscous damping matrix, the resistance matrix, the inductance matrix, the torque and the back EMF coefficient matrices, respectively, the model of three identical motors is given by:

$$\begin{cases} \mathbf{J}'\dot{\mathbf{W}}_m + \mathbf{B}'\mathbf{W}_m = \mathbf{K}_m\mathbf{I}_c - \mathbf{T}' \\ \mathbf{E} - \mathbf{K}_e\mathbf{W}_m = \mathbf{L}'\dot{\mathbf{I}}_c + \mathbf{R}'_m\mathbf{I}_c \end{cases} \tag{6}$$

where $\mathbf{J}' = J\mathbf{I}$, $\mathbf{B}' = B\mathbf{I}$, $\mathbf{L}' = L\mathbf{I}$, $\mathbf{R}'_m = R_m\mathbf{I}$, $\mathbf{K}_m = k_m\mathbf{I}$ and $\mathbf{K}_e = k_e\mathbf{I}$ are positive definite matrices accordingly. $\mathbf{W}_m = [w_1, w_2, w_3]_T$ is the 3×1 vector of the angular velocities of the motors, $\mathbf{I}_c$ is the current vector of the motors, $\mathbf{T}'$ is the driven torque vector of the motor shafts and $\mathbf{E}$ is the input voltage vector.

Since the motors employed are quite small and their rotating speeds are rather slow (due to the slow motion of the target), their inductances $\mathbf{L}'$ and back EMF's, $\mathbf{E}_{emf}$ can be neglected. Thus, (6) can be simplified as:

$$\begin{cases} \mathbf{J}'\dot{\mathbf{W}}_m + \mathbf{B}'\mathbf{W}_m = \mathbf{K}_m\mathbf{I}_c - \mathbf{T}' \\ \mathbf{E} = \mathbf{R}'_m\mathbf{I}_c \end{cases} . \tag{7}$$

Motion mapping from the motors to robot is via gear boxes. The relation between the driven torque on the motor shaft side and that of a wheel shaft side, can be illustrated in Figure 4. The torque relation is given by $\tau'_i = f_i r_m = n f_i r_w = n\tau_i$ where τ'_i is the driven torque on a motor shaft, τ_i is that on wheel shaft and $n = r_m / r_w < 1$ is the gear ratio between the two shafts. As three motors have the same gear boxes, the relation can be expressed in the vector form:

$$\mathbf{T}' = n\mathbf{T} \tag{8}$$

where $\mathbf{T} = [\tau_1, \tau_2, \tau_3]^T > 0$ is the driving torque vector.

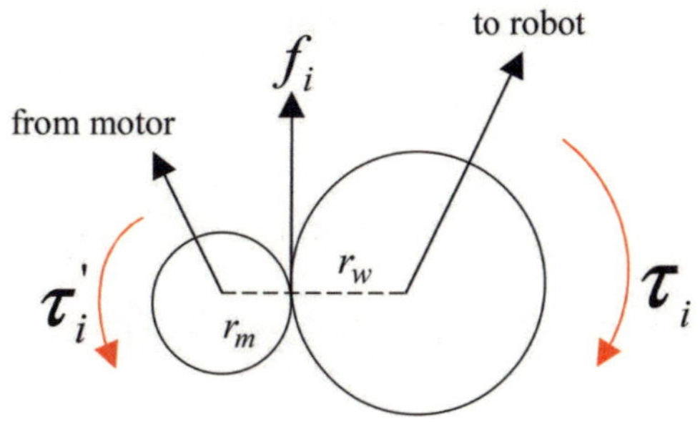

Fig. 4. Mechanism of gearing

Let v_{lx} and v_{ly} be robot's linear velocities along the x and y axes of its frame, w_r be its rotation angular velocity, and w_{m1}, w_{m2} and w_{m3} be the angular velocities of the three motors, the following relationships hold

$$\begin{cases} v_{lx} = nr'_w(w_{m1} - \frac{1}{2}w_{m2} - \frac{1}{2}w_{m3}) \\ v_{ly} = nr'_w(-\frac{\sqrt{3}}{2}w_{m2} + \frac{\sqrt{3}}{2}w_{m3}) \\ w_r = nr'_w(\frac{1}{3r}w_{m1} + \frac{1}{3r}w_{m2} + \frac{1}{3r}w_{m3}) \end{cases} \tag{9}$$

where r'_w is the radius of the wheel. (9) can be rewritten as

$$\mathbf{V} = nr'_w \mathbf{T}_v \mathbf{W}_m \tag{10}$$

where $\mathbf{T}_v$ is a constant mapping matrix given by:

$$\mathbf{T}_v = \begin{bmatrix} 1 & -1/2 & -1/2 \\ 0 & -\sqrt{3}/2 & \sqrt{3}/2 \\ 1/3r & 1/3r & 1/3r \end{bmatrix}.$$

Therefore the relationships between the linear and angular velocities of the robot and the angular velocities of the three motors are given by

$$\dot{\mathbf{q}}_r = \bar{\mathbf{R}}\mathbf{V} = nr'_w \bar{\mathbf{R}} \mathbf{T}_v \mathbf{W}_m, \tag{11}$$

or

$$\mathbf{W}_m = (nr'_w)^{-1} \mathbf{T}_v^{-1} \bar{\mathbf{R}}^{-1} \dot{\mathbf{q}}_r. \tag{12}$$

According to (12), motor motion equation (7) becomes:

$$\begin{cases} (nr'_w)^{-1}\mathbf{J}'\mathbf{T}_v^{-1}\bar{\mathbf{R}}^{-1}\ddot{\mathbf{q}}_r + (nr'_w)^{-1}\mathbf{T}_v^{-1}(\mathbf{J}'\dot{\bar{\mathbf{R}}}^{-1} + \mathbf{B}'\bar{\mathbf{R}}^{-1})\dot{\mathbf{q}}_r = \mathbf{K}_m\mathbf{I}_c - \mathbf{T}' \\ \mathbf{E} = \mathbf{R}'_m\mathbf{I}_c \end{cases}. \tag{13}$$

3.2 The overall model

As $\mathbf{f} = n^{-1}\mathbf{R}_w^{-1}\mathbf{T}'$, (3) becomes:

$$\mathbf{M}\ddot{\mathbf{q}}_r = n^{-1}\bar{\mathbf{R}}\mathbf{T}_p\mathbf{R}_w^{-1}\mathbf{T}', \tag{14}$$

where $\mathbf{R}_w = diag\{r'_w, r'_w, r'_w\} > 0$ is the wheel radius matrix. By combining (13) and (14), the overall dynamic model including both motors and robot body then is given by

$$\begin{cases} (nr'_w)^{-1}\mathbf{J}'\mathbf{T}_v^{-1}\bar{\mathbf{R}}^{-1}\ddot{\mathbf{q}}_r + (nr'_w)^{-1}\mathbf{T}_v^{-1}(\mathbf{J}'\dot{\bar{\mathbf{R}}}^{-1} + \mathbf{B}'\bar{\mathbf{R}}^{-1})\dot{\mathbf{q}}_r = \mathbf{K}_m\mathbf{I}_c - n\mathbf{M}\mathbf{R}_w\mathbf{T}_p^{-1}\bar{\mathbf{R}}^{-1}\ddot{\mathbf{q}}_r \\ \mathbf{I}_c = \mathbf{R}'^{-1}_m\mathbf{E} \end{cases}. \tag{15}$$

It can be rearranged as

$$\begin{cases} ((nr'_w)^{-1}\mathbf{J}'\mathbf{T}_v^{-1} + n\mathbf{M}\mathbf{R}_w\mathbf{T}_p^{-1})\bar{\mathbf{R}}^{-1}\ddot{\mathbf{q}}_r + (nr'_w)^{-1}\mathbf{T}_v^{-1}(\mathbf{J}'\dot{\bar{\mathbf{R}}}^{-1} + \mathbf{B}'\bar{\mathbf{R}}^{-1})\dot{\mathbf{q}}_r = \mathbf{K}_m\mathbf{I}_c \\ \mathbf{I}_c = \mathbf{R}'^{-1}_m\mathbf{E} \end{cases}. \tag{16}$$

Further let

$$\mathbf{A} = ((nr'_w)^{-1}\mathbf{J}'\mathbf{T}_v^{-1} + n\mathbf{M}\mathbf{R}_w\mathbf{T}_p^{-1})\bar{\mathbf{R}}^{-1} \tag{17}$$

and

$$\mathbf{B} = (nr'_w)^{-1}\mathbf{T}_v^{-1}(\mathbf{J}'\dot{\bar{\mathbf{R}}}^{-1} + \mathbf{B}'\bar{\mathbf{R}}^{-1}), \tag{18}$$

the combined model equation, (16), can then be simplified as

$$\mathbf{A}\ddot{\mathbf{q}}_r + \mathbf{B}\dot{\mathbf{q}}_r = \mathbf{K}_m \mathbf{R}_m'^{-1} \mathbf{E}. \tag{19}$$

In real applications, motors **E** is generated by PWM (Pulse-Width Modulation) motor driver from a motion control microprocessor. As the relationships between the PWM signal and voltages **E** are almost linear, the **E** can be controlled by the PWM signals outputted from controller by a constant gain.

4. Relative motion estimation and tracking errors

In this section we present the estimation of relative position and velocity between the robot and target using machine vision. The outcome will be used for robot to carry out tracking task.

4.1 Rotation estimation

Since the moving target tracking is performed in 2-D space, The parameters needed to be estimated are rotation angle $\theta_e = \theta_e^t - \theta_e^r$, and 2-D translation vector t_c between the robot and target, as shown in Figure 5.

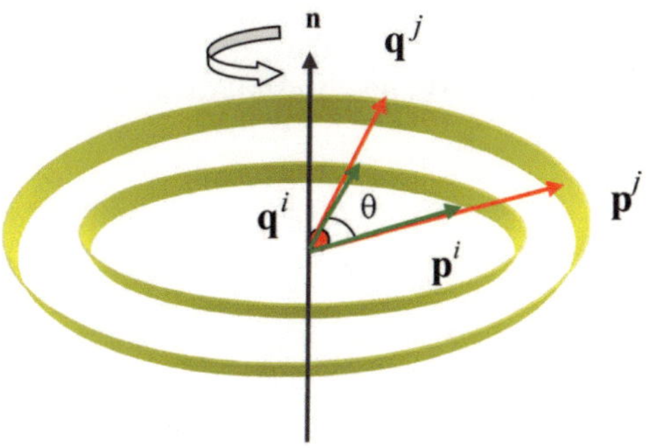

Fig. 5. The relationship between the feature vectors and the rotation angle

To solve θ_e, the feature points detected in the images are first recovered from the image coordinates back to two dimensional coordinates $\{\mathbf{q}_i\}$ expressed in the coordinate frame of the left camera by employing the reconstruction algorithm [14].

Expressed in the camera frame at the *k-th* step of the observation, the relative motion of the *i-th* feature point, $\mathbf{p}_i$, on the target is formulated by the following kinematic equation:

$$\mathbf{q}^i(k) = \mathbf{R}_c^t(\theta_e(k))\mathbf{p}^i + \mathbf{t}_c^t(i,k), \ (i = 1, 2, \cdots, n) \tag{20}$$

where $\mathbf{p}_i$ is the known coordinate of the *ith* feature point predefined in the coordinate frame of the target, $\mathbf{q}_i(k)$ is a two dimensional coordinate in the coordinate frame of the left camera, and $\mathbf{t}_c^t(k)$ is the translation between the origins of the target and the left camera viewed from C. Considering another feature point, $\mathbf{p}_j$, and subtracting the projection equation of $\mathbf{p}_j$ from (20), the rotation is decoupled from the translation, as given by the following equation:

$$\mathbf{q}^i(k) - \mathbf{q}^j(k) = \mathbf{R}_c^t(\theta_e(k))(\mathbf{p}^i - \mathbf{p}^j), \quad i \neq j. \tag{21}$$

(21) indicates that the feature vector, $(\mathbf{p}_i\text{-}\mathbf{p}_j)$, on the target becomes the one, $(\mathbf{q}_i(k)\text{-}\mathbf{q}_j(k))$, after the target rotates by angle $\theta_e(k)$ with respect to the left camera.

Given observation $\mathbf{q}_i(k)$ and $\mathbf{q}_j(k)$ and predefined feature point $\mathbf{p}_i$ and $\mathbf{p}_j$, the following relationship holds:

$$(\mathbf{p}^i - \mathbf{p}^j) \cdot (\mathbf{q}^i(k) - \mathbf{q}^j(k)) = |\mathbf{p}^i - \mathbf{p}^j||\mathbf{q}^i(k) - \mathbf{q}^j(k)| \cos\theta_e(k),$$

where $(\cdot)$ represents the inner product. Therefore rotation angle $\theta_e(k)$ can be resolved as:

$$\theta_e(k) = \arccos \frac{(\mathbf{p}^i - \mathbf{p}^j) \cdot (\mathbf{q}^i(k) - \mathbf{q}^j(k))}{|\mathbf{p}^i - \mathbf{p}^j||\mathbf{q}^i(k) - \mathbf{q}^j(k)|}.$$

As there are $s = n(n - 1) / 2$ such feature vectors. The optimal estimation of $\theta_e(k)$ can be obtained by using the following equation:

$$\theta_e(k) = \arccos \frac{1}{s} \sum_{i=1}^{n-1} \sum_{j=i+1}^{n} \frac{(\mathbf{p}^i - \mathbf{p}^j) \cdot (\mathbf{q}^i(k) - \mathbf{q}^j(k))}{|\mathbf{p}^i - \mathbf{p}^j||\mathbf{q}^i(k) - \mathbf{q}^j(k)|}. \tag{22}$$

In addition, the sign of the rotation angle is decided using the 2D polar frame.

4.2 Translation estimation

After rotation angle $\theta_e(k)$ is obtained, rotation matrix $\mathbf{R}_c^t(\theta_e(k))$ can be constructed as follows:

$$\mathbf{R}_c^t(\theta_e(k)) = \begin{bmatrix} \cos(\theta_e(k)) & -\sin(\theta_e(k)) \\ \sin(\theta_e(k)) & \cos(\theta_e(k)) \end{bmatrix}.$$

Consequently, the translation between the origins of the target and the left camera is calculated using the following equation:

$$\mathbf{t}_c^t(i,k) = \mathbf{q}^i(k) - \mathbf{R}_c^t(\theta_e(k))\mathbf{p}^i.$$

By defining the following estimation error:

$$E \stackrel{\text{def}}{=} \sum_{i=1}^{n} (\mathbf{t}_c^t(i,k) - \mathbf{t}_c^t(k))^2.$$

The optimal solution $\mathbf{t}_c^t(k)$ that minimizes E can be obtained as the solution of $dE/dt = 2\sum_{i=1}^{n}(\mathbf{t}_c^t(i,k) - \mathbf{t}_c^t(k)) = 0$ which is

$$\mathbf{t}_c^t(k) = \frac{1}{n} \sum_{i=1}^{n} \mathbf{t}_c^t(i,k). \tag{23}$$

4.3 Relative velocity estimation

In addition to the relative orientation and position, the relative velocity between the robot and target is also required by the proposed controller. Consequently, the angle velocity is given by

$$w_e(k) = \frac{\theta_e(k) - \theta_e(k-1)}{T_i} \tag{24}$$

and the translation velocity is calculated by

$$\dot{\mathbf{t}}_k = \frac{\mathbf{t}(k) - \mathbf{t}(k-1)}{T_i} \tag{25}$$

where T_i is the sampling time.

4.4 Tracking error

The relative orientation, position and velocity between the left camera and the target have been calculated by (22)-(25). In addition, tracking error e_r and its derivative between the robot and ideal configuration $\mathcal{D}$ at which the robot is expected to arrive are also calculated in this subsection. For convenience, in the sequel, step index k in (22)-(25) is removed.

4.4.1 Relative tracking error in the frame of the robot

As the orientation difference between the target and configuration $\mathcal{D}$ is a given constant angle, θ_t^d, and the orientation of the left camera is the same as that of the robot, the relative angle between the robot and configuration $\mathcal{D}$ is

$$e_\theta = \theta_e + \theta_t^d$$

in which θ_e is given in (22). Therefore, given the predefined $\mathbf{t}_t^d = [x_t^d, y_t^d]^T$ which is the position coordinate of $\mathcal{D}$ in the coordinate frame of the target, the linear position error between the origins of the coordinate frames of the robot and $\mathcal{D}$ can be obtained as follows:

$$\bar{\mathbf{e}} = \begin{bmatrix} \bar{e}_x \\ \bar{e}_y \end{bmatrix} = \mathbf{R}_r^t(\theta_e)\mathbf{t}_t^d + \mathbf{t}_r^t. \tag{26}$$

As the robot frame has the same orientation as that of the left camera, $\mathbf{t}_r^t$ is obtained by the following equation:

$$\mathbf{t}_r^t = \mathbf{t}_c^t + \mathbf{t}_r^c$$

where $\mathbf{t}_c^t$ is given by (23) and $\mathbf{t}_r^c$ is a constant, as illustrated in Figure 2.
Therefore the 3D position error in the coordinate frame of the robot, i.e. $\mathbf{e}_r$, is given by

$$\mathbf{e}_r = \begin{bmatrix} \bar{e}_x \\ \bar{e}_y \\ e_\theta \end{bmatrix} = \begin{bmatrix} \mathbf{R}_r^t(\theta_e)\begin{bmatrix} x_t^d \\ y_t^d \end{bmatrix} + \begin{bmatrix} x_r^t \\ y_r^t \end{bmatrix} \\ \theta_e + \theta_t^d \end{bmatrix}. \tag{27}$$

Since configurations $\mathcal{T}$ and $\mathcal{D}$ have the same angular velocity, the velocity error between $\mathcal{D}$ and $\mathcal{R}$ is the same as that between the target and the robot. Taking the derivative of (26), it has that

$$\begin{aligned} \dot{\bar{\mathbf{e}}} &= \dot{\mathbf{R}}_r^t(\theta_e)\mathbf{t}_t^d + \dot{\mathbf{t}}_r^t \\ &= w_e \begin{bmatrix} -\sin\theta_e & -\cos\theta_e \\ \cos\theta_e & -\sin\theta_e \end{bmatrix} \begin{bmatrix} x_t^d \\ y_t^d \end{bmatrix} + \begin{bmatrix} \dot{x}_r^t \\ \dot{y}_r^t \end{bmatrix}. \end{aligned} \tag{28}$$

where w_e is the relative angular velocity between the target and the robot given in (24). Consequently, the corresponding 3D velocity tracking error, $\dot{\mathbf{e}}_r$, is given by

$$\dot{\mathbf{e}}_r = \begin{bmatrix} \dot{\bar{\mathbf{e}}} \\ \dot{e}_\theta \end{bmatrix} = \begin{bmatrix} \dot{\bar{\mathbf{e}}} \\ w_e \end{bmatrix} \tag{29}$$

where $\dot{\bar{\mathbf{e}}}$ is given in (28).

4.5 Relative tracking error in the earth frame

In last subsection, the relative tracking error between the robot and configuration $\mathcal{D}$ expressed in the frame of the robot has been presented. Indeed, those quantities can also be expressed in other frames such as the earth frame, $\mathcal{E}$. Recalling the notations of the robot and the target configuration given in (4), ideal configuration $\mathcal{D}$ can be written as a 3D vector in $\mathcal{E}$ as

$$\mathbf{q}_d = \begin{bmatrix} \mathbf{R}_e^t(\theta_e^t)\mathbf{t}_t^d + \mathbf{t}_e^t \\ \theta_e^t + \theta_t^d \end{bmatrix}, \tag{30}$$

and its velocity and acceleration are

$$\dot{\mathbf{q}}_d = \begin{bmatrix} \dot{\mathbf{R}}_e^t(\theta_e^t)\mathbf{t}_t^d + \dot{\mathbf{t}}_e^t \\ \dot{\theta}_e^t \end{bmatrix} \tag{31}$$

and

$$\ddot{\mathbf{q}}_d = \begin{bmatrix} \ddot{\mathbf{R}}_e^t(\theta_e^t)\mathbf{t}_t^d + \ddot{\mathbf{t}}_e^t \\ \ddot{\theta}_e^t \end{bmatrix}. \tag{32}$$

Given these the relative motion errors can be expressed in earth frame $\mathcal{E}$ as follows:

$$\mathbf{e}_e = \mathbf{q}_d - \mathbf{q}_r = \begin{bmatrix} \mathbf{R}_e^t(\theta_e^t)\mathbf{t}_t^d + \mathbf{t}_e^t - \mathbf{t}_e^r \\ \theta_e^t + \theta_t^d - \theta_e^r \end{bmatrix} = \begin{bmatrix} \bar{\mathbf{e}}_e \\ e_\theta \end{bmatrix} \tag{33}$$

$$\dot{\mathbf{e}}_e = \dot{\mathbf{q}}_d - \dot{\mathbf{q}}_r = \begin{bmatrix} \dot{\mathbf{R}}_e^t(\theta_e^t)\mathbf{t}_t^d + \dot{\mathbf{t}}_e^t - \dot{\mathbf{t}}_e^r \\ \dot{\theta}_e^t - \dot{\theta}_e^r \end{bmatrix} = \begin{bmatrix} \dot{\bar{\mathbf{e}}}_e \\ \dot{e}_\theta \end{bmatrix}. \tag{34}$$

The relation between $\mathbf{e}_e$ in $\mathcal{E}$ and $\mathbf{e}_r$ in $\mathcal{R}$ is then given by

$$\mathbf{e}_e = \begin{bmatrix} \mathbf{R}_e^r\bar{\mathbf{e}} \\ e_\theta \end{bmatrix} = \bar{\mathbf{R}}\mathbf{e}_r \tag{35}$$

where

$$\bar{\mathbf{R}} = \begin{bmatrix} \mathbf{R}_e^r & 0 \\ 0 & 1 \end{bmatrix} \tag{36}$$

is a generalized 3×3 rotation matrix. The relative velocity error in $\mathcal{E}$ is:

$$\dot{\mathbf{e}}_e = \begin{bmatrix} \dot{\mathbf{R}}_e^r \bar{\mathbf{e}} + \mathbf{R}_e^r \dot{\bar{\mathbf{e}}} \\ \dot{e}_\theta \end{bmatrix} = \dot{\bar{\mathbf{R}}}\mathbf{e}_r + \bar{\mathbf{R}}\dot{\mathbf{e}}_r \tag{37}$$

which indicates the relation between the relative velocity errors in the two frames, i.e. the earth frame and the frame of the robot.

5. PD control

5.1 Control law design

To direct the robot to track the target, a motion controller is requested. For simplicity, a PD control law is designed for the tracking task. It will be shown that the proposed PD control can ensure the asymptotic stability of the tracking error if the target is still which is equivalent to the case of docking, or the bounded-input-bounded-output (BIBO) stability if the target is moving with varying but bounded rotational and linear velocities.

As $\mathbf{A}$ is obviously a nonsingular matrix given in (19), it can be rearranged to be:

$$\ddot{\mathbf{q}}_r + \mathbf{A}^{-1}\mathbf{B}\dot{\mathbf{q}}_r = \mathbf{A}^{-1}\mathbf{K}_m\mathbf{R}_m^{\prime -1}\mathbf{E}. \tag{38}$$

Changing the sign of (38) and then adding $\ddot{\mathbf{q}}_d + \mathbf{A}^{-1}\mathbf{B}\dot{\mathbf{q}}_d$ to both sides of it, it has that

$$\ddot{\mathbf{q}}_d - \ddot{\mathbf{q}}_r + \mathbf{A}^{-1}\mathbf{B}(\dot{\mathbf{q}}_d - \dot{\mathbf{q}}_r) = \ddot{\mathbf{q}}_d + \mathbf{A}^{-1}\mathbf{B}\dot{\mathbf{q}}_d - \mathbf{A}^{-1}\mathbf{K}_m\mathbf{R}_m^{\prime -1}\mathbf{E}. \tag{39}$$

Recalling Equation (33), (39) can be revised as

$$\ddot{\mathbf{e}}_e + \mathbf{A}^{-1}\mathbf{B}\dot{\mathbf{e}}_e = \ddot{\mathbf{q}}_d + \mathbf{A}^{-1}\mathbf{B}\dot{\mathbf{q}}_d - \mathbf{A}^{-1}\mathbf{K}_m\mathbf{R}_m^{\prime -1}\mathbf{E} \tag{40}$$

which can be regarded as a second order linear system in terms of error states, $\dot{\mathbf{e}}_e$ and $\mathbf{e}_e$. We need to design a control law using $\mathbf{E}$ to make the tracking errors as small as possible as $t \to \infty$. Our target is to have (40) be the form of a typical second order linear system as:

$$\ddot{\mathbf{e}}_e + \mathbf{K}_v\dot{\mathbf{e}}_e + \mathbf{K}_p\mathbf{e}_e = \mathbf{u} \tag{41}$$

where $\mathbf{K}_v$ and $\mathbf{K}_p$ are ideal controller parameter matrices to be determined, and $\mathbf{u}$ is an equivalent input signal of the system. Ideally, if we can have $\mathbf{u} \equiv 0$, then $\dot{\mathbf{e}}_e, \mathbf{e}_e \to 0$ as $t \to \infty$, that is, the asymptotic stability can be achieved. However, due to the fact that $\ddot{\mathbf{q}}_d$ and $\dot{\mathbf{q}}_d$ are unmeasurable, $\mathbf{u}$ can not be zero unless $\ddot{\mathbf{q}}_d = \dot{\mathbf{q}}_d = 0$.

Let the constant parameter matrices, $\mathbf{K}_v$ and $\mathbf{K}_p$ be the diagonal matrices, i.e. $\mathbf{K}_v = diag\{k_{v1}, k_{v2}, k_{v3}\} > 0$ and $\mathbf{K}_p = diag\{k_{p1}, k_{p2}, k_{p3}\} > 0$ where k_{vi} and k_{pi} (i = 1, 2, 3) are given positive constants, i.e. the control gains. Thus, the model, (40), can be rewritten as:

$$\ddot{\mathbf{e}}_e + \mathbf{K}_v\dot{\mathbf{e}}_e + \mathbf{K}_p\mathbf{e}_e = \ddot{\mathbf{q}}_d + \mathbf{A}^{-1}\mathbf{B}\dot{\mathbf{q}}_d - \mathbf{A}^{-1}\mathbf{K}_m\mathbf{R}_m^{\prime -1}\mathbf{E} - \mathbf{A}^{-1}\mathbf{B}\dot{\mathbf{e}}_e + \mathbf{K}_v\dot{\mathbf{e}}_e + \mathbf{K}_p\mathbf{e}_e. \tag{42}$$

As $\ddot{\mathbf{q}}_d$ and $\dot{\mathbf{q}}_d$ on the right-hand side of (42) depend on the acceleration and the velocity of the target and are unmeasurable in practice, they are diffcult to be compensated, the control algorithm can only be applied to compensate other components, expressed as:

$$\mathbf{A}^{-1}\mathbf{K}_m\mathbf{R}_m^{\prime -1}\mathbf{E} = -\mathbf{A}^{-1}\mathbf{B}\dot{\mathbf{e}}_e + \mathbf{K}_v\dot{\mathbf{e}}_e + \mathbf{K}_p\mathbf{e}_e. \tag{43}$$

Therefore, the control law can be determined as:

$$\mathbf{E} = \mathbf{R}'_m \mathbf{K}_m^{-1}(\mathbf{A}(\mathbf{K}_v \dot{\mathbf{e}}_e + \mathbf{K}_p \mathbf{e}_e) - \mathbf{B}\dot{\mathbf{e}}_e). \tag{44}$$

Substituting matrix $\mathbf{A}$ in (17) and matrix $\mathbf{B}$ in (18) back to (44), and performing manipulation, control law (44) can be revised as:

$$\mathbf{E} = \mathbf{R}'_m \mathbf{K}_m^{-1}(\mathbf{C}\bar{\mathbf{R}}^{-1}(\mathbf{K}_v \dot{\mathbf{e}}_e + \mathbf{K}_p \mathbf{e}_e) - (nr'_w)^{-1}\mathbf{T}_v^{-1}(\mathbf{J}'\dot{\bar{\mathbf{R}}}^{-1} + \mathbf{B}'\bar{\mathbf{R}}^{-1})\dot{\mathbf{e}}_e) \tag{45}$$

where $\mathbf{C} = (nr'_w)^{-1}\mathbf{J}'\mathbf{T}_v^{-1} + n\mathbf{M}\mathbf{R}_w\mathbf{T}_p^{-1}$ is a constant matrix.

Recalling $\mathbf{e}_e = \bar{\mathbf{R}}\mathbf{e}_r$ in (35) and $\dot{\mathbf{e}}_e = \dot{\bar{\mathbf{R}}}\mathbf{e}_r + \bar{\mathbf{R}}\dot{\mathbf{e}}_r$ in (37) and applying $\dot{\bar{\mathbf{R}}}\mathbf{e}_r = \bar{\mathbf{R}}\bar{\mathbf{R}}^T\dot{\bar{\mathbf{R}}}\mathbf{e}_r = \bar{\mathbf{R}}\bar{\mathbf{S}}(w_r)\mathbf{e}_r$ to (45), finally it has that

$$\mathbf{E} = \mathbf{Q}(\mathbf{D}(\bar{\mathbf{S}}(w_r)\mathbf{e}_r + \dot{\mathbf{e}}_r) - \mathbf{U}(\mathbf{e}_r + \bar{\mathbf{S}}(w_r)^{-1}\dot{\mathbf{e}}_r) + \mathbf{C}\mathbf{K}_p\mathbf{e}_r) \tag{46}$$

where $\mathbf{Q} = \mathbf{R}'_m\mathbf{K}_m^{-1}$, $\mathbf{D} = \mathbf{C}\mathbf{K}_v - (nr'_w)^{-1}\mathbf{B}'\mathbf{T}_v^{-1}$ and $\mathbf{U} = (nr'_w)^{-1}\mathbf{J}'\mathbf{T}_v^{-1}$ are constant matrices and

$$\bar{\mathbf{S}}(w_r) = \begin{bmatrix} 0 & -w_r & 0 \\ w_r & 0 & 0 \\ 0 & 0 & 0 \end{bmatrix}$$

is a skew matrix of the angular velocity of the robot.

Although the robot tracking is formulated in the earth frame as shown in (42), (46) is a controller based on the robot frame, in which $\mathbf{e}_r$ and $\dot{\mathbf{e}}_r$ are both expressed in the frame of the robot obtained by on-board vision system. w_r, the angular velocity of the robot, can be obtained by three on-board wheel encoders using (9).

5.2 Overall system stability analysis

Given designed control law (46) error model (42) can then be simplified as:

$$\ddot{\mathbf{e}}_e + \mathbf{K}_v\dot{\mathbf{e}}_e + \mathbf{K}_p\mathbf{e}_e = \ddot{\mathbf{q}}_d + \mathbf{A}^{-1}\mathbf{B}\dot{\mathbf{q}}_d. \tag{47}$$

To analyze the stability of the overall system given in (47), the similar procedure mentioned in [15] is followed. Denote $\mathbf{u} = \ddot{\mathbf{q}}_d + \mathbf{A}^{-1}\mathbf{B}\dot{\mathbf{q}}_d$, (47) can be rewritten as the typical second order linear system:

$$\ddot{\mathbf{e}}_e + \mathbf{K}_v\dot{\mathbf{e}}_e + \mathbf{K}_p\mathbf{e}_e = \mathbf{u} \tag{48}$$

where $\mathbf{u}$ is the input of the system. Certainly if $\ddot{\mathbf{q}}_d, \dot{\mathbf{q}}_d \to 0$, $\mathbf{e}_e \to 0$ and therefore $\mathbf{q}_r = \mathbf{q}_d$, i.e. the robot catches up with ideal configuration $\mathcal{D}$. This is clearly corresponding to the docking case of a still target.

However, in most cases, $\mathbf{u}$ usually is not zero as $\ddot{\mathbf{q}}_d$ and $\dot{\mathbf{q}}_d$ are not zero. Thus, it is necessary to analyze the stability of the system using BIBO stability. Since simplified error model (48) represents a linear second order decoupled differential equation with input $\mathbf{u}$, the expressions will be developed for the L_∞ gain of the operations $H : \mathbf{u} \mapsto \mathbf{e}_e$ and $G : \mathbf{u} \mapsto \dot{\mathbf{e}}_e$.

Consider the L_∞ gain of the operator $H_i : u_i \rightarrow e_{ei}$, ($i$ = 1, 2, 3), where u_i denotes the *ith* component of the input vector **u** and the same definition is given to other vectors, and write the transfer function from input u_i to output e_{ei} as:

$$\frac{e_{ei}(s)}{u_i(s)} = h_i(s) = \frac{1}{s^2 + k_{vi}s + k_{pi}}. \tag{49}$$

As such, consider the L_∞ gain of the operator $G_i : u_i \rightarrow \dot{e}_{ei}$ and write the transfer function from input u_i to output $\dot{e}_{ei}$ as:

$$\frac{\dot{e}_{ei}(s)}{u_i(s)} = g_i(s) = \frac{s}{s^2 + k_{vi}s + k_{pi}}. \tag{50}$$

Then, the two transfer functions operators can be bounded as:

$$\begin{aligned} \|H_i\|_\infty &= \int_0^\infty |h_i(t)|dt, \\ \|G_i\|_\infty &= \int_0^\infty |g_i(t)|dt \end{aligned} \tag{51}$$

where $h_i(t)$ and $g_i(t)$ are the impulse response of the transfer functions, (49) and (50), respectively.

For simplicity the critical damping case where $k_{vi}^2 = 4k_{pi}$ is considered. Evaluating the responses (51) leads to the results:

$$\begin{aligned} \|H_i\|_\infty &= k_{pi}^{-1}, \\ \|G_i\|_\infty &= 4(ek_{vi})^{-1}. \end{aligned} \tag{52}$$

Then, bounds on the multi-input multi-output (MIMO) gains are given as:

$$\begin{aligned} \|H\|_\infty &= \mathbf{K}_p^{-1} = \alpha_1, \\ \|G\|_\infty &= 4(e\mathbf{K}_v)^{-1} = \alpha_2. \end{aligned} \tag{53}$$

Therefore, for zero initial condition, the results are:

$$\begin{aligned} \|\mathbf{e}_e\|_\infty &\leqq \alpha_1\|\mathbf{u}\|_\infty, \\ \|\dot{\mathbf{e}}_e\|_\infty &\leqq \alpha_2\|\mathbf{u}\|_\infty. \end{aligned} \tag{54}$$

From (54), if $\mathbf{u} \in L_\infty$, it is clear that $\mathbf{e}_e, \dot{\mathbf{e}}_e \in L_\infty$. Then for $\mathbf{u} = \ddot{\mathbf{q}}_d + \mathbf{A}^{-1}\mathbf{B}\dot{\mathbf{q}}_d$, if assume that a smooth, bounded desired trajectory is specified so that $\mathbf{q}_d$, $\dot{\mathbf{q}}_d$ and $\ddot{\mathbf{q}}_d$ are elements of L_∞, it is clear that:

$$\|\mathbf{u}\|_\infty \leqq \|\ddot{\mathbf{q}}_d\|_\infty + \beta_1\|\dot{\mathbf{q}}_d\|_\infty \tag{55}$$

where $\beta_1 = \|\mathbf{A}^{-1}\mathbf{B}\|_\infty$. As the matrices, **A** and **B**, only involve the constant matrices and the orthogonal matrix, $\bar{\mathbf{R}}$ whose norm is 1, the following results can be obtained:

$$\begin{aligned} \|\mathbf{e}_e\|_\infty &\leqq \alpha_1(\|\ddot{\mathbf{q}}_d\|_\infty + \beta_1\|\dot{\mathbf{q}}_d\|_\infty), \\ \|\dot{\mathbf{e}}_e\|_\infty &\leqq \alpha_2(\|\ddot{\mathbf{q}}_d\|_\infty + \beta_1\|\dot{\mathbf{q}}_d\|_\infty). \end{aligned} \tag{56}$$

The results given in (56) indicate that if $\dot{\mathbf{q}}_d$ and $\ddot{\mathbf{q}}_d \in L_\infty$, $\mathbf{e}_e$ and $\dot{\mathbf{e}}_e \in L_\infty$. That is, bounded input ($\dot{\mathbf{q}}_d$ and $\ddot{\mathbf{q}}_d$) results in bounded output ($\mathbf{e}_e$ and $\dot{\mathbf{e}}_e$). From these results, it is certain that the system satisfies the requirement of BIBO stability and the control law designed is theoretically effective.

6. Experimental verification

6.1 Experimental setup

The omnidirectional robot built up for tracking experiment mainly includes three software modules, namely, the image processing module, the pose estimation module and the control module. The first two are implemented in Microsoft Visual C++ 6.0 which operates on a laptop with a CPU of Intel 7250 at 1.60GHz and system memory of 512 MB, while the control module is implemented in MPLAB v7.01 and burned into microprocessor PIC18F452.

The communications between the laptop and the PIC is via a USB-based interface. The functions of the interface are to perform the signal conversion between the USB signals and three separate serial signals, to manage the interrupts and to successfully maintain the communications. Figure 6 shows the configuration of the real-time platform on which the image processing module and the pose estimation module are run.

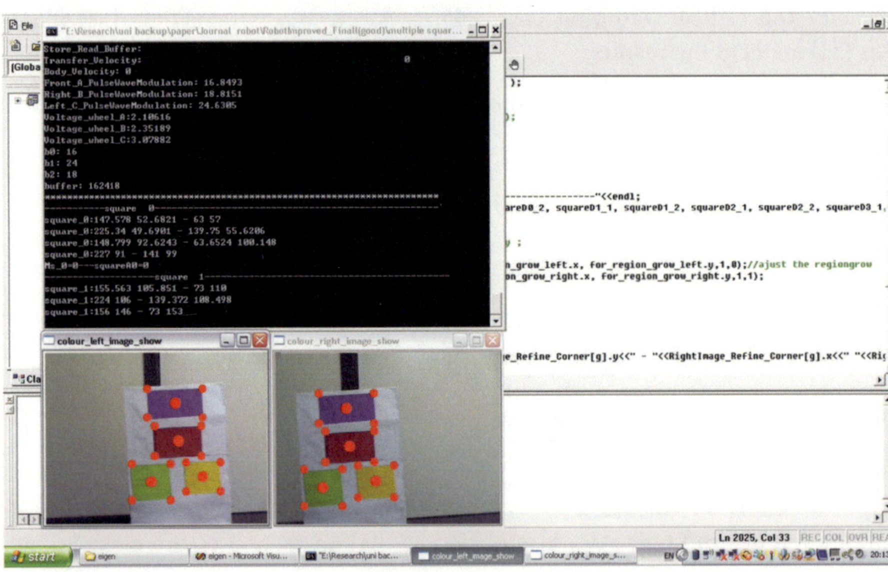

Fig. 6. The real-time platform of the experiment

The hardware of the robot can be roughly categorized into the mechanical module and the electronic module. The mechanical module includes the body of the robot which is constructed by two armor plates which have a radius of 0.21 meter, a pair of web-cameras, Logitech Pro5000, fixed on the top of the robot, and three universal wheels mounted along the edge of the robot chassis, 120° apart from each other and driven by three 12V DC geared motors. In addition, the speeds of the three wheels are measured by quadrature wheel encoders. Each encoder consists of a quadrature encoder pattern segment and a photo interrupter, ZD1901, as presented in Figure 7.

Fig. 7. The wheel encoder

The electronic module mainly involves three motor speed controllers NMIH-0050 Full H-Bridge which drive the motors to run in bidirection and vary their speeds by adjusting the PWM signals, and the PCB module which integrates all the electronic components utilized in the robot.

6.2 Image processing for target recognition

The image processing is employed for target recognition after stereo images are captured in real time. In order to simplify the image processing, a planar pattern which involves four rectangles in different colors is proposed in the tracking task, as shown in Figure 8. Prior to other parts of image processing, the captured images are first undistorted based on the lens distortion coefficients obtained by the camera calibration. Furthermore, as there are a number of existing irrelevant objects in the experimental environment, the target interested is extracted by segmentation approach. A segmentation scheme based on HSI color space (Hue, Saturation and Intensity) is applied to decouple the grey level component from the ones carrying the color information. Therefore, the effect of the light intensity is reduced, which results in a more robust segmentation.

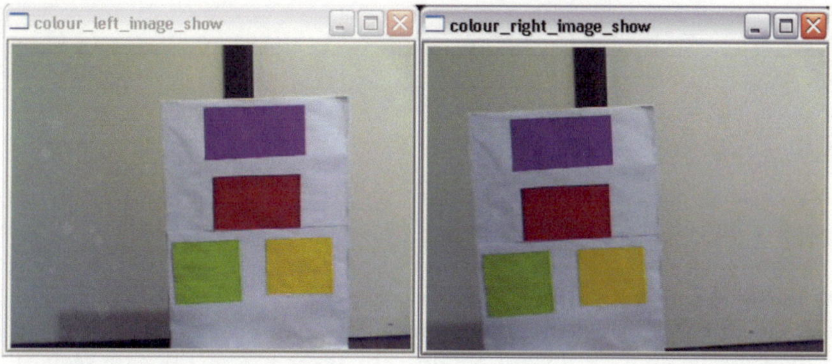

Fig. 8. The proposed target for tracking

To extract the regions of interest which include four rectangles respectively, Region Growing Segmentation, which groups pixels or subregions into larger regions with predefined criteria, is employed based on the color information of the rectangles.

Color information alone is not robust enough to recognize the target, as some objects which exclude the target may have similar colors. In order to reject the irrelevant objects which have the same colors as the target, shape information of the target is also introduced. As shown in Figure 8, the target includes four blocks which are rectangular. Based on this information, the objects which are not rectangular are rejected. By means of edge detector, the edges of the remaining objects can be directly obtained.

The edges of the four rectangles are straight lines. Therefore, the Hough transform is applied to the detected edges in order to find out the straight lines, and the contour-based shape representation approach is employed to describe the shapes of the objects constructed by the detected straight lines. Consequently, the objects whose shapes are roughly rectangular can be regarded as the targets and accepted, otherwise they are rejected.

After the four rectangles are recognized, a corner detector is applied to the interested regions for the detection of the feature points. The detected feature points in images will be recovered back to Euclidean space by using the optimal triangulation algorithm for the estimation of the relative orientation and position between the target and the robot. The Harris corner detector in OpenCV (Open Source Computer Vision Library)[16] is adopted as our corner detector to detect the corners of the four rectangles. To enhance the accuracy of the corner detection, the so called sub-pixel refinement for the corner detection is employed which iterates to find the sub-pixel accurate locations of the corners.

The complete procedure of the proposed image processing for the target detection and recognition is summarized and illustrated in Figure 9.

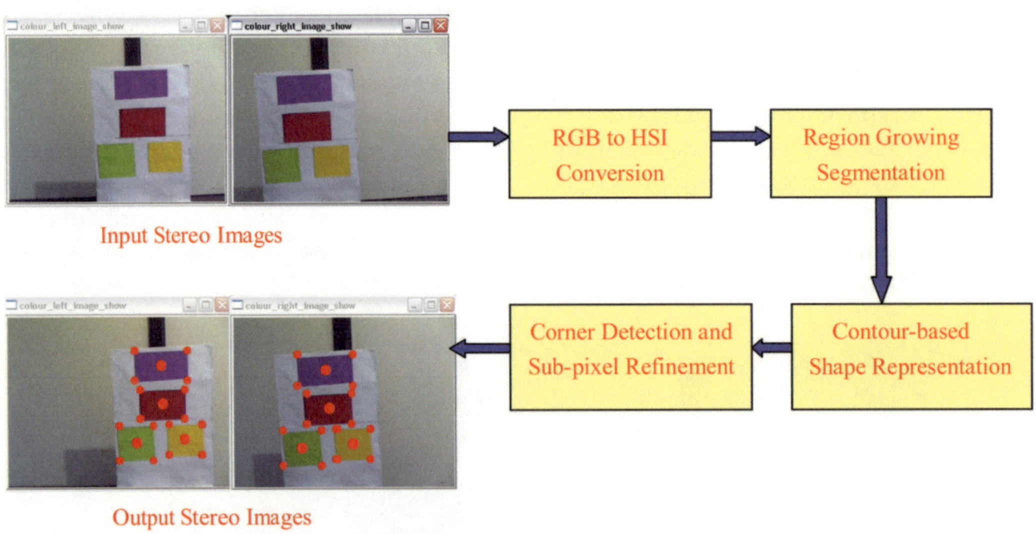

Fig. 9. The procedure of image processing

6.3 Experimental results and analysis

The experiment for the target docking and tracking is performed with the omnidirectional robot, and the experimental results are obtained by testing three docking scenarios and one tracking scenario.

The experiment is undertaken in a laboratory with a complex background. Many irrelevant objects are available in the laboratory. Natural light is used instead of using stable lighting. The planar object given in Figure 8 is employed as the target to be tracked. To record the trajectories of the robot for the visualization of the experiment, a colored chalk is fixed in the middle of the chassis during the experiment.

In addition, the relative rotation angle also called the relative orientation error, e_θ, and the position errors, $\bar{e}_x$ and $\bar{e}_y$, between the robot and ideal configuration $\mathcal{D}$ are calculated in real time and then saved into a separate file. Therefore, the change of the relative orientation error and position errors between the robot and configuration $\mathcal{D}$ during the experiment can be visualized by plotting the data from the file obtained.

The total time of one cycle through the complete control loop is about 0.05 second. The weight of the robot is around 10 Kg and its moment of inertia is about 3 Kgm^2. In addition, the relative rotation angle, θ_t^d, between the ideal configuration and the target is zero, and the linear position error between them is $\mathbf{t}_t^d = [0,-0.75]^T m$.

To avoid large overshoots or steady errors and reduce the effect of the un-modeled friction, the control gains, $\mathbf{K}_p$ and $\mathbf{K}_v$, are not fixed, that is, the control gains vary according to the relative orientation error and position errors between the robot and configuration $\mathcal{D}$. For instance, the large relative orientation error and position errors can produce large driven forces and therefore the small control gains are needed in order to avoid the large overshoot of the response or vice versa so as to reduce the steady errors of the tracking.

6.3.1 Docking scenarios

In docking scenarios, the target is still while the robot is controlled to move towards ideal configuration $\mathcal{D}$. As mentioned in the stability analysis, if the target is still, the tracking errors should satisfy the asymptotic stability, that is, the robot can finally dock itself to ideal configuration $\mathcal{D}$.

In the first docking scenario, as shown in Figure 10, the target is represented by the green star. The desired docking location, i.e. ideal configuration $\mathcal{D}$, is represented by a blue square and located at coordinate (225, 182) in the earth frame while the initial position of the robot is located at (169, 78) represented by a black circle, and the initial relative rotation angle between the robot and $\mathcal{D}$ is around zero. By observing the trajectory of the robot represented by the red line, it is clear that the position error, $\bar{\mathbf{e}}$, is driven to decrease while the movements of the robot are adjusted by the control signals. However, the position error, $\bar{e}_x$, increases while the robot goes through the range where $y = \{y_e \mid 100 \leq y_e \leq 120\}$. The reason is that the movements of the robot are affected by the friction between the wheels and the ground. Finally, the robot stops at location (221, 180) represented by the red circle and fairly close to ideal configuration $\mathcal{D}$ at (225, 182). The total experimental time is about 2.25 seconds.

For the second docking scenario, the result is illustrated in Figure 11. From the figure, it can be seen that ideal configuration $\mathcal{D}$ is at coordinate (184, 187) and the initial position of the robot is located at (275, 70). The initial relative rotation angle between the robot and ideal configuration $\mathcal{D}$ is -35°, that is, the robot needs to clockwise rotate 35° in order to coincide with the orientation of ideal configuration $\mathcal{D}$, as shown in Figure 12.

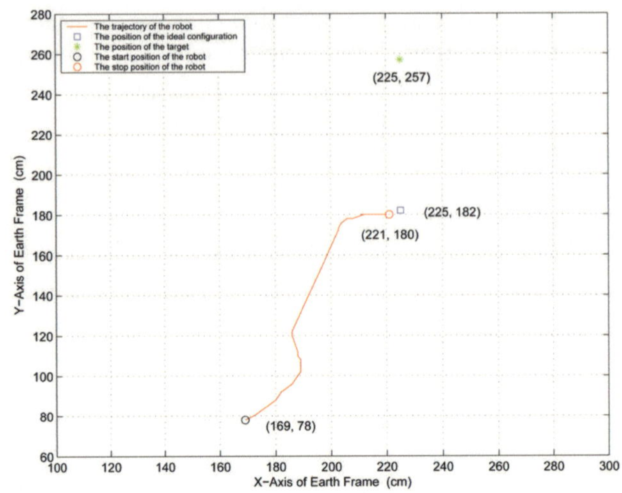

Fig. 10. The docking trajectory of the robot in Docking Scenario. 1

Furthermore, as the initial position error and relative rotation angle are big, large driven forces of the wheels are generated and therefore an overshoot comes up. The overshoot makes the robot pass ideal configuration $\mathcal{D}$ and then arrive at location (172, 186). After adjustments of control signals, the robot finally stops at location (190, 186). In terms of the orientation error between the robot and the ideal configuration, Figure 12 shows that the relative rotation angle dramatically decreases within the first second, followed by its fluctuation amongst 0° and -5°. Finally, the residual rotation angle is about -1°. The total time of the docking takes 2.5 seconds.

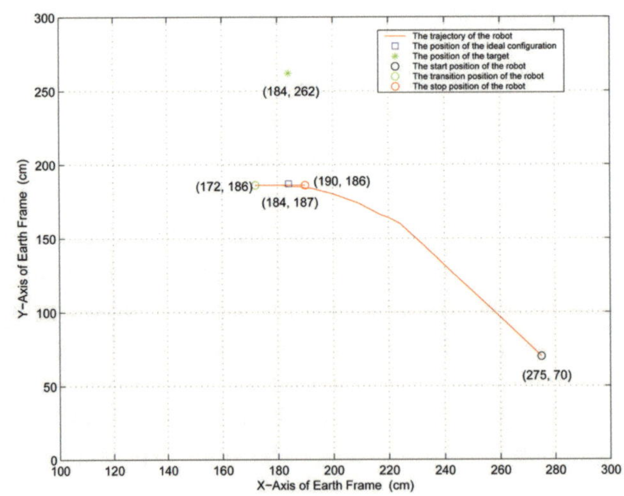

Fig. 11. The docking trajectory of the robot in Docking Scenario. 2

The third scenario of the docking is similar with the second one. The main differences are that the initial position of the robot in the third scenario is on the left-hand side of the ideal configuration and initial relative rotation angle $e_\theta = 42°$ while it is on the right-hand side and $e_\theta = -35°$ in the second one, as presented in Figure 13 and Figure 14. Figure 13 shows that the robot starts from location (63, 176), passes through transition location (101, 286), and finally stops at (155, 284) which is close to $\mathcal{D}$ at location (151, 280) represented by the blue square. In addition, the relative rotation angle remarkably drops within the first 0.6 second, and the final one is close to 0°. The total docking time is 2.75 seconds.

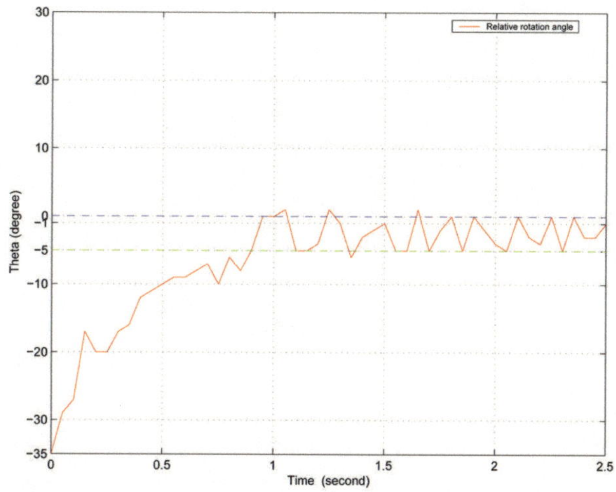

Fig. 12. The relative rotation angle between the robot and ideal configuration $\mathcal{D}$ in Docking Scenario. 2

Based on the results of the three docking scenarios reported, it is clear that the robot can not completely coincide with ideal configuration $\mathcal{D}$, that is, the residual errors exist. This is due to the influences of the un-modeled friction, the low resolution cameras employed and the unstable lighting. However, by using the proposed pose estimation algorithm and control scheme, the robot can successfully dock itself to the locations close to ideal configuration $\mathcal{D}$ at different positions and orientations. Therefore, the asymptotic stability is approximately satisfied in the docking scenarios.

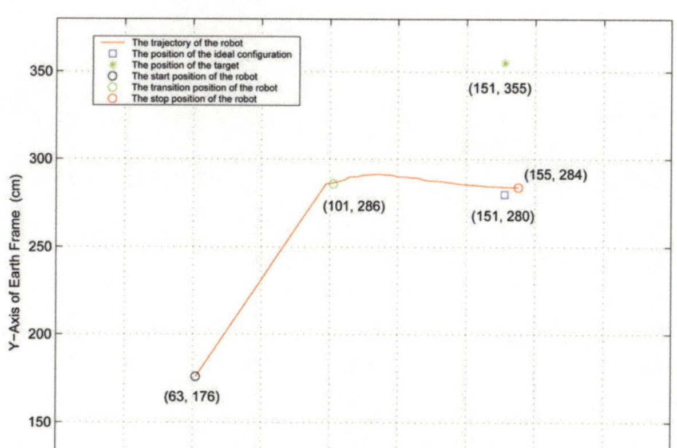

Fig. 13. The docking trajectory of the robot in Docking Scenario. 3

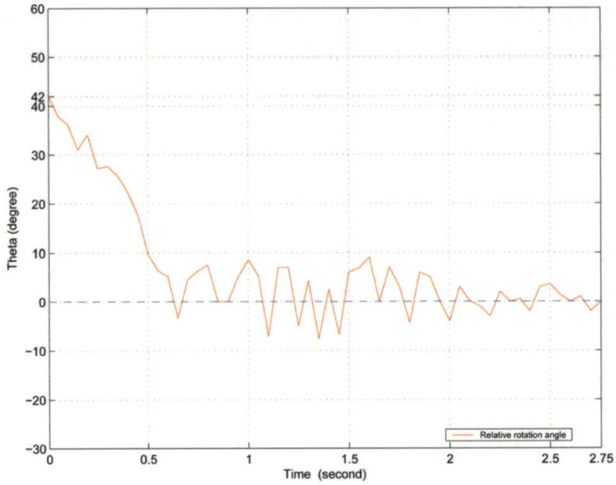

Fig. 14. The relative rotation angle between the robot and ideal configuration $\mathcal{D}$ in Docking Scenario. 3

6.3.2 Tracking

To validate the proposed robot's capability of the target tracking, a tracking experiment is performed. As shown in Figure 15, instead of being still, ideal configuration $\mathcal{D}$ clockwise moves along a predefined trajectory represented by the blue line at an average rate of $0.27m/s$. The predefined trajectory is a circle with a radius of 1 meter.

Initial location of configuration $\mathcal{D}$ is at coordinate (85, 201) represented by the blue circle while the initial position of the robot is (78, 118) represented by the red circle. Hence, the initial distance between them is about $0.833m$ and the initial relative rotation angle is about -58° as shown in Figure 15 and Figure 16. To make the tracking trajectory of the robot clear,

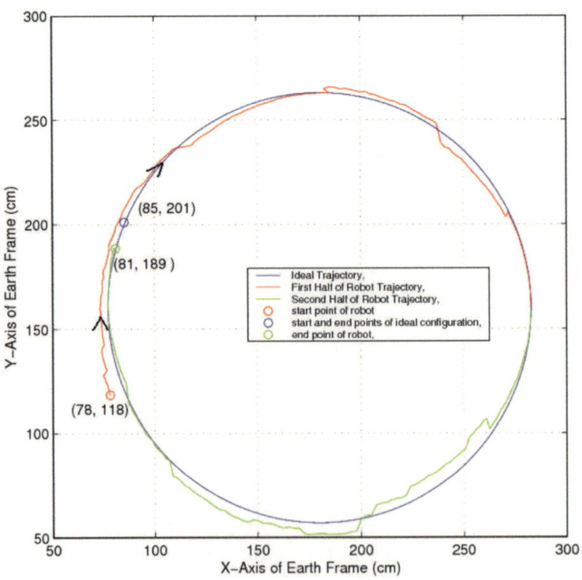

Fig. 15. The tracking trajectory of the robot

the first half of the robot trajectory is represented by the red line and the rest of it is represented by the green line. Finally, ideal configuration $\mathcal{D}$ arrives at coordinate (85, 201) again, but the robot stops at (81, 189) represented by the green circle. Thus, the residual position errors, $\bar{e}_x$ and $\bar{e}_y$, between the robot and the ideal configuration are $0.04m$ and $0.12m$, respectively, and the residual orientation error is around 0°, as shown in Figures 16-18.

Figure 15 shows that the tracking trajectory of the robot is basically consistent with that of the predefined trajectory of the ideal configuration, although some acceptable deviations between them exist. The factors which affect the tracking performance are summarized as follows:

1. Response delay: The response delay of the robot is mainly caused by the computational cost of the image and control modules and the friction. The response delay has the feedback signals, or more precisely the visual feedback and the wheel speed feedback, outdated so as to impair the tracking performance.
2. Friction: Due to the sliding motion of the wheels, there is strong friction existing between the wheels and ground. The friction plays an important role in affecting the performance of the tracking, especially the static friction which is considerably bigger than the dynamic one. In order to reduce the impacts of the friction on the tracking performance, an accurate model including friction effects should be considered and an advanced nonlinear control law should be applied.
3. Lighting condition: Lighting condition mainly exerts an impact on the image processing and therefore the pose estimation is affected. In this experiment, natural light is used rather than stable lighting. To improve the tracking performance, a laboratory with the relatively stable lighting condition is preferred.
4. Camera resolution: In this experiment, a pair of low resolution cameras are employed. Due to the limitation of the camera resolution, a small difference between the robot and ideal configuration $\mathcal{D}$ can not be detected on image planes. Therefore, the residual tracking errors exist.

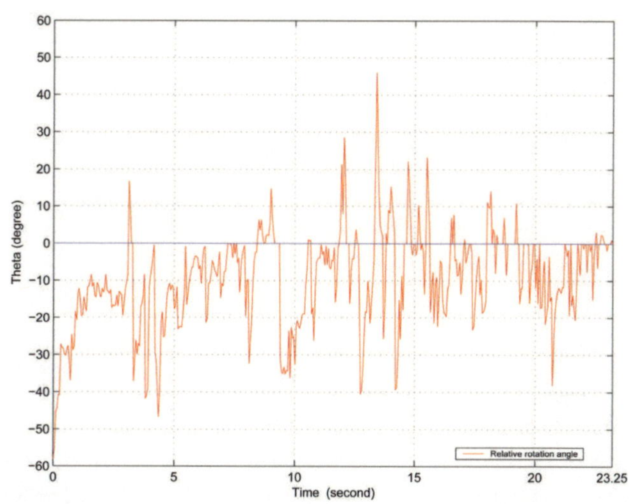

Fig. 16. The relative rotation angle between the robot and the ideal configuration expressed in the coordinate frame of the robot in the tracking scenario

Furthermore, Figure 16 shows the historic trajectory of the relative rotation angle obtained during the tracking and the total experimental time of 23.25 seconds. From the figure, it is clear that the relative rotation angle is large sometimes. This mainly results from either the sharp turn of the target, or of the robot, or of both. However, the large relative rotation angle is adjusted by the modulated control signals. Consequently, the relative rotation angle is bounded within 20° at most time, as reported in Figure 16. In terms of the position errors, Figure 17 and Figure 18 show that the position errors, $\bar{e}_x$ and $\bar{e}_y$, are bounded within 0.2m and 0.3m at most time, respectively, although the impacts of the sudden acceleration of the target and the response delay of the robot exist. Therefore, the tracking satisfies the BIBO stability.

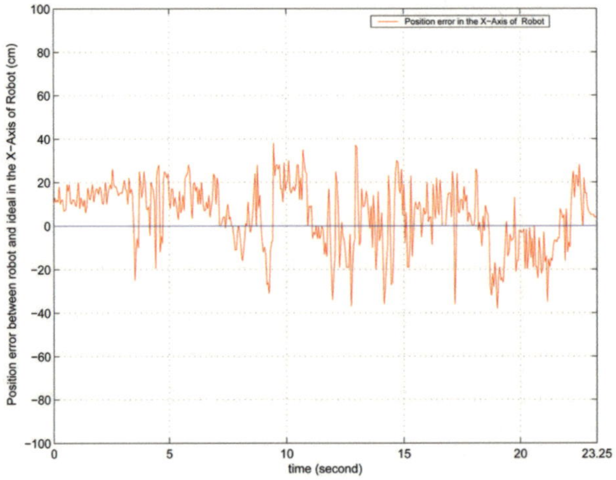

Fig. 17. The position error, $\bar{e}_x$ in the tracking scenario

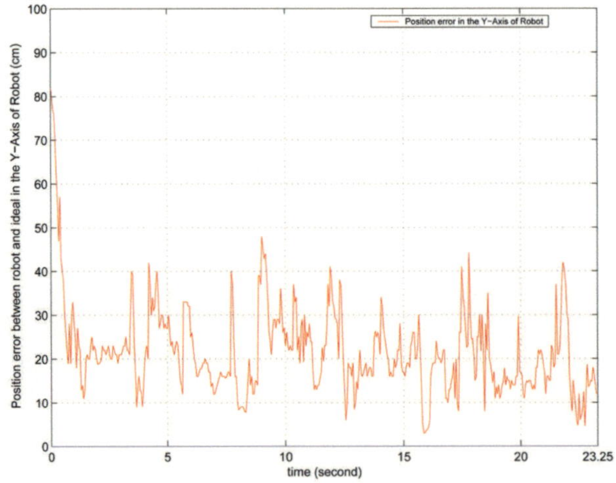

Fig. 18. The position error, $\bar{e}_y$ in the tracking scenario

The figures which include the docking and tracking trajectories of the robot are plotted with the sampled points on the physical trajectories chalked. The position errors shown in Figures 17 and 18 are expressed in the coordinate frame of the robot rather than the earth frame, and the relative rotation angle shown in Figures 12, 14 and 16 and the position errors are not the ground truth values but the estimated values obtained by the calculation in real time.

7. Conclusions and future work

In this paper, the moving target tracking of an omnidirectional robot with stereo cameras has been presented. Unlike most car-like robots which are subject to the nonholonomic constraint, the omnidirectional robot proposed is holonomic with excellent maneuverability. Furthermore, this omnidirectional feature makes the path planning, which is a tough task for the car-like mobile robots, be a straightforward task.

The dynamic model of the robot is obtained first. Based on the dynamics, a simple PD controller with visual feedback is proposed. The PD control ensures the asymptotic stability of the tracking error if the target is still which is equivalent to the case of docking, or the bounded-input-bounded-output (BIBO) stability if the target is moving with varying but bounded rotational and linear velocities. The real-time experimental results demonstrate the feasibility and effectiveness of the proposed tracking scheme.

For future work, we plan to track target with a vision system which integrates one camera and some kind of sensor. In order to suppress the noise and enhance the tracking performance, a Kalman filter will be employed in our control loop.

8. References

R. L. Williams, B. E. Carter, P. Gallina and G. Rosati, "Dynamic Model with Slip for Wheeled Omnidirectional Robots", *IEEE Trans. Robot. Automat.*, vol. 18, pp. 285-293, June, 2002.

K. Watanabe, Y. Shiraishi, S. G. Tzafestas, J. Tang and T Fukuda, "Feedback Control of an Omnidirectional Autonomous Platform for Mobile Service Robots", *Journal of Intelligent and Robotic Systems*, vol. 22, pp. 315-330, 1998.

K L. Moore and N S. Flann, "A Six-Wheeled Omnidirectional Autonomous Mobile Robot", *IEEE Control Systems Magazine* , pp. 53-66, December, 2000.

G D. Hager, "A Modular System for Robust Positioning Using Feedback from Stereo Vision", *IEEE Transactions on Robotics and Automation,* VOL.13, NO.4, pp. 582-595, AUGUST 1997.

G. Artus, P. Morin and C. Samson, "Tracking of an Omnidirectional Target with a Nonholonomic Mobile Robot", *Proc. of the 11th Int. Conf. on Advanced Robotics,* pp. 1468-1473, Portugal, June, 2003.

H. Kase, N. Maru, A. Nishikawa and S. Yamada "Visual Servoing of the Manipulator using the Stereo Vision", *Proc. of the 1993 IEEE IECON,* pp. 1791-1796, VOL.3, Maui, HI, November, 1993.

N. Andreff, B. Espiau and R. Horaud "Visual Servoing from Lines", *Proc. of the 2000 IEEE International Conference on Robotics and Automation,* pp. 2070-2075, San Francisco, April, 2000.

A. K. Das, R. Fierro, V. Kumar, B. Southall, J. Spletzer and C. J. Taylor, "Real-Time Vision-Based Control of a Nonholonomic Mobile Robot", *Proc. of the 2001 IEEE International Conference on Robotics and Automation,* pp. 1714-1719, Seoul, Korea, May 21-26, 2001.

D. Burschka and G. Hager, "Vision-Based Control of Mobile Robots", *Proc. of the 2001 IEEE International Conference on Robotics and Automation*, pp. 1707-1713, Seoul, Korea, May 21-26, 2001.

D. Xu, M. Tan and Y. Shen, "A New Simple Visual Control Method Based on Cross Ratio Invariance", *Proc. of the 2005 IEEE International Conference on Mechatronics and Automation*, pp. 370-375, Niagara Falls, Canada, July, 2005.

H. Hashimoto, T. Kubota, M. Sato and F. Harashima "Visual Control of Robotic Manipulator Based on Neural Networks", *IEEE Transactions on Industrial Electronics*, pp. 490-496, VOL.39, No. 6, December, 1992.

S. Fujisawa, M. Ryuman, T. Yamamoto, H. Sogo, Y. Suita and T. Yoshida, "Development of Path Tracking Control for Omni-Directional Mobile Robot using Visual Servo System", *The 27th Annual Conference of the IEEE Industrial Electronics Society*, pp. 2166-2170, 2001.

M. Berkeme, M. Davidson, V. Bahl, Y Q. Chen and L. Ma, "Visual Servoing of an Omni-directional Mobile Robot for Alignment with Parking Lot Lines", *Proceedings of the 2002 IEEE International Conference on Robotics and Automation*, pp. 4204-4210, May, 2002.

R. Hartley and A. Zisserman, Multiple View Geometry *Cambridge University Press 2003*

S. He, Force Sensor Based Motion Control of Wheeled Mobile Robot *Master Thesis, Monash University, 2007*

Intel Open Source Computer Vision Library www.sourceforge.net/projects/opencvlibrary

13

High School Educational Program Using a Simple and Compact Stereo Vision Robot

Takeshi Morishita[1] and Tetsuro Yabuta[2]
[1]*Toin University of Yokohama,* [2]*Yokohama National University*
Japan

1. Introduction

To promote and further encourage developments in the fields of science and technology in accordance with the Science and Technology Basic Law (hereinafter Basic Law), the Japanese Ministry of Education, Culture, Sports, Science and Technology is promoting various initiatives through measures that realize technological and scientific development according to the vision of the Basic Law. Realization of a personnel training program in science and technology is one of the primary targets of the Basic Law in order to maintain and elevate Japan's research, development capabilities and competitiveness in the international market. Various plans such as the 'Training and Development of Talented Individuals in Technology, Arts and Sciences plan', the 'I love Technology and Science Plan', and the development of digital science education teaching materials are also being carried out to realize the vision of the Basic Law. For example, in junior and senior high schools, the Japanese government has formulated and is promoting the 'Super Science High Schools' plan and the 'Science Partnership Program'. All these initiatives are aimed at providing personnel training in science and technology, which in turn promotes technical innovation and industrial competitiveness. As students are being increasingly exposed to advanced technologies, these programs intend to increase the educational incentive for students and provide them with a basic technology education (Ministry of Education, 2005).
Furthermore, the industrial education target in a high school course requires basic knowledge of each industrial field, practical and technical skills acquisition and an understanding of the role it plays in the modern society and environment. Therefore, in technical high schools, practice trainings and experiments are incorporated into the course syllabus as an educational program that encourages student interest in technology, promotes the student's practical knowledge and develops their skills positively.
Now, we see many teaching materials using REGO for studying basic technology. This REGO is the teaching materials which are easy to treat for programming education. However, students cannot acquire practical and technical skills of manufacturing by using the REGO materials (Josep et al., 2005).
On the basis of these plans and vision, the Kanagawa Comprehensive Technical and Science Senior High School (Kanagawa Sougou Sangyo H.S.) was founded in 2005. As this predecessor school was a technical high school, the learning environments are maintained such that the students are exposed to a wide range of technologies ranging from machine tools such as the NC and the milling machine to micro computer technology.

The students participating in this program, who are active in both class and extracurricular activities, study not only the individual technologies but also systematic technical synthesis by gaining exposure to the entire process from design stage to manufacturing using robotic systems. As a result, these students have participated in robot contests and have won several awards.

In this chapter, we experimented our program for high school students as an educational subject, in order to show that this program can be used effectively for the high school student. We present the results of implementing a new educational program and materials for high school students by incorporating a small stereo vision sensor module for an autonomous, simple and compact robot (Fig. 1).

Fig. 1. Compact autonomous robot with stereo vision system constructed by high school students

2. Contents research purposes of the educational program

This chapter describes a new educational program to maximise the use of a simple and compact stereo vision robot. The research purposes of this educational program are summarized as follows:

1. Development of new educational methods and materials that differ from the conventional ones for application in a new technical field.
2. Investigation of both the field of technical interest and technical knowledge of the students and the presentation of new education curricula and materials suitable for the students.
3. Measure the educational effects of the proposed educational contents such as the educational method, practical training, cost, etc.
4. Measure student responses to the program, gauge their interest levels and obtain their opinions regarding the teaching materials.

3. Educational program features

Providing the students with an opportunity to study stereo image processing technology on a computer and handmade robot is one of the specific features of this educational program. We will teach students total system integration using a new educational framework that has not been previously attempted in an educational program. The flow of this educational program is summarized as follows:

1. Manufacturing a robot body as a control object.
2. Stereo image information processing exercise by using a computer.
3. Learning of control programming and systemization of the robot system.
4. Operating the autonomous stereo vision robot that performs feature extraction and distance measurement by processing images.

4. Pre-analysis of participant students

While developing a program, a teacher always considers the relationship between the contents of the educational material and the academic aptitude of the students involved. This also holds true for technical education. Pre-analysis performed by the participating students is thus an important element for developing the lesson. In the analysis method, a questionnaire on the related technology is administered to ten students (Table 1).

This questionnaire is designed to check the student's technical skill and attitudes with regards to robots, vision sensors, and other technical elements involved in the program, in order to have a record of the technical knowledge of the students. Figure 2 shows the results of the questionnaires. All answers were either 'Yes' or 'No'. The numbers in Figure 2 denote the actual questions.

Based on the results of the questionnaire, we observed that everyone had been exposed to all or most of the technical skills related to this program. The cause of this, as shown by the results, is that the target students have received technical education for one year or more, and have received additional mechatronics education by participating in extracurricular activities. The positive responses obtained in the questionnaires revealed that the students were capable of using various technical tools; however, their technical knowledge was still not very high. Thus, it was considered that the knowledge of the students participating in this educational program was quite high as compared to that of high school students.

Further, the results of the attitude section indicated that none of the students wanted to use a vision sensor (Q1 = 0%). This shows that technical education in high schools does not typically include a study of vision sensors.

This questionnaire also revealed that the present condition of the education level in technical high schools lags behind in the field of vision technology. On the other hand, since the results indicate that all the students are interested in robots (Q2 = 100%), it is considered that the students participating in this program show interest and are highly motivated to work in this field.

5. Educational objectives

Based on the results obtained from the questionnaire, we developed the educational objectives and curriculum for this program. In Japan, the educational objectives of a technical high school curriculum comprise teaching fundamentals of both industrial mechanical science and information technology, practical training related to these fundamentals, mechanical drafting, etc. We set up the educational objectives of this program by considering the objectives of a technical high school—special knowledge, technology and ability for functional integration; these are ingrained through the fundamental study of the proposed vision technology using both experiments and training. Spontaneous study attitude and problem solving ability are also improved by this program. Also, the fundamental understanding of the relationship among mechanics, hardware, software and vision technology is increased, and the students gain interest in general technical fields, especially mechatronics.

Section	Question
Mechanism /Hardware	Q1. Can you use a saw well? Q2. Can you cut sheet metal by shirring machine? Q3. Can you file well? Q4. Can you use slide calipers and a height gage? Q5. Can you use a drilling machine? Q6. Have you used a lathe? Q7. Have you used a milling machine? Q8. Can you understand a machine design? Q9. Can you draw a machine design? Q10. Can you understand a data sheet? Q11. Can you look for a data sheet by yourself? Q12. Have you made electronic circuit? Q13. Have you soldered? Q14. Have you used cutting pliers and a nipper? Q15. Can you use an electronic millimeter? Q16. Have you used an oscilloscope? Q17. Can you understand a circuit schematic? Q18. Can you draw a circuit schematic? Q19. Do you know a polar semiconductor? Q20. Have you used a sensor? Q21. Have you used a microcomputer? Q22. Can you select parts required for an electronic circuit? Q23. Have you made an electrical system by using semiconductor module?
Software	Q1. Have you used C language? Q2. Have you used both input and output function of data in a microcomputer? Q3. Can you use condition branch of "IF" sentence? Q4. Can you use a repetition of a "For" sentence? Q5. Have you use interruption process? Q6. Do you know how to use a function? Q7. Can you perform of LED control? Q8. Have you used a Motor Driver IC? Q9. Can you write a program in a microcomputer? Q10. Have you read a book about a microcomputer?
Attitude (interest)	Q1. Have you want to use a vision sensor? Q2. Are you interested in a robot?

Table 1. Technical level questionnaire for students before program (Answer is "Yes" or "No")

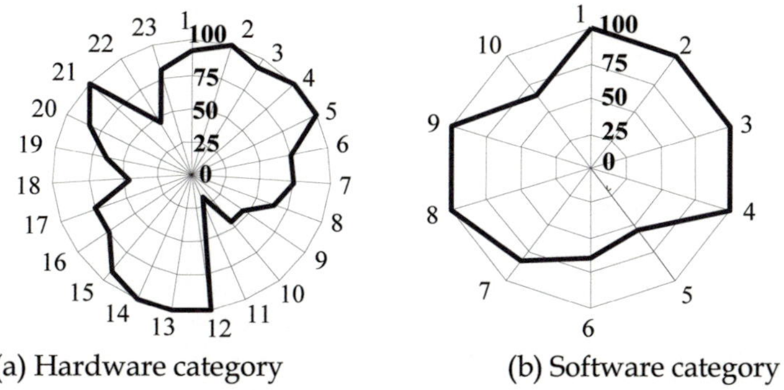

Fig. 2. Results of the technical skills questionnaires (Circled number denotes the technical question number. Radius of the circle indicates the frequency of 'Yes' responses.)

6. Educational material

6.1 Small stereo vision sensor one board module

Figure 3 shows the small Stereo Vision One board sensor module (SVO sensor module) that we developed for this program. Two sets of CMOS image sensors are mounted such that the module can obtain a stereo pair image. We can obtain the position of the three-dimensional coordinates—x, y and z—or achieve color extraction, feature extraction or distance measurement by processing the information obtained from this stereo pair image. The calculated results from this module can be output using general-purpose interface ports such as general IO, serial communication port and A/D and D/A ports. Furthermore, a 16 bit microprocessor installed in this module can perform basic visual processing calculations

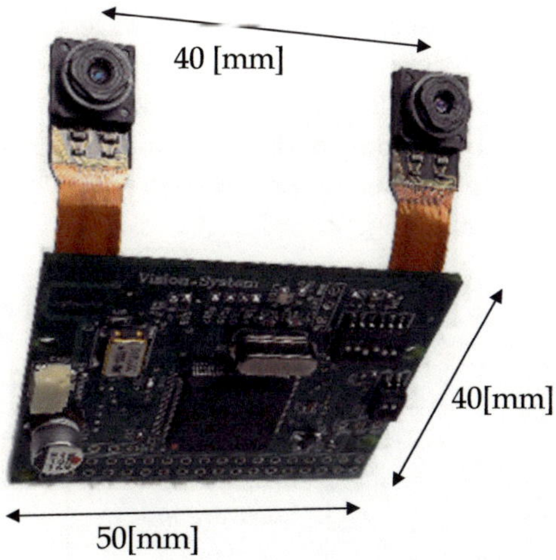

Fig. 3. Small Stereo Vision One Board Sensor Module

for image processing. We have implemented functionalities for capturing images, noise reduction, color extraction, binarization, centroid calculation, radius calculation, shape recognition, stereo image based distance calculation, pattern matching and attention area tracking. These functionalities enable us to use any image processing algorithm in order to take complete advantage of the vision sensors. In addition, when using a multiple sensor system, this SVO sensor module can be implemented easily in various robots because a general interface is employed as an output port. This module can be used as an image recognition module attached to a simple robot for this program (Okada et al., 2005; Morishita, T.& Yabuta, T., 2007).

In related works, Konolige developed a small, self-contained stereo vision module using FPGA(Kurt, 1997), which is particularly useful for depth image generation. Our sensor module provides several visual functions for multiple purposes and can be easily implemented in various types of robots—it can be attached as a visual sensor as it has a general-purpose interface.

6.2 Educational stereo vision robot

The educational robot developed for this program is a simple and compact robot with a stereo vision system. As conventional educational targets are only concerned with exposure to the manufacturing process, the contents of their corresponding exercises focus only on this process. On the other hand, the educational target of this program is aimed at covering a wide spectrum of technologies ranging from mechanical manufacturing processes to software technology. Therefore, the time spent in manufacturing the mechanism and hardware is limited to about 1/3 of the entire curriculum educational time. We selected a robot with a simple structure in order to ensure that it is manufactured within the assigned time. Figure 4 shows a simple block diagram for the stereo vision robot system; the dashed line represents the portion of the electronic circuit to be manufactured. Students fabricated this portion using a few general electric parts and under the considerations of a circuit diagram provided as educational material. In this educational program, students can attain the fundamental skills of fabricating an electronic circuit and learn to read a fundamental circuit diagram. While integrating machine and hardware parts of the electrical circuit, a technical sense of system integration is cultivated. The educational curriculum developed is shown in Table 2. All the parts and components selected for this educational system are low cost and are easily available in order to ensure that the parts can be sourced easily and are economically efficient and maintainable(Fig. 5).

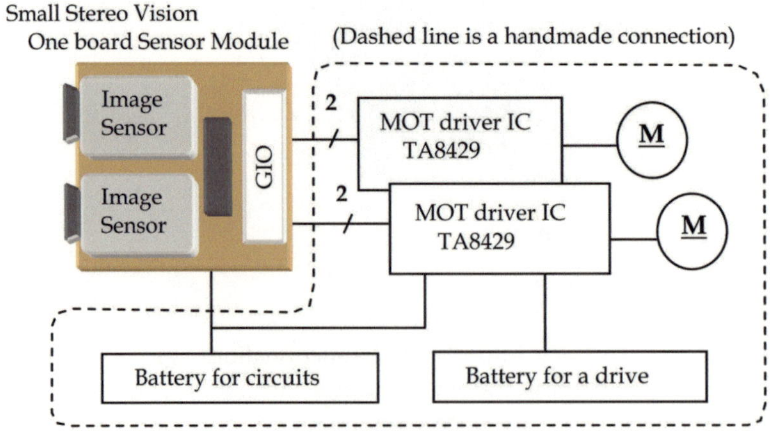

Fig. 4. Block Diagram for Educational Robot System

Fig. 5. Main parts of the robot system

Subject	Contents
Robot Manufacturing (1st - 2nd week)	1) Guidance 2) Manufacturing the robot body 3) Manufacturing the motor control circuit board
	4) Manufacturing the motor control circuit board 5) Stereo vision module mounting 6) Connection and operation check
Image information processing basics and handling(3rd week)	7) Color system, binarization, noise reduction 8) Stereo vision module handling exercise
Stereo vision basic learning (4th week)	9) Image data processing using viewer 10) Data reduction/graph (disparity - distance) 11) Visual information processing programming exercise
Experiment of vision robot (5th - 6th week)	12) Viewer handling for image processing operation 13) Object tracking control programming 14) Operation check and debugging 15) Color extraction tracking robot control experiment
	16) Distance recognition programming 17) Operation check and debugging 18) Distance recognition robot control experiment

Table 2. An example of the educational curriculum

6.3 Viewer for vision technology

Figure 6 shows the screenshots of both the sampled data image from the vision sensor and visual image analysis results for the operation. By using the viewer shown in Figure 6(a), students can obtain not only a color image of an object but also a binary image, a representative point of the object and the disparity of the stereo image. Figure 6(b) shows the real time processing results of the position, distance and size of the object. Due to the stereo vision sensor, the resultant values are three-dimensional (in x, y and z coordinates).

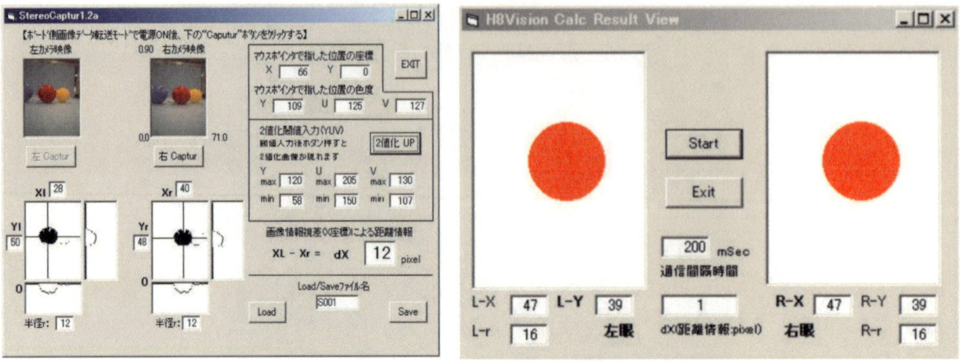

(a) Visual processing view (b) Processing result view

Fig. 6. Vision system information viewer for education

6.4 Curriculum

One subject consists of six training classes with four hours of exercise time. For example, if a four-hours class is conducted once per week, it would takes six weeks to cover all the subjects; this implies a total of 24 hours of exercise time. Moreover, this curriculum can be implemented as a conventional one. In this case, one subject is equivalent to two classes of the conventional curriculum and the 1st, 2nd & 3rd, 4th subjects can be taught under the conventional curriculum as mechatronics technology, basic image information technology and an autonomous robot exercise with vision system. These classes are considered to be taught in relation with each other. Under this arrangement, a special one subject class can be opened and an individual teacher can be assigned for each subject.

7. Educational practice

7.1 Manufacturing of the educational robot

Manufacturing the educational robot comprises the assembly of an actuator system, manufacturing the motor drive circuit and system integration with a vision sensor (Fig. 7(a)).
The time allowance planned originally allotted time for the systematization and manufacturing of the electronic circuitry module, such as manually fabricated connections and soldering. A generous time distribution is important for the composition or connection check of circuits, although the time provided in class to perform these checks is usually not sufficient. We could effectively use the time for performing important checks by simplifying the construction of the robot (Fig. 7(b)).

7.2 Basic Learning from image information processing

This section outlines the basic learning from image information processing using the viewer shown in Figure 6. The obtained information can be summarized as follows: 1) the relationship between the YIQ color specification system generated by the vision sensor module and the RGB color specification system of the viewer, 2) the method for converting YIQ to RGB, 3) image binarization method for extracting image color information, 4) noise reduction method, 5) method for calculating the representative point of an object based on its image coordinates and 6) method to use this sensor module and transmit image information to a computer.

7.3 Stereo vision basic learning

This section describes the basic learning achieved from the stereo vision. We explained that when a stereo pair image is obtained in parallel, the distance information Z can be acquired from the disparity in the image. We used two colored balls for this experiment—a yellow ball (table tennis ball) of 38 mm and a red ball(soft-ball) of 76 mm. These balls are placed at intervals of 80 to 10 cm from the vision sensor, and then the students are asked to record the disparity value of the representative point computed from a binary image Figure 8(b).

(a) Hardware manufacturing process

(b) Connection check and debugging

Fig. 7. Practice in manufacturing and an experiment

(a) Graphical experimental data

Distance [cm]	Disparity [pixel]	
	Red ball (Φ 76 mm)	Yellow ball (Φ 38 mm)
10	17	34
20	17	19
30	12	10
40	8	9
50	6	6
60	4	5
70	4	4
80	3	3

(b) Numerical experimental data

Fig. 8. Disparity vs. distance experimental results obtained by students

We then asked the students to plot these disparity values on a graph. We explained that the distance data obtained from the disparity value changes with the change in disparity. Since the two curves are in agreement, the distance measurement can be performed independent of the target size, as shown in Figure 8(a).

7.4 Operation experiment of a stereo vision robot

This exercise is an autonomous control experiment of the stereo vision robot. First, the coordinate system was explained before the experiment; this is because it is important to obtain the relationship between the object position and the results obtained from the calculation of the vision sensor data. Next, the branch control algorithm was applied to motor control in order to realize a simple robot motion. The focus of this programming exercise is on both to set up the threshold of feature extraction for the image processing and to generate a motor control program. This exercise is related to both the software and hardware fields, which makes this an essential exercise to grasp these relationships.

Therefore, we made the students verify normal signal transfer between the software and hardware before conducting the experiment. Since debugging a program is also important for acquiring programming skills, a considerable amount of time is assigned for the debugging process (Fig. 7(b)).

After all pre-exercise conditions are fulfilled, the two experiments shown in Figure 9 were conducted. In the first experiment, an object was tracked by color extraction. In other words, the autonomous robot can detect an object by color detection and can trace it by determining the direction in which the object moves. The second experiment was a stop point control experiment of the autonomous robot. In this experiment, the robot stops at a set distance from the object by using the distance measurement function; this can be realized by the comparison between the pre-assigned distance and the distance measured by the vision sensor. By using the stop experiment data, the students compared the actual stop data and the pre-experiment data of Figure 8 (section 7.3). When the actual experiment and pre-experiment data were observed to be in good agreement with each other, the students achieve a sense of surprise, admiration and technical satisfaction.

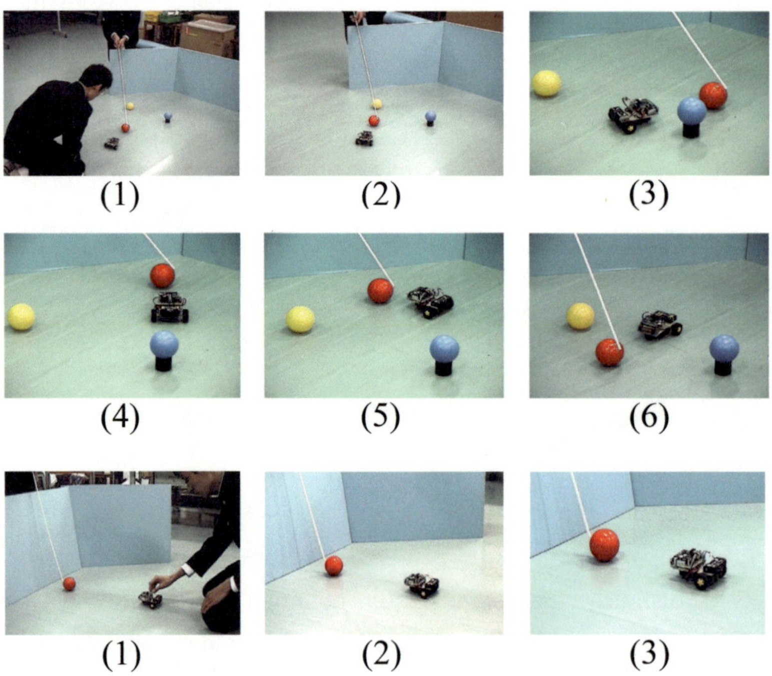

Fig. 9. Schematic figures of the color ball tracking experiment and distance measurement experiment by using the robot

8. Results after exercise

After the completion of all the exercises, we investigated both the student attitudes toward this program and the achievements of the program by using the questionnaire, in order to verify the effect of this educational program.

8.1 Skill level investigation

This questionnaire is divided into four groups in agreement with that of the curriculum. The first group comprises four questions that can be answered as either 'Yes' or 'No' (Table 3(a)). Figure 9 shows the questionnaire results; these results reveal that almost all students chose 'Yes' as an answer for each question. This educational program is planned such that it provides systematic exposure to practical technologies such as software, hardware and mechanical manufacturing process by using the vision robot. It helps students assimilate vision technology fundamentals as well as practical technologies. Thus, it can be implied that the higher the satisfaction for each a questionnaire item, the more enthusiastically the students tackled the teaching materials with interest and concern.

These results reveal that the targets of this educational program have been realized. Thus, an overall positive response to the questionnaire in relation to the basic technical education reiterates the success of this program. On the other hand, when the contents of questions marked as 'No' were investigated, it was found that they are related to either systematization or programming. Since these questions depend on the student's skill in the technical field, the answers differ based on each student's technical knowledge. Therefore, the items related to these questions must be carefully arranged from the viewpoints of teaching material, time distribution and improvement in the educational method.

8.2 Attitude survey

In order to understand student attitudes to the related technical fields of this program, we conducted the attitude survey, shown in Table 3(b). In this investigation, the students could evaluate each question by assigning one of the five grades—very satisfied, generally satisfied, normal, generally unsatisfied and very unsatisfied.

The results are shown in Table 4. Based on the pre-program questionnaire, we can see that the students were interested in robot manufacture but did not know about either vision sensors or image recognition technology until this program. On comparing the results of the questionnaire shown in Table 3, which records both academic achievements and student's attitude toward robot technology after the program, with that of the pre-program questionnaire, the following could be seen: (1) This curriculum motivated the students, as illustrated by the responses of all the students that 'making a robot is pleasant'. (2) Although there were a few differences in opinion among the students, almost all of them answered 'We were able to enjoy programming'. (3) All students also answered 'We have experienced constructing the vision robot'. This is a natural answer obtained based on the usage of these education materials. However, it is an important result for the students who did not have any prior knowledge of vision sensors, as indicated by the preliminary survey, and gained exposure to a new technical field through participation in this educational program. This result implies a significant educational result. (4) In addition, this educational program for fundamental vision technical education did not require detail learning of its contents. After the program, the students without vision technology experience felt "This technology is an important technology". This is an interesting result and has an important educational value.

(a) Academic achievements (answer in either "Yes "or "No")

Section	Question
Robot manufacture	Q1. Could you use the tools? Q2. Did you have enough time to manufacture the robot in the class? Q3. Could you solder the electrical circuit? Q4. Could you make an electrical circuit by using the electrical diagram?
Basic image information processing	Q5. Did you understand the relationship between YIQ and RGB color systems? Q6. Did you understand the binarization process? Q7. Did you understand the noise reduction method? Q8. Could you transfer the image information data into a computer?
Stereo vision technology	Q9. Could you perform color extraction on the viewer? Q10. Did you understand the calculation of the representation point by using binary information data on the viewer? Q11. Could you measure the disparity property on the viewer? Q12. Could you draw the graph by using the disparity data?
Tracking experiment	Q13. Could you generate a program by using the representation point? Q14. Could you complete the tracking program by using the representation point? Q15. Could you verify the operation of both the software and hardware systems? Q16. Could you realize the tracking experiment of a target by using color extraction results?
Distance Recognition Experiment	Q17. Could you use the distance information in programming? Q18. Could you verify program execution? Q19. Could you realize the distance recognition experiment? Q20. Could you check the correspondence of experience data with the distance data estimated by the disparity data?

(b) Attitude survey (five-grade evaluation)

Impression	Q21. Is manufacturing a robot a pleasant task? Q22. Is programming enjoyable? Q23. Do you have experience in vision robot systems? Q24. Is this technology important?

Table 3. Questionnaires of both academic achievements and student's attitude toward robot technology after program

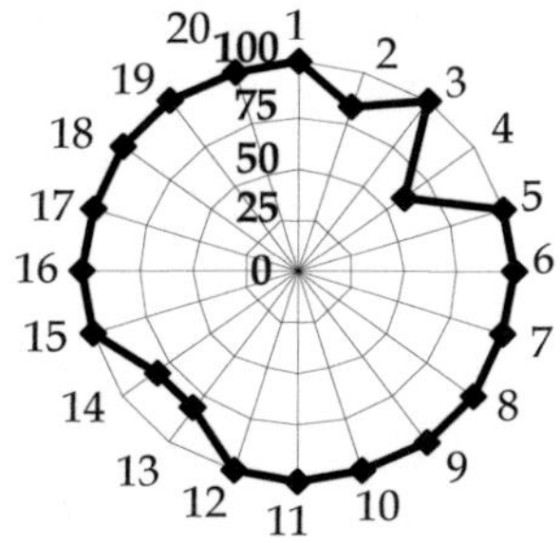

Fig. 9. Results of the academic achievements questionnaire (Circled number indicates the questionnaire number. Radius of the circle indicates the frequency of the 'Yes' responses.)

A21	Very pleasant 80%, pleasant 20%
A22	Very enjoyable 60%, Enjoyable 20%, Normal 20%
A23	Very realized 100%
A24	Significant agreement 80%, agreement 20%

Table 4. Attitude survey result

9. Conclusion

We developed practical high school educational material for a stereo vision autonomous simple and compact robot, which has not been previously reported until now in the technical education field. The results of this program are summarized as follows: The opinions of the high school students, which can be verified from the pre-program questionnaire, and the final questionnaire was 'We never had an opportunity to touch the vision technology until now, which is considered to become an important future technology'.

This opinion is almost similar to that expressed by both the first and second grade students from an undergraduate engineering university; this is because both the technical knowledge and the educational environment are almost the same as that of the technical high school students. This educational program can impart practical education including vision technology fundamentals, which have a far-reaching impact.

The salient features of this program can be summarized as follows: 1) As the educational program was formulated to encourage the pleasure of experimenting with robot technology in practical exercises, it motivated the students positively. 2) This program was intended to cultivate each basic technology and technological capability, interest and awareness among the students. 3) Furthermore, the educational materials also attained the educational goal as set in the beginning.

As for the student's opinion regarding the vision sensor, the application of vision technology in robotics is important in the field of robot research. However, the conventional educational materials use infrared sensors for basic technical education, which is not an actual vision sensor. This vision sensor module holds significant potential in education since the module is compact, has high usability and is very affordable. Visual information processing has long been researched; however, educational programs that use this technology are relatively uncommon. In order to make visual information processing technology familiar in general use, we believe that the educational material using simple, compact and low cost visual processing modules is of significant value in future educational materials.

10. References

Josep, F.; Pedro, L. & Joan, O. (2005). A Distributed Multirobot System Based On Edutainment Robots. In *Proceedings of the 2005 IEEE international Conference on Robotics and Automation*, pp. 4271-4276, April 2005, Barcelona, Spain

Okada, K.; Morishita, T.; Hayasi, M.; Inaba, M. & Inoue, H. (2005). Design and Development of Small Stereo Vision Sensor Module for Small Self-Contained Autonomous Robots. *Journal of Robotics and Mechatronics*, Vol. 17, No. 3, 2005, pp. 248-254.

Kurt, K. (1997). Small vision systems: Hardware and implementation. *In Eighth Intl. Symposium on Robotics Research,* pages 111-116, Oct 1997, Hayama, Japan

Ministry of Education, Culture, Sports, Science and Technology (2005). In: *Science and Technology white book 2005*, Basic Law on Science and Technology and Basic Plan on Science and Technology, June 2005 (in Japanese)

Morishita, T. & Yabuta, T. (2007). Development and Application of the Small Stereo Vision One Board Sensor Module using Mobile Phone CMOS Sensors for a Compact Autonomous Robot. *Transaction of JSME Series C,* Vol.73, No726, 2007, pp.363-370 (in Japanese)

14

A Sensor for Urban Driving Assistance Systems Based on Dense Stereovision

Sergiu Nedevschi, Radu Danescu, Tiberiu Marita, Florin Oniga, Ciprian Pocol, Silviu Bota and Cristian Vancea
Technical University of Cluj-Napoca, Romania

1. Introduction

Advanced driving assistance systems (ADAS) form a complex multidisciplinary research field, aimed at improving traffic efficiency and safety. A realistic analysis of the requirements and of the possibilities of the traffic environment leads to the establishment of several goals for traffic assistance, to be implemented in the near future (ADASE, INVENT, PREVENT, INTERSAFE) including: highway, rural and urban assistance, intersection management, pre-crash.

While there are approaches to driving safety and efficiency that focus on the conditions exterior to the vehicle (intelligent infrastructure), it is reasonable to assume that we should expect the best results from the in-vehicle systems. Traditionally, vehicle safety is mainly defined by passive safety measures. Passive safety is achieved by a highly sophisticated design and construction of the vehicle body. The occupant cell has become a more rigid structure in order to mitigate deformations. The frontal part of vehicles has been improved as well, e.g. it incorporates specially designed "soft" areas to reduce the impact in case of a collision with a pedestrian. In the recent decades a lot of improvements have been done in this field.

Similarly to the passive safety systems, primitive active safety systems, such as airbags, are only useful when the crash is actually happening, without much assessment of the situation, and sometimes they are acting against the well-being of the vehicle occupants. It has become clear that the future of the safety systems is in the realm of the artificial intelligence, systems that sense, decide and act.

Sensing implies a continuous, fast and reliable estimation of the surroundings. The decision component takes into account the sensorial information and assesses the situation. For instance, a pre-crash application must decide whether the situation is of no danger, whether the crash is possible or when the crash is imminent, because depending on the situation different actions are required: warning, emergency braking or deployment of irreversible measures (internal airbags for passenger protection, or inflatable hood for pedestrian protection). While warning may be annoying, and applying the brakes potentially dangerous, deploying non-reversible safety causes permanent damage to the vehicle, and therefore the decision is not to be taken lightly. However, in a pre-crash scenario it is even more damaging if the protection systems fail to act. Therefore, it is paramount that the protection systems act when needed, and only when needed, a decision that cannot be taken in the absence of reliable sensor data.

The sensorial systems for driving assistance (highway and urban) are today the focus of large, joint research projects, which combine active and passive sensors, GPS navigation, and telematics. Projects such as CARSENSE (www.carsense.org) INVENT (www.invent-online.de), PREVENT (www.prevent-ip.org), bring together car manufacturers and research partners for the common goal of solving the driving assistance problem.

In order to provide support for these applications, a sensorial system must provide an accurate and continuously updated model of the environment, fitted for high level reasoning. The environment description should include:

- Lane detection / Lane parameters estimation
- Navigable channel detection and channel parameters estimation in crowded environments
- Vehicle detection and tracking
- Detection of fixed (non-moving) obstacles
- Pedestrian detection and tracking.

There are many types of sensors that can be used for advanced driving assistance systems. The most known are:

- Long range radar: with a range of 1 to 200 m, and a response time of around 40 ms, it is a highly accurate ranging sensor, with a narrow field of view, suitable for detection of radar-reflecting targets such as vehicles in highway environments.
- Short/mid range radar: having a working range of 0-80 m, a fast response time, high accuracy and a medium width field of view, it is suitable for near range detection of vehicles in crowded urban scenarios. Both near range and far range radars have an increased reliability when detecting moving objects.
- Laser scanner: a high precision ranging sensor, working in near or far distance ranges, it is not limited to the metallic surfaces like the radar, but has considerable difficulty with low albedo objects.
- Monocular video sensors: employed in the visual or in the infrared light spectrum, the visual sensors can have a high field of view and can extract almost any kind of information relevant for driving assistance. The main problem of these sensors is that it cannot rely on accurate 3D information, having to infer it indirectly, usually with poor results.

A stereovision sensor adds the 3D information to the visual, thus becoming the most complex and complete sensor for driving assistance. It is capable of detecting any type of obstacle that falls inside its adjustable field of view, the road and lane geometry, the free space ahead, and it is also capable of visual classification, for pedestrian recognition.

The stereovision-based approaches have the advantage of directly estimating the 3D coordinates of an image feature, this feature being anything from a point to a complex structure. Stereovision involves finding correspondents from the left to the right image, and the search for correspondence is a difficult, time demanding task, which is not free from the possibility of errors. Obstacle detection techniques involving stereovision use different approaches in order to make some simplifications of the classic problem and achieve real-time capabilities. For instance, [1] uses stereovision only to measure the distance of an object after it has been detected from monocular images, [2] detects the obstacle points from their stereo disparity compared to the expected disparity of a road point, [3] detects obstacle features by performing two correlation processes, one under the assumption that the feature is part of a vertical surface and another under the assumption that it is part of a horizontal surface, and comparing the quality of the matching in each of the cases. A stereovision system that uses no correspondence search at all, but warps images instead and then performs subtraction is presented in [4].

Processing 3D data from stereo (dense or sparse) is a challenging task. A robust approach can prove of great value for a variety of applications in urban driving assistance. There are two main algorithm classes, depending on the space where processing is performed: disparity space-based and 3D space-based. Most of the existing algorithms try to compute the road/lane surface, and then use it to discriminate between road and obstacle points.

Disparity space-based algorithms are more popular because they work directly with the result of stereo reconstruction: the disparity map. The "v-disparity" [5] approach is well known and used to detect the road surface and the obstacles in a variety of applications. It has some drawbacks: it is not a natural way to represent 3D (Euclidian) data, it assumes the road is dominant along the image rows, and it can be sensitive to roll angle changes.

The 3D space algorithms have also become popular among the researchers in recent years. Obstacle detection and 3D lane estimation algorithms using stereo information in 3D space are presented in [6], [7], [8] and [9], ego pose estimation algorithms are presented in [10] and [11], and unstructured environment estimation algorithms are presented in [12], [13] and [14].

2. Stereo sensing solution for driving urban assistance

The research team of the Technical University of Cluj-Napoca has already implemented a stereovision-based sensor for the highway environment [6], [7] and [8]. This sensor was able to detect the road geometry and the obstacle position, size and speed from a pair of synchronized grayscale image pairs, using edge-based, general geometry, software stereo reconstruction.

The urban scenario required important changes in the detection algorithms, which in turn required more stereo information. Thus, the edge-based stereo engine was discarded, and replaced with a dense stereo system. A software dense stereo system being time consuming, a hybrid solution was chosen: software rectification and down sampling, followed by hardware correspondence search. The time gained by the use of a hardware board compensated the increase in complexity of the new algorithms.

The dense stereo information is vital for the new obstacle reconstruction module, which extracts oriented objects even in serious clutter, and also allows better shape segmentation for recognition of pedestrians. Dense stereo information allows us to compute and track an unstructured elevation map, which provides drivable areas in the case when no lane markings or other road delimiting features are present or visible.

Lane detection requires edges, but the vertical profile is better computed from dense stereo information. The edge based lane detection algorithms are completely changed, adapted to the limited and variable viewing distance of the urban environment. A freeform lane detection module was added, in order to solve the problem of the non-standard geometry roads.

The dense stereovision based sensor presented in this paper provides complex and accurate functionality on a conventional PC architecture, covering many of the problems presented by the urban traffic environment, and promising to be a valuable addition to a driving assistance system.

The hardware acquisition system (fig. 1) includes two grayscale digital cameras with 2/3" (1380x1030) CCD sensors and 6.5 mm fixed focal length lenses, allowing a horizontal field of view (HFOV) of 72 [deg]. The cameras are mounted on a rigid rig with a baseline of 320 [mm] (fig. 2). The images are acquired at full resolution with a digital acquisition board at a maximum frame rate of 24 fps.

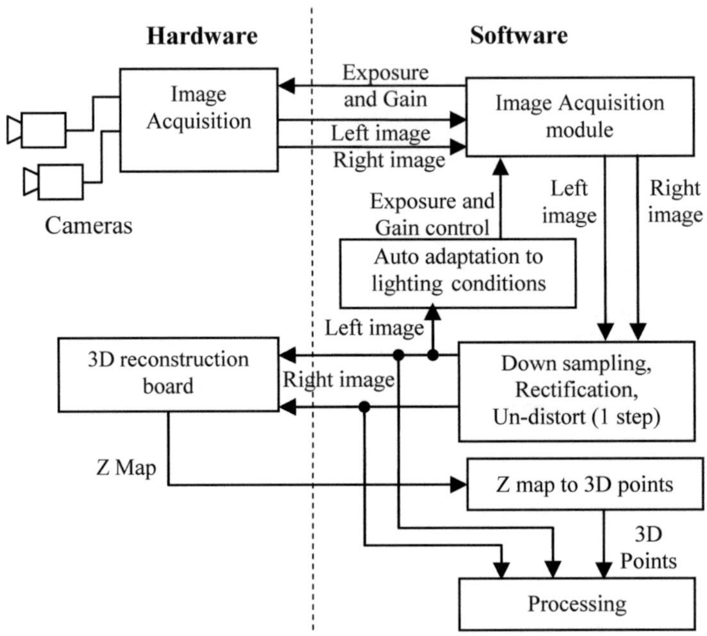

Fig. 1. The stereovision system architecture.

The camera parameters are calibrated using a dedicated method optimized for high accuracy stereovision [15],[16] and [17] using the full resolution images.

The images are further enhanced by lens distortion correction and rectified in order to fulfill the dense stereo reconstruction requirements (canonical images). A down-sampling step is used to adapt the image size to the dedicated hardware reconstruction board (512 pixels width) and to minimize the noise introduced by the digital rectification and image correction. The whole process is reduced to an image warping approach performed in a single step (fig. 1) using reverse mapping and bilinear interpolation [18]. An optimized implementation using MMX instructions and lookup tables was used in order to minimize the processing time.

The 3D reconstruction of the scene is performed using a dedicated hardware board. The input of the board consists in two rectified images and the output can be either a disparity or a Z map (left camera coordinate system). Our system uses 3D points set for scene representation; therefore the preferred output is the Z map. Using the Z coordinate value, the X and Y coordinate can be computed and then transformed into the car coordinate system.

With the current system setup a detection range optimally suited for the urban environments is obtained (fig. 2):

- minimum distance: 0.5 m in front of the ego car;

- delimiters of the current lane are visible at 1.0 m;
- reliable detection range: 0.5 ... 35 m;

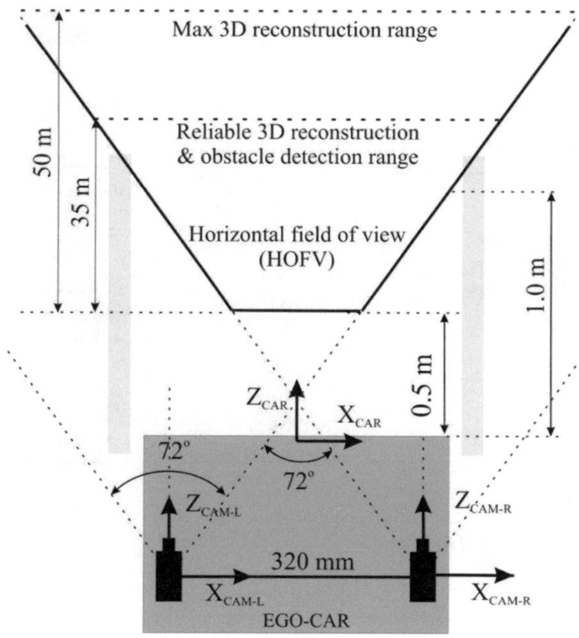

Fig. 2. Detection range of the current stereo system setup.

3. Stereovision-based lane detection

The urban lane detection system is organized as an integrator of multiple sensors, using a Kalman filter framework. Instead of having multiple physical sensors, we have multiple detection stages, which all deliver results that will be used to update the lane model state parameters. The cycle begins with the prediction, and continues with all the detection algorithms, until the final update. When one algorithm updates the lane state, the resulted estimation becomes the prediction for the next stage. In this way, we can insert any number of algorithms into the processing chain, or we can temporary disable some of them, for testing or speedup purposes.

Figure 3 shows the organization of the lane detection system, the main processing modules and the relationships between them. In what follows, we'll give a brief description for each module. A detailed description of the lane detection system is given in [19].

One of the most important advantages of stereovision-based lane detection is the possibility of direct detection of the vertical profile (pitch angle and vertical curvature). The projection of the 3D point set in the (YOZ) plane (height and distance) is analyzed by means of histograms. The pitch and the curvature range of values is divided into discrete hypotheses. First, a polar histogram counting the near range points along the lines described by each pitch angle hypothesis is built, and the lowest line having a significant number of points is selected. The vertical curvature histogram is built by taking into consideration the already detected pitch angle, and the curvature hypotheses, and counts the points in the far range. The method is somewhat similar to the Hough transform, and is described in detail in [7].

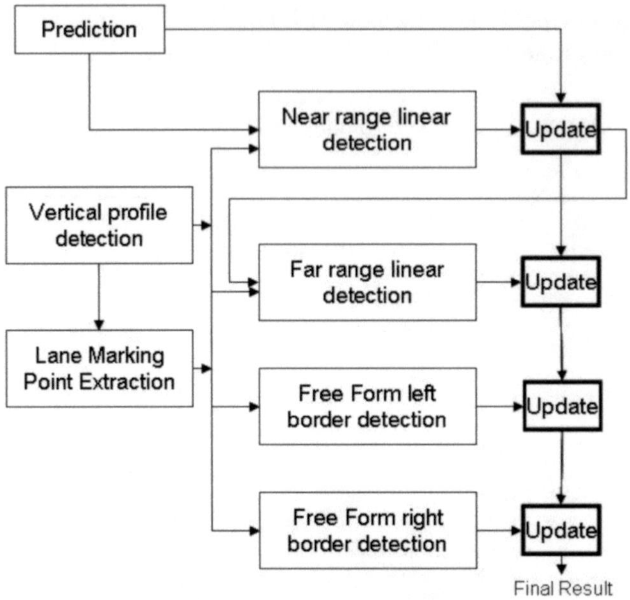

Fig. 3. Lane detection system architecture

After the vertical profile is detected, the 3D points can be labeled as belonging to the road or belonging to obstacle structures. The points belonging to the road are the main features used for lane geometry estimation. The next step is to extract, from the road surface point set, the points that have a higher relevance for lane delimitation, namely the lane markings. The highway lane detection approach required little information about lane markings, because it could rely greatly on the lane model. For the urban environment, however, we require a fast and robust lane marking extraction algorithm.

The lane marking extraction method relies on the well-known dark-light-dark transition detection [20]. We have to search for pairs of gradients of opposing sign and equal magnitude. We have improved the method by using a variable filter for computing the horizontal gradient. The size of the filter is the size of a standard width lane marking projected in the image space, and varies because of the perspective effect. The following equation shows the differentiation filter that is used:

$$G_N(x,y) = \frac{\sum_{i=x+1}^{x+D} I(i,y) - \sum_{i=x-1}^{x-D} I(i,y)}{2D} \qquad (1)$$
$$D = KernelSize(y)$$

Applying the variable width filter we preserve the level of detail in the distance while filtering the noise in the near areas. The gradient maxima and minima are paired and the DLD pairs are extracted as lane markings. The complete technique is described in [21].

Although the clothoid model is not always accurate for the urban scenario, it has several benefits, such as good results when the lane is delimited by simple edges (unmarked roads). Due to the short visibility range, we have decided to avoid matching the whole clothoid model on the image data, but to match pairs of line segments instead, in two zones: near and far.

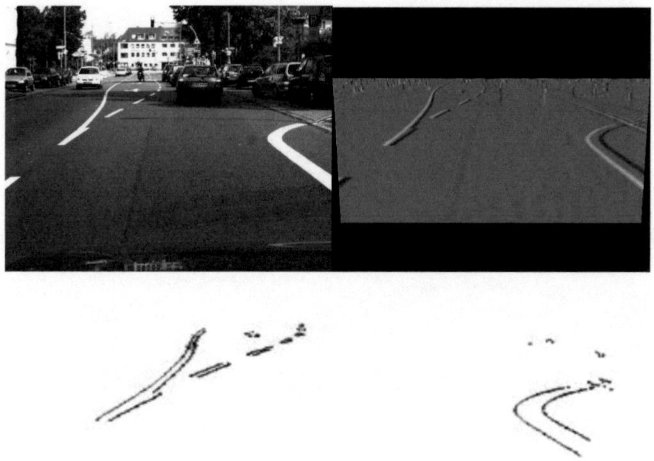

Fig. 4. Lane marking detection - top left, original image; top right, results of the adaptive gradient filter; bottom, lane marking results

First, we make an attempt for the near zone (2m to 5 m). Hough transform is performed on the image edges corresponding to the road points, and line segments are extracted. Lane markings will have a higher weight in the Hough bins, and therefore they will have a higher priority. We divide then the line segments in two sets - left and right. The segments on the left are paired with the segments on the right, and the best pair is selected as our lane measurement. The linear measurement will update the clothoidal model parameters using the Extended Kalman Filter. Depending on whether the detection has been successful on both sides, or only on one side, the measurement vector for the Kalman filter will have different configurations:

$$Y_{both} = \begin{bmatrix} xbottomleft \\ xtopleft \\ xbottomright \\ xtopright \end{bmatrix} \quad Y_{left} = \begin{bmatrix} xbottomleft \\ xtopleft \end{bmatrix} \quad Y_{right} = \begin{bmatrix} xbottomright \\ xtopright \end{bmatrix} \tag{2}$$

If the linear fit for the near zone is successful, the same is done for the far zone, and the model parameters are updated again.

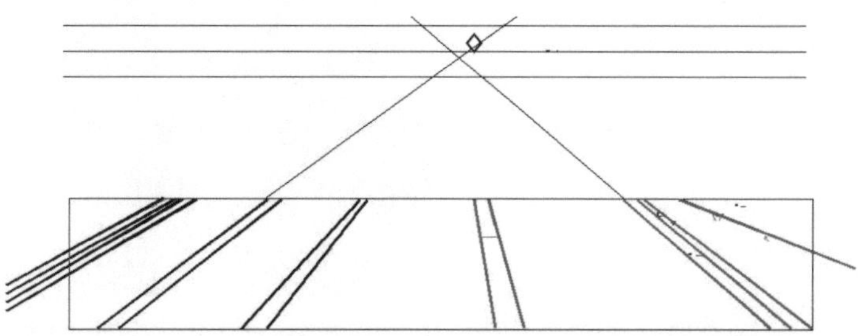

Fig. 5. The left and the right lines must intersect at the horizon line

Fig. 6. Linear lane detection result

Sometimes the clothoid lane model is not suited for the road we wish to observe, and the detection will be incorrect. For these cases, a freeform lane detection system has been implemented. Because we don't have strong models to guide our search, we have to discard the non-marking delimiters, and work with lane markings only. The markings are projected onto a top view image, and then distance transform is performed, to facilitate the fitting of a lane border. The left and the right lane borders are represented as Catmull-Rom splines with four control points. The lateral coordinates of the four control points are the model parameters, and they are found using a simulated annealing search strategy in the model space.

The result of the freeform detection module is a chain of 3D X coordinates for a set of equally spaced fixed Z coordinates. The X values will form the measurement vector for the lane state update using again the Kalman filter.

$$Y_{side} = [X_1, X_2, ..., X_n]^T \tag{3}$$

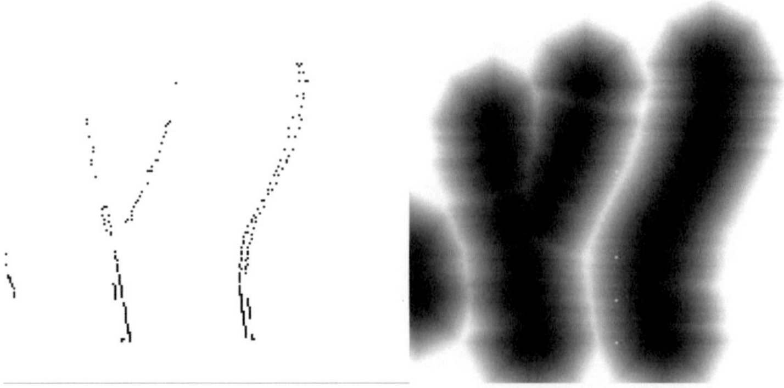

Fig. 7. Top view of the lane markings and the distance transform image used for freeform lane matching

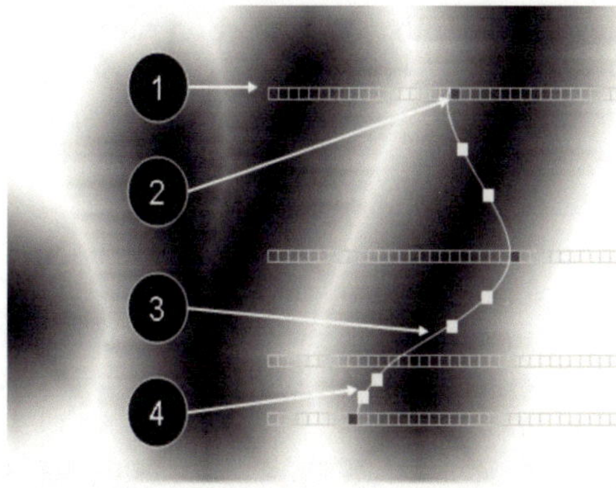

Fig. 8. The search problem for freeform detection: 1 - The search space for one of the control points. The y coordinate is fixed; 2 - A control point instance (hypothesis); 3- The Catmull-Rom spline generated by the control points hypotheses in the current iteration; 4 - The intermediate points, which, together with the control points, are used for evaluating the distance between the curve and the image

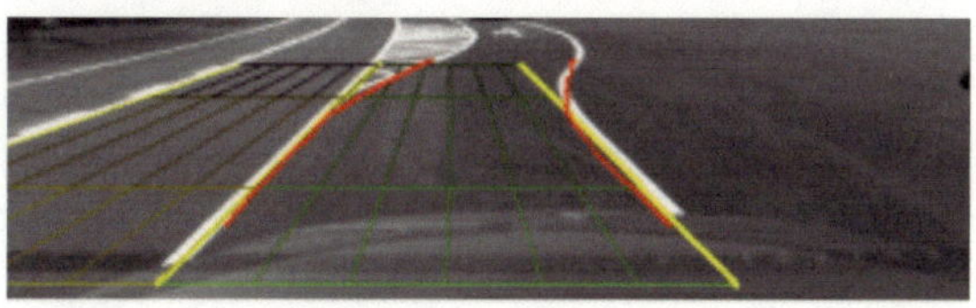

Fig. 9. Freeform lane detection succeeds in situations where the clothoid model fails

4. Stereovision-based obstacle detection and tracking

4.1 Stereovision-based obstacle detection

The obstacle detection algorithm is based on the dense stereo information, which provides a rich amount of 3D data for the observed scene and allows the possibility to carry out geometrical reasoning for generic obstacle detection regardless of the 2D appearance in images.

Starting from the obstacle points identified by the vertical profile provided by the lane detection subsystem, the target is to detect the obstacles as 3D boxes, having position, orientation and size. The confident fitting of cuboids to obstacles is achieved in several steps. By analyzing the vicinity and the density of the 3D points, the occupied areas are located. An occupied area consists of one or more cuboidal obstacles that are close to each other. By applying criteria regarding the shape, the occupied areas may get fragmented into parts that obey the cuboidal shape. The orientation of the obstacles (on the road surface) is then extracted.

The only 3D points used by the obstacle detection algorithms are those situated above the road and below the height of the ego car (fig. 10.b). It is supposed that the obstacles do not overlap each other on the vertical direction. In other words, on a top view (fig. 10.c) the obstacles are disjoint. Consequently, in what follows the elevation of the 3D points will be ignored and all the processing is done on the top view.

Due to the perspective effect of the camera, further obstacles appear smaller in our images, providing fewer pixels, and therefore, less, sparser 3D reconstructed points in the 3D space.

On the other hand, the error of the depth reconstruction increases with the distance too, which contributes to the 3D points sparseness as well. To counteract the problem of the points' density, a schema to divide the Cartesian top view space into tiles of constant density is proposed (fig. 10.c). The horizontal field of view of the camera is divided into polar slices of constant aperture, trying to keep a constant density on the X-axis. The depth range is divided into intervals, the length of each interval being bigger and bigger as the distance grows, trying to keep a constant density on the Z-axis.
A specially compressed space is created, as a matrix (fig. 11.a). The cells in the compressed space correspond to the trapezoidal tiles of the Cartesian space. The compressed space is, in fact, a bi-dimensional histogram, each cell counting the number of 3D points found in the corresponding trapezoidal tile. For each 3D point, a corresponding cell C (*Row, Column*) in the compressed space is computed, and the cell value is incremented.
The column number is found as:

$$Column = ImageColumn/c \tag{4}$$

where *ImageColumn* is the left image column of the 3D point and c is the number of adjacent image columns grouped into a polar slice as shown in fig. 10.c ($c = 6$).
The depth transformation, from the Z coordinate of the Cartesian space into the *Row* coordinate of the compressed space has a formula obtained using the following reasoning:

a. The Cartesian interval corresponding to the first row of the compressed spaces is:

$$[Z_0 \ldots Z_0 + IntervalLength(Z_0)] = [Z_0 \ldots Z_1] \tag{5}$$

where $Z_0 = Zmin$, the minimum distance available through stereo reconstruction. The length of the interval beginning at a certain Z is

$$IntervalLength(Z) = k*Z \tag{6}$$

k being empirically chosen). Thus

$$Z_0 + IntervalLength(Z_0) = Z_0+k*Z_0 = Z_0*(1+k) = Z_1 \tag{7}$$

b. The Cartesian interval corresponding to the nth row of the compressed spaces is $[Z_n \ldots Z_n + IntervalLength(Z_n)]$,
where

$$Z_n=Z_0*(1+k)^n \tag{8}$$

The above equation can be proven by mathematical induction.

c. For a certain 3D point, having depth Z, the ith interval it belongs to is $[Z_i \ldots Z_i + IntervalLength(Z_i)] = [Z_i \ldots Z_{i+1}]$.
From equation (8), we can find the row number as the index of the point's interval

$$Row = i = [\log_{1+k} \frac{Z}{Z_0}] \tag{9}$$

Each 3D point, having the (X, Z) coordinates in the top view of the Cartesian space is transformed into a cell C(*Row, Column*) in the compressed space.

a)

b)

c)

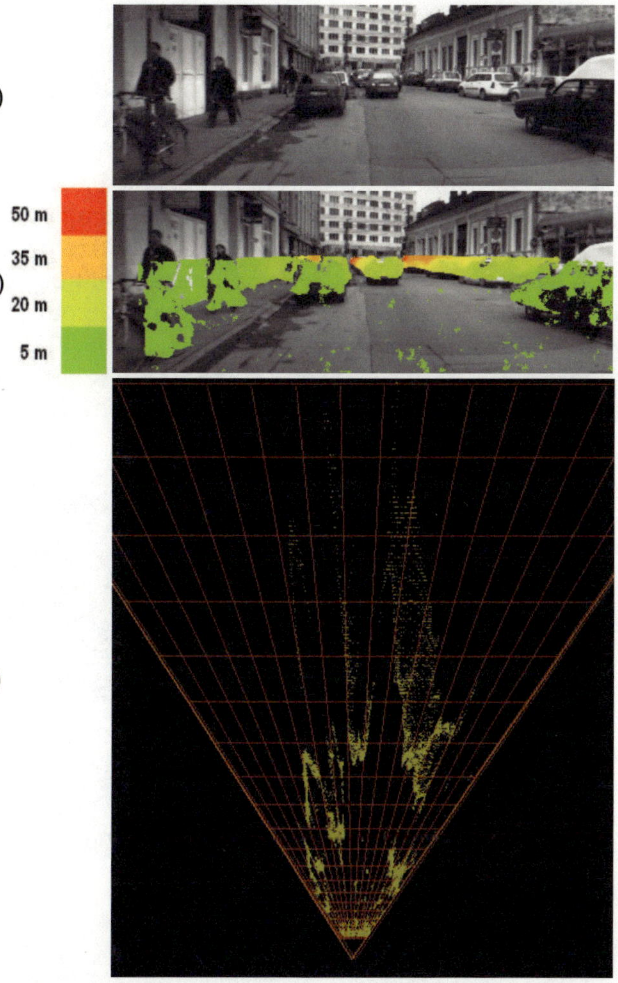

Fig. 10. Division into tiles. a) Gray scale image, b) 3D points - perspective view, c) 3D points - top view; the tiles are here considerably larger for visibility purpose; wrongly reconstructed points can be seen in random places; reconstruction error is visible as well.

The histogram cells that have a significant number of points indicate the presence of obstacle areas. On these cells, a labeling algorithm is applied, and the resulted clusters of cells represent occupied areas (fig. 11.b). The small occupied areas are discarded in this phase.

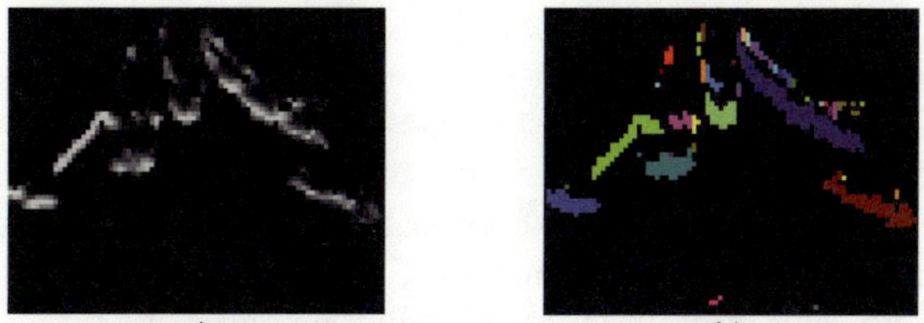

a) b)

Fig. 11. The compressed space (for scene in fig. 7) – a bi-dimensional histogram counting 3D points. Occupied areas are identified by cell labeling

The occupied areas may contain several obstacles and by consequence they may have multiple shapes. The obstacle tracking algorithms as well as the driving assistance applications need the individual cuboidal obstacles. Therefore the fragmentation of the occupied areas into the individual cuboidal obstacles is required.
An indication that an area requires fragmentation is the presence of concavities. In order to detect the concavities, the envelope of the cells of an occupied area is drawn, and then for each side of the envelope, the gap between the side and the occupied cells is determined. If the gap is significant, we consider that two or more obstacles are present, and the deepest point of the concavity gives the column where the division will be applied. The two sub-parts can be divided again and again as long as concavities are found.
In fig. 12.c the bottom side of the envelope for the cells in fig. 12.b delimits a significant concavity. For each new sub-part, the envelope of the cells has been calculated again (and painted as well), but without revealing big concavities for new divisions to be performed.

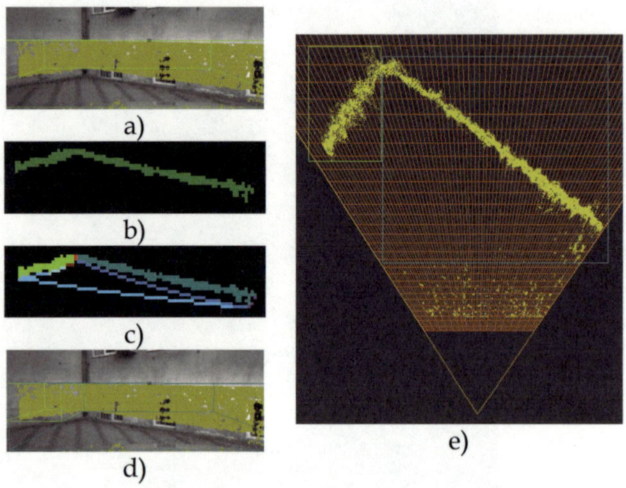

Fig. 12. Fragmentation of occupied areas into primitive obstacles. a) an occupied area, b) the labeling in the compresses space, c) sides of the envelope and the two primitive obstacles of the occupied area - compressed space, d) the two primitive obstacles - perspective view, e) the two primitive obstacles - top view

By reconsidering the coordinates (including the height) of the 3D points that have filled the cells of an obstacle, the limits of the circumscribing box are determined. Boxes are shown in fig. 12.d (perspective view) and fig. 12.e (top view). A more detailed description of the obstacle detection system is given in [22].

4.2 Stereovision-based obstacle tracking

Tracking is the final stage of the stereovision based object recognition system. The cuboids extracted using the methods described in the previous paragraphs represent the measurement data. The state of the object, composed of position, size, speed and orientation, is tracked using a Kalman filter based framework.
A measurement cuboid may initialize a track if several conditions are met: the cuboid is not associated to an existing track, the cuboid is on the road (we compare its Y position with the profile of the road), the cuboid's back side position in the image does not touch the image

limits, and the height and width of the cuboid must be consistent to the standard size of the vehicles we expect to find on the road. The classification based on size is also useful for initialization of the object's length, as in most cases the camera cannot observe this parameter directly.

The core of the tracking algorithm is the measurement-track association process. The association (matching) process has two phases: a 3D matching of the predicted track cuboid against the measurements, which is performed as a simple intersection of rectangles in the top view space, and a corner by corner matching in the image space, when each active corner is matched against the corners of the measurement cuboids that passed the 3D intersection test.

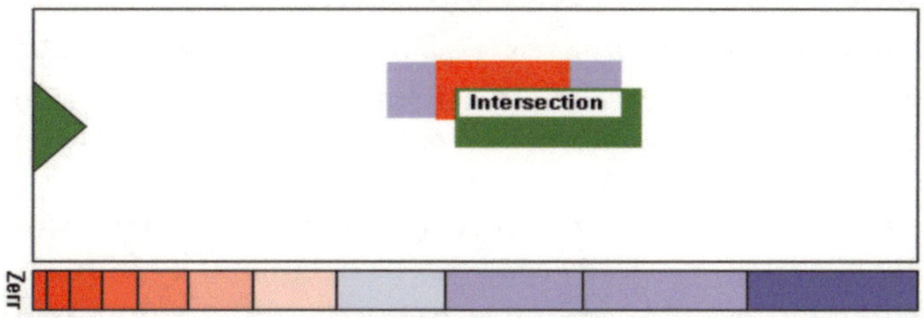

Fig. 13. Coarse association between prediction and measurement cuboids

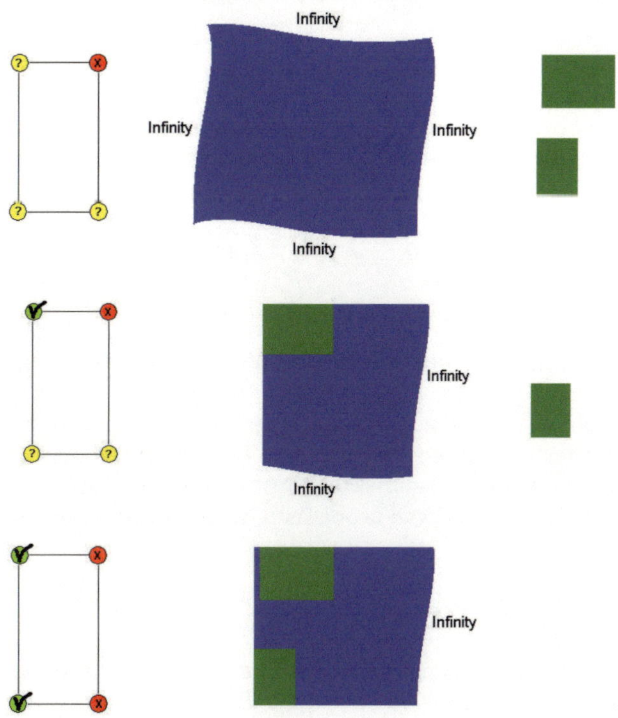

Fig. 14. Fine association at corner level and the building of a meta-measurement

The ratio between the area of the intersection and the area of the enhanced prediction is a measure of the quality of the 3D association. If a measurement intersects multiple predicted

objects, it will be associated to the one it had the best intersection measure. A predicted object may associate to multiple measurement objects, but not the other way around.
After the 3D association is completed, each track prediction is compared to each of the associated measurements corner by corner, using their 2D projection, for the visible corners only. If a relevant corner of the prediction associates to at least one relevant corner of a measurement, this corner becomes an "active" corner. The active corners form a virtual object which we'll call "meta-measurement". The meta measurement is the sum of all measurement objects associated to a predicted object (Fig. 14). The meta measurement is used in the Kalman filter for updating the track state.

5. Description of unstructured environments

The 3D lane cannot be detected in urban scenes without lane delimiters (ex. intersections). An alternative method can be used to detect elevated areas (obstacles), regions where the ego vehicle cannot be driven into. Complementary, the obstacle-free road areas can be considered as drivable.
The dense stereo engine normally reconstructs most of the road points even if lane markings are not present. Thus, the surface of the road can be computed by fitting a geometric model to the 3D data. Fitting the model to 3D data should be performed in a least-square fashion (LSQ), or, more robustly, using a statistical approach (ex. RANSAC). The model used for the road is quadratic (the vertical coordinate Y is a 2nd degree function of the depth Z and lateral displacement X).

$$Y = -a \cdot X - a' \cdot X^2 - b \cdot Z - b' \cdot Z^2 - c. \quad (10)$$

Fitting the quadratic surface to a set of n 3D points involves minimizing an error function. The error function S represents the sum of squared errors along the height:

$$S = \sum_{i=1}^{n} \left(Y_i - \overline{Y_i} \right)^2, \quad (11)$$

Where Y_i is the elevation of the 3D point i and $\overline{Y_i}$ is the elevation of the surface at coordinates (X_i, Z_i). Minimizing only along the Y-axis, instead of the surface normal, is acceptable. Even for curved roads, the normal of the surface is close to the Y-axis: for an extreme local slope of 20% (11.3 degrees), the residual of a 3D point along the vertical represents 98% of the residual along the normal. The computational complexity is highly reduced by avoiding minimization against the normal of the surface.
By replacing (10) into (11), the function S is obtained, where the unknowns are a, a', b, b', and c:

$$S = \sum_{i=1}^{n} \left(Y_i + a \cdot X_i + a' \cdot X_i^2 + b \cdot Z_i + b' \cdot Z_i^2 + c \right)^2. \quad (12)$$

For S to have a minimum value, its partial derivatives with respect to the unknowns must be 0. The following system of equations must be solved:

$$\left\{ \frac{\partial S}{\partial a} = 0, \frac{\partial S}{\partial a'} = 0, \frac{\partial S}{\partial b} = 0, \frac{\partial S}{\partial b'} = 0, \frac{\partial S}{\partial c} = 0. \right. \quad (13)$$

After writing explicitly each equation, the system (4) becomes (matrix form):

$$\begin{bmatrix} S_{X^2} & S_{X^3} & S_{XZ} & S_{XZ^2} & S_X \\ S_{X^3} & S_{X^4} & S_{X^2Z} & S_{X^2Z^2} & S_{X^2} \\ S_{XZ} & S_{X^2Z} & S_{Z^2} & S_{Z^3} & S_Z \\ S_{XZ^2} & S_{X^2Z^2} & S_{Z^3} & S_{Z^4} & S_{Z^2} \\ S_X & S_{X^2} & S_Z & S_{Z^2} & n \end{bmatrix} \begin{bmatrix} a \\ a' \\ b \\ b' \\ c \end{bmatrix} = \begin{bmatrix} -S_{XY} \\ -S_{X^2Y} \\ -S_{ZY} \\ -S_{Z^2Y} \\ -S_Y \end{bmatrix}, \quad (14)$$

Generically, each sum is computed as

$$S_\alpha = \sum_{i=1}^{n} \alpha_i \text{ (for example } S_{XZ} = \sum_{i=1}^{n} X_i \cdot Z_i\text{)}. \quad (15)$$

If weights (w) for each point are available, then the following formulas will be applied:

$$S_\alpha = \sum_{i=1}^{n} w_i \cdot \alpha_i \text{ (example } S_{XZ} = \sum_{i=1}^{n} w_i \cdot X_i \cdot Z_i\text{)}. \quad (16)$$

System (14) has 5 linear equation and 5 unknowns, therefore solving it is a trivial algebra problem. This explicit way of minimization was preferred instead of the pseudo-inverse matrix method. It allows real time re-computation (hundreds of times per frame) of the road surface during the surface growing step, as it will be explained later in section. This model allows the detection of road surfaces with non-zero pitch and roll angles relative to the ego car, and with vertical curvatures (lateral/longitudinal). The model can be extended to fit complex road surfaces, such as cubic or B-spline surfaces.

The 3D space available consists of a set of 3D points (80,000 to 120,000). Real-time fitting of the road surface to this set is not possible because it has a high computational complexity. A (bird-eye rectangular, 13x40 meters) region of interest of the 3D space can be represented similar to a digital elevation map (DEM). A DEM is formed: the value of each cell is proportional to the 3D height of the highest point (within the cell). The DEM presents poor connectivity between road points at far depths (due to the perspective effect, Fig. 15.b). Connectivity can be improved by propagating (to empty cells) the height of valid cell (with 3D points), along the depth (Fig. 15.c). The propagation amount is computed from the stereo geometry of the system, to compensate the perspective effect.

For the classification into drivable/non-drivable (road inliers/outliers, Fig. 16) areas, the depth uncertainty model was extended to a height uncertainty model (17). The expected height uncertainty *Yerr* is a function of the height *Y* and the depth *Z* of the 3D point, height of the cameras *Hcam*, and the estimated depth uncertainty *Zerr*. *Zerr* is a function of the stereo system parameters and the expected disparity uncertainty. The disparity uncertainty was chosen experimentally as 1 pixel, although a more complex model for estimating the correlation's accuracy can be developed.

$$Yerr = \left| \frac{(Y - Hcam) * Zerr}{Z} \right|. \quad (17)$$

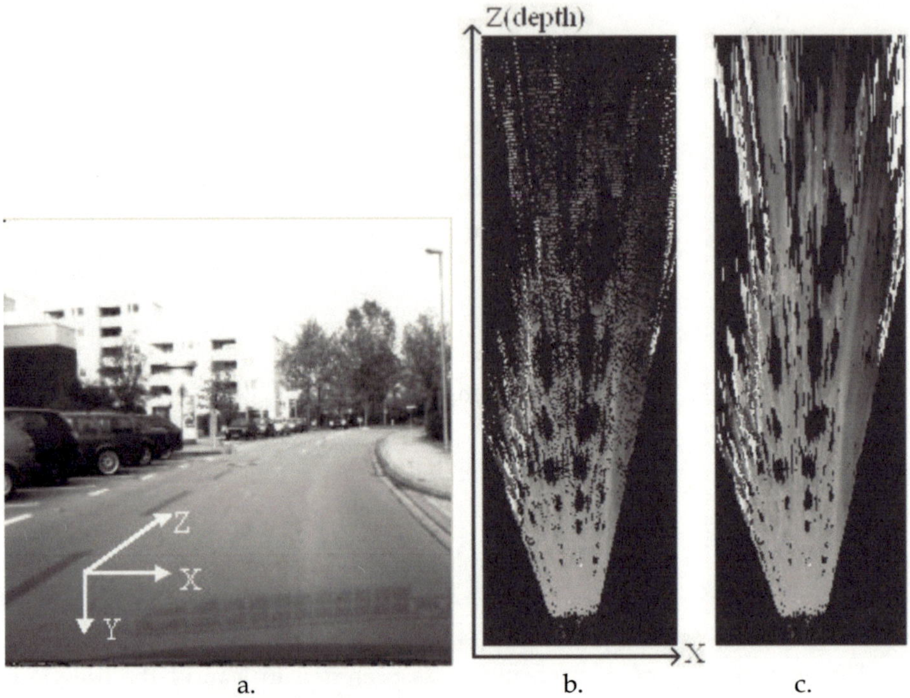

Fig. 15. The urban scenario (a) and the DEM (b. initial, c. with propagation of heights). Darker means more elevated.

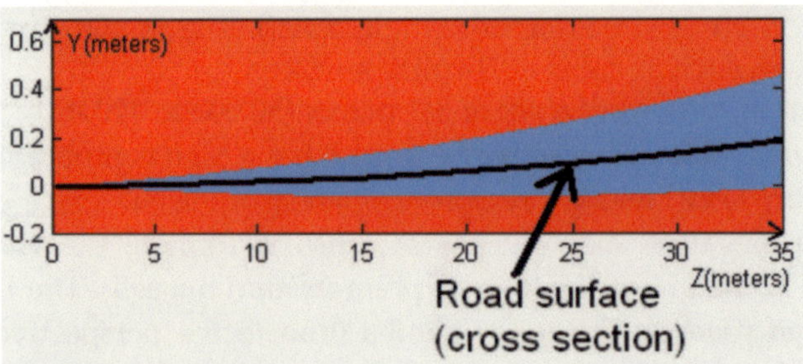

Fig. 16. Lateral view of the inliers (blue) region around a quadratic road surface, according to the proposed uncertainty model.

LSQ fitting is sensitive to rough noise in the data set. Due to road outliers, such as obstacle points, LSQ fitting is likely to fail when applied to the whole data set. We propose an optimal two-step approach for real-time detection of the road surface:

1. The road model is fitted, using RANSAC, to a small rectangular DEM patch in front of the ego vehicle. A primary road surface is extracted optimally for the selected patch.
2. The primary solution is refined through a region growing process (Fig. 17) where the initial region is the set of road inliers of the primary surface, from the initial rectangular patch. New cells are added to the region if they are on the region border and they are inliers of the current road surface. The road surface is recomputed (LSQ fitting to the current region), each time the border of the region expands with 1-2 pixels (about 100 new cells).

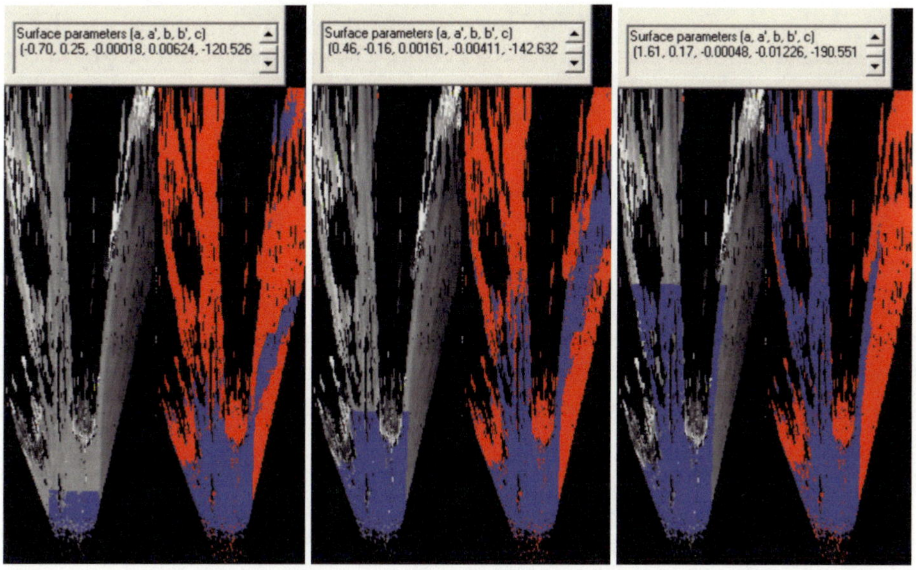

a. b. c.

Fig. 17. Road inliers (blue) / outliers (red) detected with the primary surface in a. Intermediate regions, growing from left to right (b and c). For each image, the left side shows the inliers used for surface fitting, while the right side shows the classification of the whole DEM based on the current surface parameters.

Fig. 18. The result for two scenes re-projected back as a grid onto the left image: obstacles (red) and traffic isles (yellow) are separated from the drivable road surface (blue).

Obstacle / road separation is performed based on the detected road surface. Each DEM cell is labeled as drivable if it is closer to the road surface than its estimated height uncertainty, or as non-drivable otherwise. Small clusters of non-drivable cells are rejected based on criteria related to the density of 3D points. Remaining non-drivable clusters are classified into obstacles and traffic isles, based on the density of 3D points. This approach is detailed in [13]. Some results are presented in figure 18, re-projected as a grid onto the left image.

6. Object classification

Out of the set of 3D cuboids depicting the obstacles in the urban traffic, one category of objects is of special importance: the pedestrians. The pedestrians are the most vulnerable traffic participants, and also the most undisciplined and having the most unpredictable behavior. For these reasons, the pedestrians need to be recognized as early and as reliably as possible.

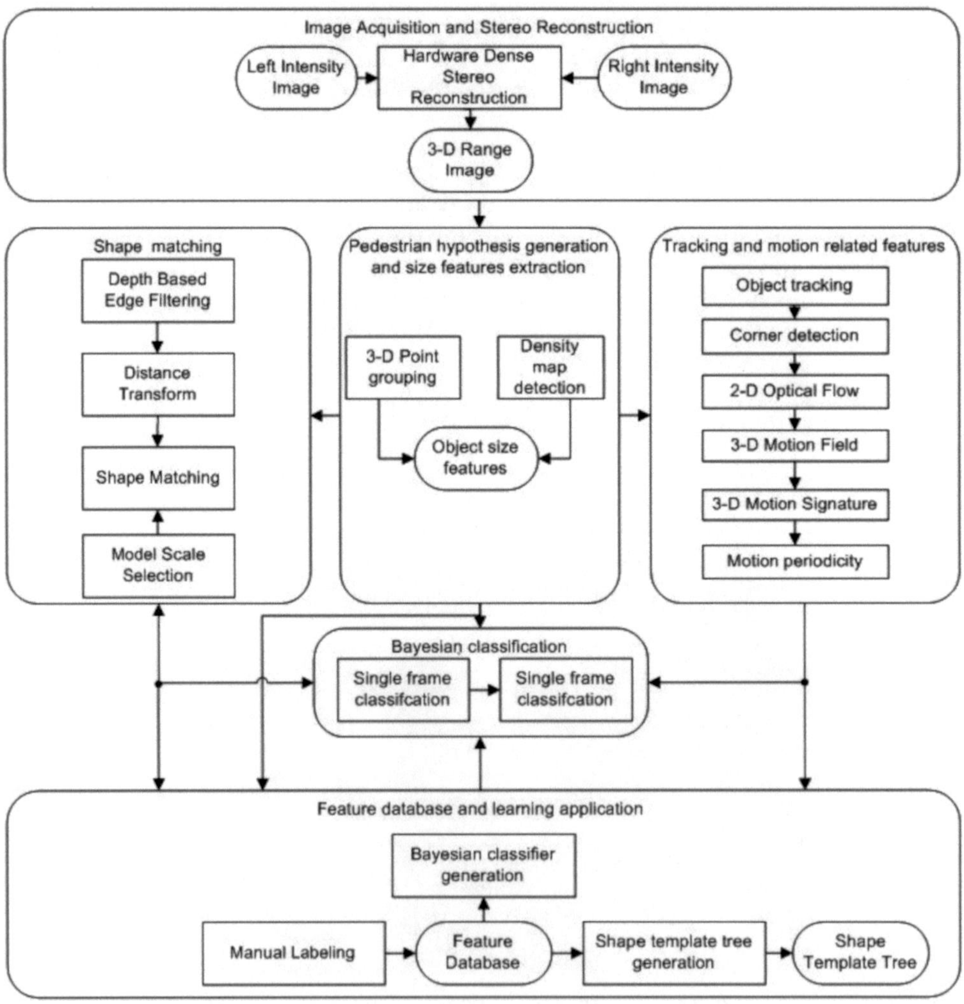

Fig. 19. Classification dataflow

In complex, urban scenarios, pedestrians are difficult to detect. This is because the variability of their appearance and the high clutter present in this type of scenarios. Therefore, as much information as possible must be used in order to correctly recognize them. A multi-feature dichotomizer was implemented [23], which can separate obstacles into pedestrians and non-pedestrians. The implementation contains the following modules (see Fig. 19):

- The *image acquisition and stereo reconstruction module,* described in section 2, supplies a dense range image, using the image pair from the stereo camera system.
- The *hypothesis generation and size features extraction* module generates pedestrian hypotheses, which will be classified into pedestrians and non-pedestrians. For short range (up to 10 meters) a specialized algorithm is used to generate reliable pedestrian hypotheses, containing only individual pedestrians. The algorithm is based on a "density map" built by accumulating 3D points, projected on the horizontal xOz plane [23]. As this algorithm does not give good results for distant objects, we use the generic obstacle detected by the algorithm described in section 4 to supply hypotheses which are further away than 10 meters. Because the objects are represented as cuboids, we can extract their height and base radius, and use them as features for classification.
- *Shape matching:* An important feature for pedestrian recognition is the specific pedestrian shape. Having a 3D cuboid, we obtain its projection as a 2D box in the image. Edges in the 2D box are extracted using a Canny filter. The extracted edges are filtered, using depth, and only those edges which have a correct depth are retained. These form the outline of the pedestrian hypothesis. We apply a distance transform on these outline edges. The shape matching algorithm uses a hierarchy (tree) of pedestrian shape templates, constructed offline. A top-down matching algorithm based on the Chamfer distance is used to select the shape template that has the best match with the outline of the pedestrian, and the matching score is supplied as a feature for classification. The scale of template is inferred from the distance of the 3D box. Our approach is similar to [24], but makes more use of the available 3D information.
- *Tracking and motion related features:* Another powerful feature for pedestrian recognition is the specific way in which pedestrians move. Pedestrians are articulated objects, as opposed to most objects present in traffic environments which are rigid. Pedestrians move their arms and legs while walking or running. By tracking the individual motions of different body parts, and observing motion variations across them, we can supply a motion based feature, called "motion signature".

The pedestrian hypothesis is tracked using the tracking algorithm described in section 4. This step is useful for both object association and to eliminate the global object motion. The next step is to select the foreground points of each object. Only points for which their 3D coordinates lie inside the tracked cuboid are considered. This step is important as it eliminates spurious background points and deals with partial occlusions. Corner points are detected, and 2D optical flow vectors computed using the approach described in [25]. The optical flow vectors are transformed from 2D to 3D using the available depth information supplied by stereo. Principal component analysis is used to find the principal direction of the 3D velocity field variation for each individual object. Variance is smoothed across frames, to increase its stability. The magnitude of this principal component represents the "motion signature". This motion signature is much smaller for non-pedestrians as compared to pedestrians, and thus it is a powerful feature for pedestrian detection. The spectrum of the motion signature variation in time is analyzed. Pedestrians display a typical periodic motion signature, while other types of objects display only impulsive noise. The cutoff frequency for the motion spectrum is much smaller for pedestrians as opposed to non-pedestrians. This makes the motion spectrum a powerful feature for pedestrian classification.

- *Bayesian classification:* A naïve Bayesian classifier is used to combine the extracted features (height, base radius, lateral and longitudinal speed, motion signature and the motion spectrum cutoff frequency). According to Bayes' formula, the probability of an object to belong to class C_1, if it displays features $F_1 \ldots F_n$ is:

$$P(C_1 \mid F_1, F_2, \ldots, F_n) = \frac{P(C_1)P(F_1, F_2, \ldots, F_n \mid C_1)}{P(F_1, F_2, \ldots, F_n)} \tag{18}$$

By assuming feature independence, switching from probability to likelihood and taking the logarithm we have:

$$\log\left(L(C_1 \mid F_1, F_2, \ldots, F_n)\right) = \frac{\log\left(P(C_1)P(F_1 \mid C_1)P(F_2 \mid C_1) \ldots P(F_n \mid C_1)\right)}{\log\left(P(\neg C_1)P(\neg F_1 \mid C_1)P(\neg F_2 \mid C_1) \ldots P(\neg F_n \mid C_1)\right)} \tag{19}$$

Finally we have:

$$\log\left(L(C_1 \mid F_1, F_2, \ldots, F_n)\right) = A + B_1 + B_2 + \ldots + B_n \tag{20}$$

Where A is the prior pedestrian likelihood and $B_1, \ldots, B_n$ are functions of feature values.

- *Feature Database and Learning Application*

In order to determine the A and B_k coefficients in equation 20, to obtain pedestrian shape templates and build the template tree, and to test our classification algorithm, a sufficiently large number of stereo image sequences had to be manually indexed. For each frame in an indexed sequence, the ground truth cuboids are stored, together with their class and extracted features. A manual labeling tool was developed, which uses the 3D cuboids detected by our hypothesis generation algorithms, and which asks the user to supply the ground truth class for each of them. Ground truth classes are tracked in order to allow more efficient labeling. From the set of indexed sequences, the subset of sequences where the ego vehicle was stopped is used to generate pedestrian shape templates for the pedestrian shape template tree. The tree is generated by an automatic clustering algorithm.
Figure 20 shows some results for pedestrian detection.

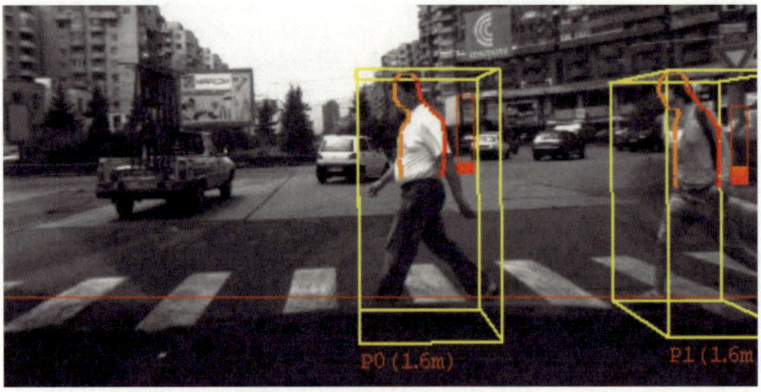

Fig. 20. Detected pedestrians

7. Possible use of stereovision in Driving Assistance Systems

The defining feature of a stereovision sensor is the capability of delivering rich meaningful data about the observed environment, combined with a reasonable precision of the 3D measurement. These features make the sensor suitable for a large variety of driving assistance applications.

The stereovision-based lane detection, capable of robust operation even when the lane markings are less than ideal or even absent, or when shadows or obstacles clutter the scene, can be used for *lane keeping and following assistance*, which can be implemented either as a form of warning or even as steering control.

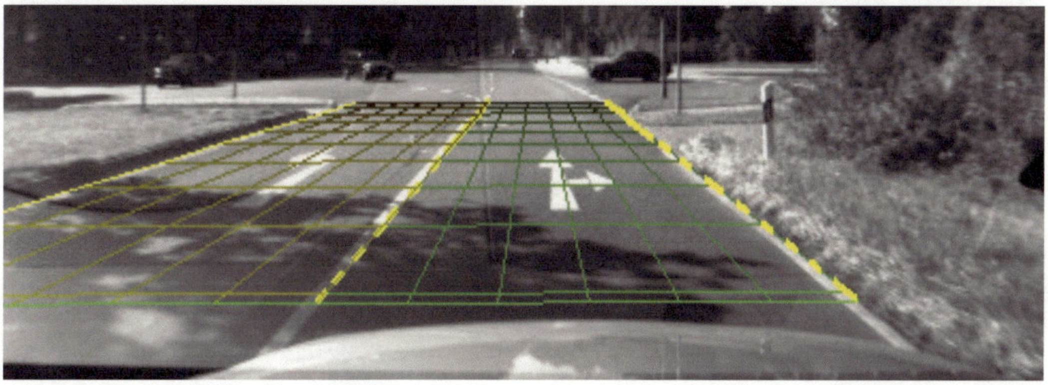

Fig. 21. Lane following assistance

Obstacle detection, in terms of 3D position, size and speed, provide the knowledge required for *collision avoidance*. The trajectory of the ego vehicle can be computed against the projected trajectory of the obstacles, the dangerous situations can be identified and appropriate measures such as emergency braking can be taken.

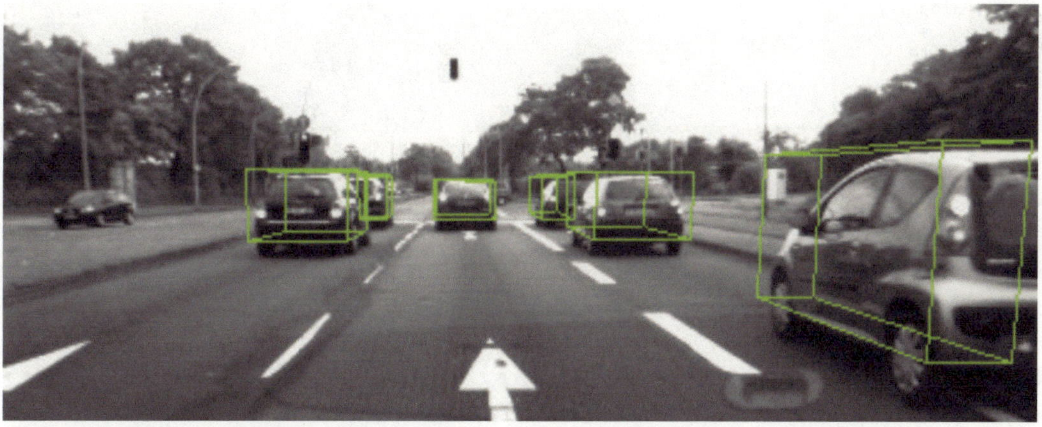

Fig. 22. Collision avoidance

A special type of collision is the one involving a pedestrian. The consequences of this collision are often very severe, and therefore it has to be avoided at (almost) all cost, even if some other type of unpleasant consequences may follow. This is the reason why the pedestrian has to be detected and classified as such as soon as it enters the field of view. The sensor we have described is thus suited for *pedestrian avoidance systems*.

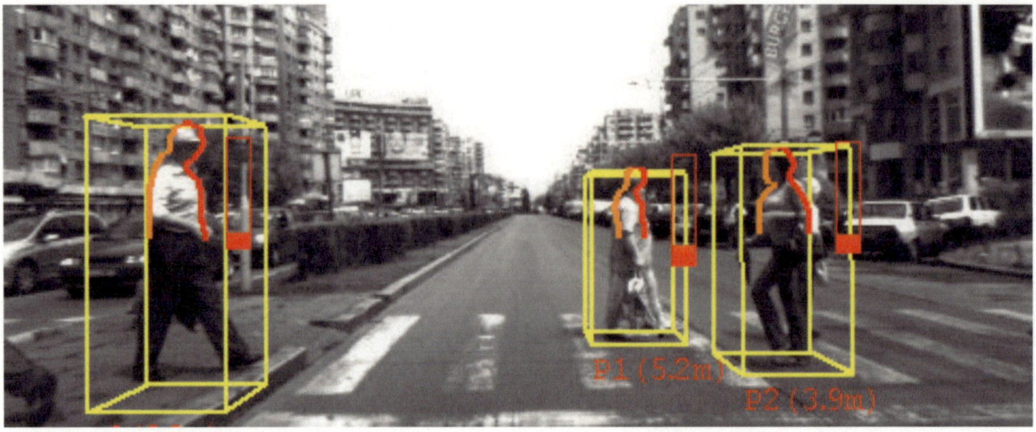

Fig. 23. Pedestrian avoidance

Accurate detection, measurement and tracking of the position, size and speed of the vehicle in front of us provides all the necessary information for a stop and go, (or follow the leader) assistance system, a helpful tool for the busy city traffic.

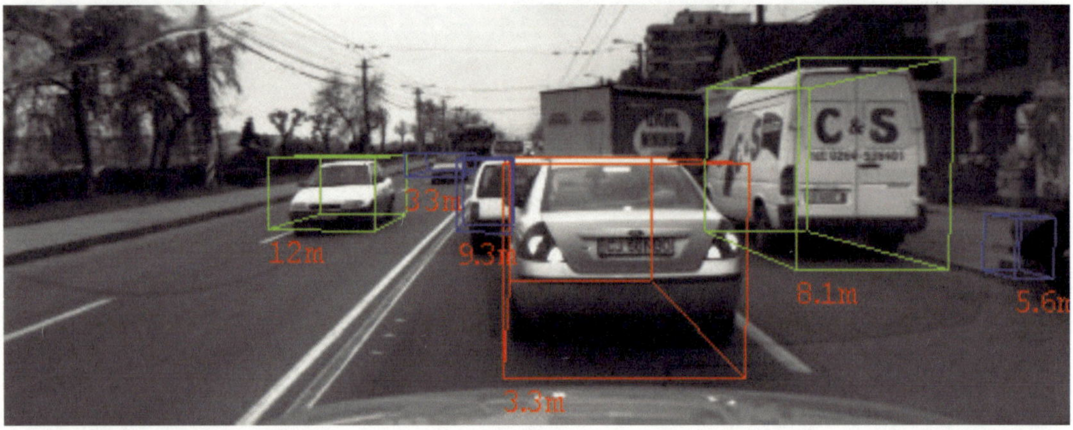

Fig. 24. Stop and go assistance

Sometimes, the environment is not easily represented in a structured way. In this case, the stereovision sensor can deliver information about areas that the vehicle is allowed to drive on, and about the forbidden areas, providing information for navigation assistance in unstructured environments.

8. References

M. Bertozzi, A. Broggi, A. Fascioli, and S. Nichele, "Stereo Vision-based Vehicle Detection", in *Proceedings of IEEE Intelligent Vehicles Symposium (IV 2000)*, October 2000, Detroit, USA, pp. 39-44.

U. Franke, D. M. Gavrila, S. Görzig, F. Lindner, F. Paetzold and C. Wöhler, "Autonomous Driving Approaches Downtown", *IEEE Intelligent Systems*, vol.13, no. 6, pp. 40-48, 1998.

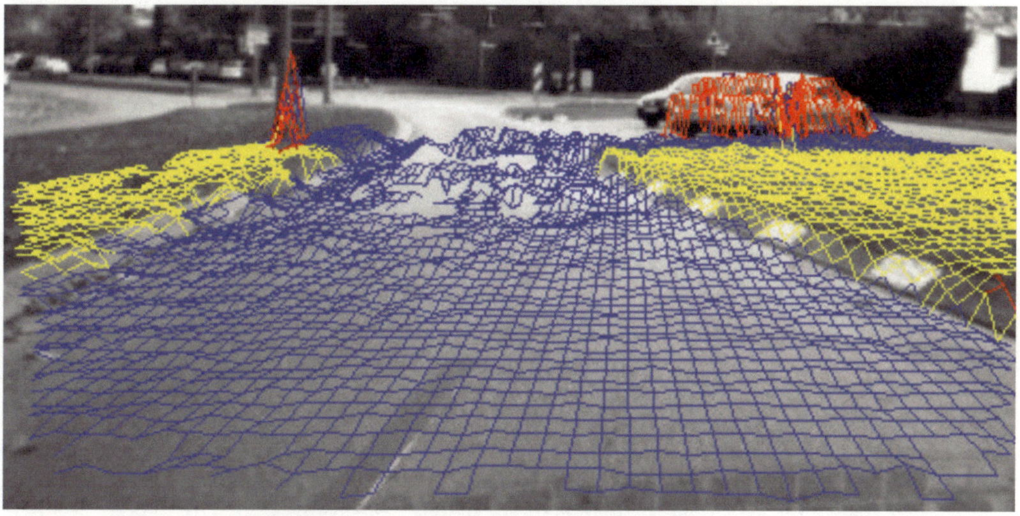

Fig. 25. Navigation assistance for unstructured environments

T. A. Williamson, "A high-performance stereo vision system for obstacle detection", CMU-RI-TR-98-24, September 25, 1998, *Robotics Institute Carnegie Mellon University*, Pittsburg, PA 15123.

M. Bertozzi and A. Broggi, "GOLD: a Parallel Real-Time Stereo Vision System for Generic Obstacle and Lane Detection", *IEEE Transactions on Image Processing*, vol. 7, no. 1, pp. 62-81, January 1998.

R. Labayrade, D. Aubert, and J.-P. Tarel, "Real time obstacle detection on non flat road geometry through V-disparity representation," in *Proceedings of IEEE Intelligent Vehicles Symposium (IV 2002)*, June 2002, Versailles, France, pp. 646–651.

S. Nedevschi, R. Danescu, D. Frentiu, T. Marita, F. Oniga, C. Pocol, R. Schmidt, T. Graf, "High accuracy stereo vision system for far distance obstacle detection", in *Proceedings of IEEE Intelligent Vehicles Symposium (IV 2004)*, June 2004, Parma, Italy, pp. 292-297.

S. Nedevschi, R..Schmidt, T. Graf, R. Danescu, D. Frentiu, T. Marita, F. Oniga, C. Pocol, "3D Lane Detection System Based on Stereovision", in *Proceedings of IEEE Intelligent Transportation Systems Conference (ITSC'04)*, October 2004, Washington, USA, pp. 161-166.

S. Nedevschi, R. Danescu, T. Marita, F. Oniga, C. Pocol, S. Sobol, T. Graf, R. Schmidt, "Driving Environment Perception Using Stereovision", in *Proceedings of IEEE Intelligent Vehicles Symposium, (IV 2005)*, June 2005, Las Vegas, USA, pp.331-336.

S. Nedevschi, F. Oniga, R. Danescu, T. Graf, R. Schmidt, Increased Accuracy Stereo Approach for 3D Lane Detection, *Proceedings of IEEE Intelligent Vehicles Symposium*, (IV2006), June 13-15, 2006, Tokyo, Japan, pp. 42-49.

A. Sappa, D. Gerónimo, F. Dornaika, and A. López, "Real Time Vehicle Pose Using On-Board Stereo Vision System", Int. Conf. on Image Analysis and Recognition, *LNCS*, Vol. 4142, September 18-20, 2006, Portugal, pp. 205-216.

M. Cech, W. Niem, S. Abraham, and C. Stiller, "Dynamic ego-pose estimation for driver assistance in urban environments", in *Proceedings of IEEE Intelligent Vehicles Symposium (IV 2004)*, Parma, Italy, 2004, pp. 43-48.

Britta Hummel, Soeren Kammel, Thao Dang, Christian Duchow, Christoph Stiller, Vision-based Path Planning in Unstructured Environments, in *Proceedings of IEEE Intelligent Vehicles Symposium (IV 2006)*, June 13-15, 2006, Tokyo, Japan, pp. 176-181.

F. Oniga, S. Nedevschi, M-M. Meinecke, T-B. To, "Road Surface and Obstacle Detection Based on Elevation Maps from Dense Stereo", in *Proceedings of the 10th International IEEE Conference on Intelligent Transportation Systems (ITSC'07)*, Sept. 30 - Oct. 3, 2007, Seattle, Washington, USA.

F. Oniga, S. Nedevschi, M-M. Meinecke, "Curb Detection Based on Elevation Maps from Dense Stereo", in *Proceedings of the 3rd International IEEE Conference on Intelligent Computer Communication and Processing (ICCP 2007)*, Sept. 6-8, 2007, Cluj-Napoca, Romania, pp.119-125,.

T. Marita, F. Oniga, S. Nedevschi, T. Graf, R. Schmidt, Camera Calibration Method for Far Range Stereovision Sensors Used in Vehicles, in *Proceedings of IEEE Intelligent Vehicles Symposium (IV 2006)* , June 13-15, 2006, Tokyo, Japan, pp. 356-363.

S. Nedevschi, C.Vancea, T. Marita, T. Graf, "On-line calibration method for stereovision systems used in vehicle applications", in *Proceedings of 2006 IEEE Intelligent Transportation Systems Conference (ITSC'06)*, September 17-20, 2006, Toronto, Canada, pp. 957 - 962.

S. Nedevschi, C. Vancea, T. Marita, T. Graf, "On-Line Calibration Method for Stereovision Systems Used in Far Range Detection Vehicle Applications", *IEEE Transactions on Intelligent Transportation Systems*, vol. 8, no. 4, pp. 651-660, December, 2007.

C. Vancea, S. Nedevschi, "Analysis of different image rectification approaches for binocular stereovision systems", in *Proceedings of IEEE 2nd International Conference on Intelligent Computer Communication and Processing (ICCP 2006)*, September 1-2, 2006, Cluj-Napoca, Romania, pp. 135-142.

R. Danescu, S. Nedevschi, M.M. Meinecke, T.B. To, "Lane Geometry Estimation in Urban Environments Using a Stereovision System", in *Proceedings of the IEEE Intelligent Transportation Systems Conference (ITSC'07)*, Sptember 30th- October 3rd, 2007, Seattle, USA, pp. 271-276.

J. Goldbeck, B. Huertgen, "Lane Detection and Tracking by Video Sensors", in *Proceedings of IEEE International Conference on Intelligent Transportation Systems (ITSC'99), October 5-8, 1999, Tokyo Japan, pp. 74–79.*

R. Danescu, S. Nedevschi, "Robust Real-Time Lane Delimiting Features Extraction", in Proceedings of *2nd IEEE International Conference on Intelligent Computer Communication and Processing (ICCP 2006)*, September 1-2, 2006, Cluj Napoca, Romania, pp. 77-82.

C. Pocol, S. Nedevschi, M. M. Meinecke, "Obstacle Detection Based on Dense Stereovision for Urban ACC Systems", in *Proceedings of 5th International Workshop on Intelligent Transportation (WIT 2008)*, March 18-19, 2008, Hamburg, Germany, pp. 13-18.

S. Nedevschi, C. Tomiuc, S. Bota, "Stereo Based Pedestrian Detection for Collision Avoidance Applications", in *Proceedings of Workshop on Planning, Perception and Navigation for Intelligent Vehicle, at IEEE ICRA 2007*, April 10-14, 2007, Roma, Italy, pp. 39-44.

D. M. Gavrila, J. Giebel, and S. Munder, "Vision-based pedestrian detection: the PROTECTOR system", in *Proceedings of Intelligent Vehicles Symposium (IV 2004)*, June, 2004, Parma, Italy, pp. 13–18.

J.-Y. Bouguet, "Pyramidal implementation of the Lucas Kanade feature tracker", Available: http://mrl.nyu.edu/ bregler/classes/vision spring06/bouget00.pdf

15

Stereo Algorithm with Reaction-Diffusion Equations

Atsushi Nomura[1], Makoto Ichikawa[2], Koichi Okada[1] and Hidetoshi Miike[1]
[1]Yamaguchi University, [2]Chiba University
Japan

1. Introduction

Three-dimensional depth reconstruction from a pair of stereo images needs to find reliable stereo correspondence between left and right images. Stereo disparity refers to difference of positions between two corresponding points on the stereo images. Although there are many possibilities in finding the stereo correspondence, the human vision system successfully reconstructs three-dimensional depth distribution with binocular vision. Thus, we can expect that solving the stereo correspondence problem brings a key mechanism to understand the human vision system.

The human vision system can solve the stereo correspondence problem from random-dot stereograms, which refer to a pair of stereo images consisting of only a randomly dotted pattern. Julesz demonstrated that the randomly dotted pattern is enough for the human vision system to solve the stereo correspondence problem (Julesz, 1960). When the human vision system is exposed to random-dot stereograms, it perceives three-dimensional structure emerging spontaneously. This implies that the human vision system has a module being able to detect stereo disparity from only a randomly dotted pattern and not requiring key information such as edges and feature points, which generally appear in natural scenes.

Several researchers presented stereo algorithms to solve the stereo correspondence problem from random-dot stereograms in the early period of computer vision research. In particular, Marr and Poggio presented a representative stereo algorithm named the cooperative algorithm (Marr & Poggio, 1976). Their motivation to approach the stereo correspondence problem exists in the biological aspect of the human vision system; their algorithm consists of multi-layered cell networks. Their most important proposal to the stereo algorithm is the two famous constraints: continuity and uniqueness.

The authors have been approaching several subjects in image processing and computer vision research, by utilizing reaction-diffusion equations as a biologically motivated tool (Nomura et al., 2007). We call the group of algorithms utilizing reaction-diffusion equations the reaction-diffusion algorithm. The previous research done by the authors presented several algorithms for edge detection, segmentation (grouping) and stereo disparity detection by utilizing the FitzHugh-Nagumo type reaction-diffusion equations.

This chapter presents a stereo algorithm utilizing multi-sets of the FitzHugh-Nagumo type reaction-diffusion equations. We associate each set of the equations with each of possible disparity levels; a particular grid point of the set represents existence or non-existence of its associated disparity level. The filling-in process originally built in the reaction-diffusion

equations realizes the continuity constraint; the multi-sets mutually connected via a mutual-inhibition mechanism realize the uniqueness constraint.
Although the human vision system can detect stereo disparity from only random-dot stereograms, the authors believe that feature points such as edges detected from image brightness distribution help stereo algorithms to achieve more reliable stereo disparity detection. An integration mechanism of edge information into the stereo perception is interesting from the scientific point of view. According to this, the authors furthermore propose to integrate edge information into the stereo algorithm. As mentioned above, the authors have also presented an edge detection algorithm utilizing the reaction-diffusion equations (Nomura et al., 2008). Thus, in this chapter the stereo algorithm with edge information is fully realized with the reaction-diffusion equations biologically motivated.
Finally, from the engineering point of view we need to evaluate quantitative performance of the stereo algorithms presented here. We apply the stereo algorithms to test stereo image pairs provided on the Middlebury website (http://vision.middlebury.edu/stereo/). Results of the performance evaluations show that the stereo algorithm utilizing the reaction-diffusion equations with edge information achieves better performance in areas having depth discontinuity, in comparison to the stereo algorithm without edge information. However, in other areas not having depth discontinuity, edge information obtained for image brightness distribution is useless. Thus, the performance of the stereo algorithm is not improved. In addition, other state-of-the-art stereo algorithms achieve much more performance. Thus, future work needed for the reaction-diffusion algorithm is to improve its performance also in other areas not having depth discontinuity.

2. Previous stereo algorithms

It is essentially difficult to detect stereo disparity from only a pair of stereo images, since image brightness distribution does not provide enough information to identify a particular point in image. There exists ambiguity in finding stereo correspondence between stereo images; the stereo correspondence problem is the typical ill-posed problem. Researchers have devoted their efforts to obtain reliable stereo correspondence with several different algorithms, such as, the template-matching algorithm and the cooperative algorithm.
In the template-matching algorithm that finds stereo correspondence with a cross-correlation function, we need to extend the size of its correlation window for reducing the ambiguity lying on the stereo correspondence problem. However, the larger size of the correlation window causes unreliable solution to the stereo correspondence problem, in particular, in areas having depth discontinuity. Thus, on the one hand, the larger correlation window size assuming the spatial uniformity of disparity distribution is necessary to reduce the ambiguity; on the other hand, such the spatial uniformity can not be assumed, in particular, in areas having depth discontinuity. The two requirements of reducing ambiguity and preserving depth discontinuity is mutually exclusive.
Marr and Poggio imposed the two constraints: uniqueness and continuity on the cooperative algorithm (Marr & Poggio, 1976). The uniqueness constraint states that a particular point on a stereo disparity map has only one stereo disparity level except for transparent objects; the continuity constraint states that neighboring points on a stereo disparity map share same or similar disparity level(s) except for object boundaries. They tried to detect stereo disparity by taking account of the two constraints with a biologically motivated multi-layered network model, each grid point of which was considered as a cell

activated by neighboring cells locating on the same layer and inhibited by other cells locating on other layers of disparity levels. The continuity constraint causes error of stereo disparity in areas having depth discontinuity. Thus, the problem arising from the depth discontinuity is one of the most significant issues not only in the template-matching algorithm, but also in other stereo algorithms including the cooperative algorithm.

Since the original cooperative algorithm is designed for random-dot stereograms, it is not applicable to natural stereo images. More recently, Zitnick and Kanade presented a cooperative algorithm designed for natural stereo images (Zitnick & Kanade, 2000). In addition, the cooperative algorithm can detect occlusion areas in which an object occludes another object in either of two stereo images; it is generally difficult for stereo algorithms to detect occlusion areas.

Other algorithms such as the belief-propagation algorithm and the graph-cuts algorithm are well known as state-of-the-art stereo algorithms. The Middlebury website is providing the ranking table showing which stereo algorithm achieves the best performance for test stereo image pairs with respect to the bad-match-percentage error measure. According to the ranking table, those state-of-the-art algorithms achieve much better performance, in comparison to the cooperative algorithm.

The authors are interested in biologically motivated algorithms and thus contribute to the stereo correspondence problem by utilizing the reaction-diffusion algorithm. The algorithm presented by the authors is rather similar to the cooperative algorithm and is linked directly with mathematical models of information transmission and pattern formation observed in biological systems.

3. Previous research on reaction-diffusion equations

3.1 Modeling pattern formation processes with reaction-diffusion equations

Biological systems self-organize spatio-temporal patterns for information transmission (Murray, 1989). Typical examples of self-organized patterns are impulses propagating along a nerve axon and a spiral pattern and a target pattern observed in a two-dimensional system of slime mould. These self-organized patterns have been explained with a mathematical model of reaction-diffusion equations, which are generally described with time-evolving partial differential equations. Each of the equations consists of a diffusion term coupled with a reaction term describing its corresponding phenomenon such as, for example, a response to stimuli given by other neighboring cells or to external stimuli in biological systems.

The next set of equations describes reaction-diffusion equations with the two variables $u(x,y,t)$ and $v(x,y,t)$ in two-dimensional space (x,y) and time t, as follows:

$$\partial_t u = D_u \nabla^2 u + f(u,v)\,,\quad \partial_t v = D_v \nabla^2 v + g(u,v)\,, \tag{1}$$

where D_u and D_v are diffusion coefficients and $f(u,v)$ and $g(u,v)$ are reaction terms. The symbol ∂_t refers to the partial differential operator $\partial/\partial t$ and ∇^2 refers to the Laplacian operator $\nabla^2=\partial^2/\partial x^2+\partial^2/\partial y^2$. The FitzHugh-Nagumo type reaction-diffusion equations describe impulses propagating along a nerve axon in a biological system (FitzHugh, 1961; Nagumo et al., 1962); the equations have the reaction terms $f(u,v)$ and $g(u,v)$, as follows:

$$f(u,v) = \frac{1}{\varepsilon}[u(u-a)(1-u)-v]\,,\quad g(u,v) = u - bv\,, \tag{2}$$

where a and b are constants and ε is a positive small constant (0<ε<<1). A set of ordinary differential equations derived from Eqs. (1) and (2) under D_u=D_v=0 indicates two different types of system behavior such as the bi-stable system or the mono-stable system as shown in Fig. 1. When the set of the equations is bi-stable, it has the two stable steady states, one of which locates at the origin (0,0) and the other of which locates at another point (u_s,v_s) having u_s close to one.

Turing found that a condition causes a spatial static pattern such as a spot or stripe pattern with a set of reaction-diffusion equations (Turing, 1952), even if diffusion terms in the equations induce spatial homogeneity of the two variables u and v. When the diffusion coefficient D_v is much larger than the diffusion coefficient D_u, such the spatial static pattern appears. The Turing condition refers to the condition that causes the spatial static pattern in reaction-diffusion equations and the Turing pattern refers to the static pattern. Several self-organized patterns observed in biological systems have been explained as the Turing pattern and with the Turing condition (Kondo & Asai, 1995). We need to emphasize that the Turing pattern is due to the rapid diffusion on the inhibitor variable v, in comparison to the slow diffusion on the activator variable u (D_u<<D_v).

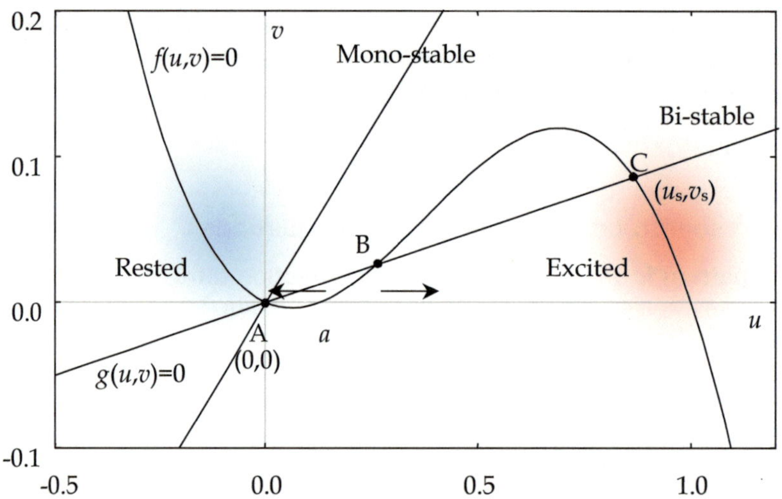

Fig. 1. Plots of $f(u,v)=u(u\text{-}a)(1\text{-}u)\text{-}v=0$ and $g(u,v)=u\text{-}bv=0$. The steady states A locating at (0,0) and C locating at (u_s,v_s) are stable; the steady state B is unstable. Depending on the parameter values a and b, the system of $f(u,v)$=0 and $g(u,v)$=0 becomes mono-stable or bi-stable. When the system is bi-stable, a solution converges either of the two stable states. The excited state denotes that the system is in the red area; the rested state denotes that the system is in the blue area.

3.2 Modeling and realizing visual functions with reaction-diffusion equations

Kuhnert et al. found that a chemical reaction system can realize visual functions such as edge detection, segmentation and image memory (Kuhnert, 1986; Kuhnert et al., 1989). The chemical reaction system is described with a set of reaction-diffusion equations having activator and inhibitor variables. Thus, their experimental reports made a strong impact on us and highly motivated us to develop image processing and computer vision research with reaction-diffusion equations.

After the reports by Kuhnert et al., many other researchers have also done much efforts on modeling and realizing visual functions with reaction-diffusion equations. Asai et al. realized a reaction-diffusion system with the large-scale integrated circuits technology for image processing; Adamatzky et al. furthermore proposed a novel computer architecture and named it the reaction-diffusion computer (Adamatzky et al., 2005). Ueyama et al. proposed a reaction-diffusion system as a mathematical model of the human visual perception (Ueyama et al., 1998). They tried to explain a visual phenomenon observed in motion perception with a reaction-diffusion equation.

In image processing and computer vision research, we usually utilize the Gaussian filter or a diffusion equation for reducing noise or selecting scales in patterns. For example, the edge detection algorithm proposed by Marr and Hildreth utilizes the difference of two Gaussians (Marr & Hildreth, 1980), each of which is alternatively expressed by a diffusion equation. Perona and Malik proposed an edge detection algorithm by utilizing an anisotropic diffusion (Perona & Malik, 1990).

The authors have tried to build models of several visual functions such as edge detection, grouping and stereo disparity detection by utilizing reaction-diffusion equations. The authors believe that reaction-diffusion equations become a basic tool for realizing visual functions from the engineering point of view, and a basic model for understanding the human visual functions from the scientific point of view. This is the motivation for the present research work done by the authors by utilizing reaction-diffusion equations in image processing and computer vision research.

This chapter presents a stereo algorithm by utilizing multi-sets of the FitzHugh-Nagumo type reaction-diffusion equations (1) and (2), on which the authors impose the Turing like condition ($D_u << D_v$). The grouping mechanism built in the reaction-diffusion equations works as the continuity constraint required in the stereo algorithm. In addition, the Turing like condition helps to prevent over-smoothing around corners in structure of stereo disparity distribution.

4. Stereo algorithm with reaction-diffusion equations

Upon acceptance of the two constraints: continuity and uniqueness proposed by Marr and Poggio (1976), the authors present a stereo algorithm that utilizes the FitzHugh-Nagumo type reaction-diffusion equations (1) and (2). In order to realize the stereo algorithm, we consider multi-sets of the equations, each set of which is associated with a possible disparity level, and each set of which governs areas having the associated disparity level. In addition, each set inhibits other sets from having other disparity levels via a mutual-inhibition mechanism.

A set of the reaction-diffusion equations organizes propagating waves, which are applicable to the realization of the continuity constraint (Fig. 2). A wave triggered by an external stimulus at a point in space begins to propagate. When the wave collides with another wave triggered at another point in the space, the two waves become one, if the system of the equations is bi-stable. This is a filling-in process that fills in undefined areas between the two triggered positions. Let us give an output of a matching cost function to the set of the reaction-diffusion equations as its external stimulus. Then, we can understand that the set realizes the continuity constraint.

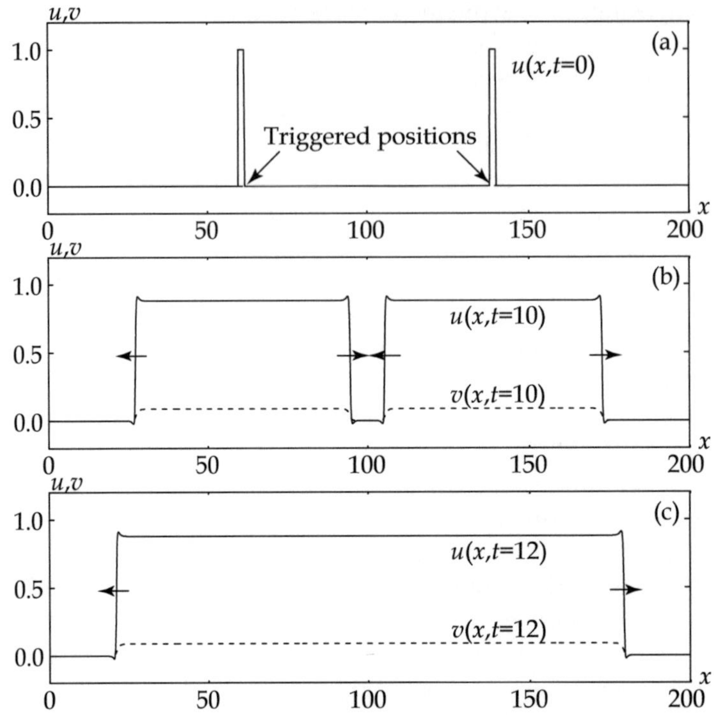

Fig. 2. Numerical result of the FitzHugh-Nagumo type reaction-diffusion equations (1) and (2) in one-dimensional space x, in which waves triggered at two different positions propagate and become one after the collision of the two waves. (a) Initial condition on $u(x,t=0)$; (b) one-dimensional spatial distributions of u and v at t=10 before the collision; (c) one-dimensional spatial distributions of u and v at t=12 after the collision. Parameter values chosen here for the numerical computation are D_u=1.0, D_v=3.0, a=0.05, b=10 and $\varepsilon=1.0\times10^{-2}$ (the bi-stable system); finite differences for the numerical computation are δx=1/5 and δt=1/100. An initial condition for $v(x,t=0)$ is zero over the one-dimensional space. In each figure, the solid line indicates the one-dimensional distribution of $u(x,t)$ and the broken line indicates the one-dimensional distribution of $v(x,t)$.

The mutual-inhibition mechanism built in the multi-sets of the reaction-diffusion equations realizes the uniqueness constraint. Each set organizes propagating waves, for which the parameter a in Eq. (2) controls their propagating speed. Thus, when we change the parameter value a, we can observe waves propagating at different speeds. If we choose a large value of a, we can inhibit the waves from propagating. Thus, each set of the reaction-diffusion equations utilized in the stereo algorithm is as follows:

$$\partial_t u_n = D_u \nabla^2 u_n + f(u_n, v_n, u_{\max}) + \mu s_n, \quad \partial_t v_n = D_v \nabla^2 v_n + g(u_n, v_n), \tag{3}$$

$$f(u_n, v_n, u_{\max}) = \frac{1}{\varepsilon}\left[u_n(u_n - a(u_{\max}))(1-u_n) - v_n\right], \quad g(u_n, v_n) = u_n - b v_n, \tag{4}$$

where n=0,1,⋯,N-1 denotes the index of possible disparity levels and μ is a constant. We provide an output of a matching cost function for $s_n(x,y)$ as an external stimulus to the set; the matching cost function returns one for a stereo image pair exactly matched with a disparity level d_n and zero for an un-matched pair. The original reaction term $f(u,v)$ in Eq. (2) has a constant parameter a. In contrast to this, the above modified version of the reaction term in Eq. (4) depends on the state of the other set having the maximum value u_{max} as follows:

$$a(u_{\max}) = a_0 + \frac{u_{\max}}{2}\left[1 + \tanh\left(|d| - a_1\right)\right],\; u_{\max} = \max_{(x',y',n')\in\Omega_i} u_{n'}(x',y',t),\; |d| = \left| d_n - d_{\underset{(x',y',n')\in\Omega_i}{\arg\max}\, u_{n'}(x',y',t)} \right|, \quad (5)$$

where a_0 and a_1 are constants and Ω_i denotes an inhibition domain for the uniqueness constraint (Zitnick & Kanade, 2000).

For preserving sharp corners existing in stereo disparity distribution, we impose the Turing like condition (D_u<<D_v) on each of the reaction-diffusion equations in Eq. (3). The condition induces rapid diffusion of the inhibitor variable v_n, in comparison to diffusion of the activator variable u_n. The rapid diffusion of the inhibitor prior to the diffusion of the activator prevents a wave from propagating into undefined areas. The phenomenon preventing the propagation of the wave becomes remarkable for a curved wave front having negative curvature. Thus, the wave propagation inhibited by itself in a disparity level helps a sharp corner governed by another disparity level to survive. The authors name this phenomenon due to the Turing like condition the self-inhibition mechanism.

By computing Eqs. (3), (4) and (5) during enough finite duration of time needed for convergence, we finally obtain a stereo disparity map $M(x,y,t)$ at the time t with

$$M(x,y,t) = \underset{n\in[0,1,\cdots,N-1\}}{\arg\max}\, u_n(x,y,t)\,. \quad (6)$$

5. Integration of edge information into the stereo algorithm

Areas having depth discontinuity cause much error in stereo disparity detection. If the areas are given in advance of the stereo disparity detection, information of the areas helps stereo algorithms to achieve better performance around the areas. However, the areas are generally unknown. Although we can not directly link edge areas existing in image brightness distribution with the areas having depth discontinuity, we can expect that information of the edge areas becomes a key to success in stereo disparity detection in the areas having depth discontinuity.

The present chapter furthermore proposes the integration of edge information into the stereo algorithm presented above. We obtain edge information from either of stereo images with an edge detection algorithm and then feed the edge information to the stereo algorithm. The authors previously proposed the edge detection algorithm by utilizing the reaction-diffusion equations (Nomura et al., 2008). Let the two distributions $u_e(x,y)$ and $v_e(x,y)$ be the results of the edge detection algorithm; they represent spatial distributions of the activator variable $u(x,y,t)$ and the inhibitor variable $v(x,y,t)$ obtained at a point (x,y) and at time t after enough finite duration of time needed for convergence. In edge areas, a set of solutions $u_e(x,y)$ and $v_e(x,y)$ is in the excited state nearly equal to the stable steady state (u_s,v_s) (see also Fig. 1). Thus, we integrate the edge information (u_e,v_e) into the stereo algorithm as follows:

$$\partial_t u_n = D_u \nabla \cdot [(1-u_e)\nabla u_n] + f(u_n, v_n, u_{max}) + \mu s_n, \quad \partial_t v_n = D_v \nabla \cdot [(1-v_e)\nabla v_n] + g(u_n, v_n), \tag{7}$$

where the terms $\nabla \cdot [(1-u_e)\nabla u_n]$ and $\nabla \cdot [(1-v_e)\nabla v_n]$ describe anisotropic diffusion, according to the results of the edge detection. Since ($1-u_e$) becomes almost zero in edge areas, slow diffusion across an edge line prevents disparity information from propagating into other areas having other disparity levels. Thus, we can expect that the slow diffusion brings better performance around areas having depth discontinuity.

6. Experimental results

This experimental section reports performance of stereo algorithms. The stereo algorithms evaluated here are as follows: the reaction-diffusion algorithm without the integration of edge information (RD_s), the reaction-diffusion algorithm with the integration of edge information (RD_{s+e}) and the cooperative algorithm proposed by Zitnick and Kanade (ZK) (Zitnick & Kanade, 2000). The two algorithms RD_s and RD_{s+e} utilize a normalized cross-correlation function as a matching cost function; the normalized cross-correlation function computes similarity between two areas, each of which consists of five points including the target point and its nearest four points. In order to show how much the two reaction-diffusion algorithms improve initial disparity maps, we also evaluated the initial disparity maps (CC_5).

We utilized the four pairs of test stereo images: MAP, TSUKUBA, SAWTOOTH and VENUS for the performance evaluation of the stereo algorithms (Scharstein & Szeliski, 2002). The Middlebury website provides all of the four image pairs as well as their ground truth data of stereo disparity maps and depth discontinuity areas. Figure 3 shows the four pairs of left and right stereo images and their initial disparity maps obtained by only the cross-correlation function (CC_5).

The two error measures: the root-mean-square error measure and the bad-match-percentage error measure quantitatively evaluate stereo disparity maps obtained by the stereo algorithms. The root-mean-square error measure R (pixel) evaluates absolute difference between the true disparity map $M_t(x,y)$ and an obtained disparity map $M(x,y,t)$ in a domain F as follows:

$$R = \left[\frac{1}{|F|} \sum_{(x,y) \in F} \{M_t(x,y) - M(x,y,t)\}^2 \right]^{1/2} \text{ (pixel)}, \tag{8}$$

where $|F|$ denotes the number of pixel sites belonging to the domain F. The bad-match-percentage error measure B (%) evaluates the ratio of the number of bad match pixel sites, which have absolute error larger than δd (pixel), to the number of total pixel sites belonging to the domain F as follows:

$$B = \frac{1}{|F|} \sum_{(x,y) \in F} \sigma(|M_t(x,y) - M(x,y,t)|, \delta d) \times 100 \text{ (\%)}, \tag{9}$$

where the function $\sigma(S,\delta d)$ is the threshold function which returns one for $S \geq \delta d$ and zero for $S < \delta d$. The domain F in Eqs. (8) and (9) refers to the domain consisting of all pixel sites except for borders and occlusion areas (all), the domain consisting of untextured areas (untex.) or the domain consisting of depth-discontinuity areas (disc.). The Middlebury website also provides definitions of the three domains. Borders of 10 pixels (18 pixels for TSUKUBA) and occlusion areas were ignored in the present evaluation.

The reaction-diffusion algorithms RD_s and RD_{s+e} utilize time-evolving partial-differential equations, for which the finite difference method provides a set of linear equations with the finite differences δx, δy and δt for space (x,y) and time t. The parameter values $\delta x=\delta y=1/5$, $\delta t=1/100$, $D_u=1.0$, $D_v=3.0$, $a_0=0.13$, $a_1=1.5$, $b=10$, $\varepsilon=10^{-2}$, $\mu=3.0$ were chosen for the present experiments in RD_s and RD_{s+e}. The reaction-diffusion algorithm with edge information (RD_{s+e}) requires edge detection results. Figure 4 shows the edge detection results obtained by the reaction-diffusion algorithm designed for edge detection. See the literature (Nomura et al., 2008) for the edge detection algorithm and its parameter values utilized here. These results were fed to the algorithm RD_{s+e} through the anisotropic diffusion terms, as described in Eq. (7).

Figure 5 shows the ground truth disparity maps, disparity maps obtained by RD_{s+e} and their absolute error maps. Table 1 shows error values evaluated for the disparity maps obtained by RD_s, RD_{s+e} and ZK, as well as for the initial disparity maps denoted by CC_5.

With respect to the bad-match-percentage error measure B (%), we can confirm that the algorithm RD_{s+e} achieves better performance than RD_s in areas having depth discontinuity for all of the four stereo image pairs. In particular, for the stereo image pair SAWTOOTH, RD_{s+e} achieves better performance for all of the three evaluation domains: "all", "untex." and "disc.". However, RD_s is better than RD_{s+e} in the evaluation domains: "all" and "untex." for the stereo image pair TSUKUBA. The algorithm RD_{s+e} assumes that edge areas obtained for brightness distribution coincide with areas having depth discontinuity. When a stereo image pair does not satisfy the assumption, RD_{s+e} provides rather much error in stereo disparity detection. When comparing the reaction-diffusion algorithms RD_s and RD_{s+e} with the cooperative algorithm ZK, the reaction-diffusion algorithms achieve better performance for SAWTOOTH and VENUS and the cooperative algorithm achieves better performance for TSUKUBA. That is, we confirm that overall performance is almost the same.

Future work required for the stereo algorithm RD_{s+e} is to improve its performance not only for the areas having depth discontinuity, but also for other areas including untextured areas. In order to do this, we need to feed edge information obtained for an initial disparity map instead of image brightness distribution to the stereo algorithm RD_{s+e}. Since initial disparity maps have much noise, we need to confirm how much the edge detection algorithm works correctly for the noisy maps. Furthermore, the authors are considering a new vision system that dynamically integrates the edge detection algorithm with the stereo algorithm. First, we obtain edge information from temporarily detected stereo disparity distribution; second, we compute one time step of the stereo algorithm with the integration of the edge information. By iterating these two steps alternately, we expect to achieve better performance of stereo disparity detection in the reaction-diffusion algorithm.

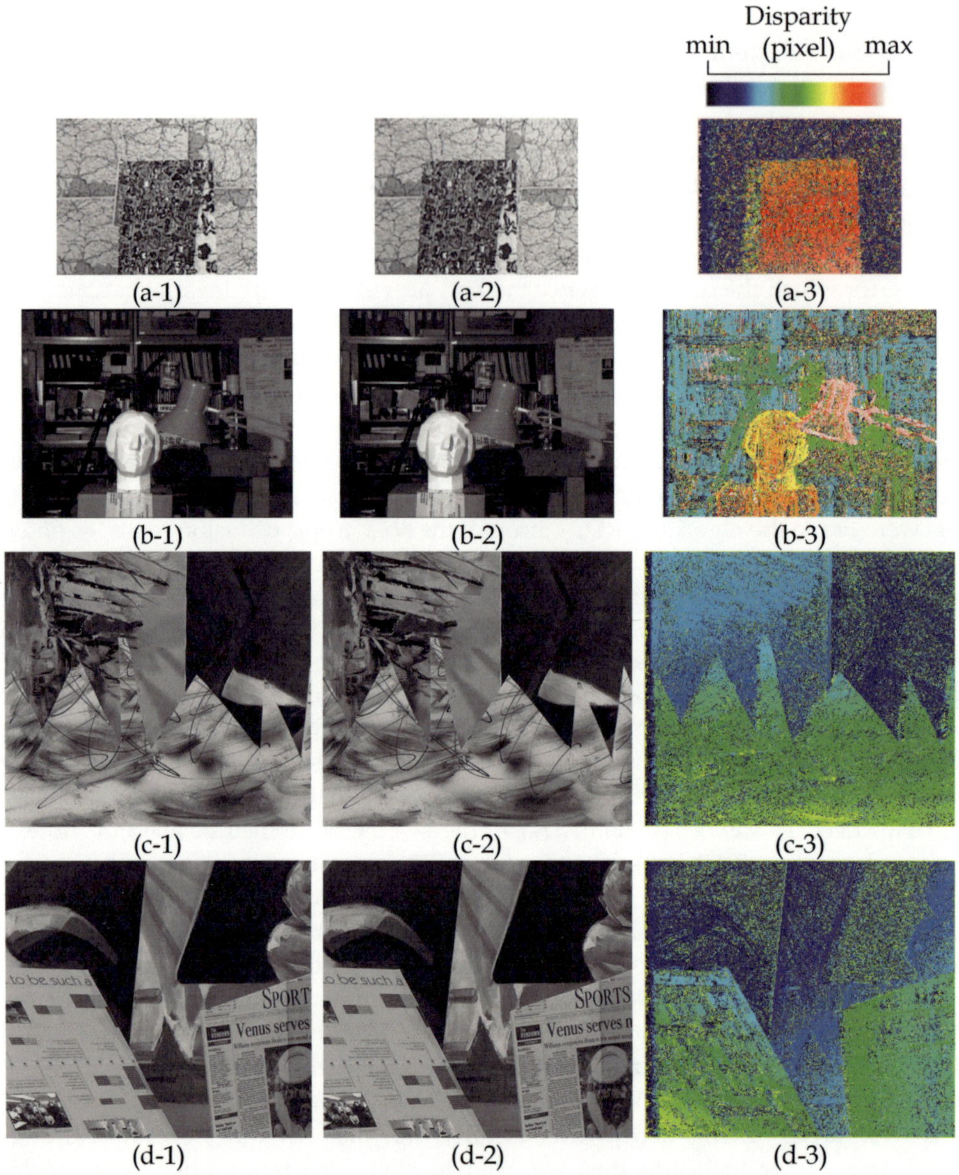

Fig. 3. Stereo image pairs (a) MAP, (b) TSUKUBA, (c) SAWTOOTH and (d) VENUS provided on the Middlebury website, and their initial disparity maps provided for the reaction-diffusion stereo algorithms RD_s and RD_{s+e}. (a-1), (b-1), (c-1) and (d-1) are left images; (a-2), (b-2), (c-2) and (d-2) are right images. Initial disparity maps shown in (a-3), (b-3), (c-3) and (d-3) were obtained by the normalized cross-correlation function CC_5. Image sizes are (a) 284×216 (pixels), (b) 384×288 (pixels), (c) 434×380 (pixels) and (d) 434×383 (pixels); possible disparity levels are (a) {0,1,···,29} (pixels), (b) {0,1,···,15} (pixels), (c) {0,1,···,19} (pixels), and (d) {0,1, ···,19} (pixels). (a-3), (c-3) and (d-3) visualize the initial disparity maps in the disparity range of min=0 (pixel) and max=31 (pixels); (b-3) visualizes that in the disparity range of min=0 (pixel) and max=15 (pixels).

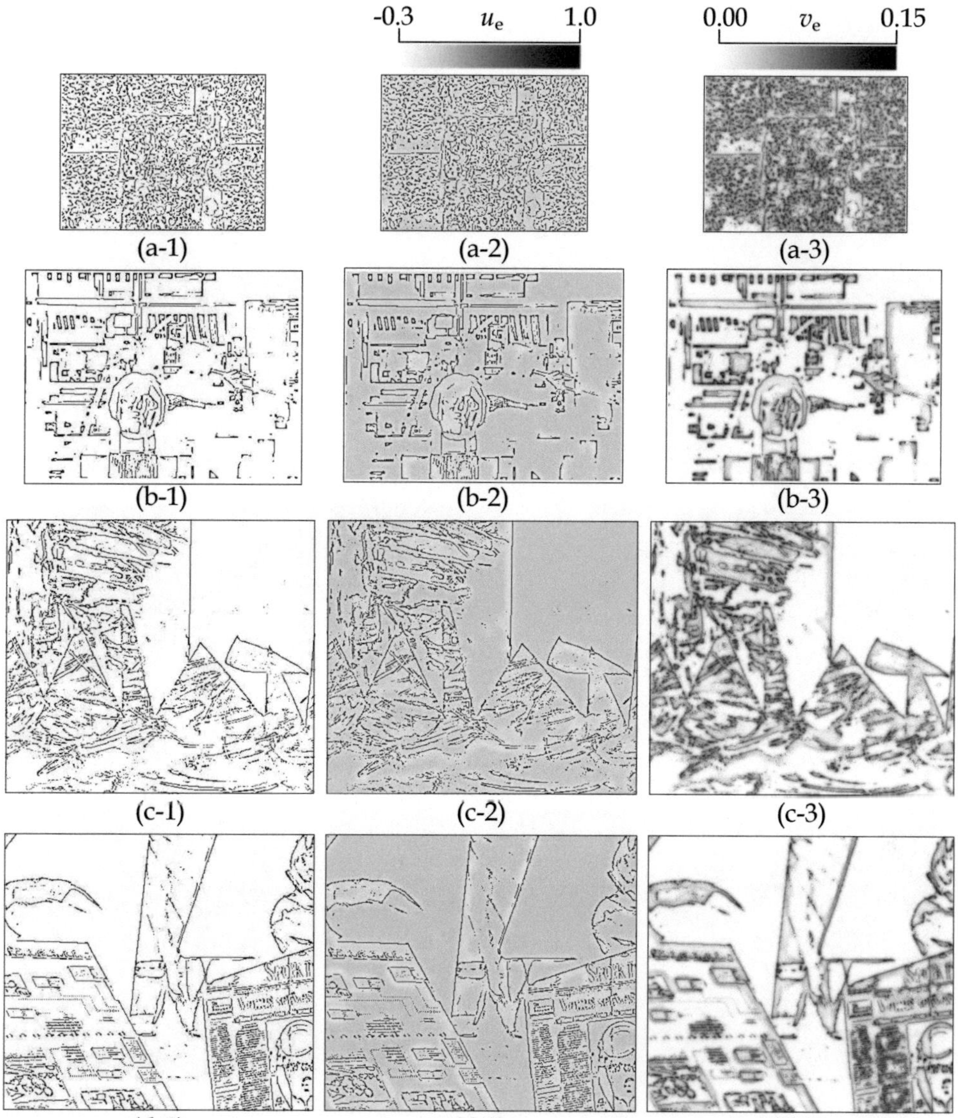

Fig. 4. Edge information obtained by the edge detection algorithm utilizing reaction-diffusion equations from the left images of (a) MAP, (b) TSUKUBA, (c) SAWTOOH and (d) VENUS (see Fig. 3 for the images). Refer to the literature (Nomura et al., 2008) for the edge detection algorithm and its parameter values utilized here. (a-1), (b-1), (c-1) and (d-1) show edge detection results, in which black dots and lines denote detected edges. (a-2), (b-2), (c-2) and (d-2) show spatial distributions of the activator variable $u_e(x,y)$; (a-3), (b-3), (c-3) and (d-3) show spatial distributions of the inhibitor variable $v_e(x,y)$. All of the results were obtained at t=5.0 in the reaction-diffusion equations.

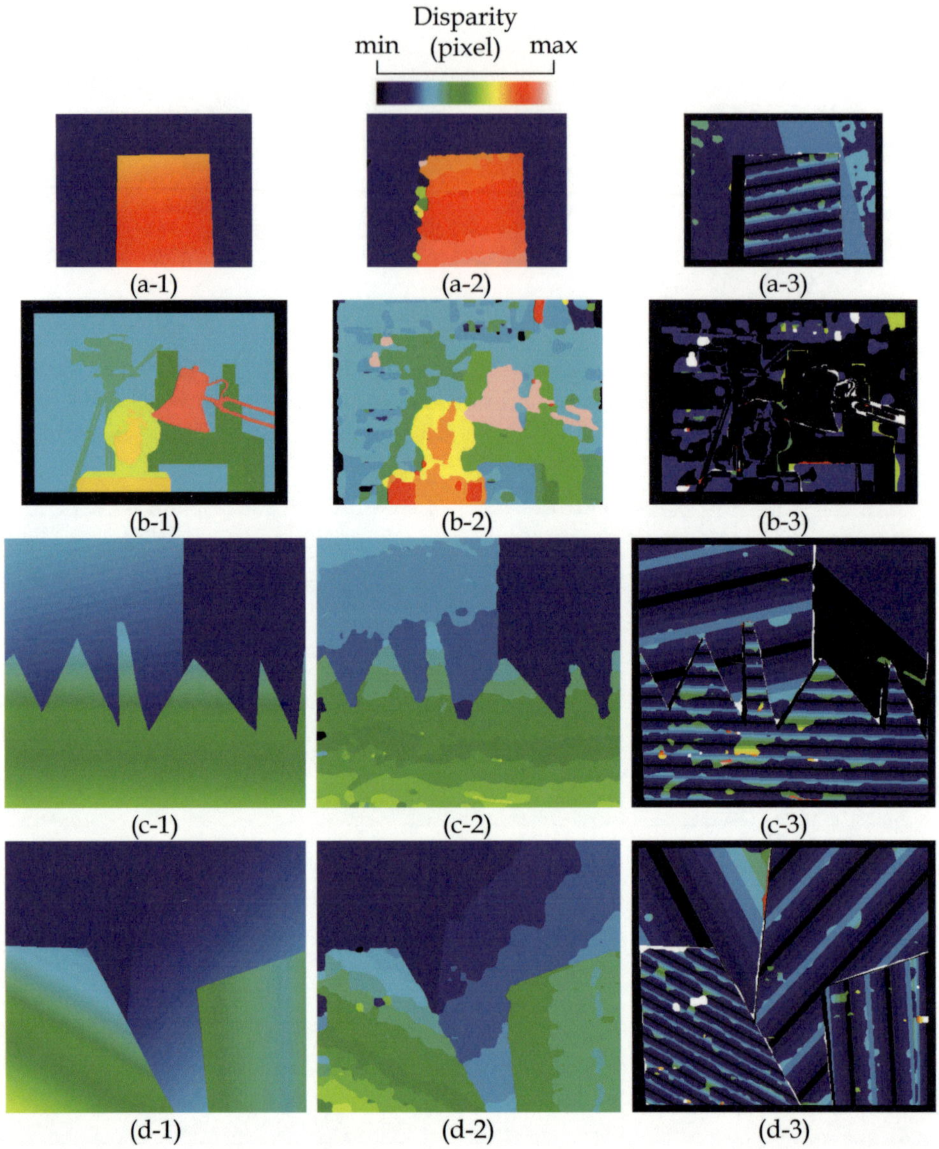

Fig. 5. Stereo disparity maps obtained by the proposed stereo algorithm RD_{s+e}. (a-1), (b-1), (c-1) and (d-1) show the ground truth data of disparity maps provided on the Middlebury website; (a-2), (b-2), (c-2) and (d-2) show disparity maps obtained by RD_{s+e} at t=10; (a-3), (b-3), (c-3) and (d-3) show absolute error distributions evaluated for the disparity maps. Borders and occlusion areas of disparity maps were ignored in the performance evaluation. See Fig. 3 for the stereo image pairs and Fig. 4 for the edge detection results utilized here. (a-1), (a-2), (c-1), (c-2), (d-1) and (d-2) visualize the stereo disparity maps in the disparity range of min=0 (pixel) and max=31 (pixels); (b-1) and (b-2) visualize those in the disparity range of min=0 (pixel) and max=15 (pixels). (a-3), (c-3) and (d-3) visualize the error distribution maps in the range of min=0.0 (pixel) and max=2.0 (pixels); (b-3) visualize that in the range of min=0.0 (pixel) and max=5.0 (pixels).

		CC_5		RD_s		RD_{s+e}		ZK	
		R(pixel)	B(%)	R(pixel)	B(%)	R(pixel)	B(%)	R(pixel)	B(%)
MAP	all	8.43	43.22	1.02	0.25	1.02	0.29	0.87	0.22
	untex.	12.32	76.19	0.39	0.00	0.39	0.00	1.48	0.95
	disc.	9.41	50.81	3.58	3.53	3.59	3.30	2.95	2.37
TSUKUBA	all	3.42	39.82	1.23	3.89	1.35	4.85	0.96	3.49
	untex.	3.97	57.94	1.27	4.54	1.41	6.28	0.87	3.65
	disc.	3.49	38.51	2.20	14.90	2.23	14.61	2.05	14.77
SAWTOOTH	all	4.36	36.35	0.70	1.74	0.68	1.72	0.79	2.03
	untex.	5.77	61.46	0.51	0.35	0.48	0.32	0.69	2.29
	disc.	3.90	35.16	2.23	12.50	1.82	8.94	2.20	13.41
VENUS	all	4.67	45.66	0.68	1.53	0.71	1.78	0.81	2.57
	untex.	6.01	68.99	0.61	1.05	0.69	1.28	0.93	3.52
	disc.	4.17	43.46	2.11	18.58	2.05	16.74	2.44	26.33

Table 1. Results of quantitative performance evaluations obtained for the reaction-diffusion stereo algorithms RD_s and RD_{s+e} and the cooperative algorithm ZK as well as for the initial disparity maps denoted by CC_5. We evaluated the algorithms in all areas (all), un-textured areas (untex.) and depth discontinuity areas (disc.); we ignored borders and occlusion areas in the performance evaluations. Underlined scores denote the best performance among the algorithms. See Eq. (8) for the root-mean-square error measure R (pixel) and Eq. (9) for the bad-match-percentage error measure B (%). The scores evaluated for ZK were resulting from the disparity maps provided on the Middlebury website.

7. Conclusion

This chapter presented a stereo algorithm utilizing multi-sets of the FitzHugh-Nagumo type reaction-diffusion equations. The stereo algorithm realizes the two constraints: continuity and uniqueness, as follows. Each set of the reaction-diffusion equations self-organizes propagating waves, the collision of which works as the continuity constraint. The mutual inhibition mechanism connecting the multi-sets works as the uniqueness constraint. In addition, the authors imposed the Turing like condition on each set of the reaction-diffusion equations; the condition helps the stereo algorithm to preserve small features such as sharp corners in stereo disparity distribution.

In order to improve the performance of the stereo algorithm in areas having depth discontinuity, the authors additionally proposed the integration of edge information into the stereo algorithm. That is, the algorithm weakens the diffusion processes of the reaction-diffusion equations in areas having edges, which are detected by also the reaction-diffusion algorithm designed for edge detection from image brightness distribution.

We evaluated quantitative performance of the two stereo algorithms with/without the integration of edge information, by applying them to the well known test stereo image pairs. Results of the performance evaluations show that the algorithm with the integration of edge information achieves better performance than the stereo algorithm without the integration of edge information in the areas having depth discontinuity. Overall performance of the stereo algorithms presented here is almost equivalent to that of the cooperative algorithm proposed by Zitnick and Kanade. According to the Middlebury website, other state-of-the-art stereo algorithms achieve much better performance, in comparison to the reaction-diffusion algorithm. The authors believe that it is possible to improve the performance of the reaction-diffusion algorithm by dynamically integrating the edge detection algorithm into the stereo algorithm. Thus, we need to improve the performance of the reaction-diffusion algorithm designed for stereo disparity detection.

8. References

Adamatzky, A.; Costello, B. D. L. & Asai, T. (2005). *Reaction-Diffusion Computers*, Elsevier, Amsterdam

FitzHugh, R. (1961). Impulses and physiological states in theoretical models of nerve membrane. *Biophysical Journal*, Vol. 1, pp. 445-466

Julesz, B. (1960). Binocular depth perception of computer-generated patterns. *The Bell System Technical Journal*, Vol. 39, pp. 1125-1162

Kondo, S. & Asai, R. (1995). A reaction-diffusion wave on the skin of the marine angelfish *Pomacanthus*. *Nature*, Vol. 376, pp. 765-768

Kuhnert, L. (1986). A new optical photochemical memory device in a light-sensitive chemical active medium. *Nature*, Vol. 319, pp. 393-394

Kuhnert, L.; Agladze, K. I. & Krinsky, V. I. (1989). Image processing using light-sensitive chemical waves. *Nature*, Vol. 337, pp. 244-247

Marr, D. & Poggio, T. (1976). Cooperative computation of stereo disparity. *Science*, Vol. 194, pp. 283-287

Marr, D. & Hildreth, E. (1980). Theory of edge detection. *Proceedings of the Royal Society of London. Series B, Biological Sciences*, Vol. 207, pp. 187-217

Murray, J. D. (1989). *Mathematical Biology*, Springer-Verlag, Berlin

Nagumo, J.; Arimoto, S. & Yoshizawa, S. (1962). An active pulse transmission line simulating nerve axon. *Proceedings of the I.R.E.*, Vol. 50, pp. 2061-2070

Nomura, A.; Ichikawa, M.; Sianipar, R. H. & Miike, H. (2007) Reaction-diffusion algorithm for vision systems, In: *Vision Systems: Segmentation & Pattern Recognition*, Goro Obinata and Ashish Dutta (Ed.), pp.61-80, i-Tech Education and Publishing, Vienna

Nomura, A.; Ichikawa, M.; Sianipar, R. H. & Miike, H. (2008) Edge detection with reaction-diffusion equations having a local average threshold. *Pattern Recognition and Image Analysis*, Vol. 18, pp. 289-299

Perona, P. & Malik, J. (1990). Scale-space and edge detection using anisotropic diffusion. *IEEE Transactions on Pattern Analysis and Machine Intelligence*, Vol. 12, pp. 629-639

Scharstein, D. & Szeliski, R. (2002). A taxonomy and evaluation of dense two-frame stereo correspondence algorithms. *International Journal of Computer Vision*, Vol. 47, pp. 7-42

Turing, A. M. (1952). The chemical basis of morphogenesis. *Philosophical Transactions of the Royal Society of London. Series B, Biological Sciences*, Vol. 237, pp. 37-72

Ueyama, E.; Yuasa, H.; Hosoe, S. & Ito, M. (1998) Figure-ground separation from motion with reaction-diffusion equation – The front of the generated pattern and a subjective contour –. *IEICE Transactions on Information and Systems D-II*, Vol. J81-D-II, pp.2767-2778 [in Japanese]

Zitnick, C. L. & Kanade, T. (2000). A cooperative algorithm for stereo matching and occlusion detection. *IEEE Transactions on Pattern Analysis and Machine Intelligence*, Vol. 22, pp. 675-684

16

Point Cloud Streaming for 3D Avatar Communication

Masaharu Kajitani, Shinichiro Takahashi and Masahiro Okuda
Faculty of Environmental Engineering, The University of Kitakyushu
Japan

1. Introduction

Real-time 3D imaging has gained special attention in many fields such as computer vision, tele-imaging, and virtual reality. Especially real-time 3D modeling is one of emerging topics. During the last years, many systems that reconstruct 3D images using image sequences obtained from several viewpoints have been proposed. In the work (Saito et al., 1999), the 3D image sequences are acquired by multiple images of about 50 cameras. They accomplish high quality 3D images. In the method, a modeling step is done offline. In (Wu & Matsuyama, 2003), (Ueda et al., 2004), (Würmlin et al., 2002), and (Würmlin et al., 2004), the 3D modeling is employed in real-time using parallel processing with PC clusters. All of them use Silhouette Volume Intersection (SVI) method to approximate a shape by a visual hull. In order to make the surface finer and map textures on it, some extra tasks are required, which are in principle time-consuming.

On the other hand, some methods have been established to generate range images in realtime, for example (Kanade et al., 1996). Some of them have already been commercialized (Pointgrey). These methods achieve range images at 5-30 fps by the stereo matching algorithm.

In this paper, we propose a client-server system in which the server reconstructs, encodes, and transmits the 3D images to client in real-time. We do not apply the polygonal modeling but transmit a point cloud of multiple range images and render it directly, due to the following two reasons:

1. To reconstruct mesh model, a triangulation is required. Moreover, one needs to extract textures from colour images, and map them onto the meshes, which are complicated and time-consuming. The computational cost of the processes depends on complexity of the object shapes. The point rendering does not need these.
2. In general, the polygonal meshes have connectivity information as well as the 3D coordinates of the vertices. The connectivity has to be encoded without any loss, which consumes many bits. Moreover, only one bit error on the connectivity causes catastrophic damage on its quality. This is not suitable for transmission over wireless network or UDP.

The point cloud can easily be divided and packetized to individual blocks.

The client receives and decodes a bit-stream, and then renders the point cloud quickly by using a strip-based rendering. As a rendering method for the point cloud, Point-Sampled-Geometry (Rusinkiewicz & Levoy, 2000) is well known, which is used to represent high resolution mesh data. Our rendering scheme is similar to the method. The client divides point cloud obtained from range image into strips, and renders it with triangle strips.

In certain situations at the modeling stage, it is hard to capture the 3D shape perfectly. Failure to obtain precise camera-calibration or Region of Interest (ROI) extraction results in overlaps and annoying artifacts between different range images. To address these problems, we apply a cylindrical method to each frame of 3D sequence before rendering stage. The cylindrical method maps each point of the Point Cloud on the inner wall of a virtual cylinder, which is then warped to a 2D plane. Mapping of the 3D points on a plane facilitates handling of the 3D Point Cloud efficiently, and makes the process work in realtime.

In the next section, we review conventional real-time 3D imaging methods. In Section 3, we introduce our client-server system that reconstructs, encodes and displays the 3D images. In Section 4, we verify the validity of our method with some experimental results. And then we conclude the chapter in Section 5.

2. Past work

In this section, we review conventional methods that reconstruct 3D mesh models and explain the drawback and advantage with comparison to our method.

Some papers (Wu & Matsuyama, 2003), (Ueda et al., 2004), (Würmlin et al., 2002), (Würmlin et al., 2004), have already presented systems reconstructing 3D images in real-time. In many of these methods, first object silhouettes are extracted from multiple images obtained from several viewpoints. Next, these silhouettes are re-projected to a 3D space, and the intersectional space is assumed as Visual Hull of the object. The procedure is called Silhouette Volume Intersection (SVI). Fig. 1 illustrates the SVI method. This method successfully reconstructs the rough shape of an object without holes, while the stereo method often fails to make a continuous surface due to mismatch of corresponding points. Thus the shape without holes can be reconstructed robustly if the image is divided exactly into regions and the silhouettes are faithfully extracted. However, SVI method has a drawback that it is difficult to reconstruct curved or concave surfaces in principle. In (Wu & Matsuyama, 2003), to overcome this drawback, the Visual Hull is divided into voxels and the 3D mesh is estimated by Marching Cube method, finally, high-resolution shape is obtained by relocating vertices of 3D mesh appropriately.

On the contrary, our method reconstructs a 3D point cloud using depth images obtained from several viewpoints by the stereo method. In general, reconstructing the 3D mesh model by the stereo method requires the following five processes: (1) taking depth image, (2) integration of several depth images, (3) polygonal meshing of a point cloud, (4) simplification of meshes, (5) generating texture and mapping it to the 3D mesh. As mentioned above, the computational cost of the stereo method is basically high. The feature of our method is to treat point cloud with the 3D coordinates and the colour information directly without any 3D meshes. So, calculation is reduced drastically because above (3), (4) and (5) processes are not required. Moreover, each depth image can be processed separately. Thus it can be implemented with simple parallel processing. In (Waschbüsch, et al., 2007), a 3D imaging system based on the stereo matching method has been proposed for real-time application.

In the stereo method, estimating depth is unstable because of the mismatch of corresponding points. The precision of corresponding points depends on lighting condition or complexity of colour, etc. And it is impossible to reconstruct flat colour surface with low reflectance and specular reflection. And, many points are required to represent the colour information of an object surface precisely because the colour is assigned to each point. And, additional process is necessary before rendering point cloud because the overlapping of the range images is caused.

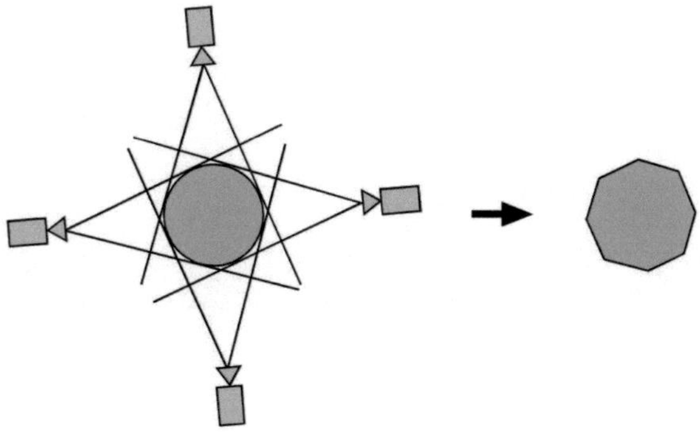

Fig. 1. Silhouette Volume Intersection

3. Client-server system

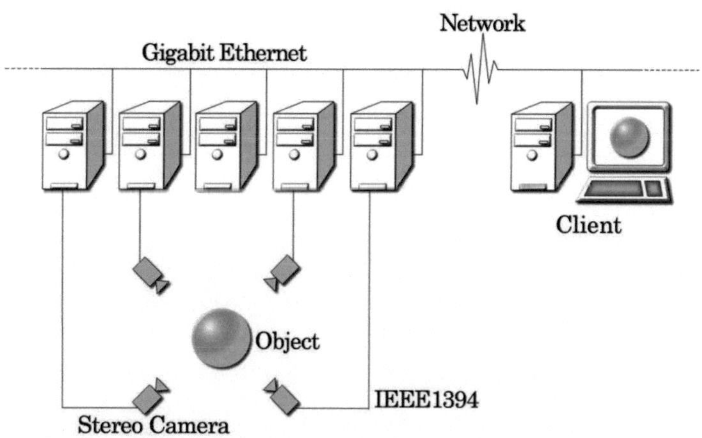

Fig. 2. 3D image transmission system

3.1 Outline

Our Client-Server system is illustrated in Fig.2. The server consists of N+1 Personal Computers (PCs) and N stereo cameras. The PCs in the server are connected through Gigabit switch. We use the commercial stereo cameras [10]. This camera consists of three CCD cameras and transfers three RGB colour images at the same time. The maximum frame rate is 15 fps. Each stereo camera is allocated to surround the object and connected with a PC through IEEE 1394. The PC takes a depth image and encodes it. Thus the entire server takes N depth images. The encoded bitstream is transmitted to the remaining PC, which is responsible for integration of data, packetization, transmission to the client, and sending out a command of taking images to each PC. We call the PC *ParentPC*. The relative position of each stereo camera is calculated by calibration in advance and informed to the client. The stereo cameras are synchronized with each other by special synchronization equipments. *ParentPC* generates synchronized signal, so each PC can take an image at the same time. Hereinafter, we explain the details of the server and the client.

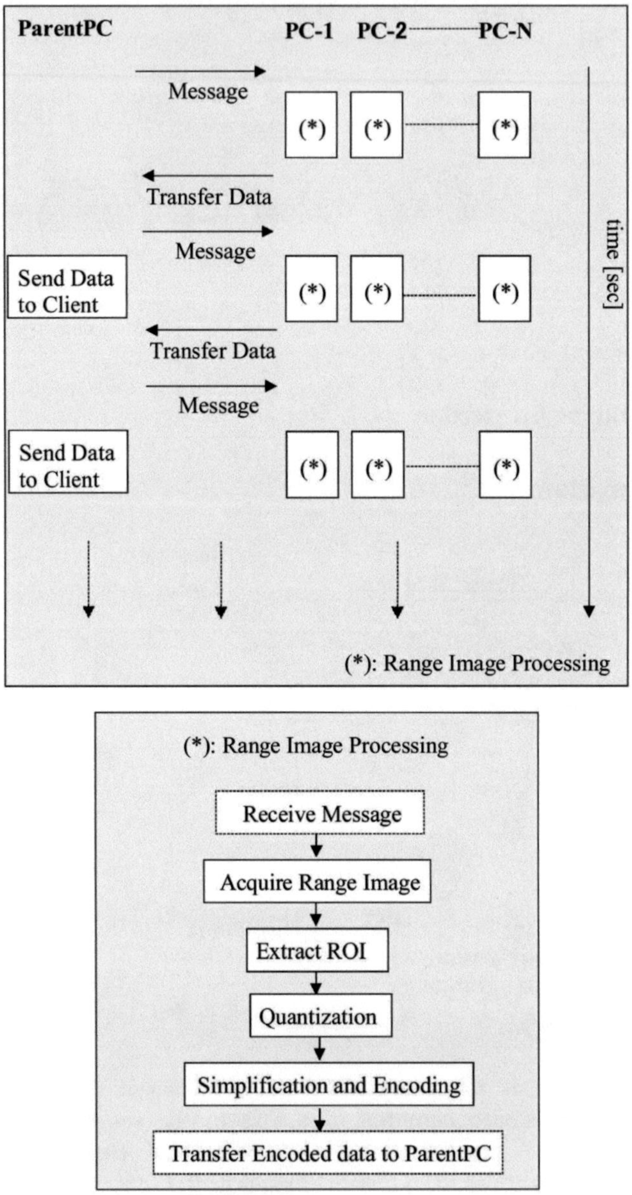

Fig. 3. Processing of depth image for each PC

3.2 Tasks in server

Flowchart of tasks in the server is illustrated in Fig.3. *ParentPC* tells the rest of PCs to take colour images. The *N* PCs does same process in parallel. Here, we explain about the process of the PCs.

Range Image Acquisition, Extraction of ROI, and Quantization

Each PC acquires colour images from its connected stereo camera. Each camera has three CCD units and can capture three colour images at the same time. All PCs are synchronized based on a signal sent from *ParentPC*. At the second step, range images are generated using these colour images by the stereo matching method. In this paper, we do not refer the detail of method of acquiring depth image and we use a conventional method. The next step is ROI extraction based on background subtraction. Here, a background image is taken previously without any object in front of the cameras. This method removes the pixels corresponding to the background based on colour and depth values of the obtained images. In our case, the ROI is a bounding box including remaining pixels and we only transmit the ROIs. Each pixel in ROI is quantized by linear quantization.

Simplification and Encoding

For streaming 3D images obtained in real-time, the simplification and the compression of the 3D images are essential. In general, a 3D mesh model includes coordinate, colour and connectivity information of vertexes. Since this connectivity information must be encoded losslessly and it is represented by a set of integers, it makes coding efficiency relatively low. Most of conventional 3D mesh simplification methods, for example (Hoppe, 1996) and compression methods such as (Khodakovsky et al., 2000) have high computational complexity and it can hardly be executed in real-time.

There exist many compression methods for equally sampled signals and data, such as DPCM (Gersho & Grey, 1992) and an integer type orthogonal transformation (Pratt et al., 1969). These encoding methods can be implemented with low computational complexity and can reduce its entropy after quantization by removing correlation with surrounding pixels. However since in the range images, some pixels may be blank due to failure in finding depth value, the above methods, which basically handles regularly sampled signals, cannot be applied directly. Moreover some extra computation is required for the entropy coding. Our simplification proposed here is simple. After all the vertices are quantized, one first finds the difference to neighbours. And then if the difference is small, completely replace the pixel value with it. The method does not need the entropy coding. We explain the detail of the algorithm.

First, the server scans the pixels in ROI of the range image along the horizontal line from upper left to lower right. For the depth values and the colour of RGB, we calculate the differences between the current pixel and each of four neighbours, the left, and upper left, upper, and upper right pixels. And then, if the absolute value of the difference between the current value and one of the four pixels is smaller than a threshold, the pixel value is replaced with the neighbor's. Fig. 4 illustrates the example. The left pixel is given priority if there were more than one pixel that can be replaced, in order to improve the rendering efficiency with triangle strip, the detail of which is explained later. Note that the pixel values of four compared neighbors are quantized in advance to enable to decode at the client. Four codes are prepared to identify pixels in the neighbourhood. If a pixel is replaced, then one of the four codes is allocated to the pixel. After all, the depth and RGB of the pixel is described by one code. The simplification procedure is applied with the valid pixels only. So, the valid pixels that have only invalid neighbours are not simplified. Fig. 4 (upper right) is the range image after the simplification. The arrows show that the pixel values are replaced with other neighbours indicated by the arrows.

As a result, the encoded bit stream is divided into packets and transferred to the decoder. The packets are well aligned according to a rule so that the client can determine which range images the packets come from.

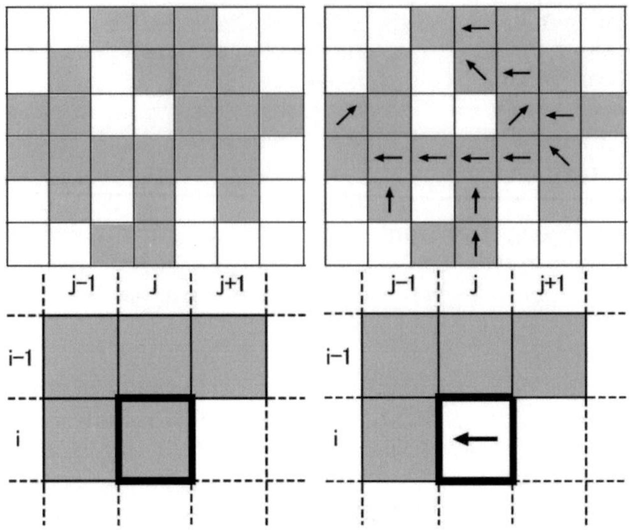

Fig. 4. The range image before (upper left) and after (upper right) the simplification. Current pixel (i, j) is compared with its for neighbors (lower left) and if the difference between the current pixel value and left pixel is smaller than a threshold, the pixel value is replaced with the left (lower right). A special code is allocated to the (i, j) pixel.

3.3 Tasks in client

The client receives the bit-stream of the point cloud from *ParentPC*. The received bit-stream consists of the spatial coordinates and colour information of the points. The 3D shape is represented by the point based geometry. At this time, overlaps and annoying artefacts are caused due to failure of camera calibration or the ROI extraction. In order to remove the invalid points, we apply the cylindrical method to the 3D model before rendering. The detail is explained below.

Post Processing

In the 3D image, annoying artifacts often appears due to overlaps between different range images. Fig.13 (a) shows an example of the 3D image with the overlaps. Due to the influence of the shadow of the object itself, the annoying artifacts are often left around the boundary at the ROI extraction stage. That results in image degradation. Since the 3D point cloud does not have connectivity information, it is not straightworward to decide the relationships among points. This makes the determination of the unnecessary points more difficult. To handle this, we employ a method which projects the vertices of the point cloud onto the inner wall of the cylinder. It can treat whole the points in a single 2D plane. Note that the cylindrical method is used to determine the unnecessary points, and that does not actually transform the coordinates of points for rendering.

We assume that the 3D object is surrounded by a virtual open cylinder (see Fig.5). The axis of the cylinder lies along the principal axis of the object. The cylinder and the object are cut into thin horizontal slices. In each slice a 2D point set are mapped from the surface of the object surrounded by a circle corresponding to the circumference of the cylinder (see Fig.6 (a)). In each slice, each point of the 2D point set is mapped on an intersection of a ray and the circle. This ray is emitted from a center of the circle and passes through that point.

Therefore, a point can be represented by magnitude r and an angle ϕ based on coordinates x and z of that point (see Fig.6 (b)). The values of r and ϕ are calculated by following simple mathematical formula.

$$r = \sqrt{x^2 + z^2}, \qquad \phi = \tan^{-1}(z/x) \tag{1}$$

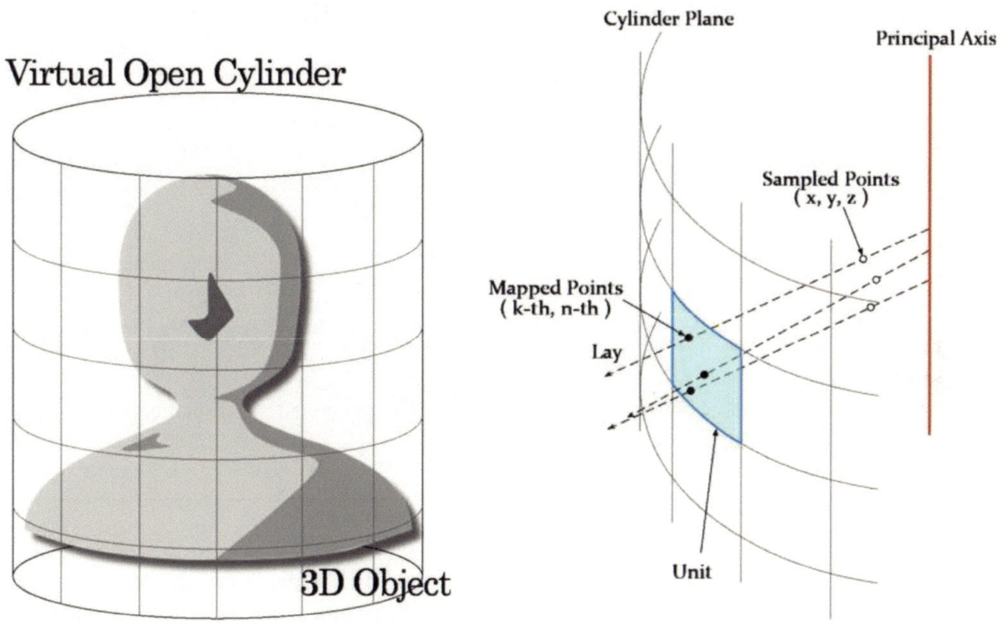

(a) (b)

Fig. 5. Mapping on the wall of the Cylinder: (a) 3D object with virtual open cylinder. (b) Image of mapping on the inner wall of the cylinder. Sampled points in 3D space are converted to mapped points in 2D plane.

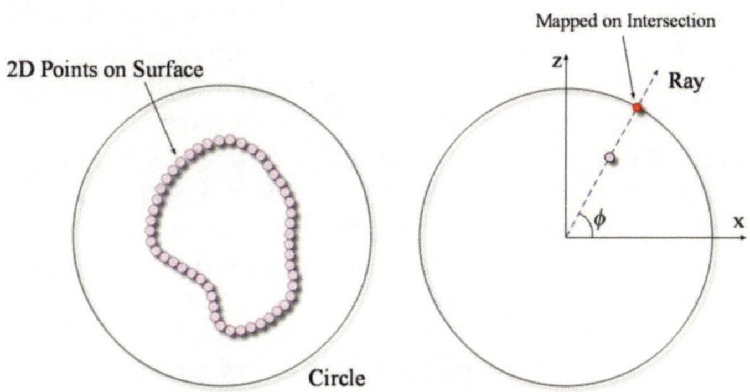

(a) (b)

Fig. 6. Mapping based on angle: (a) One of horizontal slice. 2D point set are surrounded by a circle corresponding to circumference of cylinder. (b) Image of mapping based on angle. Each point of 2D point set is mapped on intersection of ray and circle. Ray is decided angle ϕ calculated by x and z components of that point.

By repeating this process for all points in a slice, all of 2D point set is projected on the circle. When this procedure has been done for all the slices, the entire surface of the object is mapped on the inner wall of the cylinder.
After the mapping process, the cylinder is opened and warped to a plane (Fig.7). Then this plane is divided by a regular grid based on the angle ϕ and the y components. These divided small domains are named "Units". Here, one of the rows corresponds to the slice. If the plane has M columns, points stored in the k-th unit have angle ϕ satisfying the following condition.

$$\frac{2\pi(k-1)}{M} \le \phi < \frac{2\pi k}{M} \tag{2}$$

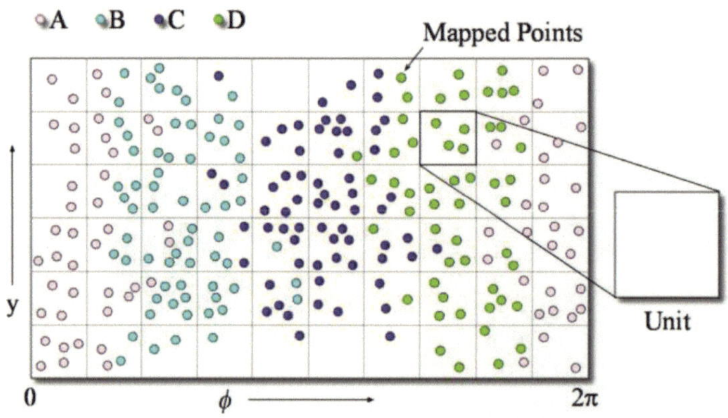

Fig. 7. Unit Plane: Unit Plane warped from the cylinder. This plane is divided by a regular grid. These divided small domains are called "Unit". The units store the mapped points.

We consider the points assigned to a unit as one segment. We can control quality and speed of processing by changing the size of the grid.
When there are points from more than one camera in a segment, we take it as the overlap. Fig.8 shows the image observed from the vertical direction. In Fig.8 (lower left), one unit stores the points from only one camera. However in (lower right), the points acquired from two cameras are stored in the unit. In this case, one of these surfaces should be removed. First, we find a tangent vector of each surface by calculating the differences of each point from the same camera. In the case of Fig.9, we have two vectors. These vectors are normalized. Secondly, the weighted average of each tangent vector is computed using number of points from each camera. Thirdly, the inner products of the average tangent vector and view axis of each camera are calculated. In case of Fig.9 (b), we have two inner products. Finally, the points coming form a camera with the minimum inner product are left, and others are removed (see Fig.9 (c)).

Rendering

The client receives only the XYZ coordinates and its colour of each point. Although the polygonal representation reconstructs a contiguous surface, computational complexity for the triangulation is relatively high. Our method renders the point cloud without reconstructing polygonal meshes to enhance the rendering speed. In our method, the texturing is not needed, since the colour is provided as attributes of the points. To reduce the load of rendering and to make up for the drawback of the stereo method, we adopt the strip-based rendering. First, we explain about a simple point-based rendering method using a rectangle primitive. Next, we explain about strip-based rendering using triangle strip as primitive, and then we compare the two rendering methods.

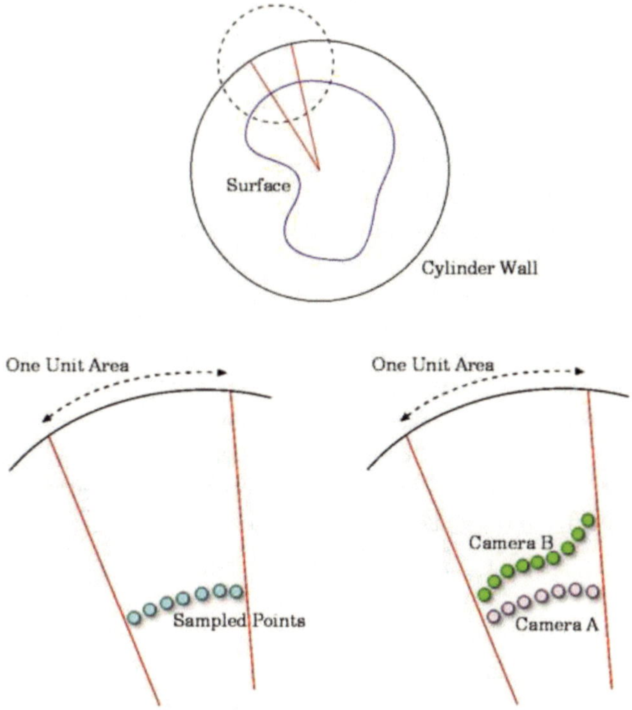

Fig. 8. (top) Overlap in One Unit: These are images observed from y direction. (lower left) One unit stores points that came from only one camera. (lower right) There are points came from two cameras.

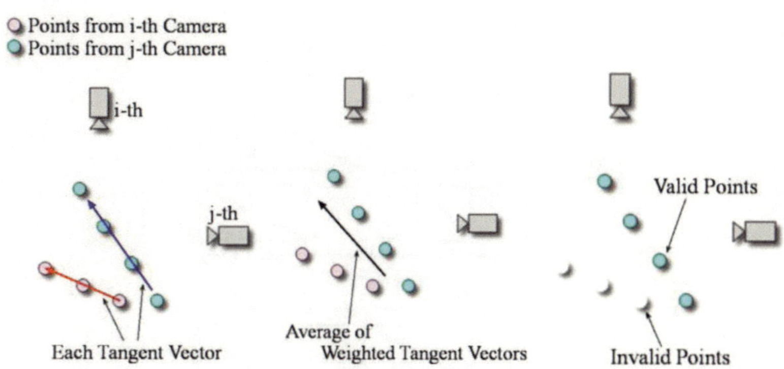

Fig. 9. Procedure of Unit Process: Blue points come from i-th camera and pink points from j-th one. (a) Each tangent vector is acquired. (b) The average of weighted tangent vectors is estimated. After that, inner products are calculated. (c) Points came form camera with minimum inner product are left.

1. Point-Based Rendering

The point-based geometry is proposed for example in (Rusinkiewicz & Levoy, 2000). In our implementation, the client renders the received point could with the rectangle (see Fig.

10(a)). The client estimates the normal vector using the neighbors. By using the estimation of the normal vector and the viewpoint of a user, the size and direction are changed adaptively with respect to rotation by the user. Note that although in the conventional point rendering method it is possible to control the size of the rectangles and avoid holes and it is efficient especially for large scale meshes, in our framework the client avoids the time-consuming step and linearly changes the size of the rectangles based on its depth.

2. Strip-Based Rendering.

Since high accuracy is not required in geometry and computational complexity in the client should be low, we employ the strip-based rendering. The client treats the successive valid pixels in the horizontal line as a band and expresses the pixels by a set of the bands (see Fig.10(b)). The client renders it with the triangle strip. Since the normal vector of each triangle is determined automatically in the steps, we can skip the step of estimating the normals. To enhance the efficiency of rendering, if the valid pixel is replaced by the left pixel in the above simplification algorithm, the pixel is integrated with the left pixel and the pixels are rendered by one strip (Fig.10(c)). And, when scanning the range image in the horizontal direction, if there are invalid pixels, the client calculate the distance of its left and right valid pixels. If the distance is within the threshold, the client fills the blank with its valid pixel. The isolated valid point is rendered by a rectangle. For example, if the pixel is Fig.4(upper left), the rendering results in the strips shown in Fig.10(d).

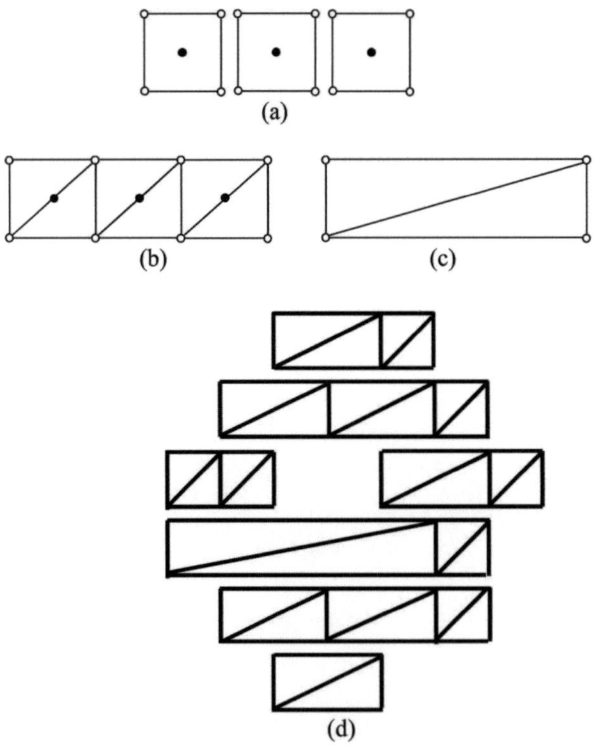

Fig. 10. (a) Rectangle as primitive. The number of vertices is four times as many as the number of the points. (b)Triangle strip as primitive. The number of vertices is about two times as many as the number of the points. (c) Example of merged triangle strip. (d) The example of rendering Fig.4 (upper left).

4. Experimental results

The server consists of five PCs (Pentium IV) and four stereo cameras. We use the commercial stereo cameras (PointGrey). The resolution of colour image and depth image obtained from the camera is 240 x 320. The PCs in the server are connected through Gigabit switch. The server acquires four range images and *ParantPC* integrates them and transmits them to the client. For the client PC, we tested several types such as high end PCs to laptops. The client is located in the same LAN via a router. The example of a man's upper body obtained under these experiment conditions is presented in Fig.11. The system worked at about 9-12fps(frame per sec).

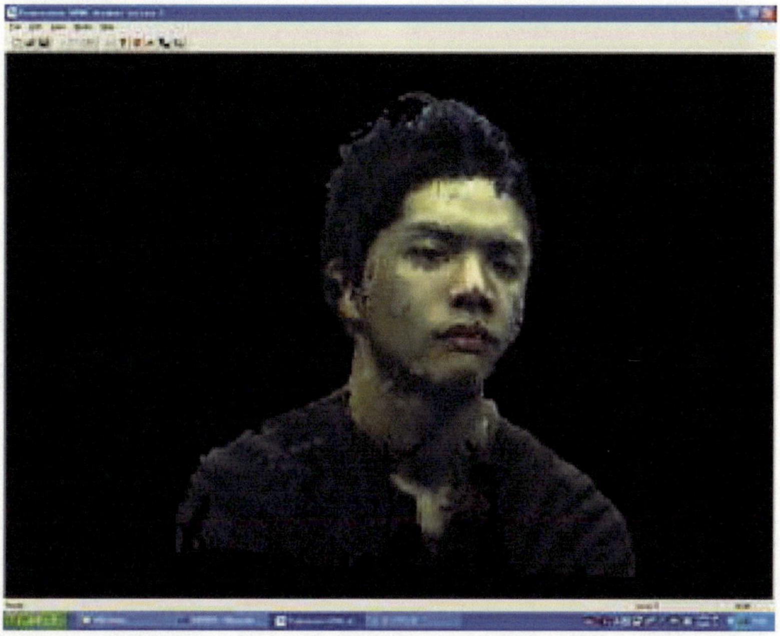

Fig. 11. 3D Viewer

Fig.12 shows the comparison of the point based and strip based rendering. In this experiment, the total data size of 100 frames of the human face was 3.86M bytes without the proposed simplification. On the contrary, our method achieves 1.46M bytes. The data is decreased by 38 %. The average rendering time for one frame was 0.036 sec by the rectangle rendering. On the contrary, with the triangle strip, the rendering time was 0.022 sec. It can be seen from the Fig. 12 that our method performs well without much degradation.

Fig.13 (a) illustrates the 3D model reconstructed simply by integrating four range images. On the other hand, Fig.13 (b) is the model obtained by the integration with our cylindrical method. Blue points from right and left range images are the annoying artifacts on the front of the face in (a). However in (b), these annoying artifacts are removed and the facial expression is finer than (a). The validity of cylindrical method is shown in Fig.14. These pictures are horizontal slice of the face. Each line represents the surface acquired from a camera. Fig.14 (b) is the result by our cylindrical method and (a) is the result of the conventional method. One can see that some overlaps are removed by the cylindrical method in (b). By applying the cylindrical method, the number of points is reduced by 25%.

The computation time to apply the cylindrical method was less than 0.1 second and it did not almost affect the sequences visually.

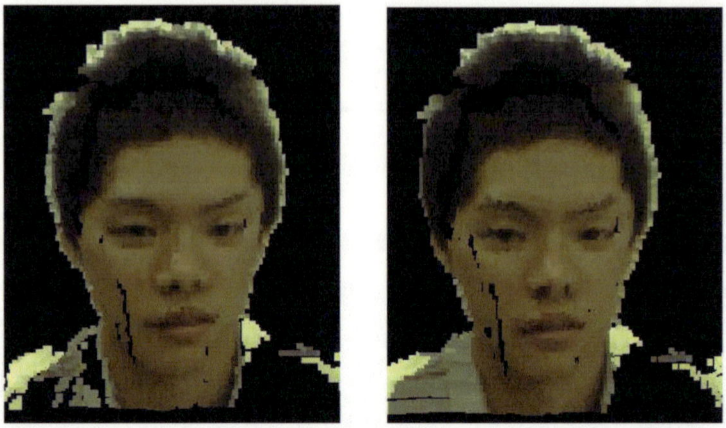

Fig. 12. Rendering example of rectangle as primitive (left). Rendering example of triangle strip as primitive using proposal simplification (right).

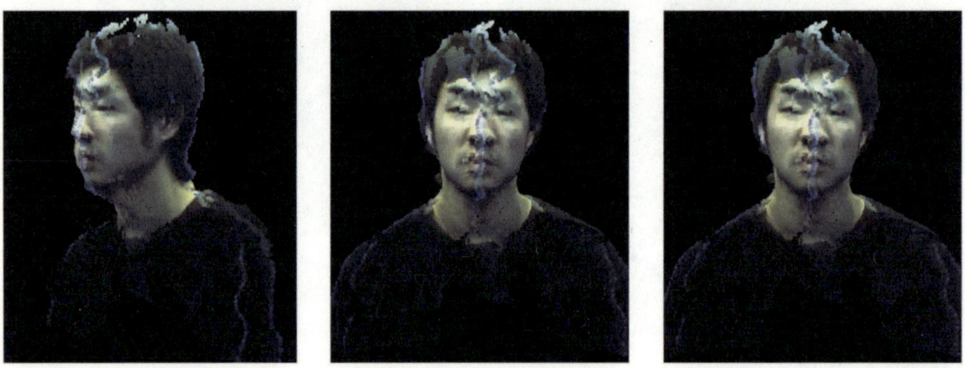

(a)

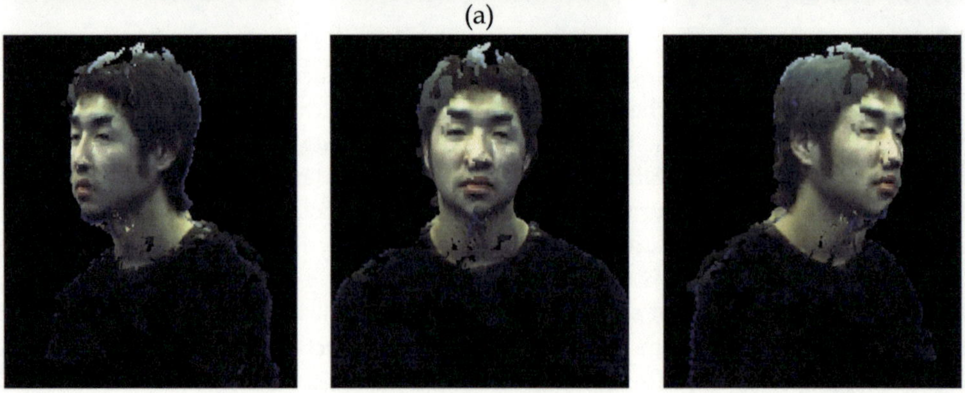

(a)

Fig. 13. Result of Cylindrical Method: These show one frame of the 3D sequence. (a) Integrated four range images. (b) After applying the cylindrical method. Annoying artifacts on the front of face are removed in (b).

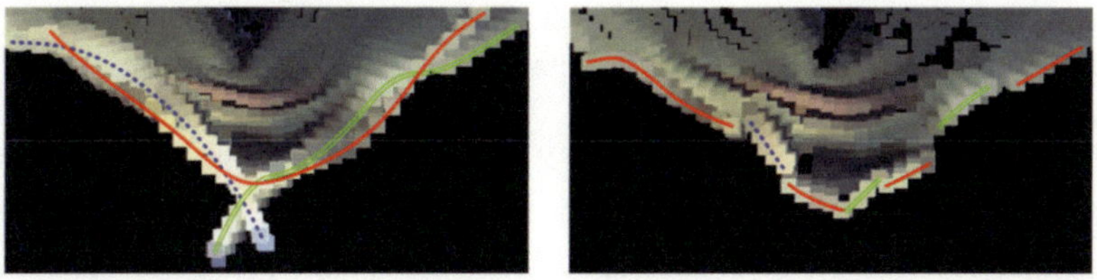

Fig. 14. Horizontal Cut Surface of Nose: (left) without the cylindrical method. Overlapping area cannot be in sight like blue line. However overlaps are removed especially front side of face in (right).

5. Conclusion

In this paper, we introduced a real-time 3D imaging system by handling the 3D shape and the colour information as Point Cloud. It is an advantage of our system that it enhances the transmission efficiency by eliminating the connectivity information. Moreover, by applying the cylindrical method to the real-time 3D imaging system, the overlaps and the annoying artifacts are removed efficiently in real-time.

6. References

Gersho, A. & Gray, R. M. (1992) "Vector Quantization and Signal Compression", Kluwer Academic Publishers.

Hoppe, H. (1996) "Progressive Meshes.", SIGGRAPH '96, pp.99-198.

Kanade, T.; Yoshida, A.; Oda, K.; Kano H. & Tanaka, M. (1996) "A Stereo Machine for Video-rate Dense Depth Mapping and Its New Application", the 15th Computer Vision and Pattern Recognition Conference, pp. 196-202.

Khodakovsky, A., Schröder, P., and Sweldens, W., (2000) "Progressive Geometry Compression.", SIGGRAPH '00, pp.271-278.

PointGrey Research, Digiclops, http://www.ptgrey.com/products/stereo.html

Pratt, W. K.; Andrews, H. C. & Kane, J. (1969) "Hadamard Transform Image Coding", Proc. IEEE, Vol.57, pp.58-68, Jan, 1969.

Rusinkiewicz, S., & Levoy, M., (2000) "Qsplat: A Multiresolution Point Rendering System for Large Meshes", SIGGRAPH '00, pp.343-352.

Saito, H.; Baba, S.; Kimura, M.; Vedula, S. & Kaneda, T. (1999) "Appearance-Based Virtual View Generation of Temporally-Varying Events from Multi-Camera Image in the 3D Room", Second International Conference on 3-D Digital Imaging and Modeling (3DIM99), pp.516-525.

Ueda M.; Arita, D. & Taniguchi, R. (2004) "Real-time Free-viewpoint Video Generation Using Multiple Cameras and a PC-cluster", Pacific-Rim Conference on Multimedia, pp.418-425.

Wu, X. & Matsuyama, T, (2003) "Real-time Active 3D Shape Reconstruction for 3D Video", Proc. of 3rd International Symposium on Image and Signal Processing and Analysis, September, pp.186-191.

Waschbüsch, M.; Würmlin, S.; Cotting D., & Gross, M. (2007) "Point-sampled 3D video of real-world scenes", Journal of Signal Processing: Image Communication, vol. 22, no. 2, 2007, pp. 203-216

Würmlin, S.; Lamboray, E.; Staadt, O. G. & Gross, M. H. (2002) "3D video recorder.", Proceedings of Pacific Graphics '02, pages 325-334. IEEE Computer Society Press.

Würmlin, S.; Lamboray, E.; Gross, M. (2004) "3D Video Fragments: Dynamic Point Samples for Real-time Free-Viewpoint Video", Computers & Graphics, Special Issue on Coding, Compression and Streaming Techniques for 3D and Multimedia Data.

17

A Self Navigation Technique using Stereovision Analysis

Sergio Nogueira[1], Yassine Ruichek[2] and François Charpillet[3]
[1,2] *Systems and Transportation laboratory, University of Technology of Belfort Montbéliard,*
[3] *LORIA Laboratory INRIA Lorraine*
France

1. Introduction

One of the most common issues in developing intelligent vehicle concepts is self localization and autonomous navigation. In this chapter, we are particularly interested in global localization of urban autonomous vehicles. To get a global position of a vehicle navigating in urban areas, localization systems must be adapted to different kinds of environments in presence of dynamics objects. In order to achieve global localization, there are two approaches. The first one is based on sensors data fusion. The second approach uses knowledge on the environment of navigation.

In most data fusion based methods, the environment is represented using a GIS (Geographic Information System). The environment representation is augmented by introducing structured elements from sensors. In [1], Chen proposes a method that estimates the camera movements using edge matching. The initial position is given by fusing data coming from a D-GPS (Differential Global Positioning System), a GIS and other sensors. Kais and al. [2] propose to fuse vertical features recorded into a GIS with images provided by an embedded camera. The principle is to construct regions around vertical features in order to compute the position and the orientation of the camera. The localization process is achieved by determining correspondences between virtual and real features.

The environment knowledge based approach is interesting when sensors unreliability is considered. In spite of GPS sensors can be accurate (for example by using RTK-GPS systems) GPS information is not adapted into dense urban areas. Indeed, because of the difficulty to detect satellites (due to the presence of buildings) and reflection of GPS signals, the GPS system may lose in his accuracy and may even provide false positions. This situation is known by the urban cayonning problem [3].

In order to discard this problem, the environment knowledge approach consists in creating an image key database. The camera position and orientation are computed for each image referenced during the learning phase. In [4], Katsura et al. propose a method, which adds a region analysis to distinguish different region types. The aim is to process specifically each region that evolves differently through the time. Based also on image key learning sequence, the method proposed by Royer and al. [5] computes 3D specific points. Bundle adjustments [6] are used in order to increase the model precision. For each movement of the camera, the position is determined using the closest 3D features.

Between these two main approaches, there are some hybrid localization methods. For vehicle navigation in urban areas, Gerogiev and Allen propose a technique, which consists in fusing data coming from several sensors [7]. When the sensors information is unavailable, a camera makes a visual servoing between the images provided by the camera and the images coming from a geo-referenced image database.
In this chapter, the presented approach is based on image key learning sequence. The aim is to achieve self localization using only stereoscopic information. As Royer and al. in [5], the proposed stereovision based method constructs a 3D model. In the approach, the 3D model is built from 3D features reconstructed using SIFT based stereo matching and tracked using temporal matching.
This chapter is organized as follows. Section 2 presents the concept and fundament to construct a 3D model based on image SIFT features. Section 3 explains how to localize a camera using a database containing 3D points, reconstructed from the matching of SIFT features. Before concluding, results in various conditions are presented in section 4.

2. Model construction

In order to compute the global position of a camera, one needs to construct a 3D model composed by features that can be matched under image changes like translation, rotation and scaling. In external conditions like in urban environments, it is important that image features are partially invariant to illumination changes. Most of the existing approaches use Harris corner detector [8], which is sensitive to the scale of images. As a consequence, building a map requires a lot of images and a considerable number of points extracted in each image. Moreover, in the localization process, the vehicle exploration must be close to the learned trajectory.
In the proposed method, Scale Invariant Feature Transform (SIFT) features are used. Introduced by Lowe [9], SIFT features have detailed characteristics that make them suitable landmarks for robust Self Localization And Mapping (SLAM), because when mobile robots are moving in an environment, landmarks are observed from different angles, distances and under different illumination changes. Each reconstruction or localization step is based on the matching process of SIFT features.
Figure 1 shows an example of SIFT features extracted from the left and right images taken by a virtual simulated stereoscopic sensor with two cameras. The resolution of the images is 640x480. In this example, eight levels of scale are used to extract SIFT features. There are about 2000 features in each image. This quantity of features is generally sufficient for the considered task, but if desired, the number can be increased by increasing the scales and the image resolution.
At each reconstruction step-frame, the SIFT features are extracted from the two rectified stereo images and are then matched. The images are rectified using the Zhang method for stereo camera calibration [10]. Matched SIFT features are stable and serve as better landmarks for detecting and tracking the objects observed in the environment. The stereo matching of the SIFT features provides 3D points that serve to construct the 3D model.

2.1 Stereo matching

Considering the stereovision sensor, the right camera serves as the reference camera. The two cameras are separated by a distance $E=12cm$ and have the same focal length f (see Figure 2). The stereo matching of the SIFT features uses the following constraints:

Fig. 1. SIFT features extraction

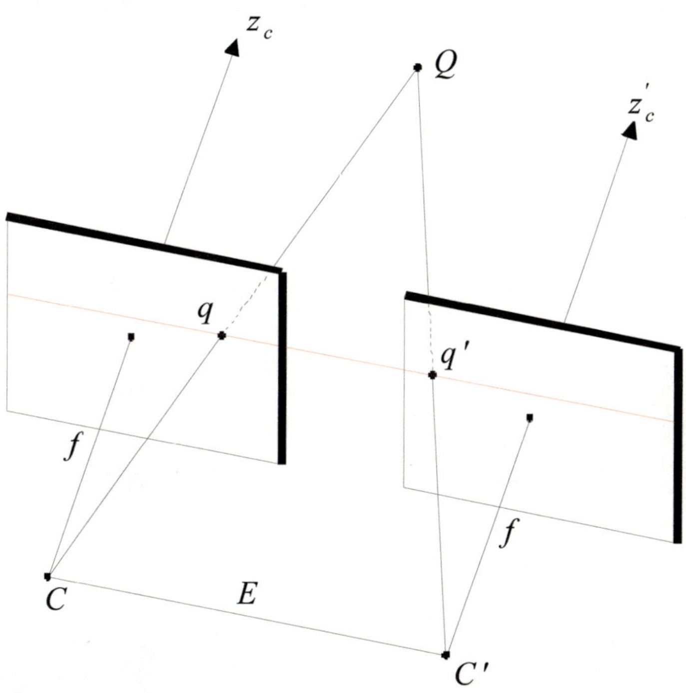

Fig. 2. Stereoscopic configuration

Disparity constraint. Considering a candidate match composed with a left feature and a right feature, the difference between the horizontal image coordinates of the features must be within a predefined disparity range.

Epipolar constraint. The vertical image coordinates of the left and right features must be within 1 pixel of each other, as the images are aligned and rectified.
Orientation constraint. The difference of the orientations of the left and right features must be within a predefined range.
Scale constraint. One scale must be at most one level higher or lower than the other. Adjacent scales differ by a factor of 1.5 in the proposed SIFT extraction procedure.
Uniqueness constraint. If a feature has more than one match satisfying the above constraints, the matches are considered as ambiguous and are discarded so that the resulting matches are more consistent and reliable.
After matching the SIFT features, a list of matched couples is obtained. For each couple, an image horizontal disparity is computed and then a 3D point is reconstructed by considering the intrinsic and extrinsic parameters of the cameras. The 3D point has its coordinates in a reference associated to the stereovision sensor. All the reconstructed 3D points are added to the database after computing their global position (see section 3). In order to use them as landmarks, the orientations and scales of the corresponding SIFT features are set respectively to the average of the orientations and the scales computed in the left and right images.

2.2 Model computation

Let's define an axis-aligned stereoscope *S* where the left and the right cameras are defined respectively by their optical centers *C* and *C′*, with the same focal length *f*, and separated with a distance *E* (cf. figure 2). Let $W = \{W_x, W_y, W_z\}$ be a 3D point from the world coordinate system, $q = \{q_x, q_y\}$ and $q' = \{q'_x, q'_y\}$ are the projections of the point *W* into the left and right cameras, respectively. Using the stereo triangulation technique, the point *W* can be computed given the image projections *q* and *q′* and knowing the intrinsic and extrinsic parameters:

$$W_x = \frac{W_z * q'_x}{f}, W_y = \frac{W_z * q_y}{f}, W_z = \frac{f * E}{|q_x - q'_x|} \quad (1)$$

Figure 3 shows the projection on the left and right images of the reconstructed 3D points after the stereo matching process.

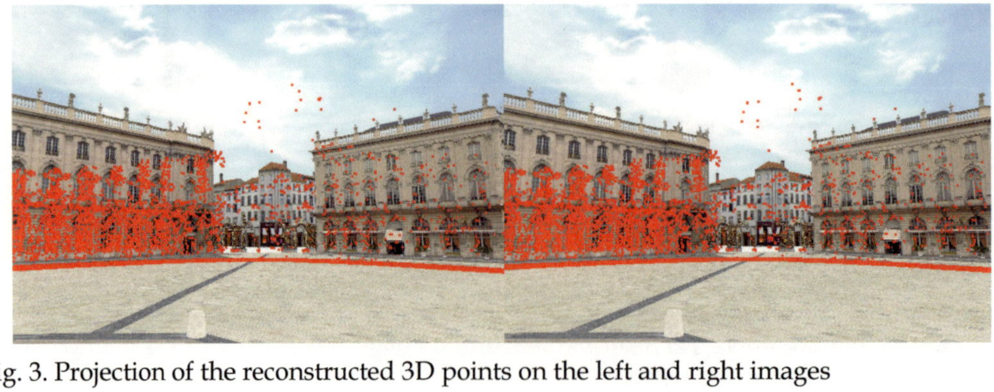

Fig. 3. Projection of the reconstructed 3D points on the left and right images

2.3 3D model construction

The global 3D construction is based on an incremental method. At the beginning, the first stereoscopic acquisition is used to initialize the 3D reconstruction process.

Let's $C_1, C_2 \ldots C_N$, be the camera positions computed respectively from the pose 1 to the pose $N(N \geq 3)$. The goal is to determine the position C_{N+1} of the camera, associated to the pose *N+1*. After image acquisition I_{N+1} from the unknown position C_{N+1}, the SIFT features are extracted. Let q_{N+1}^{i} be the ith extracted SIFT feature from the image I_{N+1}. The extracted points are then matched with the SIFT points extracted from the image I_N. A list of couples of matched points $\left(q_N^i, q_{N+1}^i\right)$ is finally obtained. Note that the point q_{N+1}^{i} may have any corresponding point in the image I_{N-1}. This means that from one image to the other, identified points may disappear and new points may appear. The 3D global position of the new points is determined by using a robust construction method described in the next section.

The main problem of the path reconstruction with temporal matching is that the estimation of each point position is based on the previous computed ones. Consequently, the computation errors increase throughout the reconstruction process. The bundle adjustment process [6] limits the error computation. This process increases the construction precision by using multiple views. It consists on a minimization process based on the Levenberg-Marquardt algorithm [11]. The function $f\left(C_E^1,\ldots,C_E^N,Q^1,\ldots Q^M\right)$ to be minimized is defined from the extrinsic cameras parameters C_E^i and the 3D points $\{Q^j\}$ extracted from stereo 3D reconstruction using multiple views. This function is expressed as follows:

$$f\left(C_E^1,\ldots,C_E^N,Q^1,\ldots Q^M\right) = \sum_{i=1}^{N}\sum_{j=1}^{M}\left\| q_i^j - \pi\left(P_i Q^j\right)\right\|^2 \tag{2}$$

Where $\left\| q_i^j - \pi\left(P_i Q^j\right)\right\|^2$ is the square Euclidian distance between the SIFT point q_i^j and the point $\pi\left(P_i Q^j\right)$, which is the projection of the 3D point Q^j on the camera at the ith position, called also the retro-projection of the point Q^j. P_i is the projection matrix obtained from the extrinsic $\left(C_E^i\right)$ and intrinsic parameters of the camera at the ith position.

In order to make the algorithm more robust, false matches are removed. The minimization process begins by keeping the points satisfying the condition $\left\| q_i^j - \pi\left(P_i Q^j\right)\right\|^2 < \varepsilon$, where ε is a constant fixed empirically to 9. The minimization algorithm converges when the number of the retro-projection points becomes stable.

Figure 4 shows an example of an environment 3D model construction. This figure represents a model view in two different render types. In the middle of the right sub-image, the circular shape represents the vehicle trajectory. The other shapes around the trajectory represent the 3D reconstructed points, corresponding to the environment 3D model.

3. Localization

After completing the 3D model construction of the environment of navigation, the camera position can be computed by matching the SIFT features extracted from the acquired images

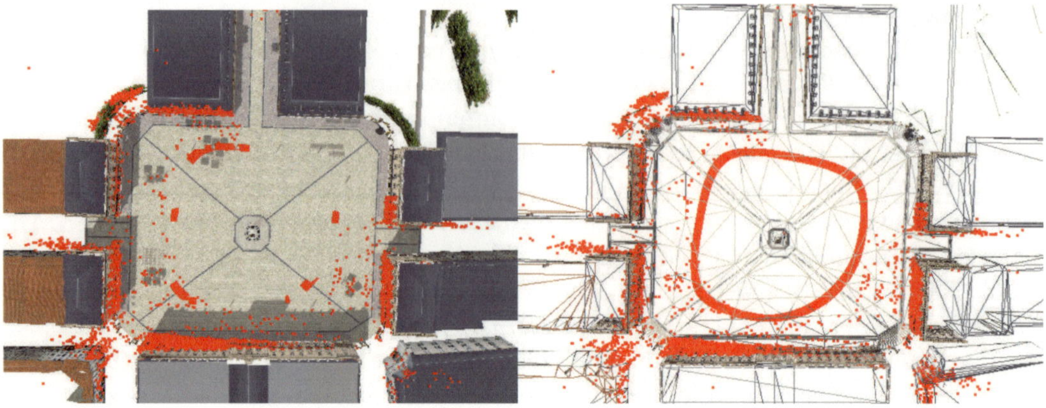

Fig. 4. Environment 3D model construction

with those stored in the database, corresponding to the learned trajectory. When the initial camera position is unknown, the extracted SIFT features are matched with all features from the database. This starting procedure necessitates a considerable computation time. However, when started, the localization process uses the last computed position in order to compute the current one. This allows filtering the features from the database in order to consider only those corresponding to the current view. This filtering stage reduces the complexity computation by ignoring more than 90% of the features during the matching procedure, allowing thus real time localization.

Let $\{Q^i\}$, $i = 0 \ldots n$, be the 3D points extracted from the stereoscopic sensor defined as C (see figure 5). The points Q^i are expressed into the stereoscopic coordinate system. In order to compute the global position of the stereoscopic sensor, the SIFT matching process is used to associate to each point Q^i a point W^i of the environment 3D model.

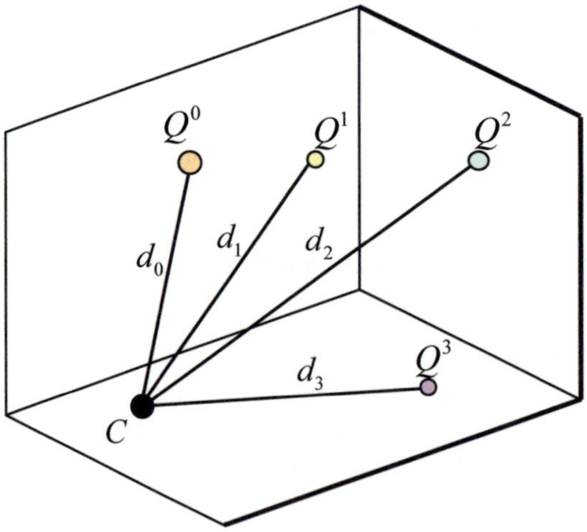

Fig. 5. Localization scheme

The Euclidian distances between the stereoscopic sensor and the points Q^i are computed and defined as d_j. The relation giving the camera position into the global coordinate system is expressed as follows:

$$\begin{cases} \left(C_x - Q_x^0\right)^2 + \left(C_y - Q_y^0\right)^2 + \left(C_z - Q_z^0\right)^2 = d_0^2 \\ \vdots \\ \left(C_x - Q_x^i\right)^2 + \left(C_y - Q_y^i\right)^2 + \left(C_z - Q_z^i\right)^2 = d_i^2 \\ \vdots \\ \left(C_x - Q_x^n\right)^2 + \left(C_y - Q_y^n\right)^2 + \left(C_z - Q_z^n\right)^2 = d_n^2 \end{cases} \quad (3)$$

In order to resolve the equations system (3), it is necessary to minimize the equation (4) using the Newton-Gauss algorithm.

$$S(\beta) = \sum_{i=0}^{n} r_i^2 \quad (4)$$

where $r_i = d_i^2 - f_i(\beta)$, $f_i(\beta) = (C_x - Q_x^i)^2 + (C_y - Q_y^i)^2 + (C_z - Q_z^i)^2$ and $\beta = (C_x, C_y, C_z)$.

The Newton-Gauss algorithm is particularly interesting when it starts with an initial value of β closed to the desired solution. This initialization step can be achieved using different techniques like Kalman filtering, fusion from several sensors like odometer or simply using the previous position. The minimization process can be expressed as follows:

$$\beta^{k+1} = \beta^k + \left(\mathrm{J}^{\mathrm{T}}\mathrm{J}\right)^{-1} \mathrm{J}^{\mathrm{T}} r \quad (5)$$

where J is the Jacobian matrix of $f(\beta)$:

$$\mathrm{J} = 2\begin{pmatrix} C_x - Q_0^x & C_y - Q_0^x & C_z - Q_0^z \\ \vdots & \vdots & \vdots \\ C_x - Q_n^x & C_y - Q_n^x & C_z - Q_n^z \end{pmatrix} \quad (6)$$

r and $f(\beta)$ are respectively the vectors composed with r_i and $f_i(\beta)$ $(i=1\ldots n)$

In addition, it is possible to increase the localization robustness by using the RANSAC technique [12]. This technique consists first in choosing randomly three points Q^0, Q^1 and Q^2 from the 3D reconstructed points, provided by the stereo matching of the SIFT features. Using these chosen points, the global position is then computed. These two steps are repeated until convergence, i.e., when the current calculation result is close to the previous one. Taking into account the RANSAC scheme, equation (3) (with n=4) can be rewritten as:

$$t^2 \cdot H_1 + t \cdot H_2 + H_3 = 0 \quad (7)$$

This equation has two solutions t_1, t_2:

$$t_1 = \frac{-H_2 - \sqrt{H_2^2 - 4H_1 \cdot H_3}}{2H_1} \text{ or } t_2 = \frac{-H_2 + \sqrt{H_2^2 - 4H_1 \cdot H_3}}{2H_1} \quad (8)$$

$$\text{where}: \begin{cases} H_1 = N_4^2 + N_2^2 + 1 \\ H_2 = 2\left(N_4 M_1 + N_2 M_2 - Q_z^0\right), \\ H_3 = M_1^2 + M_2^2 + \left(Q_z^0\right)^2 - d_0^2 \end{cases} \begin{cases} M_1 = N_3 - Q_x^0 \\ M_2 = N_1 - Q_y^0 \end{cases},$$

$$\text{with}: \begin{cases} N_1 = \dfrac{A_1 D_2}{B_1 A_2} - \dfrac{D_1}{B_1} \\ N_2 = \dfrac{A_1 C_2}{B_1 A_2} - \dfrac{C_1}{B_1} \end{cases}, \begin{cases} N_3 = -\dfrac{D_1 + B_1 P_1}{A_1} \\ N_4 = -\dfrac{B_1 P_2 + C_1}{A_1} \end{cases}$$

$$\text{and}: \begin{cases} A_1 = 2\left(Q_x^1 - Q_x^0\right); A_2 = 2\left(Q_x^2 - Q_x^0\right) \\ B_1 = 2\left(Q_y^1 - Q_y^0\right); B_2 = 2\left(Q_y^2 - Q_y^0\right) \\ C_1 = 2\left(Q_z^1 - Q_z^0\right); C_2 = 2\left(Q_z^2 - Q_z^0\right) \\ D_1 = \left(Q_x^0\right)^2 - \left(Q_x^1\right)^2 + \left(Q_y^0\right)^2 - \left(Q_y^1\right)^2 + \left(Q_z^0\right)^2 - \left(Q_z^1\right)^2 - \left(d_0\right)^2 + \left(d_1\right)^2 \\ D_2 = \left(Q_x^0\right)^2 - \left(Q_x^2\right)^2 + \left(Q_y^0\right)^2 - \left(Q_y^2\right)^2 + \left(Q_z^0\right)^2 - \left(Q_z^2\right)^2 - \left(d_0\right)^2 + \left(d_2\right)^2 \end{cases}$$

The camera (left or right) position estimated from equation (8) is considered to compute the retro-projection (see section 2.3) of each point Q^i. A retro-projection error can be thus computed for each point Q^i. If the error is less than 2 pixels, then the point Q^i associated to this error is considered as an inlier point with respect to the RANSAC procedure. The final estimation of the camera position is then computed using equation (3) and by considering only the inliers.

4. Experimental results

The proposed global localization method is tested and evaluated considering different environment conditions (cf. figure 6). The tests are achieved using a specific software simulation platform, which allows to generate 3D virtual environments and to construct mobile vehicles equipped with different video sensors. The evaluation consists to compare the real trajectory and the estimated one, (during the self global localization process). This comparison can be achieved by computing the deviation between the two trajectories. Recall that the estimated trajectory is based on the learned one, which corresponds to the constructed environment 3D model.

To construct the environment 3D model, the learning process is realized using a stereo sequence of 80 couples. The constructed model is composed with 9452 3D points, which represents a circular trajectory of about 142.42 meters (see figure 4).

Graphs 1–4 represent the evolution of the error (in meters) between the real and estimated trajectories in different environment conditions. The error represents the distance between the estimated camera position and the closest point of the real trajectory.

Fig. 6. upper left image: with normal illumination condition; upper right image: with dark illumination; lower left image: with high illumination; lower right image: with presence of objects in the environment

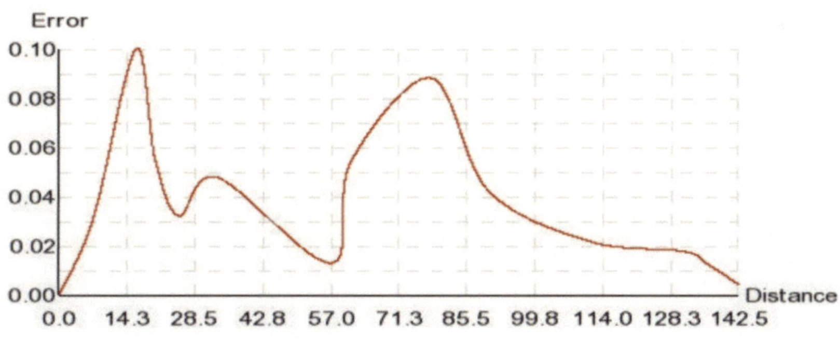

Graph 1. Error (in meter) in normal illumination conditions

The graph 1 shows that the error is less than 10 cm. Compared to classic GPS systems, the proposed localization process is more reliable.

Most global localization techniques based on feature extraction are often sensitive to the environment illumination problem [13]. Graphs 2 and 3 show respectively the error obtained by the proposed method in dark and high illumination conditions. The localization results remain accurate. The pose of the camera is estimated with a maximum error of 15 cm. This performance is due principally to the using of the SIFT feature extractor, which is not considerably influenced by the illumination changes, and, as a consequence, the SIFT matching process provide enough correct matched points to get an efficient result.

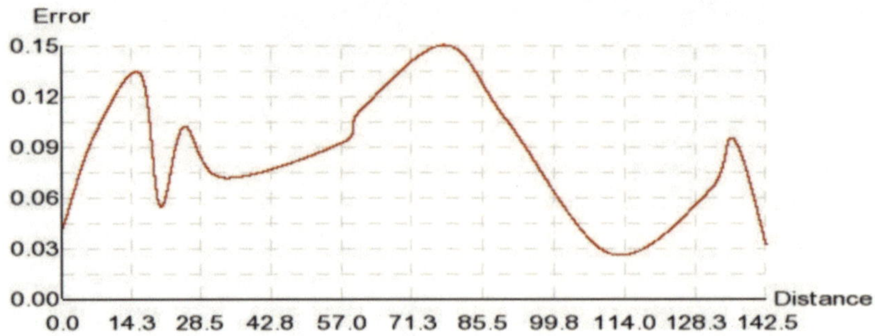

Graph 2. Error (in meter) in dark illumination conditions

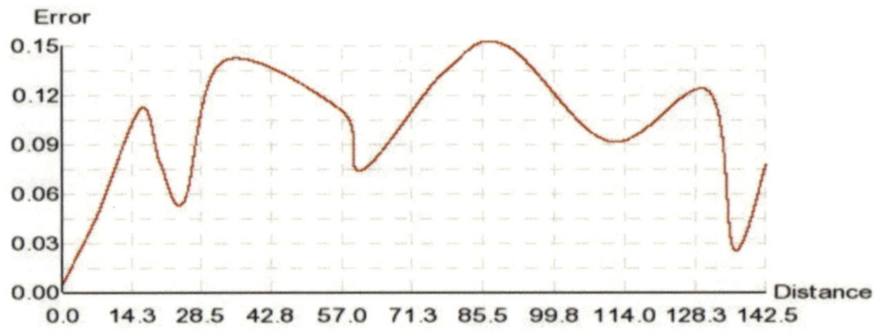

Graph 3. Error (in meter) in high illumination conditions

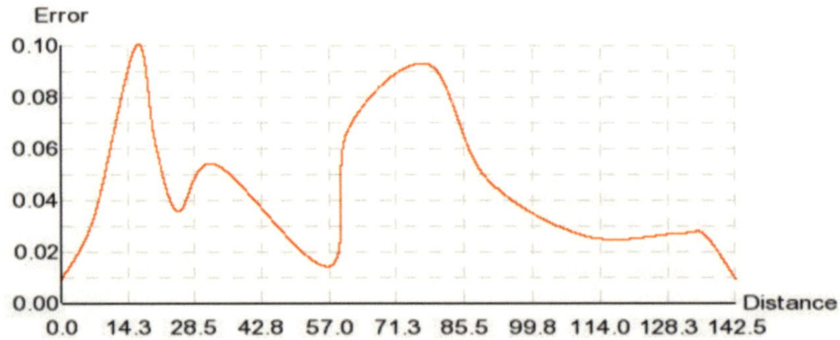

Graph 4. Error (in meter) in the case of presence of objects near to the learned trajectory

The last test consists in changing statically the learned environment by adding objects near to the trajectory, with normal illumination conditions. The error evolution shown in the graph 4 is close to the one associated to the case without presence of objects near to the learned trajectory (see graph 1). The presence of objects in the navigation environment decreases the number of matched SIFT features. This does not influence the precision of localization, while the illumination conditions remain unchanged. Indeed, when graphs 1 and 4 are compared, one can see that, globally, the precision is approximately identical.

5. Conclusion

This chapter proposes a robust method to achieve global localization. This method is based on a learning process, which consists to construct an environment 3D model from spatial and temporal matching of the SIFT features. Having the environment 3D model, the localization step is performed by searching correspondences between 3D reconstructed points and the 3D points belonging to the 3D model throughout the utilization of the SIFT features. The reliability of the proposed method is improved by using the RANSAC technique.

The proposed method is tested and evaluated in different environment conditions, using a software simulation platform. The tests show that the method is robust and provides reliable results. In deed, the error between the estimated and the real trajectories is less than 10 cm in normal illumination conditions. Thanks to the SIFT extractor, the maximum error does not exceed 15 cm, when considering illumination changes.

In terms of computation time, the localization process runs at 1.2 Hz for images with a resolution of 640x480 images, using a PC machine with a Core Duo running at 2.2 GHz. The SIFT extraction procedure consumes about 95% of the processing time. This is due to the high number of scales considered during the SIFT extraction. More tests are in progress in order to reduce the number of the SIFT points without loss of precision. Indeed, a relation can be established between the number of SIFT points, the processing time and the desired global precision.

Thanks to the using of stereovision and SIFT extractor, the proposed localization method is more interesting in terms of computation time and reliability. The work is in progress to compare the method with other ones using simulation and evaluation in real conditions through an experimental automated vehicle platform.

6. References

T. Chen. "*Development of a vision-based positioning system for high density area*". In Asian Conference on Remote Sensing (ACRS'99), Hong Kong, China, Nov 22-25 1999

M. Kais, S. Morin, A. de la Fortelle, and C. Laugier. "*Geometrical model to drive vision systems with error propagation*". In 8th International Conference on Control, Automation, Robotics and Vision (ICARCV'04), Kunming, China, Dec. 3-9 2004.

Cui, Sheyhi. "*Autonomous vehicle positioning with GPS in urban canyon environment*". IEEE Transaction on Robotics and Automation, 2003. Monocular Vision for Mobile Robot Localization and Autonomous Navigation,

H. Katsura, J. Miura, M. Hild, and Y. Shirai. "*A view-based outdoor navigation using object recognition robust to changes of weather and seasons*". In IEEE RSJ/International conference on Intelligent Robot and System (IROS'03), pages 2974-2979, Las Vegas, Nev., USA, Oct. 27-31 2003.

E Royer, M Lhuilier, M. Dhome, and T. Chateau. "*Towards an alternative gps sensor in dense urban environment from visual memory*". In 15th British Machine Vision Conference (BMVC'04), London, U.K., Sept. 7-9 2004.

Bill Triggs Philip F.Lauchian Richard I. Hartley and Andrew W. Fitzgibbon, "*Bundle Adjustment a Modern Synthesis*", "*Vision Algorithms: Theory and Practice*" vol. 1883, pp. 298—372. Springer Verlag, LNCS (2000)

A. Georgiev and P.K. Allen. "*Localization methods for a mobile robot in urban environments*". In IEEE Transactions on Robotics and Automation (ICRA'04), 2004.

C. Harris, and M.J. Stephens, "*A combined corner and edge detector*". In Alvey Vision Conference, pages 147-152, 1988.

David G. Lowe, "*Distinctive image features from scale-invariant keypoints*", International Journal of Computer Vision, 60, 2 (2004), pp. 91-110

Z. Zhang. "*A flexible new technique for camera calibration*". IEEE Transactions on Pattern Analysis and Machine Intelligence, 22(11):1330-1334, 2000.

W.H. Press, S.A. Teukolsky, W.T. Vetterling, and B.P. Flannery. Numerical Recipies in C. Cambridge University Press, 1988. pp. 681-689.

Martin A. Fischler and Robert C. Bolles. "*Random Sample Consensus: A Paradigm for Model Fitting with Applications to Image Analysis and Automated Cartography*". Comm. of the ACM 24: 381–395

K. Mikolajczyk and C. Schmid. "*A Performance Evaluation of Local Descriptors*". IEEE Trans. on Pattern Analysis And Machine Intelligence, Vol. 27, No. 10, Oct. 2005

18

Real Time Stereo Image Registration for Planar Structure and 3D Sensor Pose Estimation

Fadi Dornaika[1] and Angel D. Sappa[2]
[1]*Institut Géographique National, 2 avenue Pasteur, 94165 Saint-Mandé,*
[2] *Computer Vision Center, UAB, 08193 Bellaterra, Barcelona,*
[1]*France,*
[2]*Spain*

1. Introduction

In recent years, several techniques to on-board vision pose estimation have been proposed (Zhu et al., 1998; Labayrade & Aubert, 2003; Liu & Fujimura, 2004; Stein et al., 2000; Suzuki & Kanade, 1999). Vision system pose estimation is required for any advanced driver assistance application. The real-time estimation of on-board vision system pose-position and orientation- is a challenging task since i) the sensor undergoes motions due to the vehicle dynamics and the road imperfections, and ii) the viewed scene is unknown and continuously changing.

Of particular interest is the estimation of on-board camera's position and orientation related to the 3D road plane. Note that since the 3D plane parameters are expressed in the camera coordinate system, the camera's position and orientation are equivalent to the 3D plane parameters. Algorithms for fast road plane estimation are very useful for driver assistance applications as well as for augmented reality applications. The ability to use continuously updated plane parameters (camera pose) will considerably make the tasks of obstacle detection more efficient (Viola et al., 2005; Sun et al., 2006; Toulminet et al., 2006). However, dealing with an urban scenario is more diffcult than dealing with highways scenario since the prior knowledge as well as visual features are not always available in these scenes (Franke et al., 1999).

In general, monocular vision systems avoid problems related to 3D Euclidean geometry by using the prior knowledge of the environment as an extra source of information. Although prior knowledge has been extensively used to tackle the driver assistance problem, it may lead to wrong results. Hence, considering a constant camera's position and orientation is not a valid assumption to be used in urban scenarios, since both of them are easily affected by road imperfections or artifacts (e.g., rough road, speed bumpers), car's accelerations, uphill/downhill driving, among others.

The use of prior knowledge has also been considered by some stereo vision based techniques to simplify the problem and to speed up the whole processing by reducing the amount of information to be handled (Bertozzi & Broggi, 1998; Bertozzi et al. 2003; Nedevschi et al., 2006). In the literature, many application-oriented stereo systems have been proposed. For instance, the edge based *v*-disparity approach proposed in (Labayrade et al., 2002), for an automatic estimation of horizon lines and later on used for applications such as obstacle or pedestrian detection (e.g., (Bertozzi et al., 2005; Labayrade & Aubert,

2003)), only computes 3D information over local maxima of the image gradient. A sparse disparity map is computed in order to obtain a real time performance. Recently, this *v*-disparity approach has been extended to a *u-v*-disparity concept in (Hu & Uchimura, 2005). In this work, dense disparity maps are used instead of only relying on edge based disparity maps. Working in the disparity space is an interesting idea that is gaining popularity in on-board stereo vision applications, since planes in the original Euclidean space become straight lines in the disparity space.

In (Sappa et al., 2006), we have proposed an approach for on-line stereo camera pose estimation. Although the proposed technique does not require the extraction of visual features in the images, it is based on dense depth maps and on the extraction of a dominant 3D plane that is assumed to be the road plane. This technique has been tested on different urban environments. The proposed algorithm took, on average, 350 ms per frame.

As can be seen, existing works adopt the following main stream. First, features are extracted either in the image space (optical flow, edges, ridges, interest points) or in the 3D Euclidean space (assuming the 3D data are built online). Second, an estimation technique is then invoked in order to recover the unknown parameters.

In this chapter, we propose a novel paradigm for on-board camera pose tracking trough the use of image registration (Romero & Calderon, 2007). Since we do not rely on features, the image registration should be featureless. We solve the featureless registration by using two optimization techniques: the Differential Evolution algorithm (a stochastic search) and the Levenberg-Marquardt algorithm (a directed search). Moreover, we propose two tracking schemes based on these optimizations. The advantage of our proposed paradigm is twofold. First, it can run in real-time. Second, it provides good results even when the road surface does not have reliable features. We stress the fact that our proposed methods are not restricted to the estimation of on-board camera pose/roads. Indeed, the proposed methods can be used for extracting any planar structures using stereo pairs.

The rest of the chapter is organized as follows. Section 2 describes the problem we are focusing on as well as some backgrounds. Section 3 presents the proposed approach in details. Section 4 gives some experimental results and method comparisons. Section 5 concludes the chapter.

2 Problem formulation and background

2.1 Experimental setup

A commercial stereo vision system (Bumblebee from Point Grey[1]) was used. It consists of two Sony ICX084 color CCDs with 6mm focal length lenses. Bumblebee is a pre-calibrated system that does not require in-field calibration. The baseline of the stereo head is 12cm and it is connected to the computer by an IEEE-1394 connector. Right and left color images can be captured at a resolution of 640×480 pixels and a frame rate near to 30 fps. This vision system includes a software able to provide the 3D data. Figure 1(a) shows an illustration of the on-board stereo vision system as well as its mounting device.

The problem we are focusing on can be stated as follows. Given a stream of stereo pairs provided by the on-board stereo head we like to recover the parameters of the road plane for every captured stereo pair. Since we do not use any feature that is associated with road structure, the computed plane parameters will completely define the pose of the on-board vision sensor. This pose is represented by the 3D plane parameters, that is, the height d and

---

[1] [www.ptgrey.com]

the plane normal $\mathbf{u} = (u_x, u_y, u_z)^T$ from which two independent angles can be inferred (see Figure 1(b)). In the sequel, the pitch angle will refer to the angle between the camera's optical axis and the road plane; and the roll angle will refer to the angle between the camera horizontal axis and the road plane (see Figure 1(b)). Due to the reasons mentioned above, these parameters are not constant and should be estimated online for every time instant. Note that the three angles (pitch, yaw, and roll) associated with the stereo head orientation can vary. However, only the pitch and roll angles can be estimated from the 3D plane parameters.

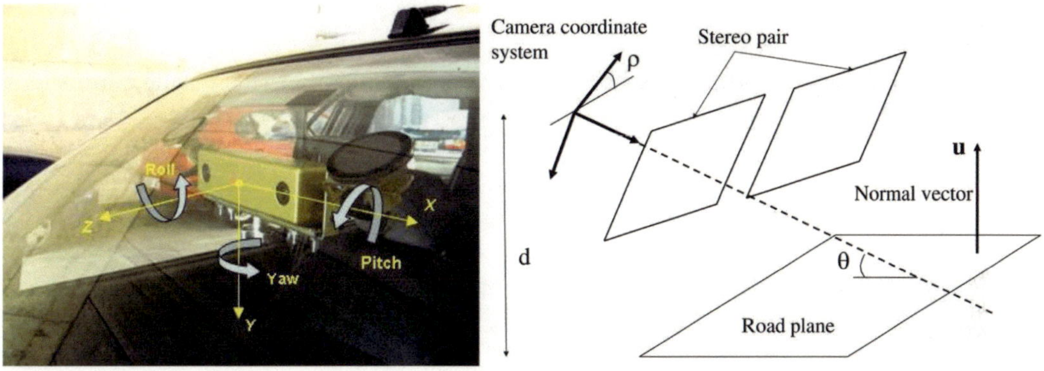

(a) (b)

Fig. 1. (a) On-board stereo vision sensor. (b) The time-varying road plane parameters d and $\mathbf{u}$. θ denotes the pitch angle and ρ the roll angle.

2.2 Image transfer function

Before going into the details of the proposed approach, this section will describe the geometric relation between road pixels belonging to the same stereo pair—the left and right images. It is well-known (Faugeras & Luong, 2001) that the image coordinates of the projections of 3D points belonging to the same plane onto two different images are related by a 2D projective transform having 8 independent parameters-*homography*. In our setup, the right and left images are horizontally rectified[2]. Let $p_r(x_r, y_r)$ and $p_l(x_l, y_l)$ be the right and left projection of an arbitrary 3D point P belonging to the plane (d, u_x, u_y, u_z). In the case of a rectified stereo pair where the left and right cameras have the same intrinsic parameters, the right and left coordinates of corresponding pixels belonging to the road plane are related by the following linear transform (the homography reduces to a linear mapping):

$$x_l = h_1\, x_r + h_2\, y_r + h_3 \quad (1)$$

$$y_l = y_r \quad (2)$$

where h_1, h_2, and h_3 are function of the intrinsic and extrinsic parameters of the stereo head and of the plane parameters. For our setup (rectified images with the same intrinsic parameters), those coefficients are given by:

---

[2] The use of non-rectified images will not have any theoretical impact on our developed method. However, the image transfer function will be given by a general homography.

$$h_1 = 1 + b\,\frac{u_x}{d} \tag{3}$$

$$h_2 = b\,\frac{u_y}{d} \tag{4}$$

$$h_3 = -b\,u_0\frac{u_x}{d} - b\,v_0\,\frac{u_y}{d} + \alpha\,b\,\frac{u_z}{d} \tag{5}$$

where b is the baseline of the stereo head, α is the focal length in pixels, and (u_0, v_0) is the image center (principal point). Let **w** be the 3-vector encapsulating the 3D plane parameters, that is, $\mathbf{w} = \frac{\mathbf{u}}{d}$.

$$\mathbf{w} = (w_x, w_y, w_z)^T = (\frac{u_x}{d}, \frac{u_y}{d}, \frac{u_z}{d})^T \tag{6}$$

Note that the vector **w** fully describes the current road plane parameters. The problem can be stated as follows. Given the current stereo pair estimate the corresponding 3D road plane parameters d and **u** or equivalently the vector **w**.

3. Approach

Since the goal is to estimate the road plane parameters **w** for every stereo pair (equivalently the 3D pose of the stereo head), the whole process is invoked for every stereo pair. Figure 2 illustrates the tracking of the stereo head pose over time. The inputs to the algorithm are the current stereo pair as well as the estimated road plane parameters associated with the previous frame. The algorithm is split into two consecutive stages. First, a rough road region segmentation is preformed for the right image. Let $\mathcal{R}$ denotes this region—a set of pixels. Second, recovering the plane parameters from the rawbrightness of a given stereo pair will rely on the following fact: *if the parameter vector* **w** *corresponds to the actual plane parameters—the distance d and the normal* **u**—*then the registration error between corresponding road pixels in the right and left images over the region* $\mathcal{R}$ *should correspond to a minimum.* In our work, the registration error is set to the Sum of Squared Differences (SSD) between the right image and the corresponding left image computed over the road region $\mathcal{R}$. The registration error is given by:

$$e(\mathbf{w}) = \sum_{(x_r,y_r)\in\mathcal{R}} \left(I_{r(x_r,y_r)} - I_{l(h_1\,x_r+h_2\,y_r+h_3,y_r)}\right)^2 \tag{7}$$

The corresponding left pixels are computed according to the linear transform given by (1) and (2). The computed $x_l = h_1\,x_r + h_2\,y_r + h_3$ is a non-integer value. Therefore, the grey-level, $I_l(x_l, y_l)$, is set to a linear interpolation of the grey-level of two neighboring pixels—the ones whose horizontal coordinates bracket the value x_l.

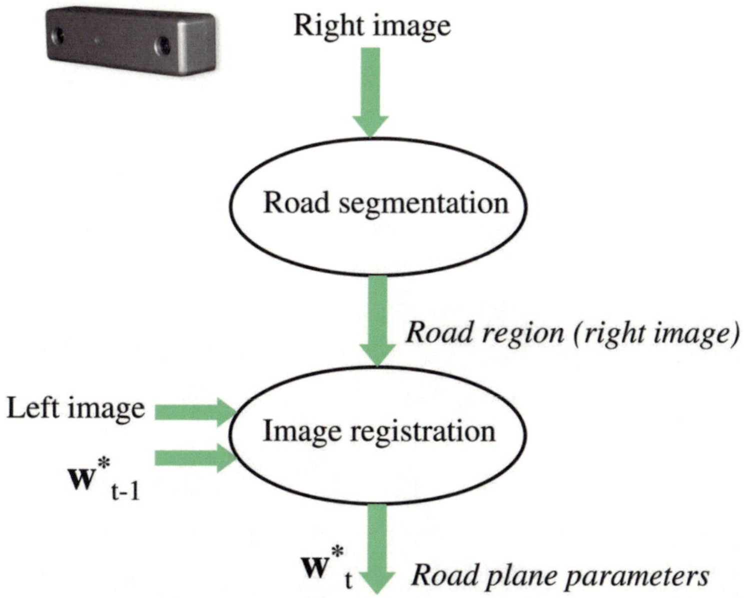

Fig. 2. The proposed approach consists of two stages: A rough road segmentation followed by image registration.

3.1 Road segmentation

In this section, we briefly describe how the road region $\mathcal{R}$ is detected in the right images. Road segmentation is the focus of many research works (Lombardi et al., 2005; Jansen et al., 2005; Guzman & Parra, 2007; Alvarez et al., 2008). In our study, the sought segmentation should meet two requirements: (i) it should be as fast as possible, and (ii) it should be as generic as possible (both urban roads and highways). Thus our segmentation scheme will be a color-based approach which works on the hue and saturation components. The segmentation stage is split into two phases. The first phase is only invoked every T frames for updating the color model and for obtaining a real-time performance. The second phase exploits the road color consistency over short time. The first phase consists of a classical K-means algorithm that is applied on the hue and saturation values of the pixels belonging to a predefined region of interest (ROI) that is centered at the bottom of the image. The number of classes can be between 3 and 5. The cluster having the largest number of pixels will be assumed to belong to the road. Once the cluster is identified, the mean and the covariance of its color (hue and saturation components) can be easily computed. In the second phase (invoked for every frame), by assuming that the color distribution of the detected cluster is Gaussian, we can quantify the likelihood of an arbitrary pixel to be a road pixel. Thus, the pixels within the ROI are labelled as road pixels if their *Mahalanobis* distance to the mean is below a certain threshold. Figure 3 shows the segmentation results obtained with the proposed scheme. Detected road pixels are shown in white within the ROI of two different frames. As can be seen, all detected pixels belong to the road plane.

3.2 Image registration

The optimal current road parameters are given by:

$$\begin{aligned}\mathbf{w}^\star &= arg\min_{\mathbf{W}} e(\mathbf{w}) \\ &= arg\min_{\mathbf{W}} \sum_{(x_r,y_r)\in\mathcal{R}} \left(I_{r(x_r,y_r)} - I_{l(h_1\,x_r+h_2\,y_r+h_3,y_r)}\right)^2\end{aligned} \tag{8}$$

where $e(\mathbf{w})$ is a non-linear function of the parameters $\mathbf{w} = (w_x, w_y, w_z)^T$. In the sequel, we describe two minimization techniques: i) the Differential Evolution minimization, and ii) the Levenberg-Marquardt minimization. The first one is a stochastic search method and the second one is a directed search method. Moreover, we present two tracking schemes.

Fig. 3. Rapid road segmentation associated with two frames.

3.2.1 Differential evolution minimization

The Differential Evolution algorithm (DE) is a practical approach to global numerical optimization that is easy to implement, reliable and fast (Price et al., 2005). We use the DE algorithm (Das et al. 2005; Storn & Price, 1997) in order to minimize the error (8). This is carried out using generations of solutions-population. The population of the first generation is randomly chosen around a rough solution. We point out that even the exact solution for the first frame is not known, the search range for the camera height as well as for the plane normal can be easily known. For example, in our experiments, the camera height and the normal vector are assumed to be around $1m$ and $(0, 1, 0)^T$, respectively.
The optimization adopted by the DE algorithm is based on a population of N solution candidates $\mathbf{w}_{n,i}$ (n = 1, ..., N) at iteration (generation) i where each candidate has three components. Initially, the solution candidates are randomly generated within the provided intervals of the search space. The population improves by generating new solutions iteratively for each candidate.
Calibration. Since the stereo camera is rigidly attached to the car, the differential evolution algorithm can also be used as a calibration tool by which the camera pose can be estimated off-line. To this end, the car should be at rest and should face a flat road. Whenever the car moves, the off-line calibration results can be used as a starting solution for the whole tracking process. Note that the calibration process does not need to run in real-time.

3.2.2 Levenberg-Marquardt minimization

Minimizing the cost function (8) can also be carried out using the Levenberg-Marquardt technique (Fletcher, 1990; Press et al., 1992) —a well-known non-linear minimization technique. One can notice that the Jacobian matrix only depends on the horizontal image gradient since the right and left images are rectified.

3.3 Tracking schemes

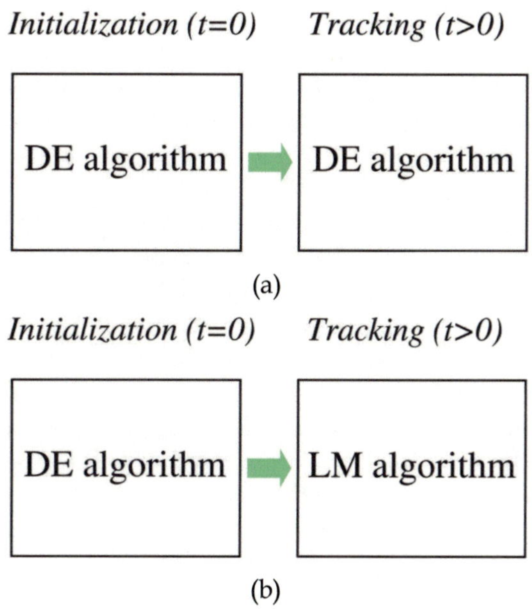

Fig. 4. Parameter tracking using two strategies. (a) The tracking is only based on the Differential Evolution search. (b) The tracking is based on the Differential Evolution search and on the Levenberg-Marquardt search.

The unknown parameters (road parameters/camera pose) should be estimated for every stereo pair. Thus, we will adopt a tracking strategy in which the estimated parameters $\mathbf{w}_{t-1}^{\star}$ associated with the previous frame will be handed over to the current frame.

Since the unknown parameters (road parameters/camera pose) are estimated by two optimization techniques, we propose two tracking schemes which are illustrated in Figure 4. The first scheme (Figure 4(a)) is only based on the Differential Evolution minimization. In other words, the solution for every stereo frame is computed by invoking the whole algorithm where the first generation is generated by diffusing the previous solution using a normal distribution. A uniform distribution is used for the first stereo frame.

The second scheme (Figure 4(b)) uses the Differential Evolution minimization for the first stereo frame only. It utilizes the Levenberg-Marquardt for the rest of the frames where the initial solution for a given frame is provided by the solution $\mathbf{w}_{t-1}^{\star}$ associated with the previous frame.

Although the first scheme might have better convergence properties than the second scheme, the latter one is better suited for real-time performance since the Levenberg-Marquardt algorithm is faster than the Differential Evolution search (the corresponding CPU times are illustrated in Section 4.2). In both tracking schemes, the pose parameters associated with the first stereo pair are estimated by the DE search. The Differential

Evolution algorithm performs a global search whereas the Levenberg-Marquardt performs a directed and local search.

4. Experimental results

The proposed technique has been tested on different urban environments since they correspond to the most challenging scenarios. In this section, we provide results obtained with two different videos associated with different urban road structures. Moreover, we provide a performance study using synthetic videos with ground-truth data.

4.1 Tracked road parameters

The first experiment has been conducted on a sequence corresponding to an uphill driving. The stereo pairs are of resolution 320 × 240. Figure 5(a) depicts the estimated camera's height as a function of the sequence frames. Figures 5(b) and 5(c) depict the estimated pitch and roll angles as a function of the sequence frames, respectively. The dotted curves correspond to the first scheme that is based on the Differential Evolution minimization. The solid curves correspond to the second scheme which is based on both the Differential Evolution algorithm and the Levenberg-Marquardt algorithm. As can be seen, the estimated parameters are almost the same for the two proposed schemes. However, as we will show, the second scheme is much faster than the first scheme (the stochastic search).

Differential Evolution convergence. Figure 6 illustrates the behavior of the Differential Evolution algorithm associated with the first stereo pair of the above stereo sequence. This plot depicts the best registration error (SSD per pixel) obtained by every generation. The three curves correspond to three different population sizes. The first generation (iteration 0) has been built using a uniform sampling around the solution $d = 1m$ and $\mathbf{u} = (u_x, u_y, u_z)^T = (0, 1, 0)^T$. The algorithm converged in five iterations (generations) when the population size was 30 and in two iterations when the population size was 120. At convergence the solution was $d = 1.25m$ and $\mathbf{u} = (u_x, u_y, u_z)^T = (-0.03, 0.99, -0.02)^T$. Note that even the manually provided initial camera's height has 25cm discrepancy from the current solution, the DE algorithm has rapidly converged to the actual solution. Also, we have run the Levenberg-Marquardt algorithm with the same starting solution but we get at convergence $d = 1.09m$ and $\mathbf{u} = (u_x, u_y, u_z)^T = (0.01, 0.99, -0.02)^T$.

Horizon line. In the literature, the pose parameters—plane parameters—can be used to compute the horizon line. In our case, since the roll angle is very small, the horizon line can be represented by an horizontal line in the image. Once the 3D plane parameters d and $\mathbf{u} = (u_x, u_y, u_z)^T$ are computed, the vertical position of the horizon line will be given by:

$$v_h = v_0 + \frac{\alpha\, d}{u_y\, Z_\infty} - \frac{\alpha\, u_z}{u_y} \approx v_0 - \frac{\alpha\, u_z}{u_y} \tag{9}$$

The above formula is derived by projecting a 3D point $(0, Y_p, Z_\infty)$ belonging to the road plane and then taking the vertical coordinate $v = \alpha\, \frac{Y_p}{Z_\infty} + v_0$. Z_∞ is a large depth value. The right-hand expression is obtained by using the fact that u_y is close to one and Z_∞ is very large. Figure 7 illustrates the computed horizon line for frames 10 and 199. The whole video illustrating the computed horizon line can be found at *www.cvc.uab.es/~asappa/HorizonLine.avi.*

Approach behavior in the presence of road segmentation error. In order to study the algorithm behavior in the presence of significant segmentation errors or non-road objects, we conducted the following experiment. We used a video sequence corresponding to a flat road (see Figure 3). We run the proposed technique described in Section 3 twice. We used the second tracking scheme (DE-LM). The first run was a straightforward run. In the second run, the right images were corrupted to simulate a significant registration error (road segmentation error). To this end we set the vertical half of a set of 20 right images to a fixed color. The left images were not modified.

Figure 8 compares the pose parameters obtained in the two runs. The solid curves were obtained with the non corrupted images. The dotted curves were obtained when the right images of the same sequence are artificially corrupted. The simulated corruption starts at frame 40 and ends at frame 60. The upper part of the Figure illustrates the stereo pair 40. As can be seen, the only significant discrepancy has affected the camera height. Moreover, one can see that the correct parameters have been recovered once the perturbing factor has disappeared. Figure 9 shows the registration error obtained at convergence as a function of the sequence frames. As can be seen, the obtained registration error has suddenly increased, which can be used for validating the estimated parameters.

Figure 10(a) illustrates the registration error (8) as a function of the camera's height while the orientation is kept fixed. Figure 10(b) illustrates the registration error as a function of the camera's pitch angle for four different camera's height. In both figures, the depicted error is the SSD per pixel. From the slop of the error function we can see that the camera height will not be recovered with the same accuracy as the plane orientation. This will be confirmed in the accuracy evaluation section (see Section 4.3).

4.2 Method comparison

The second experiment has been conducted on a short sequence of stereo pairs corresponding to a typical urban environment (see Figure 3). The stereo pairs are of resolution 320 × 240. Here the road is almost flat and the changes in the pose parameters are mainly due to the car's accelerations and decelerations. Figures 11(a) and 11(b) depict the estimated camera's height and orientation as a function of the sequence frames using two different methods. The solid curves correspond to the developed direct approach (DE-LM) and the dashed curves correspond to a 3D data based approach (Sappa et al., 2006). This approach uses a dense 3D reconstruction followed by a RANSAC-based estimation of the dominant 3D plane—the road plane. One can see that despite some discrepancies the proposed direct method is providing the same behavior of changes.

On a 3.2 GHz PC, the proposed approach processes one stereo pair in about 20 ms assuming that the ROI size is 190 × 90 pixels and the number of the detected road pixels is 11000 pixels (3 ms for the fast color-based segmentation and about 17 ms for the Levenberg-Marquardt minimization). One can notice that this is much faster than the 3D data based approach, which needs 350 ms. Moreover, the Levenberg-Marquardt algorithm is faster than the DE algorithm which needs 120 ms assuming that the number of iterations is 5 and the population number is 30 (the number of pixels is 11000). Obviously, devoting a very small

CPU time for estimating the road parameters/camera pose is advantageous for real-time systems since the CPU power can be used for extra tasks such as pedestrian or obstacle detection.

4.3 Accuracy evaluation

The evaluation of the proposed approach has been carried out on real video sequences, including a comparison with a 3D data based approach (Section 4.2). However, it is very challenging to get ground-truth data for the on-board camera pose. In this section, we propose a simple scheme giving the ground-truth data for the road parameters through the use of synthetic stereo sequences. To this end, we use a 1000-frame real video captured by the on-board stereo camera. For each stereo pair, we set the distance (camera height) and the plane normal—the ground-truth 3D plane road parameters. Those ones can be constant for the whole sequence or can vary according to a predefined trajectory. In our case, we keep them constant for the whole synthesized sequence. Each left image in the original sequence is then replaced with a synthesized one by warping the corresponding right image using the image transfer function encapsulating the road parameters. The obtained stereo pairs are then perturbed by adding Gaussian noise to their grey levels.

Figure 12 depicts a perturbed stereo pair. The Gaussian noise standard deviation is set to 20. Here the grey-level of the images has 256 values. The noise-free left image is synthesized using the ground-truth road parameters. The proposed approach is then invoked to estimate the road parameters from the noisy stereo pair. The performance can be directly evaluated by comparing the estimated parameters with the ground-truth parameters. The camera height error is simply the absolute value of the relative error. The orientation error is defined by the angle between the direction of the ground-truth normal and the direction of the estimated one.

Figure 13 summarizes the obtained errors associated with the synthetic stereo pairs. Figure 13(a) depicts the distance error and Figure 13(b) the orientation error. Here one percent error corresponds to $1.2cm$. Each point of the curves—each noise level—corresponds to 10000 stereo pairs corresponding to 10 realizations, each of which is a sequence of 1000 perturbed stereo pairs. The solid curves correspond to the global average of errors over the 10000 stereo pairs and the dashed curves correspond to the maximum error. As can be seen, the performance of the method downgrades gracefully with the image noise. Moreover, one can appreciate the orientation accuracy.

4.4 Convergence study

In order to study the convergence behavior of the two optimization techniques we run the following experiment. We used the same synthetic stereo sequence containing 1000 stereo frames. The standard deviation of the added image noise is kept fixed to 4. For every stereo frame in the sequence the starting solution was shifted from the ground-truth solution by 20 cm for the camera height and by 10 degrees for the plane normal. This shifted solution is used as the starting solution for the Levenberg-Marquardt technique and as the center of the first generation for the Differential Evolution technique. Table 1 depicts the average height and orientation errors obtained with the LM and DE minimizations. As can be seen, the DE minimization has better convergence properties than the LM minimization which essentially looks for a local minimum.

5. Conclusion

A featureless technique for real time estimation of on-board stereo head pose has been presented. The method adopts a registration scheme that uses images' brightness. The advantages of the proposed technique are as follows. First, the technique does need any specific visual feature extraction neither in the image domain nor in 3D space. Second, the technique is very fast compared to almost all proposed stereo-based techniques. The proposed featureless registration is carried out using two optimization techniques: the Differential Evolution algorithm (a stochastic search) and the Levenberg-Marquardt algorithm (a directed search).

A good performance has been shown in several scenarios—uphill, downhill and flat roads. Although it has been tested on urban environments, it could be also useful on highways scenarios. Experiments on real and synthetic stereo sequences have shown that the accuracy of the orientation is better than the height accuracy, which is consistent with all 3D pose algorithms. The provided experiments tend to confirm that (i) the Differential Evolution search was crucial for obtaining an accurate parameter estimation, and (ii) the Levenberg-Marquardt technique was crucial for obtaining a real-time tracking. As a consequence, the DE optimization can be used as a complementary tool to the LM optimization in the sense that it provides the initialization as well as the recovery solution from a tracking discontinuity adopting the Levenberg-Marquardt algorithm.

We stress the fact that our proposed framework is not restricted to the estimation of on-board camera pose/roads. Indeed, the proposed methods can be used for extracting any planar structures using stereo pairs.

6. Acknowledgment

This work was partially supported by the Government of Spain under MEC project TRA2007-62526/AUT and research programme Consolider Ingenio 2010: MIPRCV (CSD2007-00018).

1000 stereo frames	*LM minimization*	*DE minimization*
Ave. height error (%)	26.6	3.5
Ave. orientation error (degrees)	10.9	0.41

Table 1. Average camera pose errors. The first column corresponds to the Levenberg-Marquardt minimization and the second column to the Differential Evolution minimization.

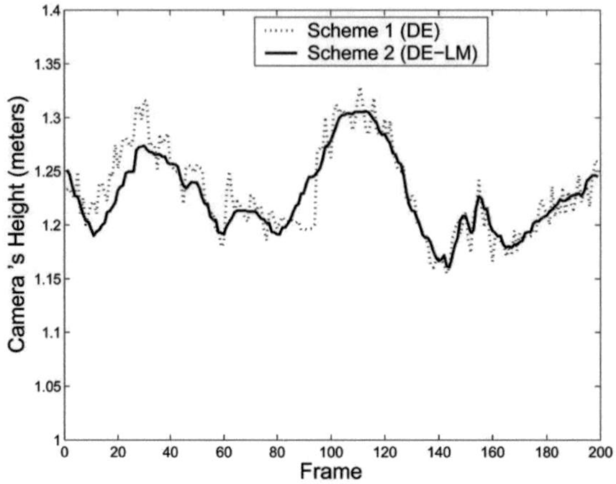

(a)

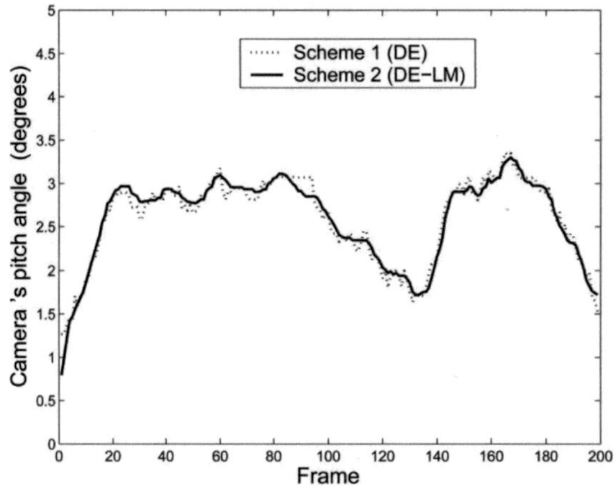

(b)

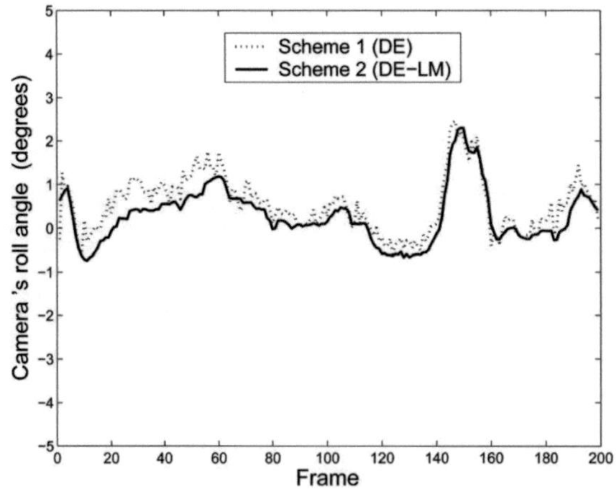

(c)

Fig. 5. Camera's height and orientation computed by the proposed tracking schemes.

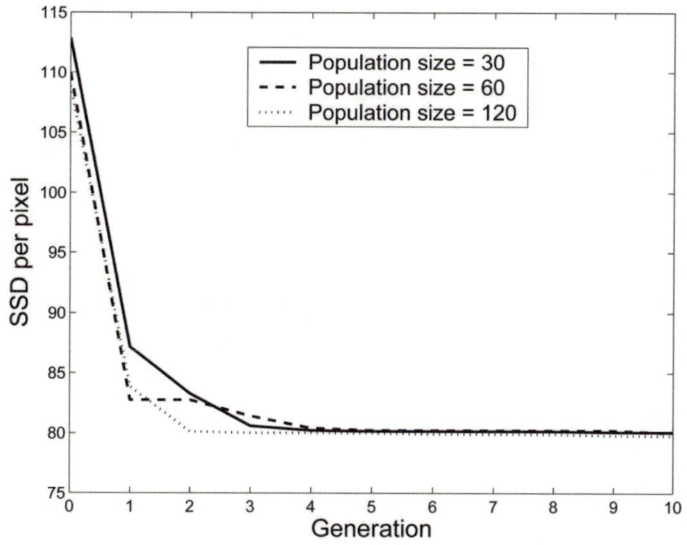

Fig. 6. The evolution of the best registration error obtained by the Differential Evolution algorithm associated with the first stereo pair. The algorithm has converged in 5 iterations (generations) when the population size was 30 and in two iterations when the population size was 120.

Fig. 7. The estimated horizon line associated with frames 10 and 199. The sequence corresponds to an uphill driving.

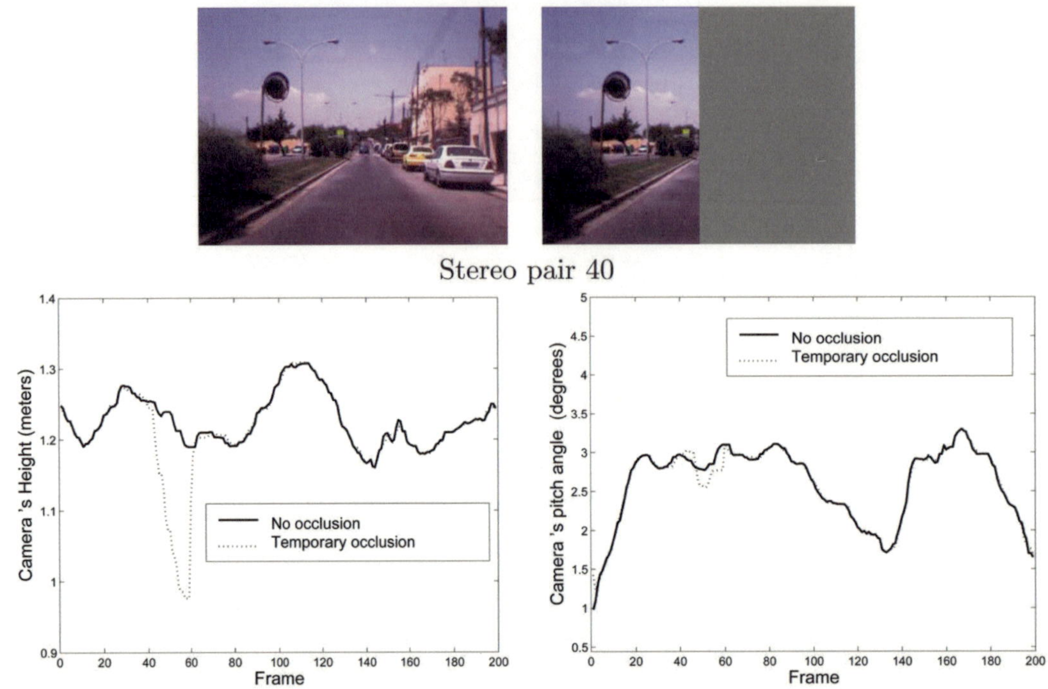

Fig. 8. The camera pose parameters in the presence of a significant corruption (road segmentation errors). The solid curves are obtained with the non corrupted images. The dotted curves are obtained when 20 frames of right images of the same sequence are artificially corrupted. The corruption is simulated by setting the vertical half of the right images to a fixed color. This corruption starts at frame 40 and ends at frame 60.

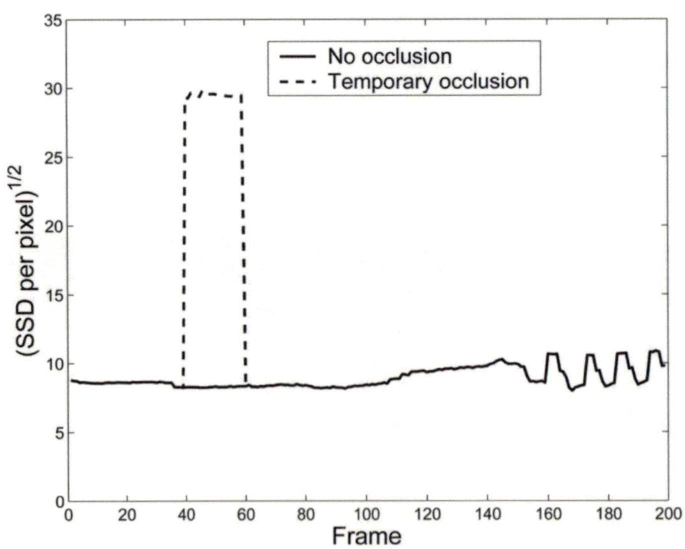

Fig. 9. The registration error obtained at convergence as a function of the sequence frame. The second tracking scheme is used.

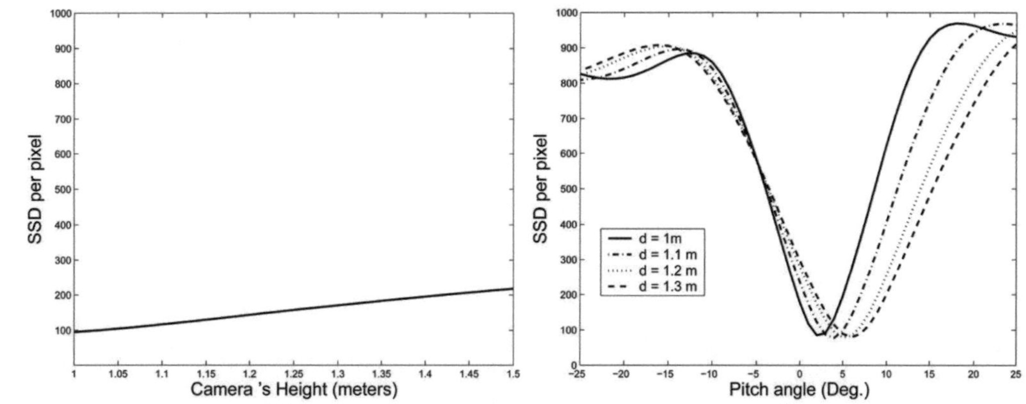

(a) (b)

Fig. 10. The registration error as a function of the camera pose parameters. (a) Depicts the error as a function of the camera height with a fixed orientation. (b) Depicts the error as a function of the camera's pitch angle associated with four different camera heights.

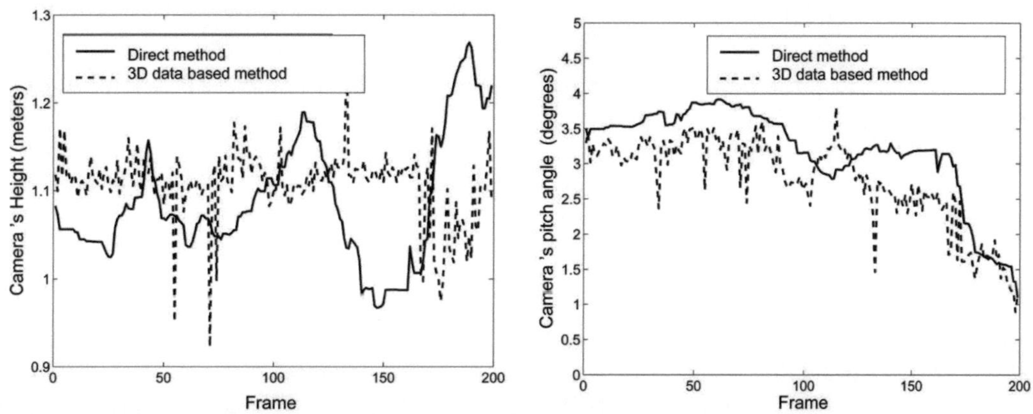

(a) (b)

Fig. 11. Camera's height and orientation using two different methods.

Fig. 12. A stereo pair from a perturbed 1000-frame video. The standard deviation of the added Gaussian noise is 20. The left images are synthesized using the ground-truth road parameters.

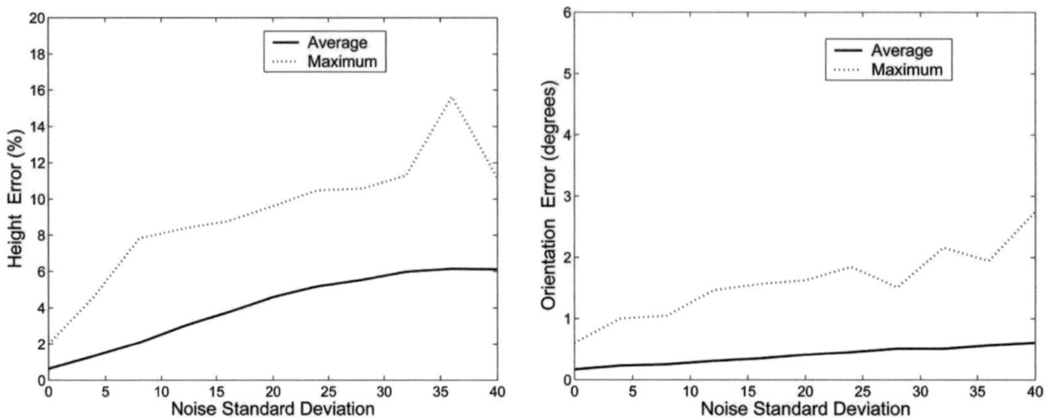

(a) (b)

Fig. 13. The errors associated with the plane parameters as a function of the noise standard deviation using synthesized video sequences. (a) Depicts the height errors. (b) Depicts the plane orientation errors. Each point of the curves—each noise level—corresponds to 10000 stereo pairs corresponding to 10 realizations, each of which is a sequence of 1000 perturbed stereo pairs.

7. References

J. M. Alvarez, A. López, and R. Baldrich. Illuminant-invariant model-based road segmentation. In *IEEE Intelligent Vehicules Symposium*, 2008.

M. Bertozzi, E. Binelli, A. Broggi, and M. Del Rose. Stereo vision-based approaches for pedestrian detection. In *Procs. Computer Vision and Pattern Recognition*, San Diego, USA, June 2005.

M. Bertozzi and A. Broggi. GOLD: A parallel real-time stereo vision system for generic obstacle and lane detection. *IEEE Trans. on Image Processing*, 7(1):62-81, January 1998.

M. Bertozzi, A. Broggi, R. Chapuis, F. Chausse, A. Fascioli, and A. Tibaldi. Shape-based pedestrian detection and localization. In *Procs. IEEE Intl. Conf. on Intelligent Transportation Systems*, pages 328-333, Shangai, China, October 2003.

S. Das, A. Konar, and U. Chakraborty. Two improved differential evolution schemes for faster global search. In *Genetic and Evolutionary Computation*, 2005.

O. Faugeras and Q.T. Luong. *The Geometry of Multiple Images*. The MIT Press, 2001.

R. Fletcher. *Practical Methods of Optimization*. Wiley, New York, 1990.

U. Franke, D. Gavrila, S. Görzig, F. Lindner, F. Paetzold, and C. Wöhler. Autonomous driving approaches downtown. *IEEE Intelligent Systems*, 13(6):1-14, 1999.

A. Guzmán and C. Parra. Extraction of roads from outdoor images. In *Vision Systems: Applications*, pages 101-112. 2007.

Z. Hu and K. Uchimura. U-V-Disparity: An efficient algorithm for stereovision based scene analysis. In *Procs. IEEE Intelligent Vehicles Symposium*, pages 48-54, Las Vegas, USA, June 2005.

P. Jansen, W. van der Mark, J.C. van der Heuvel, and F.C.A. Groen. Colour based off-road environment and terrain type classification. In *IEEE Intelligent Transportation Systems*, 2005.

R. Labayrade and D. Aubert. A single framework for vehicle roll, pitch, yaw estimation and obstacles detection by stereovision. In *Proc. IEEE Intelligent Vehicles Symposium, Columbus, OH, USA*, pages 31-36, June 2003.

R. Labayrade, D. Aubert, and J. Tarel. Real time obstacle detection in stereovision on non flat road geometry through "V-disparity" representation. In *Proc. IEEE Intelligent Vehicles Symposium, Versailles, France*, pages 646-651, June 2002.

X. Liu and K. Fujimura. Pedestrian detection using stereo night vision. *IEEE Trans. on Vehicular Technology*, 53(6):1657-1665, November 2004.

P. Lombardi, M. Zanin, and S. Messelodi. Switching models for vision-based on-board road detection. In *IEEE Intelligent Transportation Systems*, 2005.

S. Nedevschi, F. Oniga, R. Danescu, T. Graf, and R. Schmidt. Increased accuracy stereo approach for 3D lane detection. In *IEEE Intelligent Vehicles Symposium*, 2006.

W. H. Press, S. A. Teukolsky, W. T. Wetterling, and B. P. Flannery. *Numerical Recipes, The Art of Scientific Computing*. Cambridge University Press, New York, 1992.

K. V. Price, J. A. Lampinen, and R. M. Storn. *Differential Evolution: A Practical Approach To Global Optimization*. Springer, 2005.

L. Romero and F. Calderón. A tutorial on parametric image registration. In *Scene Reconstruction, Pose Estimation and Tracking*, pages 167-184. 2007.

A. Sappa, D. Gerónimo, F. Dornaika, and A. López. On-board camera extrinsic parameter estimation. *Electronics Letters*, 42(13):745-747, June 2006.

G. Stein, O. Mano, and A. Shashua. A robust method for computing vehicle ego-motion. In *IEEE Intelligent Vehicles Symposium*, 2000.

R. Storn and K. Price. Differential evolution - A simple and efficient heuristic for global optimization over continuous spaces. *Journal of Global Optimization*, 11:341-359, 1997.

Z. Sun, G. Bebis, and R. Miller. Monocular precrash vehicle detection: Features and classifiers. *IEEE Trans. on Image Processing*, 15(7):2019-2034, 2006.

T. Suzuki and T. Kanade. Measurement of vehicle motion and orientation using optical flow. In *IEEE Intelligent Vehicles Symposium*, 1999.

G. Toulminet, M. Bertozzi, S. Mousset, A. Bensrhair, and A. Broggi. Vehicle detection by means of stereo vision-based obstacles features extraction and monocular pattern analysis. *IEEE Trans. on Image Processing*, 15(8):2364-2375, 2006.

P. Viola, M.J. Jones, and D. Snow. Detecting pedestrians using patterns of motion and appearance. *International Journal of Computer Vision*, 63(2):153-161, 2005.

Z. Zhu, S. Yang, G. Xu, X. Lin, and D. Shi. Fast road classification and orientation estimation using omni-view images and neural networks. *IEEE Trans. on Image Processing*, 7(8):1182-1197, 1998.

19

Robust Visual Correspondence: Theory and Applications

Federico Tombari, Luigi Di Stefano and Stefano Mattoccia
DEIS/ARCES – University of Bologna
Italy

1. Introduction

Visual correspondence represents one of the most important tasks in computer vision. Given two sets of pixels (i.e. two images), it aims at finding corresponding pixel pairs belonging to the two sets (*homologous* pixels). As a matter of fact, visual correspondence is commonly employed in fields such as stereo correspondence, change detection, image registration, motion estimation, pattern matching, image vector quantization.
The visual correspondence task can be extremely challenging in presence of disturbance factors which typically affect images. A common source of disturbances can be related to photometric distortions between the images under comparison. These can be ascribed to the camera sensors employed in the image acquisition process (due to dynamic variations of camera parameters such as auto-exposure and auto-gain, or to the use of different cameras), or can be induced by external factors such as changes of the amount of light emitted by the sources or viewing of non-lambertian surfaces at different angles.
All of these factors tend to produce brightness changes in corresponding pixels of the two images that can not be neglected in real applications implying visual correspondence between images acquired from different spatial points (e.g. stereo vision) and/or different time instants (e.g. pattern matching, change detection). In addition to photometric distortions, differences between corresponding pixels can also be due to the noise introduced by camera sensors. Finally, the acquisition of images from different spatial points or different time instants can also induce occlusions. Evaluation assessments have also been proposed which compared visual correspondence approaches for tasks such as stereo correspondence (Chambon & Crouzil, 2003), image registration (Zitova & Flusser, 2003) and image motion (Giachetti, 2000).

2. Literature review

Let *Ir, It* be respectively the reference image patch vector and the target image patch vector, that have to be matched together. Traditional matching measures can be subdivided into two categories: correlation-based or distance-based. Between the first, the most commonly adopted are the *Normalized Cross-Correlation* (NCC) and the *Zero-mean Normalized Cross-Correlation* (ZNCC):

$$NCC(I_r, I_t) = \frac{I_r \circ I_t}{\|I_r\|_2 \cdot \|I_t\|_2} \quad (1)$$

$$\mathrm{ZNCC}(\mathrm{I_r},\mathrm{I_t}) = \frac{(\mathrm{I_r}-\bar{\mathrm{I}}_r)\circ(\mathrm{I_t}-\bar{\mathrm{I}}_t)}{\left\|\mathrm{I_r}-\bar{\mathrm{I}}_r\right\|_2\cdot\left\|\mathrm{I_t}-\bar{\mathrm{I}}_t\right\|_2} \tag{2}$$

with $\circ$ denoting the dot product, $||\cdot||_p$ the L_p norm, $\bar{\ }$ the mean value over the patch. Thanks to normalization with regards to the magnitude of the vectors and to the mean intensity value of the image patch, NCC and ZNCC are invariant, respectively, to linear and affine transformation between *Ir* and *It*. Efficient techniques for exhaustive template matching based on NCC and ZNCC matching measures have been proposed in (Mattoccia et al., 2008-1) and (Mattoccia et al., 2008-2).
On the other side, commonly used dissimilarity measures are those derived from the L_p-distance between *Ir* and *It*. Between this class, two popular measures are the *Sum of Absolute Differences* (SAD) and the *Sum of Squared Differences* (SSD):

$$\mathrm{SAD}(\mathrm{I_r},\mathrm{I_t}) = \left\|\mathrm{I_r}-\mathrm{I_t}\right\|_1 \tag{3}$$

$$\mathrm{SSD}(\mathrm{I_r},\mathrm{I_t}) = \left\|\mathrm{I_r}-\mathrm{I_t}\right\|_2^2 \tag{4}$$

These two measures showed experimentally good robustness towards noise (Aschwanden & Guggenbuhl, 1992), (Martin & Crowley, 1995). An efficient technique for exhaustive template matching based on L_p-distance has been proposed in (Tombari et al., 2008). A similar approach was also adopted in (Mattoccia et al., 2007) for motion estimation.
While all these measures are usually computed directly on the pixel intensities of the images, in (Martin & Crowley, 1995) it was shown that by computing these measures on the gradient norm of each pixel a higher robustness is attained, i.e. for what concerns insensitivity to illumination changes the SSD and the NCC applied on gradient norms (referred to here respectively as G-SSD and G-NCC) showed to perform well. In particular, if we denote with $G_r(i, j)$ the gradient of *Ir* at pixel (i, j):

$$\mathrm{G}_r(i,j) = \left[\frac{\partial I_r(i,j)}{\partial i}, \frac{\partial I_r(i,j)}{\partial j}\right]^T = \left[G_i^r(i,j), G_j^r(i,j)\right]^T \tag{5}$$

and similarly with $G_t(i, j)$ the gradient of *It* at pixel (i, j):

$$\mathrm{G}_t(i,j) = \left[\frac{\partial I_t(i,j)}{\partial i}, \frac{\partial I_t(i,j)}{\partial j}\right]^T = \left[G_i^t(i,j), G_j^t(i,j)\right]^T \tag{6}$$

the gradient norm, or magnitude, in both cases is defined as:

$$\left\|\mathrm{G}_r(i,j)\right\|_2 = \sqrt{G_i^r(i,j)^2 + G_j^r(i,j)^2} \tag{7}$$

$$\left\|\mathrm{G}_t(i,j)\right\|_2 = \sqrt{G_i^t(i,j)^2 + G_j^t(i,j)^2} \tag{8}$$

Hence the G-NCC function can be defined as:

$$\mathrm{G\text{-}NCC}(\mathrm{I_r},\mathrm{I_t}) = \frac{\sum_{(i,j)\in I_r} \left\|\mathrm{G_r}(i,j)\right\|_2 \cdot \left\|\mathrm{G_t}(i,j)\right\|_2}{\sqrt{\sum_{(i,j)\in I_r} \left\|\mathrm{G_r}(i,j)\right\|_2^2} \cdot \sqrt{\sum_{(i,j)\in I_t} \left\|\mathrm{G_t}(i,j)\right\|_2^2}} \tag{9}$$

and, analogously, the G-SSD function as:

$$\text{G-SSD}(I_r, I_t) = \sum_{(i,j)\in I_r} \left(\left\| G_r(i,j) \right\|_2 - \left\| G_t(i,j) \right\|_2 \right)^2 \tag{10}$$

In addition to these measures, many alternatives have been proposed in literature with the specific aim of deploying robust image matching. The *Gradient Correlation* (GC) measure, proposed in (Crouzil et al., 1996) and derived from a measure originally introduced in (Scharstein, 1994), is based on two terms, referred to as *distinctiveness* (D) and *confidence* (C), both computed from intensity gradients:

$$D(I_r, I_t) = \sum_{(i,j)\in I_r} \left\| G_r(i,j) - G_t(i,j) \right\|_2 \tag{11}$$

$$C(I_r, I_t) = \sum_{(i,j)\in I_r} \left(\left\| G_r(i,j) \right\|_2 + \left\| G_t(i,j) \right\|_2 \right) \tag{12}$$

The GC measure is then defined as:

$$GC(I_r, I_t) = \frac{D(I_r, I_t)}{C(I_r, I_t)} \tag{13}$$

Its minimum value is 0, corresponding to the highest similarity between *Ir* and *It*. For any other positive value, the greater the value, the higher the dissimilarity between the two vectors. In order to compute the partial derivatives, (Crouzil et al., 1996) proposes to use either the Sobel operator or the Shen-Castan ISEF filter (Shen & Castan, 1992).

The *Orientation Correlation* (OC) measure (Fitch et al., 2002) is based on the correlation of the orientation of the intensity gradient. In particular, for each gradient $G_r(i,j)$ a complex number representing the orientation of the gradient vector is defined as:

$$O_r(i,j) = \operatorname{sgn}\left(G_i^r(i,j) + iG_j^r(i,j) \right) \tag{14}$$

with *i* denoting the imaginary unit and where:

$$\operatorname{sgn}(x) = \begin{cases} 0 & \text{if } |x| = 0 \\ \dfrac{x}{|x|} & \text{elsewhere} \end{cases} \tag{15}$$

Analogously, a complex number representing the orientation of the gradient vector $G_t(i,j)$ is defined as:

$$O_t(i,j) = \operatorname{sgn}\left(G_i^t(i,j) + iG_j^t(i,j) \right) \tag{16}$$

As proposed in (Fitch et al., 2002), the partial derivatives for the gradient computation should be calculated by approximating them with *central differences*. Hence, the OC measure between *Ir* and *It* is defined as the real part of the correlation between all gradient orientations belonging to *Ir* and *It*:

$$\mathrm{OC}(\mathrm{I_r},\mathrm{I_t}) = \mathrm{Re}\left\{ \sum_{(i,j)\in I_r} O_r(i,j)\cdot O_t^*(i,j) \right\} \tag{17}$$

with * indicating the conjugate of the complex vector. In (Fitch et al., 2002) it is proposed to exploit the correlation theorem to compute the correlation operation in the frequency domain by means of the FFT in order to achieve computational efficiency.

Another class of measures concerns the so-called order-consistency or order-preservation hypothesis, that is the assumption that the considered distortions do not violate the ordering between the intensities of neighbouring pixels. This assumption includes a more general class of transformations compared to the linear or affine case. These measures are called *ordinal* and a typical example of this class is represented by the *Rank transform*. As for this measure, both *Ir* and *It* are transformed into two novel images where each pixel stores the number of points in the patch whose intensity is less than that of the central point of the patch:

$$\mathrm{R_r}(\mathrm{i,j}) = \left|\left\{(u,v)\in I_r \mid I_r(u,v) < I_r(i,j)\right\}\right| \tag{18}$$

$$\mathrm{R_t}(\mathrm{i,j}) = \left|\left\{(u,v)\in I_t \mid I_t(u,v) < I_t(i,j)\right\}\right| \tag{19}$$

where | · | represents the cardinality operator. Once the two transforms are computed, a matching measure is deployed to compare *Rr* and *Rt*, e.g. (Zabih & Woodfill, 1994) proposes to use the SAD.

A typical example of this class is represented by the *Rank transform* (Zabih & Woodfill, 1994), and the measure proposed in (Bhat & Nayar, 1998). Further approaches of robust visual correspondence measures specifically conceived for change detection are (Ohta, 2001), (Xie at al., 2004), (Mittal & Ramesh, 2006).

Finally, other robust approaches have been proposed in (Seitz, 1989), (Lai, 2000), (Odone et al., 2001), (Ullah et al., 2001), (Kaneko et al., 2003).

3. The MF measure

This section describes a novel approach, referred to here as *Matching Function* (MF), which is implicitly based on the ordering assumption. In particular, MF aims at quantifying how well the order is preserved between corresponding pairs of neighbouring pixels in the two images. A simple and effective approach for evaluating the order-consistency is to evaluate the difference between the intensities of pairs of neighbouring pixels. As an example, let *Ir* be a 3×3 patch. In order to evaluate the order preservation between neighbouring elements within this window, many pairs (e.g. 72) should be considered, as each of the 9 pixels has to be put in correspondence with each other. In order to simplify the problem, we propose to consider only a subset of the whole neighbouring pairs set by evaluating only horizontal and vertical neighbouring pixels. Hence, the considered pairs are reduced to 18, as shown in Fig. 1.

In particular, in order to quantify how well the ordering is preserved between the two image patches *Ir* and *It* we propose to correlate the differences between the considered corresponding pairs within the 3×3 window. If the ordering is preserved for a given pair, the result of the pointwise correlation is a positive coefficient regardless of the sign of the intensity difference, which tends to increase the correlation score associated with the 3 × 3 window. Conversely, if the order is not preserved the correlation coefficient is negative, and the correlation score is decreased. Moreover, since horizontal and vertical differences may be thought as the discrete approximation of the horizontal and vertical derivatives of the image, the proposed measure can also be interpreted as the cross-correlation between two vectors made out of derivatives computed within the two 3×3 patches.

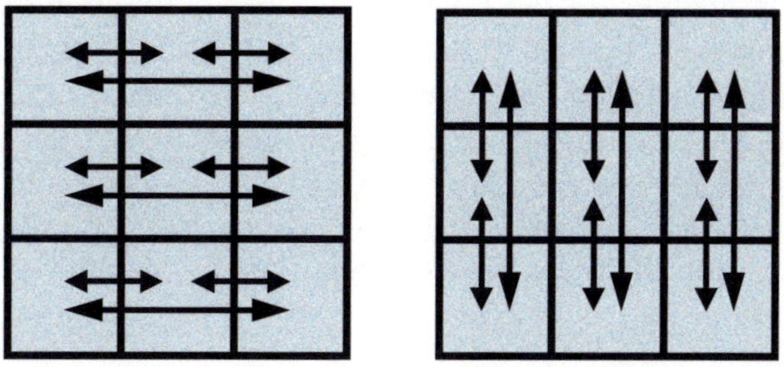

Fig. 1. considered subset of horizontal and vertical pairs of neighbouring pixels in a 3x3 patch

In the general case of two MxN patches, the considered pairs of pixels in each set include all pixels at distance 1 and 2 along horizontal and vertical directions. In order to compute this set, we define a vector of pixel differences computed at a point (*i, j*) on *Ir*:

$$\delta_{1,2}^{r}(i,j)=\begin{bmatrix} I_r(i-1,j)-I_r(i,j) \\ I_r(i,j-1)-I_r(i,j) \\ I_r(i-1,j)-I_r(i+1,j) \\ I_r(i,j-1)-I_r(i,j+1) \end{bmatrix} \tag{20}$$

and, similarly, at a point (*i, j*) on *It*:

$$\delta_{1,2}^{t}(i,j)=\begin{bmatrix} I_t(i-1,j)-I_t(i,j) \\ I_t(i,j-1)-I_t(i,j) \\ I_t(i-1,j)-I_t(i+1,j) \\ I_t(i,j-1)-I_t(i,j+1) \end{bmatrix} \tag{21}$$

Hence, the MF function consists in correlating these two vectors for each point of *Ir, It*, and in normalizing the correlation with the *L*2 norm of the vectors themselves:

$$MF_{1,2}(x,y)=\frac{\sum_{(i,j)\in I_r}\delta_{1,2}^{r}(i,j)\circ\delta_{1,2}^{t}(i,j)}{\sqrt{\sum_{(i,j)\in I_r}\delta_{1,2}^{r}(i,j)\circ\delta_{1,2}^{r}(i,j)}\cdot\sqrt{\sum_{(i,j)\in I_r}\delta_{1,2}^{t}(i,j)\circ\delta_{1,2}^{t}(i,j)}} \tag{22}$$

It is worth noticing that the normalization allows the measure to range between [−1,1]. It is a peculiarity of this method that, because of the correlation between differences of pixel pairs, intensity edges tend to determine higher correlation coefficients (in magnitude) with respect to low-textured regions. Thus, this can be seen as if the measure mostly relies on the patch edges. For this reason, MF can be usefully employed also in presence of high levels of noise, as this disturbance factor can typically violate the ordering constraint on low-textured regions, but seldom along intensity edges. Similar considerations can be made in presence of partially occluded patches.

The set of pixel pairs in (21, 22) can be seen as made out of two subsets: the set of horizontal and vertical lateral derivatives (i.e. all pixels at distance 1 one to another along horizontal and vertical directions), and the set of horizontal and vertical central derivatives (i.e. all pixels at distance 2 one to another along same directions). Theoretically, the former should benefit of the higher correlation given by adjacent pixels, while the latter should be less influenced by quantization ("sampling") noise that is introduced by the camera sensor. We will refer to two additional measures of the MF class applied on each of these two subsets as, respectively, *MF*1 and *MF*2. For these last two cases, we define the vector of pixel differences at distance 1 pixel:

$$\delta_1^r(i,j) = \begin{bmatrix} I_r(i-1,j) - I_r(i,j) \\ I_r(i,j-1) - I_r(i,j) \end{bmatrix} \tag{23}$$

$$\delta_1^t(i,j) = \begin{bmatrix} I_t(i-1,j) - I_t(i,j) \\ I_t(i,j-1) - I_t(i,j) \end{bmatrix} \tag{24}$$

and the pixel differences relative to the case of distance 2 pixels:

$$\delta_2^r(i,j) = \begin{bmatrix} I_r(i-1,j) - I_r(i+1,j) \\ I_r(i,j-1) - I_r(i,j+1) \end{bmatrix} \tag{25}$$

$$\delta_2^t(i,j) = \begin{bmatrix} I_t(i-1,j) - I_t(i+1,j) \\ I_t(i,j-1) - I_t(i,j+1) \end{bmatrix} \tag{26}$$

Then, *MF*1 and *MF*2 are defined respectively as:

$$MF_1(x,y) = \frac{\sum_{(i,j)\in I_r} \delta_1^r(i,j) \circ \delta_1^t(i,j)}{\sqrt{\sum_{(i,j)\in I_r} \delta_1^r(i,j) \circ \delta_1^r(i,j)} \cdot \sqrt{\sum_{(i,j)\in I_r} \delta_1^t(i,j) \circ \delta_1^t(i,j)}} \tag{27}$$

and:

$$MF_2(x,y) = \frac{\sum_{(i,j)\in I_r} \delta_2^r(i,j) \circ \delta_2^t(i,j)}{\sqrt{\sum_{(i,j)\in I_r} \delta_2^r(i,j) \circ \delta_2^r(i,j)} \cdot \sqrt{\sum_{(i,j)\in I_r} \delta_2^t(i,j) \circ \delta_2^t(i,j)}} \tag{28}$$

A graphical representation of the 3 different pixel pair sets used by $MF_{1,2}$, MF_1 and MF_2 is shown in Fig. 2.

Fig. 2. The 3 considered sets of neighbouring pixel pairs.

4. Application to pattern matching

This section shows the application of the class of measures referred to as MF in a typical pattern matching scenario. Pattern matching aims at finding the most similar instances of a given pattern, *P*, within an image. In particular, in this section MF measures are compared against traditional general purpose approaches as well as against proposals specifically conceived to achieve robustness. One goal of the proposed comparison is to determine which measure is more suitable to deal with the aforementioned disturbance factors represented by photometric distortions, noise and occlusions.

More precisely, in the comparison with MF we will consider the following matching measures: GC (Crouzil et al., 1996), OC (Fitch et al., 2002), G-NCC, G-SSD (Martin and Crowley, 1995). Considered traditional measures are NCC, ZNCC and SSD. All the considered measures are tested on 3 datasets which represent a challenging framework for what regards the considered distortions. These datasets, which are publicly available at www.vision.deis.unibo.it/pm-eval.asp, are characterized by a significant presence of the disturbance factors discussed previously, and are now briefly described.

Guitar. In this dataset, 7 patterns were extracted from a picture which was taken with a good camera sensor (3 MegaPixels) and under good illumination conditions given by a lamp and some weak natural light. All these patterns have to be sought in 10 images which were taken with a cheaper and more noisy sensor (1.3 MegaPixels, mobile phone camera). Illumination changes were introduced in the images by means of variations of the rheostat of the lamp illuminating the scene (*G*1–*G*4), by using a torch light instead of the lamp (*G*5–*G*6), by using the camera flash instead of the lamp (*G*7– *G*8), by using the camera flash together with the lamp (*G*9), by switching off the lamp (*G*10). Furthermore, additional distortions were introduced by slightly changing the camera position at each pose and by the JPEG compression.

Mere Poulard - Illumination Changes. In dataset *Mere Poulard - Illumination Changes* (MP-IC), the picture on which the pattern was extracted was taken under good illumination conditions given by neon lights by means of a 1.3 MegaPixels mobile phone camera sensor. This pattern is then searched within 12 images which were taken either with the same camera (prefixed by *GC*) or with a cheaper, 0.3 VGA camera sensor (prefixed by *BC*). Distortions are due to slight changes in the camera point of view and by different illumination conditions such as: neon lights switched off and use of a very high exposure time (*BC* – *N*1, *BC* – *N*2, *GC* – *N*), neon lights switched off (*BC* – *NL*, *GC*–*NL*), presence of structured light given by a lamp light partially occluded by various obstacles (*BC*–*ST*1, ...,

BC−ST5), neon lights switched off and use of the camera flash (*GC−FL*), neon lights switched off, use of the camera flash and of a very long exposure time (*GC−NFL*). Also in this case, images are JPEG compressed.

Fig. 3. *Guitar* dataset

Mere Poulard - Occlusions. In the dataset *Mere Poulard - Occlusions* (MP-Occl) the pattern is the same as in dataset MP-IC, which now has to be found in 8 images taken with a 0.3 VGA camera sensor. In this case, partial occlusion of the pattern is the most evident disturbance factor. Occlusions are generated by a person standing in front of the camera (*OP*1, …, *OP*4), and by a book which increasingly covers part of the pattern (*OB*1, …, *OB*4). Distortions due to illumination changes, camera pose variations, JPEG compression are also present.

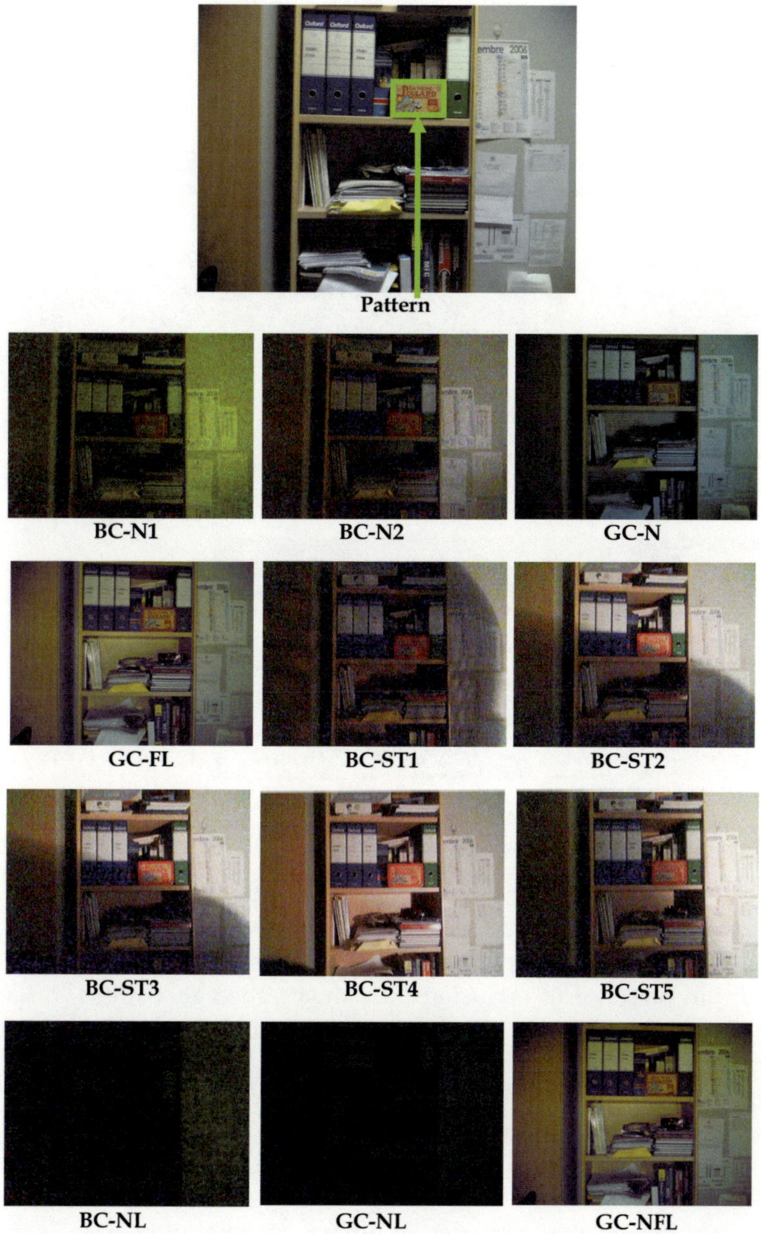

Fig. 4. *MP-IC* dataset

The number of pattern matching instances is thus 70 for the *Guitar* dataset, 12 for the *MP-IC* dataset and 8 for the *MP-Occl* dataset, for a total of 90 instances overall. The result of a pattern matching process is considered erroneous when the coordinates of the best matching subwindow found by a certain measure are further than ±5 pixel from the correct ones.

O-P1 O-P2 O-P3

O-P4 O-B1 O-B2

O-B3 O-B4

Fig. 5. *MP-Occl* dataset

Figures 6 and 7 report the matching errors yielded by the considered measures respectively on each of the 3 datasets and overall. As it can be seen, approaches specifically conceived to achieve robustness generally outperform classical measures, apart from the ZNCC which performs badly in presence of occlusions but shows good robustness in handling strong photometric distortions. The two measures which yield the best performance are MF and GC, with a number of total errors respectively equal to 6 and 8. In particular, MF performs better on datasets characterized by strong photometric distortions, conversely GC seems to perform better in presence of occlusions.

For what regards the 3 MF measures themselves, it seems clear that the use of differences relative to adjacent pixels suffers of the sampling noise introduced by the camera sensor, hence they appear less reliable compared to differences computed on a distance equal to 2. Moreover, as a consequence of the fact that $MF_{1,2}$ and MF_2 yield the same results on all datasets, MF_2 seems the more appropriate measure of the class since it requires only 2 correlation terms instead of the 4 needed by $MF_{1,2}$. Finally, for what regards traditional approaches, it is interesting to note that the application of NCC and SSD on the gradient norms rather than on the pixel intensities allows for a significantly higher robustness throughout all the considered datasets.

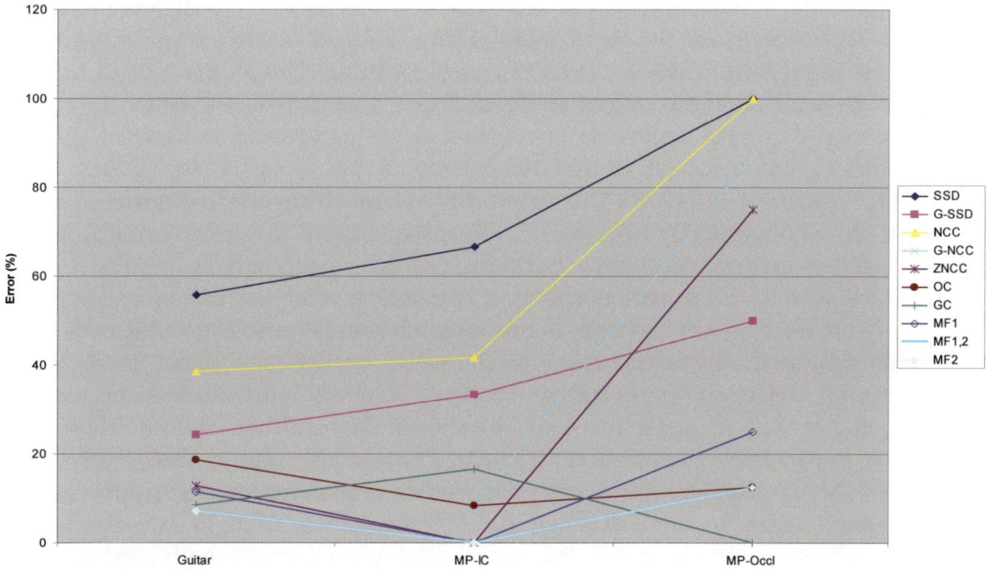

Fig. 6. results of the comparison on the 3 datasets.

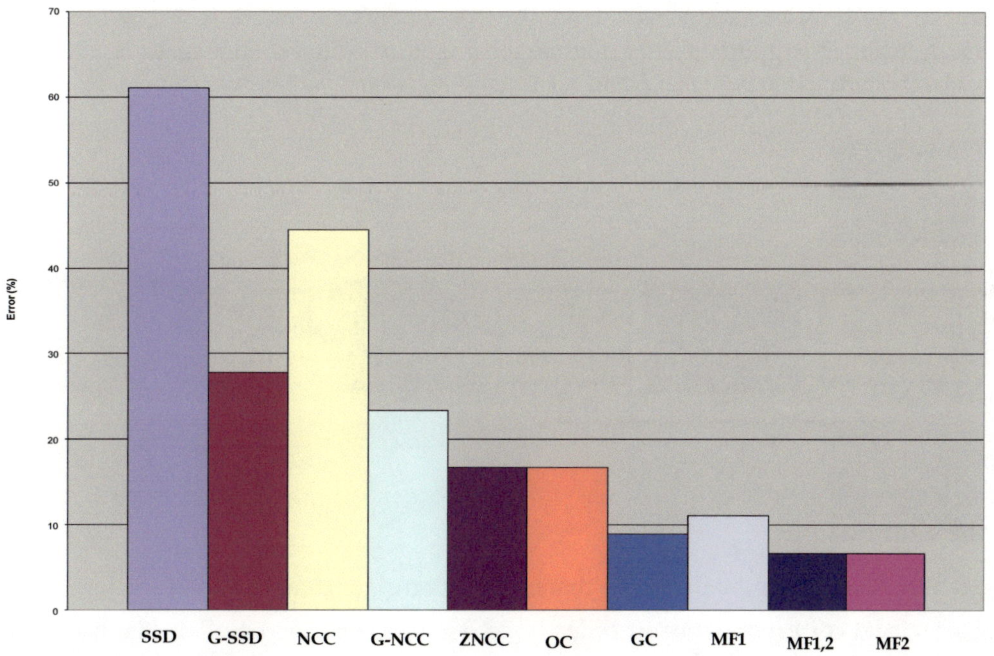

Fig. 7. overall results of the pattern matching evaluation.

5. Application to change detection

In this section we present the application of the proposed MF measures to the change detection task. Change detection aims at detecting structural changes occurring in time in a scene by analyzing a sequence of frames. This is a key task in most advanced video-surveillance applications, for the mask highlighting changed pixels (*change mask*) typically represents the input data to higher level vision algorithms. This is the case of traditional single view as well as more recent and advanced multiple-views systems. The most common change detection approach is referred to as *background subtraction*: given the current frame, F, and a model of the background of the scene, B, the change mask is obtained by comparing F and B. This approach assumes that the background model is available or can be obtained by processing a short sequence of frames at initialization time. A wide variety of change detection algorithms has been proposed in literature, so as to address issues such as illumination changes, camouflage and vacillating background. A recent survey providing good coverage of this research area is given by (Radke et al., 2005).

Sudden illumination changes occurring in the scene represent a major issue for most practical change detection applications. Properly dealing with such a problem is a challenging task for change detection algorithms since the resulting photometric variations can be easily misinterpreted as structural changes, leading to many false positives in the change mask. As depicted in Fig. 8, the proposed change detection algorithm consists of three processing stages. In the first stage, the MF measure is used to extract a subset of pixels in the current frame that can be marked as background with a high confidence level. Once such a subset, referred to as F_B, is obtained, it can be usefully employed to remove the photometric distortion between F and B. To this purpose, in the second stage the algorithm computes the transformation that aligns tonally the current frame, F, to the background image, B, using as support subset F_B. In the third stage, the final change mask is achieved by a pixelwise subtraction between F and the tonally registered background image, B_R.

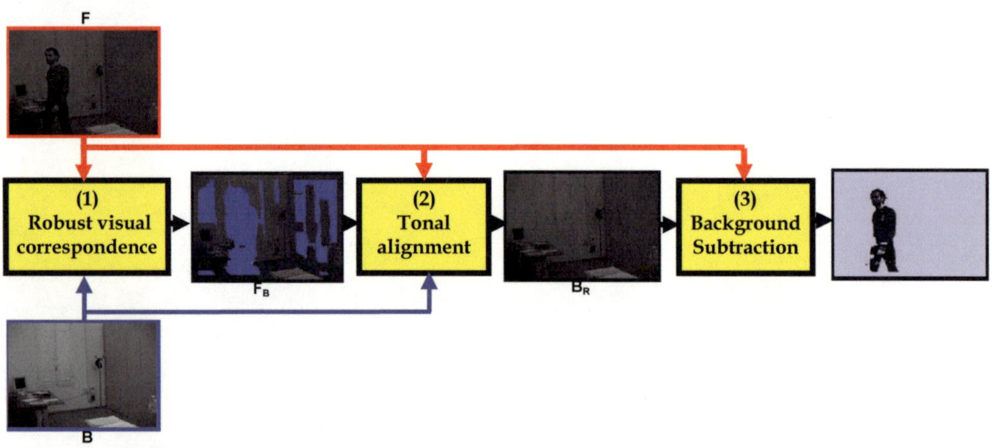

Fig. 8. Flow diagram of the proposed change detection algorithm

Robust visual correspondence In order to get F_B we match the points in the background image to the current frame. To achieve robustness with respect to outliers and noise, a block-based approach is used: that is, for each pair of correspondent points in B and F, a $M \times M$ surrounding block is considered, and the MF measure is computed between the two blocks. Points having a score higher than a given threshold are included into F_B . To explain the usefulness of the MF measure, let's discuss Fig. 9 where, for the sake of simplicity, we consider only two kind of regions, i.e. uniform and highly-textured. When dealing with a uniform region in both *F* and *B* (case *a* in Fig. 9), photometric differences between *F* and *B* can occur due to either variations of the illumination conditions of the background scene as well as to structural changes induced by a uniform foreground object. Thus, in this case the required matching measure should yield a low score, for nothing can be said reliably on whether the point belongs to the background or not. As for cases *b,c,d,* it is easy to observe that the matching score should be low too, since there's evidence of the presence of a foreground object. Finally, when the background is highly textured and the texture pattern does not change in spite of possible photometric changes (case *e*), it is reasonable to flag the point as background with a high confidence level. Hence, in case *e* we should get a high score from the required matching measure. Based on the above considerations, we adopt the MF measure which, as previously mentioned, matches corresponding blocks of two images by implicitly checking an ordering constraint. Since photometric variations tend not to violate the ordering of intensities in a neighbourhood of pixels, MF allows handling sudden and strong illumination variations between the background scene and the current frame. As previously discussed, MF tries to match the high contrast regions (i.e. the intensity edges) of the two blocks under comparison, since only high intensity differences can provide high contributions to the correlation score. Hence, MF behaves exactly as pointed out in Fig. 9. In fact, only two highly textured and highly correlated patterns can provide a high matching score (case *e*), while the presence of at least one *untextured* region (cases *a,b,c*) or of two textured but uncorrelated patterns (case *d*) yields a low score.

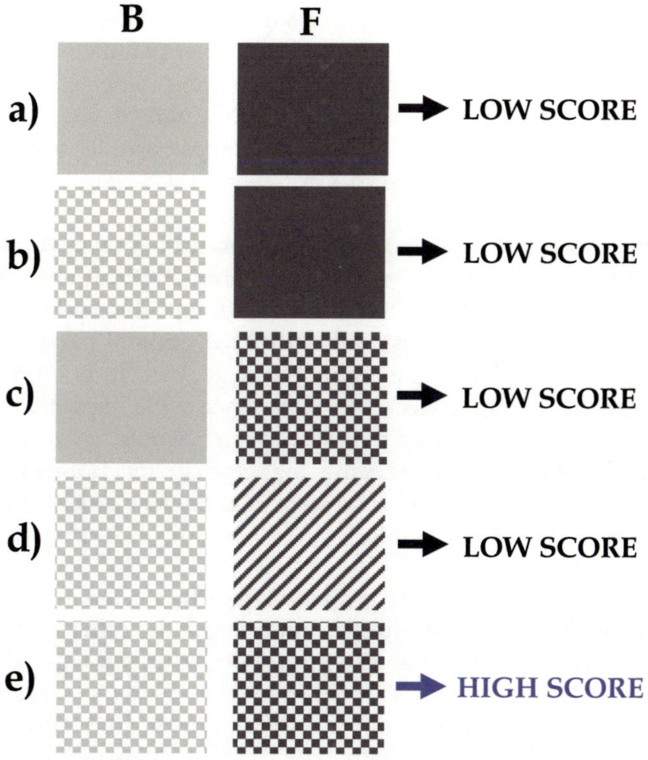

Fig. 9. Reasoning concerning the robust visual correspondence stage

Tonal alignment At this point of the algorithm, F_B represents a subset of F denoting pixels that reliably belong to the current background. Hence, B is tonally aligned to F by applying the *histogram specification* method (Gonzalez and Wood, 2002). In the evaluation of the IMF (*Intensity Mapping Function*) that aligns B to F only the set of corresponding points that belong to the mask F_B is taken into account. By applying the IMF obtained from the histogram specification method to B we get a novel background, B_R, where the photometric distortions have been removed.

Background subtraction Finally, a simple pixelwise difference between B_R and F highlights structural changes by correctly extracting foreground regions. It is worth pointing out that since background subtraction is carried out pixelwise, it is not affected by the aperture problem and allows for accurately detecting the borders as well as interior parts of foreground objects. Obviously, false negatives can still be found due to the possible camouflage between the tonally registered background and the foreground objects.

Experimental Results We now show the results dealing with a quantitative comparison between our approach and other proposals. The testing sequence was a synthetic sequence (available at: *http://muscle.prip.tuwien.ac.at/data here.php*) which comes together with groundtruth. For what regards the comparison, as representative of change detection algorithms that model false image changes according to a linear relation we consider the Normalised Cross Correlation (NCC) between pixel intensities. As for algorithms relying on checking the order preservation of intensities we consider the Rank transform (Zabih and Woodfill, 1994). We also consider as baseline for comparison the basic pixelwise background subtraction approach (BBS). For a fair comparison, we used the same block side for each algorithm (i.e. equal to 7). Then, for what regards the other parameters of each algorithm (in particular, the threshold for the final change mask), in order to determine the best parameter set of each algorithm we selected as a measure of comparison the *Precision*, i.e. the ratio between the true positives (TP) and the sum between true positives and false positives (FP), and the *Recall*, i.e. the ratio between the true positives and the sum between true positives and false negatives (FN).

In order to obtain experimental results, we started from the observation that most change detection algorithms, especially for video-surveillance applications, require to have a minimum guaranteed value of Recall. Hence, for different thresholds of minimum Recall (i.e. 70%, 80%, 90%, we selected for each algorithm the optimal parameter set maximizing the Precision value. Such results are shown in Tab. 1. It is worth pointing out that we fixed the maximum constraint value of Recall to 90%, since with higher values all algorithms would provide Precision values lower than 50%, which would result in very poor change masks (the number of false positives being higher than the number of true positives). Moreover, it is worth noting that also for these results no post processing was added to the output of the evaluated algorithms, similarly no morphology operator was used at any stage of the evaluated algorithms.

From the Table it is easy to infer that the proposed algorithm is the most robust and accurate between the evaluated ones, since it always outperforms the other approaches in terms of Precision for all different constraint values of Recall. In addition, Fig. 10 shows, for a single frame of the evaluated testing sequence, the outputs of the various algorithms at the different constraint values of Recall. In addition, in the first row of the Figure the background model as well as the current frame together with the correspondent ground truth frame are shown. These results qualitatively confirm the trend shown in Tab. 1, proving that our approach provides overall the most accurate results.

	> 70%	> 80%	> 90%
Proposed	87.3	81,7	52,2
NCC	59,6	57,2	43,0
Rank	24,5	18,8	13,1
BBS	2,2	1,9	1,7

Table 1. Best values of Precision yielded by the evaluated algorithms with different constraint values on Recall.

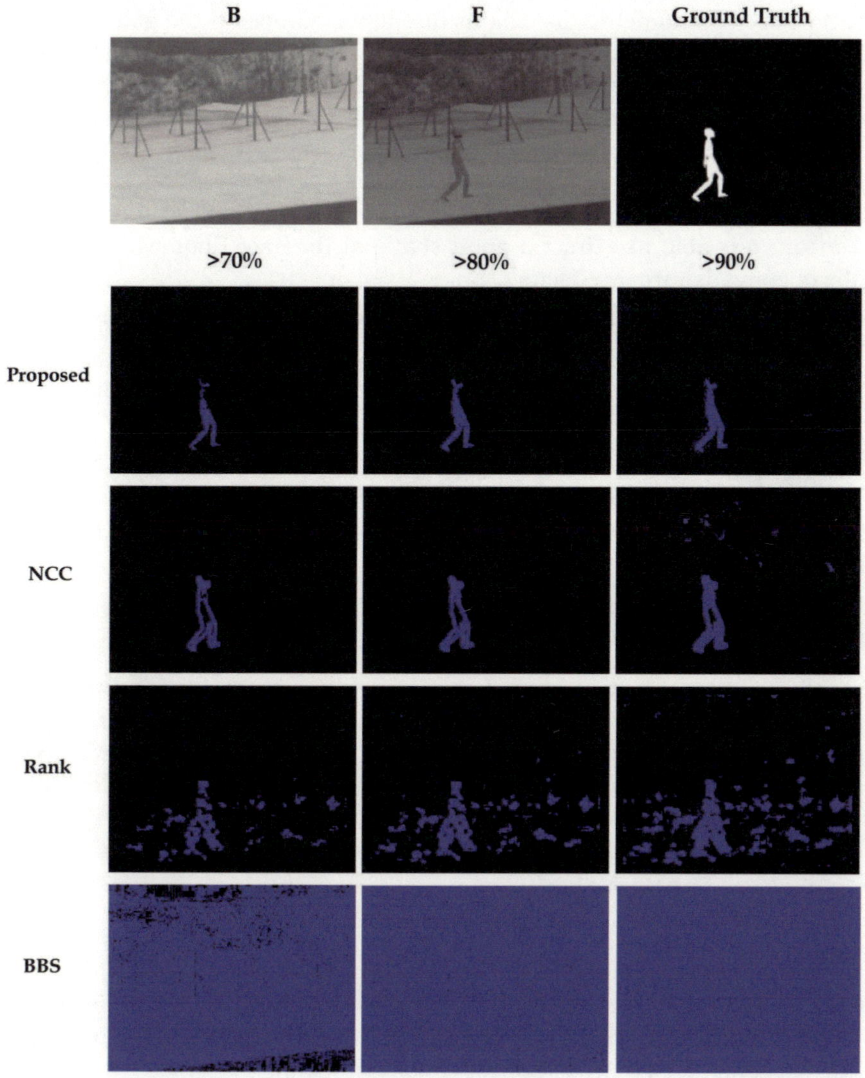

Fig. 10. Comparison of outputs yielded by the evaluated algorithms on the same sequence and with the same constraint values on Recall used for results in Tab. 1. First row, from left to right: background model *B*, current frame *F*, Ground Truth

6. Application to video-surveillance

In this section a case study is presented where access to a high security gate has to be monitored to assess for the presence/absence of people as well as to ensure that only one person occupies the gate at a given time (anti-tailgating). To accomplish this surveillance task, first of all we apply a change detection algorithm in order to discriminate between foreground (i.e. occupants) and background (i.e. the gate floor) regions of incoming frames. The change detection algorithm relies on the MF measure: at each point of the current frame and background image MF is computed on the window centred on the current point, then a threshold is used to discriminate background points from foreground points. The images are subject to heavy photometric distortions due to reflections on the gate floor, changes in indoor illumination and unpredictable light coming from outside. Fig. 11 shows the results where 3 frames acquired in different moments and with different subjects are compared with the same background image acquired previously (shown on the left). The 3 images on the right show the shape of the region detected as the gate floor using a fixed set of parameters (window side = 15, threshold = 0.2) and, as post-processing, a fixed sequence of simple morphological operators such as erosion and dilation. Results show that the proposed measure is able to extract a good shape of the gate floor with good robustness towards the ongoing disturbance factors.

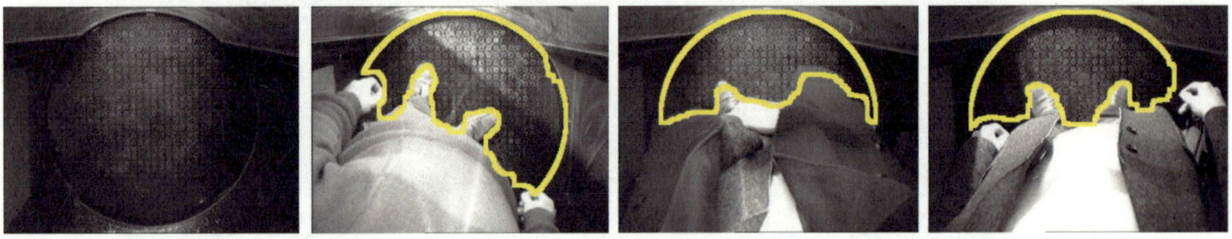

Fig. 11. The background (*left*) and 3 examples dealing with the presented video-surveillance application.

7. Conclusions

A review of the state of the art and a novel class of measures for robust visual correspondence under disturbance factors such as photometric distortions, noise and occlusions have been proposed. The proposed approach is based on the order preservation hypothesis, and aims at measuring how well the ordering constraint between neighbouring pixels is preserved. The novel measures were demonstrated to perform effectively in task such as pattern matching and change detection, as well as in a challenging surveillance scenario considered as a case study.

8. References

P. Aschwanden & W. Guggenbuhl (1992). *Experimental results from a comparative study on correlation-type registration algorithms.* In W. Forstner and S. Ruwiedel, editors, Robust computer vision, pages 268–289. Wichmann.

D. Bhat & S. Nayar (1998). *Ordinal measures for image correspondence.* IEEE Trans. Pattern Recognition and Machine Intelligence, 20(4): 415–423.

S. Chambon & A. Crouzil (2003). *Dense matching using correlation: new measures that are robust near occlusions.* Proceedings of British Machine Vision Conference, volume 1, pages 143–152.

A. Crouzil; L. Massip-Pailhes & S. Castan (1996). *A new correlation criterion based on gradient fields similarity.* In Proceedings Int. Conf. on Pattern Recognition, pages 632–636.

A. J. Fitch; A. Kadyrov; W. J. Christmas & J. Kittler (2002). *Orientation correlation.* Proceedings of British Machine Vision Conference, volume 1, pages 133–142.

S. Giachetti (2000). *Matching techniques to compute image motion.* Image and Vision Computing, 18: 247-260.

R. Gonzalez & R.Woods (2002). *Digital Image Processing.* Prentice Hall, 2nd edition.

S. Kaneko; Y. Satoh & S. Igarashi (2003). *Using selective correlation coefficient for robust image registration.*Journal of Pattern Recognition, 36(5):1165–1173.

S. Lai (2000). *Robust image matching under partial occlusion and spatially varying illumination change.* Computer Vision and Image Understanding, 78:84–98.

J. Martin & J. Crowley (1995). *Experimental comparison of correlation techniques.* Proceedings of Int. Conf. on Intelligent Autonomous Systems, volume 4, pages 86–93.

S. Mattoccia, F. Tombari & L. Di Stefano (2008-1). *Fast full-search equivalent template matching by Enhanced Bounded Correlation.* IEEE Trans. on Image Processing, 17(4): 528-538

S. Mattoccia, F. Tombari & L. Di Stefano (2008-2). *Reliable rejection of mismatching candidates for efficient ZNCC template matching.* Proceedings of IEEE Int. Conference on Image Processing (*in press*)

S. Mattoccia, F. Tombari, L. Di Stefano & M. Pignoloni (2007). *Efficient and optimal block matching for motion estimation.* 14th IAPR Int. Conference on Image Analysis and Processing, volume 1, pages 705-710

A. Mittal & V. Ramesh (2006). *An intensity-augmented ordinal measure for visual correspondence.* Proceedings of IEEE Conf. on Computer Vision and Pattern Recognition, volume 1, pages 849–856.

F. Odone; E. Trucco & A. Verri (2001). *General purpose matching of grey level arbitrary images.*Proceedings of 4th Int. Workshop on Visual Form, pages 573–582.

N. Ohta (2001). *A statistical approach to background subtraction for surveillance systems.* In Proceedings of Int. Conf. on Computer Vision, volume 2, pages 481–486.

R. Radke; S. Andra; O. Al-kofahi & B. Roysam (2005). *Image change detection algorithms: a systematic survey.* IEEE Trans. Image Processing, 14(3):294–307.

D. Scharstein (1994). *Matching images by comparing their gradient fields.* Proceedings of Int. Conf. on Pattern Recognition, volume 1, pages 572–575.

P. Seitz (1989). *Using local orientational information as image primitive for robust object recognition.* Proceedings SPIE, Visual Communication and Image Processing IV, volume 1199, pages 1630–1639.

J. Shen & S. Castan (1992). *An optimal linear operator for step edge detection.* Graphical Models and Image Processing, 54(2):112–133.

F. Tombari, S. Mattoccia & L. Di Stefano (2008). *Full search-equivalent pattern matching with Incremental Dissimilarity Approximations.* IEEE Trans. Pattern Analysis and Machine Intelligence (*in press)*

F. Ullah; S. Kaneko & S. Igarashi (2001). *Orientation code matching for robust object search.* IEICE Trans. Information and Systems, E-84-D(8):999–1006.

B. Xie; V. Ramesh & T. Boult (2004). *Sudden illumination change detection using order consistency.* Image and Vision Computing, 2(2):117–125.

R. Zabih & J. Woodfill (1994). *Non-parametric local transforms for computing visual correspondence.* In Proc. European Conf. on Computer Vision, pages 151–158.

B. Zitova & J. Flusser (2003). *Image registration methods: a survey.* Image and Vision Computing, 21(11): 977–1000.

20

Using Optical Flow as an Additional Constraint for Solving the Correspondence Problem in Binocular Stereopsis

Yeon-Ho Kim[1] and Soo-Yeong Yi[2]

[1]Purdue University

[2]Seoul National University of Technology

[1]USA

[2]Korea

1. Introduction

A stereo matching algorithm for 3D structure reconstruction relies on the correlation of image features in a pair of images. Usually, two calibrated cameras are used to capture scenes, and the left and right images are rectified to reduce the search space from 2D image region to 1D line. Then, the disparity is defined by the horizontal difference between the corresponding points on the search line. To solve the correspondence problem, various image features such as image intensity, edge, color, infrared light, pixel motion, etc. are employed either separately or integrated together. Among these image features, pixel motion or optical flow has been rarely used as one of the matching features (A. Scheuing & H. Niemann, 1986, G. Sudhir et al, 1995, A. M. Waxman & J. H. Duncan, 1986). In this chapter, we focus on the use of optical flow as an additional constraint in solving the correspondence problem.

The stereo matching algorithm for 3D structure reconstruction can be divided into two groups based on the matching primitives: intensity based matching (also referred to as area based matching) and feature based matching. The intensity based matching method searches the best matching points using only intensity values of pixels. The method can further be divided into two groups depending on the smoothness constraints in minimization of matching cost function: local (window-based) method and global method. More details on the intensity based method can be found in a survey by Scharstein (D. Scharstein & R. Szeliski. 2002, M. Z. Brown et al, 2003). The intensity based matching method produces a dense disparity map without any additional post-processing, but usually needs an exhaustive search. Since the intensity based method uses the intensity value at each pixel directly, this method may suffer from the varying illumination problem.

The feature based matching method searches the best matching points using some special symbolic feature points, such as line, contour, corner, etc. Since this method calculates disparity values only on the pixels corresponding to the feature point, this method does not require an exhaustive search. Also, symbolic features are less sensitive to illumination changes than pixel intensity, so the calculated disparity is more reliable than that of the intensity based matching method in the case of varying illumination. However, this method requires an additional process for extracting image features, and the produced disparity map is not as dense compared to that produced by the intensity based matching method unless a surface-fitting step is applied. More details on the feature based matching method can be found in a survey by Dhond (U. R. Dhond & J. K. Aggarwal, 1989).

In this chapter, we propose to use optical flow as an additional constraint in solving stereo correspondence problem in both intensity and feature based stereo matching methods. For the intensity based matching method, optical flow is used to reduce mismatching, which is described in Section 2.1. For the feature based matching method, optical flow simplifies the matching procedure as described in Section 2.2. In Section 3, we present some preliminary results on the surface reconstruction of a human hand in stereo image sequences using our proposed methods.

2. Modification of two matching techniques

2.1 Modification of an intensity based matching technique

In this section, we restrict the intensity based method to the local (window-based) method. The local intensity based matching technique calculates the disparity based on two geometric constraints. The first is the epipolar constraint that reduces two-dimensional search space to one-dimensional one. The second constraint is the assumption that the disparity is constant in small image regions (O. Faugeras et al, 1993). Based on these two constraints, small image regions are matched between two images along the epipolar line. If left and right images are rectified, then the vertical position of the search line in the right image is the same as that of the corresponding point in the left image. Then the disparity is defined as the horizontal distance between the corresponding points. To find the best matching pixel, various matching criteria are used. One of the most basic criteria is the sum of squared distances (SSD) defined as:

$$SSD_{Intensity}(x,y,d) = \sum_{i,j}[I_L(x+i,y+j) - I_R(x+i+d,y+j)]^2, \qquad (1)$$

where I_L and I_R are left and right images, (x,y) is the pixel coordinates where the disparity d is calculated, and the indices i and j are used to cover the entire pixels within a rectangular window. The best matching pixel has the minimum SSD value. This method is very simple but requires an assumption that the illumination conditions are same for the left and right images. But in real stereo images, due to the different setting of each camera or relative distance from each camera to the light sources, this assumption is not always satisfied. To ameliorate this problem, we use another matching criterion that is more robust to the variance of the illumination. This criterion is referred to as the normalized sum of squared distance (NSSD)(O. Faugeras et al, 1993) and defined as follows:

$$NSSD_{Intensity}(x,y,d) = \frac{\sum_{i,j}[(I_L(x+i,y+j) - \overline{I_L(x,y)}) - (I_R(x+i+d,y+j) - \overline{I_R(x+d,y)})]^2}{\sqrt{\sum_{i,j}(I_L(x+i,y+j) - \overline{I_L(x,y)})^2} \times \sqrt{\sum_{i,j}(I_R(x+i+d,u+j) - \overline{I_R(x+d,y)})^2}}, \qquad (2)$$

where $\overline{I_L(x,y)}$ is the mean of $I_L(x+i,y+j)$ within the window in the left image and $\overline{I_R(x+d,y)}$ is the mean of $I_R(x+i+d,y+j)$ within the window in the right image.

In addition to the intensity based matching criterion, we use another matching criterion that uses optical flow. For each left and right image sequence, the optical flow is calculated using our robust motion estimation technique presented in (Y.-H. Kim et al, 2005). The optical flow corresponding to a pixel in the left image is matched to that corresponding to a pixel in the right image along the epipolar line using the following motion correspondence matching criterion in a similar way of the intensity based matching technique described above:

$$NSSD_{Motion}(x,y,d) = \frac{\sum_{i,j}[(I_L(x+i,y+j)-\overline{I_L(x,y)})-(I_R(x+i+d,y+j)-\overline{I_R(x+d,y)})]^2}{\sqrt{\sum_{i,j}(M_L(x+i,y+j)-\overline{M_L(x,y)})^2}\times\sqrt{\sum_{i,j}(M_R(x+i+d,u+j)-\overline{M_R(x+d,y)})^2}}, \quad (3)$$

where $M(x,y)=u^2(x,y)+v^2(x,y)$ is the squared magnitude of optical flow vector which consists of the horizontal (u) and vertical (v) components and other notations are same as defined in Eqs. (1) and (2).

Finally, we integrate the intensity based matching criterion and the motion based matching criterion in a single matching criterion as follows:

$$NSSD_{Integrated}(x,y,d) = NSSD_{Intensity}(x,y,d)\times(1+NSSD_{Motion}(x,y,d)), \quad (4)$$

2.2 Modification of a feature based matching algorithm

In this section, we first describe the Marr-Poggio-Grimson (MPG) algorithm that is one of the most popular feature based matching methods. Then explain how we modify this algorithm using optical flow.

The MPG algorithm, implemented by Grimson (W. E. L. Grimson, 1981), solves the stereo correspondence problem based on the Marr and Poggio's computational model of human stereopsis (D. Marr & T. Poggio. 1979). The MPG algorithm consists of following three main steps: feature extraction, feature matching, and matching analysis.

The first step of the MPG algorithm is to extract features in images. For this purpose, the Laplacian of Gaussian $(\nabla^2 G)$ operator (referred to as the primal sketch operator) is first applied to the image where the operator is formed by:

$$\nabla^2 G(x,y) = \frac{r^2-2\sigma^2}{\sigma^4}\exp(-r^2/(2\sigma^2)), \quad (5)$$

where (x,y) is the pixel coordinate and $r^2 = x^2+y^2$. The width of the function is represented by the distance between the first zeros on either side of the origin and denoted by $w_{2D} = 2\sqrt{2}\sigma$. From the image filtered by this operator, zero crossing points are then detected as final image features.

The second step of the MPG algorithm is feature matching. Based on the Marr and Poggio's hierarchical model of human stereopsis, the MPG algorithm employs a coarse-to-fine approach for feature matching. The disparity values obtained at a coarse level of images are used for the disparity calculation at the next finer level of images. At each level, for each zero-crossing point in the left image, its matching candidates in the right image are searched within the scan line where the center of the line is the coordinates of the zero-crossing point in the left image and the width of the line is $2w_{2D}$. Within this search line, the points with similar local orientations are selected as its matching candidates. Instead of using the local orientation at zero-crossings as originally proposed by Grimson (W. E. L. Grimson, 1981), we use binary neighborhood comparison -- implemented by Tanaka and Kak (S. Tanaka & A. C. Kak, 1990) - as the candidate selection criterion in our implementation.

The third step of the MPG algorithm is matching analysis. If just only one matching is found in the scan line, then the distance between the corresponding points will be the final disparity. However, within the search line, more than one matching candidates can be found and, in that case, further consideration is required to determine the final disparity. For this purpose, Marr and Poggio used two constraints. One is referred as to *uniqueness*: each zero-crossing point in the left image must have only one disparity value. The other one is referred as to *continuity*: disparity value must vary smoothly. To implement these two constraints of Marr and Poggio's theory, Grimson grouped matching candidates by three types depending on the position of the matching candidates in the search line: convergent, divergent, or zero. If more than one matching candidates are found, then the one having the same type of matching as the dominant matching type in the neighbourhood is accepted.

In our modification of the MPG algorithm, we simplify the final matching analysis step by employing optical flow. First, we calculate squared magnitude of the optical flow vectors in the left image and the right image separately. Then in the multiple matching candidates, we choose the one having the minimum difference of the squared magnitude of the optical flow vectors between the corresponding points.

3. Results

In this section, we present some preliminary results of the 3D structure reconstruction using our modified matching algorithms described in the previous section. Fig. 1 shows left and right image sequences we used here. These image sequences include a human hand that is moving at the front of the left and right cameras. To enrich the texture of the hand, we used a glove covered by random-dot pattern.

For each image sequence, optical flow is obtained frame by frame using our robust optical flow estimation method (Y.-H. Kim et al, 2005). Generally, differential optical flow techniques based on the image derivatives are able to estimate only the small motion. The small optical flow does not provide significant differences between neighbour pixels and is not adequate for the matching criterion. To ameliorate this problem, we accumulate optical flow vectors calculated from several image frames to get larger optical flow. Fig. 2 shows the squared magnitude of the accumulated optical flow vectors in three-dimensional space. The width and depth of the three-dimensional space represent the axes of the horizontal and vertical direction of the image plane, and the height of the three dimensional space represents the axis of the squared magnitude of the optical flow vectors. Figs. 2(a, b) show the squared magnitude of the optical flow vectors from the 1st image frames to the 2nd image frames for each left and right image sequence respectively. Figs. 2(c, d) show the squared magnitude of the accumulated optical flow vectors from the 1st image frames to the 3rd image frames for each left and right image sequence respectively. Figs. 2(e, f) show the squared magnitude of accumulated optical flow vectors from the 1st image frames to the 5th image frames for each left and right image sequence respectively. As shown in this figure, the squared magnitude of optical flow and the difference of these values between neighbour pixels increase as the number of image frames used for the optical flow calculation increases.

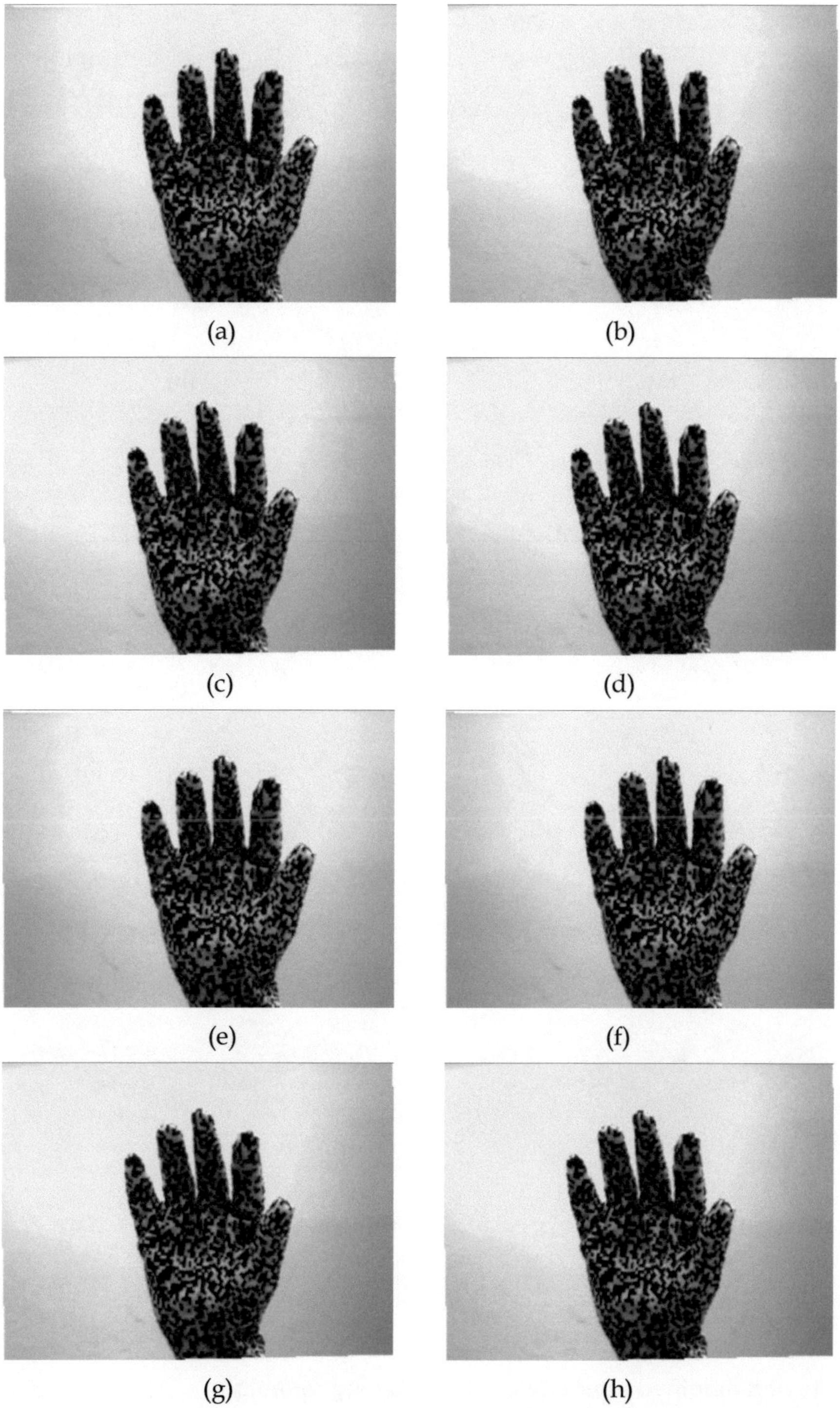

Fig. 1. Left and right image sequences: (a, b) The 1st image frames, (c, d) The 2nd image frames, (e, f) The 3rd image frames, and (g, h) The 5th image frames.

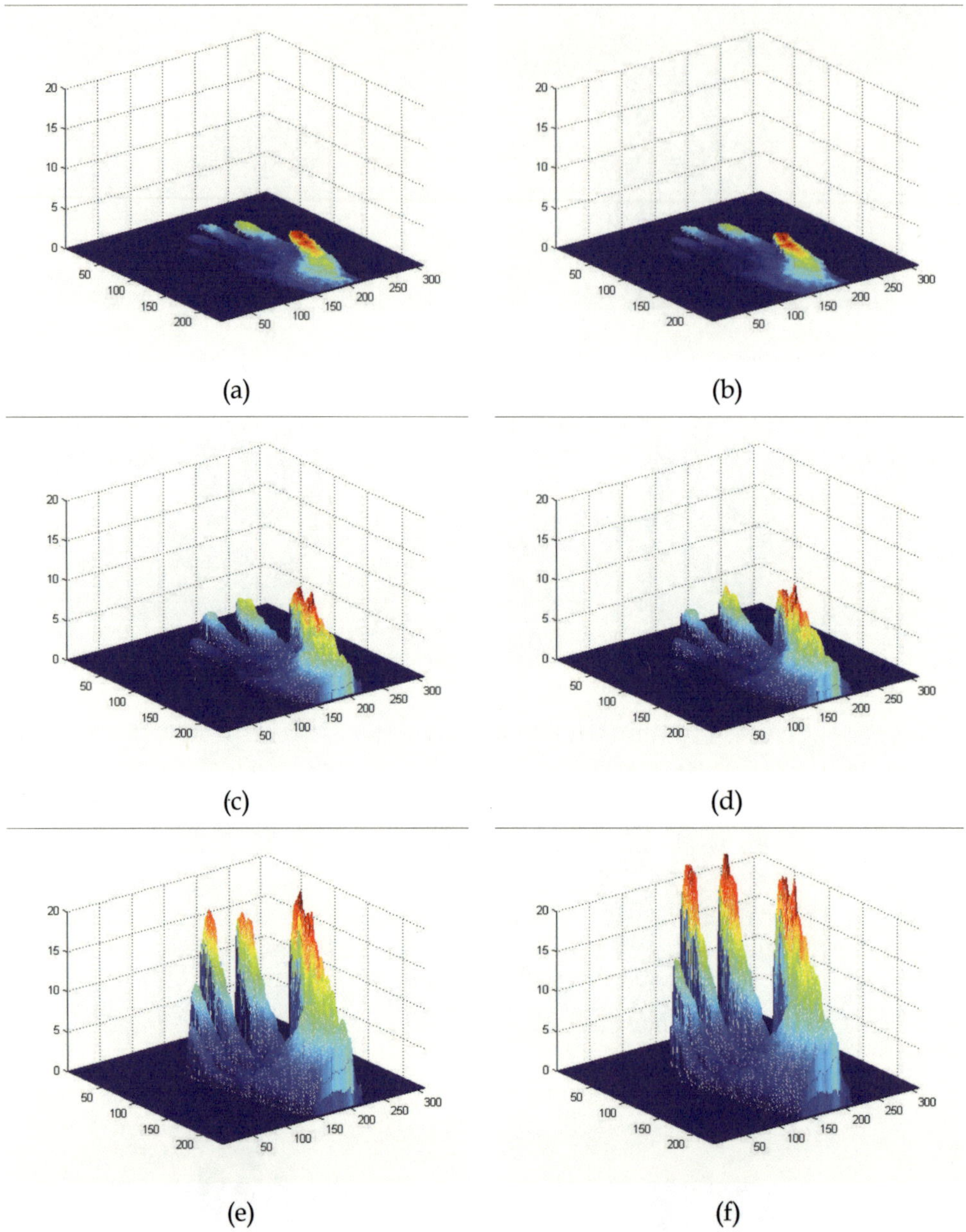

Fig. 2. Squared magnitude of (a, b) optical flow from the 1st image frames to the 2nd image frames, (c, d) accumulated optical flow from the 1st image frames to the 3rd image frames, (e, f) accumulated optical flow from the 1st image frames to the 5th image frames for the left and the right image sequences respectively.

3.1 Results of a modified intensity based matching technique

Fig. 3(a) shows the disparities obtained from the original intensity based matching algorithm using the first image frames of the left and right image sequences. No motion information is used for this result. The width and the depth of the three dimensional space represent the axes of the vertical and horizontal direction of the image plane and the height of the three dimensional space represents the axis of the disparity. Fig. 3(b) shows the disparities obtained from the modified intensity based matching algorithm using the intensity matching criterion and additional motion matching criterion. The intensity matching criterion uses pixel intensities in the first image frames of the left and right image sequences. The motion matching criterion uses the optical flow from the 1st image frames to the 2nd image frames for each left and right image sequence respectively. For the integrated matching criterion, Eq. (3) is used. Fig. 3(c) shows the disparities obtained from the modified matching algorithm in which the intensity based matching criterion uses the first image frames of the left and right image sequences and the motion based matching criterion uses accumulated optical flow from the 1st image frames to the 5th image frames for each left and right image sequence respectively. We can see the number of mismatches decreases by using the motion matching criterion and also by increasing the number of image frames to calculate optical flow. The mismatches in Fig. 3(c) are separated as shown in Fig. 4(a) using our de-nosing technique (that counts the pixels in a certain cubic area around a pixel in interest and if the number of count is less than a threshold then the pixel in interest is discarded). Fig. 4(b) shows the finally reconstructed surface of hand using a surface fitting function (spaps) in MATLAB (C. de Boor, 2004).

3.2 Results of a modified feature based matching algorithm

Fig. 2 shows different levels of zero-crossing points for the test image sequences filtered by the Laplacian of Gaussian operator with different w_{2D} values. In Fig. 2, the value of w_{2D} for (a, b) is 32, for (c, d) 16, for (e, f) 8, and for (g, h) 4. Fig. 6(a) shows disparities obtained using the original feature based matching algorithm from the first image frames of the left and right image sequences. No motion information is used for this result. The width and the depth of the three dimensional space represent the vertical and horizontal axes of the image plane and the height of the three dimensional space represents the axis of the disparity. Fig. 6(b) shows the disparities obtained using our modified feature based matching algorithm. This algorithm uses zero-crossing features obtained from the first image frames of the left and right image sequences and optical flow obtained from the 1st image frame and the 2nd image frame for each left and right image sequence respectively. Although our modified feature based matching algorithm uses simpler matching procedure than the original feature based matching algorithm as described in Section 2.2, we can see no significant difference between the disparity results in Fig. 6(a) and those in (b). Fig. 6(c) shows the disparities obtained using our modified feature based matching algorithm using the accumulated optical flow from the 1st image frame to the 5th image frame for each left and right image sequence respectively. There is no significant difference between the result in Figs. 6(b) and that of (c). Fig. 7(a) shows the disparities after separating noisy disparities using the same de-nosing algorithm used in Section 3.1. The final reconstructed hand surface result is shown in Fig. 7(b).

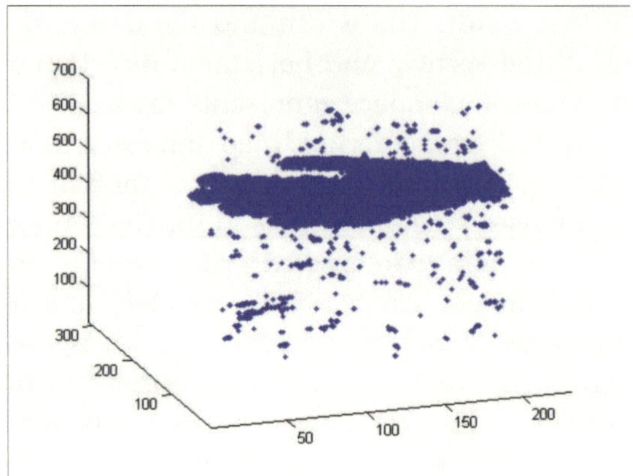

(a)

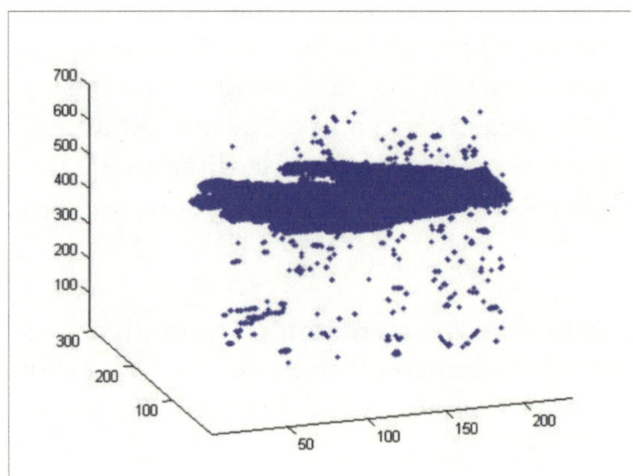

(b)

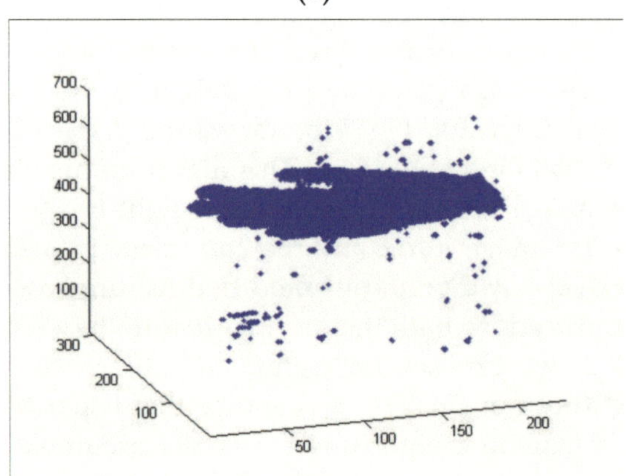

(c)

Fig. 3. Calculated disparities: (a) using the original intensity based stereo matching technique, (b) using the modified intensity based stereo matching technique with optical flow from the 1st to the 2nd image frames, and (c) using the modified intensity based stereo matching technique with accumulated optical flow from the 1st to the 5th image frames.

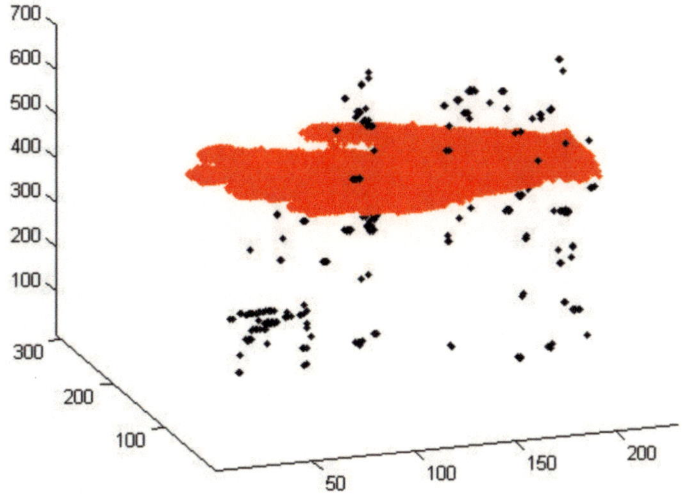

(a)

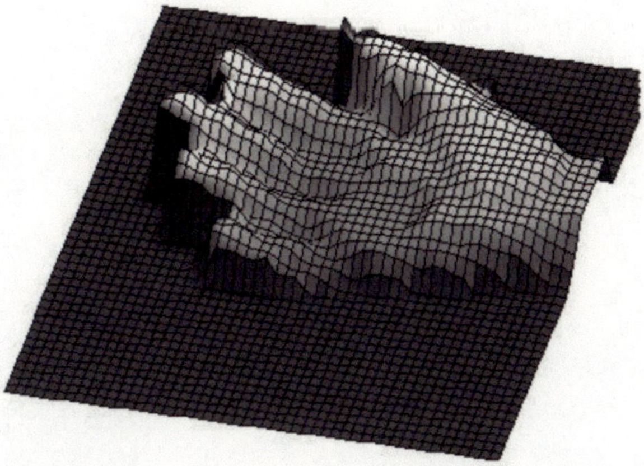

(b)

Fig. 4. (a) Noise separated disparities, (b) A human hand surface reconstruction result using our modified intensity based stereo matching method.

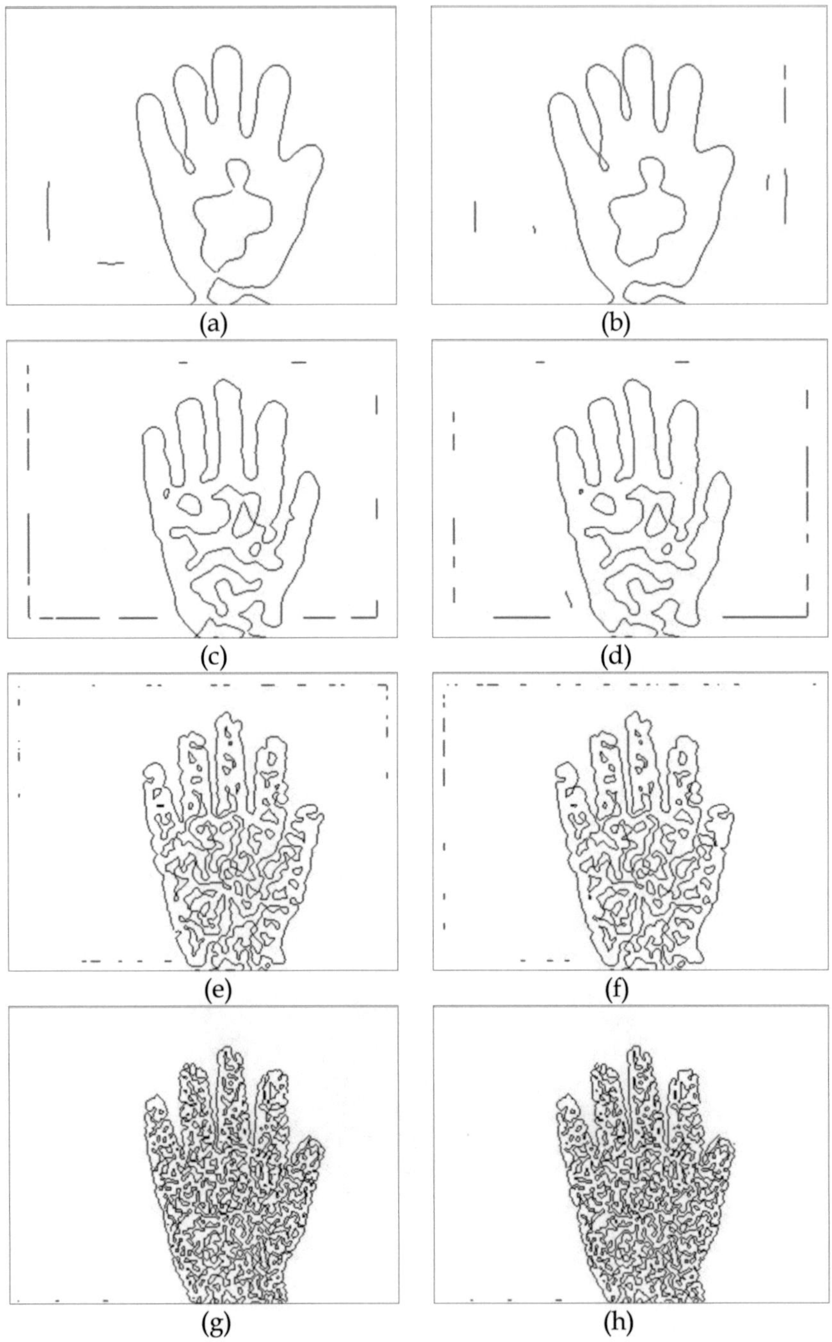

Fig. 5. (a) Zero-crossing feature points in the left and right images which are filtered by LoG operator with different w_{2D}: The value of w_{2D} is (a, b) 32, (c, d) 16, (e, f) 8, and (g, h) 4, respectively.

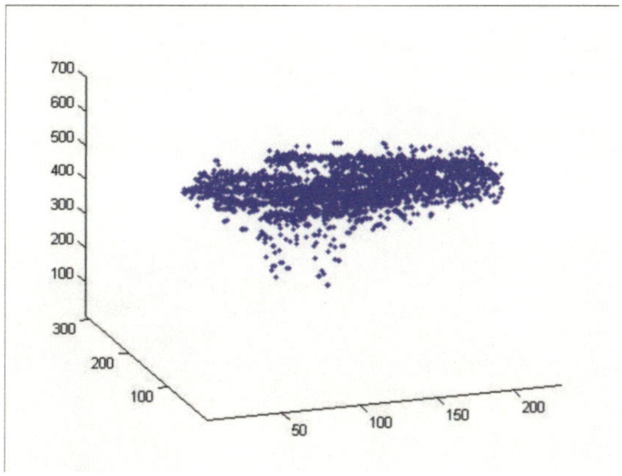

(a)

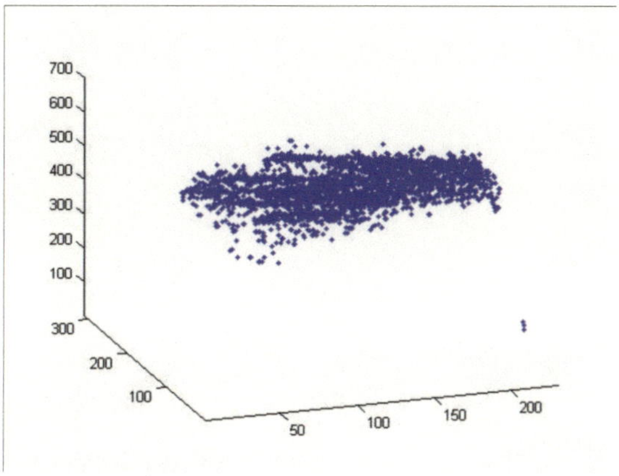

(b)

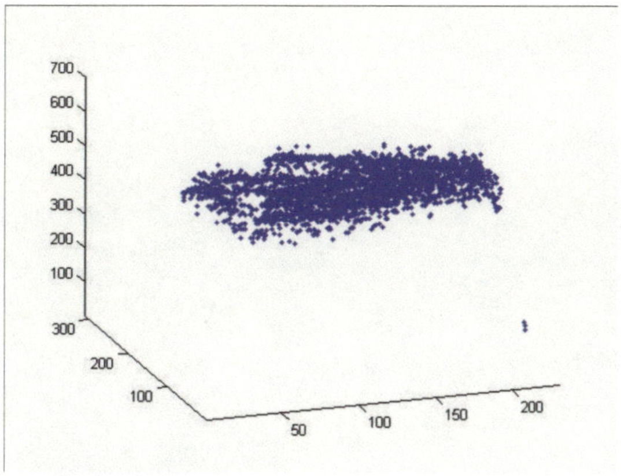

(c)

Fig. 6. (a) Calculated disparities: (a) using the original MPG algorithm, (b) using our modified MPG algorithm with optical flow from the 1st to the 2nd image frame, and (c) using our modified MPG algorithm with accumulated optical flow from the 1st to the 5th image frames.

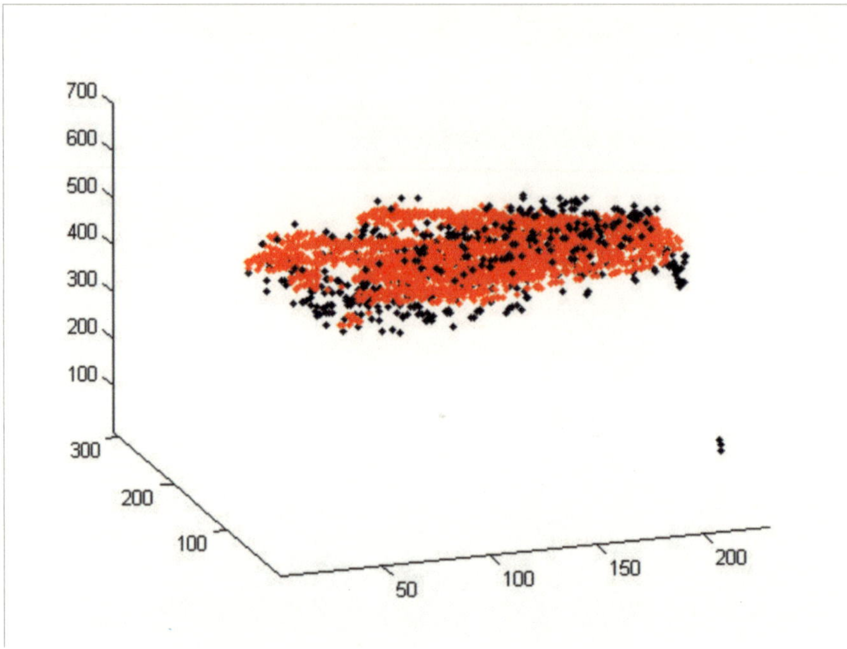

(a)

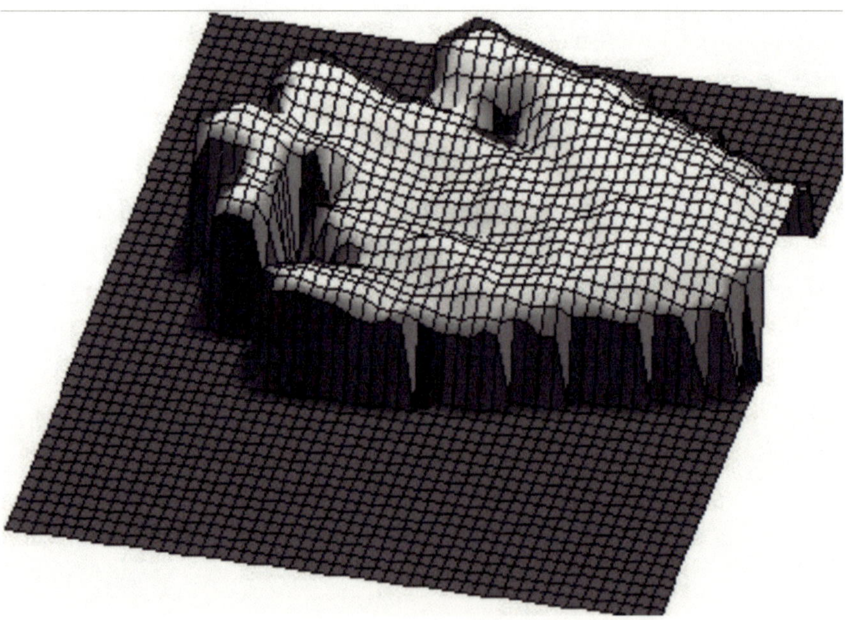

(b)

Fig. 7. (a) Noise separated disparities, (b) A hand surface reconstruction result using our modified MPG algorithm.

3. Conclusion

In this chapter, we proposed two modified stereo matching algorithms. One is the modification of an intensity based matching algorithm and the other one is the modification of a feature based matching algorithm. For the modification of an intensity based matching algorithm, we employed an additional matching criterion using optical flow to reduce the number of mismatching disparities. For the modification of a feature based matching algorithm, we simplified the matching procedure in the original MPG algorithm using optical flow. We presented some preliminary experimental results of the 3d structure reconstruction obtained by each modified matching algorithm for a pair of image sequence including a human hand.

8. References

A. M. Waxman & J. H. Duncan. (1986). Steps toward stereo-motion fusion. *IEEE Trans. on Pattern Analysis and Machine Intelligence*, Vol. 8, No. 6, November, (715-729)

A. Scheuing & H. Niemann. (1986). Computing depth form stereo images by using optical flow. *Pattern Recognition Letters*, 4, (205-212)

C. de Boor. (2004). *Spline Toolbox for use with MATLAB*. The Math Works, Inc.

D. Marr & T. Poggio. (1979). A computational theory of human stereo vision. *Proc. Royal Soc. London*, B204

D. Scharstein & R. Szeliski. (2002). A taxonomy and evaluation of dense two-frame stereo correspondence algorithms. *International Journal of Computer Vision*, 47, (7-42)

G. Sudhir; S. Banerjee; K. K. Biswas & R. Bahl. (1995). A cooperative integration of stereopsis and optic flow computation. *Journal of the Optical Society of America A*, Vol. 12, No. 12, December, (2564-2572)

M. Z. Brown; D. Burschka & G. D. Hager. (2003). Advances in computational stereo. *IEEE Trans. on Pattern Analysis and Machine Intelligence*, Vol. 25, No. 8, (993-1008)

O. Faugeras; B. Hotz; H. Mathieu; T. Viéville; Z. Zhang; P. Fua; E. Théron; L. Moll; G. Berry; J. Vuillemin; P. Bertin & C. Proy. (1993). Real time correlation-based stereo: algorithm, implementations and applications. *INRIA Technical Report*, (2013)

S. Tanaka & A. C. Kak. (1990). *Analysis and Interpretation of Range Images: Chapter 2. A Rule-Based Approach to Binocular Steropsis*. Springer-Verlag, New York

U. R. Dhond & J. K. Aggarwal. (1989). Structure from stereo - a review. *IEEE Transactions on Systems, Man, and Cybernetics*, Vol. 19, No. 6, (November-December), (1489-1510)

W. E. L. Grimson. (1981). A computational implementation of a theory of human stereo vision. *Phil. Trans. Royal Soc. London*, B292

Y.-H. Kim, A. M. Martínez, & A. C. Kak. (2005). Robust motion estimation under varying illumination. *Image Vision Computing*, Vol. 23, No. 4, (365-375)

Stereo Matching and Graph Cuts

Ayman Zureiki [1,2], Michel Devy [1,2] and Raja Chatila [1,2]
[1]*CNRS; LAAS; 7 avenue du Colonel Roche, F-31077 Toulouse,*
[2]*Université de Toulouse; UPS, INSA, INP, ISAE; LAAS-CNRS : F-31077 Toulouse, France*

1. Introduction

The stereo matching aims to find corresponding entities between two (or more) images, i.e. entities that are projections of the same 3D object in the scene. Constraints used in stereo matching can be classified into two categories: local constraints, which rely only on a pixel and on some pixels in its surrounding, and global constraints, which must be verified by the whole pixels of a line or of the image. The local methods aim to find a matching for a given pixel without taking into account neighbour pixels correspondences. Global methods try to define a global model of the observed scene and to minimize a global cost function. They try to find the correspondences once for all pixels in one line or for all pixels in the image.
Graph cuts techniques have been recently used to solve the stereo matching problem involving global constraints. These methods transform the matching problem to a minimization of a global energy function. The minimization is achieved by finding out an optimal cut (of minimum cost) in a special graph. Different methods were proposed to construct the graph. However, when applied to minimize a global cost function in stereo vision, all of them consider for each pixel, all possible disparities between minimum and maximum values. Our contribution is a new method for constructing a reduced graph: only some potential values in the disparity range are selected for each pixel. These values can be provided by a preliminary matching process using only local constraints. We will detail how this method allows to make wider the disparity range, and at the same time to limit the volume of the graph, and therefore to reduce the computation time.
In this chapter, the stereo matching problem is defined. A brief state of the art about stereo matching is presented. The stereo matching can be solved by either local methods or global ones. Among the global methods, we give a short introduction of dynamic programming techniques to be a logic introduction of the Graph Cuts methods. We recall the definition of Graph, flow, cut, and the different algorithms to solve the problem of maximum flow. A general formalism of Relabelling problem is used to express the stereo matching as a minimization of an energy function. The implementation of both complete graph (Roy & Cox, 1998) and reduced graph (our contribution) are detailed. The two methods are compared from experimental results.

2. Stereo matching

The stereo vision is a tool to find 3D information on a scene perceived by two images (or more) acquired at the same moment from different points of view. The stereo reconstruction is based on the aptitude to find in each image the projection of the same object in the scene. In fact, depth information of an object is related from one side to the disparity (difference of projections of the object in both images), and on the other side, to the relative position of both cameras (baseline) and to the image resolution focal length, etc.). Thus, the stereoscopic reconstruction raises two different problems. The first is the disparity calculation, which is attached to the problem of stereo correspondence (matching). The second problem is the ability to inverse the projective geometry problem. In other words, the tridimensional reconstruction, or how to exploit disparity knowledge and relative position of the two sensors to find 3D information. The works of (Faugeras, 1993), on the projective geometry have established a solid basis for the 3D reconstruction problem. For the matching problem, there is no method sufficiently reliable, robust and effective that allows a simple use of stereo vision as a sensor of depth measurement.

Binocular stereo vision uses two images acquired by two cameras. A preliminary phase of calibration is needed to estimate the different parameters of a stereo rig: the parameters of the projection model for each camera (pinhole geometric model) and the spatial relationship between the two cameras. This knowledge allows us to calculate the 3D coordinates of a point from its projections in the two images by a simple triangulation.

Stereo matching is one of the most studied topics in computer vision since more than half a century (Julesz, 1962). Stereo matching aims at finding in the left and right images, features, which are the projections of the same entity in the 3D scene. In general, the matching problem can be seen as a minimization problem. Local approaches try to minimize separately many energy functions, representing local matching costs supposed independent between different entities to be matched: a local cost depends on similarity constraints between these entities. Global approaches try to minimize a unique energy function taking into account all matching costs: this global cost integrates local matching costs, and also compatibility costs expressing how consistent are matchings computed on a line or on the whole image. A detailed taxonomy of stereo correspondence algorithms is proposed in (Scharstein & Szeliski, 2002). The authors classify stereo matching methods with respect to four criteria: (i) The local matching cost, (ii) The aggregation area while computing the local cost, (iii) The optimization method, (iv) The method performed to refine matching results.

The method developed at LAAS since 1995 (Grandjean & Lasserre, 1995) is a modified version of the dense or pixel-based algorithm described by (Faugeras et al., 1993). This method is suitable for robotic applications, especially because it satisfies real-time constraints. It can be classified under local methods, because there is no a global optimization step to make consistent matchings of adjacent pixels. Here we propose to apply such a global optimization method as a second step of our matching algorithm, to take advantage of the global minimization while remaining within real-time constraints as much as possible. We consider that left and right images are already rectified: it allows to limit the research area in the right image, for the match of a pixel in the left image.

2.1 Local stereo matching methods

Local stereo matching methods search separately for the best match of each pixel strating from one image (e.g. the left one) without taking into account the matches of other ones. The matching cost between two pixels is based on similarity measurements of the local intensity function. Intuitively, the projections of the same 3D point will have (naturally) similar intensities in the two images. In fact, the Lambertian model (Horn, 1986) assumes that the object surface reflects uniformly the light in all directions. Using this model, we can suppose that the corresponding pixels in both images are similar, and indeed, their neighbours are also similar, assuming that view fields between the two cameras are very close (no occlusions). A correlation measurement can calculate a degree of similarity between two point sets. Local methods try to find a match p2 in the right image for a point p1 in the left one. The correlation measurement uses information given by p1, p2 and their neighbour pixels. The pixel p1 and its neighbours form a first point set, and the point p2 and its neighbours constitute the other point set. A correlation score evaluates the similarity between these data sets.

Many local approaches use comparison windows centred on the considered pixel. Among the most known measurements, we can find: Sum of Squared Differences (SSD) (Cox et al., 1996), Sum of Absolute Differences (SAD) (Hirschmüller, 2001), Zero-mean Normalized Cross-Correlation (ZNCC) (Chen & Medioni, 1999), (Sára, 2002), etc.

Local stereo correspondence methods are in general fast algorithms, so can be used for real-time applications. However, they are exposed to many failure sources, in particular occlusions or variations of intensity between the two images. In fact, these situations can produce many false matches. In addition, because of the absence of any constraint between matches, adjacent pixels can have very different disparities, which can be particularly remarkable in scenes having vertical lines (edges of an open door for example). Global methods try to overcome these problems.

2.2 Global stereo matching methods

Global approaches try to define a global model of observed scene and to minimize a global cost function. Matches for pixels of one line or or pixels of the whole image, are searched at the same time. In a global method, the matching between a pixel in the left image and a pixel in the right image does not depend only on their neighbours, but also on the matches of their neighbours. Hence, the match of a pixel influences the matches of its neighbour pixels. This influence is modelled by regularization constraints on the matches set. Some methods are based only on the epipolar constraint to transform the bidimensional matching problem into one-dimensional problem, as in dynamic programming (Belhumeur, 1996), (Cox, 1992). Other methods address the bidimensionnal problem by taking into account, inter-lines constraints, i.e. compatibility constraints between matchings provided on every epipolar lines, as in graph cuts algorithms (Boykov et al., 1998), (Ishikawa & Geiger, 1998).

The global regularization aims to reduce the sensibility of stereo correspondence to ambiguities caused by occlusions, poor local texture or fluctuation of illumination. This improvement has a cost, which is the increasing of algorithms complexity, and in consequence, a longer execution time, in addition to some secondary effects due to this regularization (smoothing). We detail two global methods: dynamic programming and graph cuts.

Dynamic Programming:

Dynamic programming, introduced by Richard Bellman (Bellman, 1957), allows resolving optimization problems having an objective function as a sum of monotone non-decreasing functions of resources. In practice, this means that we can infer the optimal solution of a problem using optimal solutions of sub-problems. The dynamic programming applied to stereo matching searches for a path of minimal cost through a matrix composed of possible matches. To reduce the complexity, this technique is applied on two sets of points of the same epipolar line. Thus, the stereo correspondence is applied successively to find matchings for all pixels of a line of one image with pixels located on its epipolar line in the other image (Ohta & Kanade, 1985).

To obtain a global path cost equal to the sum of the partial-paths costs, it is mandatory to use additive costs. We define the local cost for each point in the research zone as the cost of a local stereo matching (SAD, SSD, etc.). Occlusions can be taken into account, making possible to link a set of image pixels with the same pixel in the other image; penalties are considered for these relations (occlusion costs), which will be added to the global cost of any path in the matrix (Ohta & Kanade, 1985), (Cox et al., 1996). This formulation presents many inconvenient as the sensibility to the occlusion cost, the difficulty to guarantee inter-lines consistency (Bobick & Intille, 1999), and the weak application of constraints on order and continuity, that could be not satisfied.

The dynamic programming can help in finding matchings in poorly textured zones, and in solving some occlusion problems. But this method brings also some weak points, as complexity of calculation, possibility of propagation of a local error through all the research line, and non-consistency of disparity between lines.

Graph Cuts:

The majority of stereo matching dynamic programming methods try to match pixels between epipolar lines in both images without taking into account inter-lines consistency. Hence, they do not use the bidimensionnal nature of the problem. To overcome this drawback, and to take into consideration bidimensionnal continuity constraint, a solution has been proposed using the graph theory. Initially, graph cuts applied to stereo matching were proposed by (Roy & Cox, 1998), and then reformulated by (Veksler, 1999) in which the matching problem is considered as a minimization of an energy function. Afterwards, the iterative graph-cuts algorithms were introduced in (Kolmogorov & Zabih, 2001), (Kolmogorov & Zabih, 2002a), (Kolmogorov & Zabih, 2002b).

The first global method based on graph cuts for stereo correspondence (Roy & Cox, 1998), (Roy, 1999), started from the 1D formulation of the order constraint used by the dynamic programming applied separately to each image line. They tried to find a more general 2D formulation for this constraint to be applied to all lines together. They proposed a *local coherence* constraint, which suggests that the disparity function is locally smooth, which means that neighbour pixels in all directions have similar disparities. They applied this local coherence constraint by defining a disparity matching cost, which depends on intensity variations of matched pixels. In the case of two cameras, the matching cost is the squared difference of intensities.

The next step in the method proposed by Roy and Cox is to estimate the optimal disparity map over the entire image. The matching constraints are expressed in a 3D mesh composed of planes, themselves composed of an image of nodes. There is a plane for each level of disparity, and each node represents a matching between two pixels in original images. The 3D mesh is then transformed into a graph of maximal flow by connecting each node to its four neighbours in the same plane by edges called occlusion edge, and with the two nodes in the neighbour planes with edges called disparity edges. Edges are not oriented. We add two special nodes: a source connected to all nodes in the plane of minimum disparity, and a sink connected to all nodes in the plane of maximum disparity. The weight of a disparity edge is equal to the mean value of matching costs of the two nodes. For occlusion edges, the weight is multiplied by a constant to control the smoothness of the optimal disparity map. A graph cut will separate the nodes in two sub-sets: the optimal disparity map is constructed by the assignment of each pixel with the bigger value of disparity for which the corresponding node is still connected to the source.

(Ishikawa & Geiger, 1998) pointed out that the method of Roy and Cox can deal only with convex maps. Thereby, it can only take into account linear penalties on disparity, which may lead to not very good results due to an over smoothing of disparities. They proposed a novel graph with oriented edges, and then it is possible to reinforce the constraint of uniqueness and order. However, their method is also weak at discontinuities because of linear penalties.

Sub-optimal algorithms: (Boykov et al., 1998; Boykov et al., 1999) proposed another method to resolve the stereo corresponding using graph cuts. The authors showed that the problem of stereo corresponding could be formulated by a *Markov Random Field* (MRF). They showed that the estimate Maximum A Priori (MAP) of such MRF, could be obtained by a minimal multi-way cut, using a maximum flow. The advantage of such a method is that it accepts non linear penalties for discontinuities, and then it gives more precise disparity maps especially near objects' edges. As the general problem of minimal multi-way cut is NP-complete (Dahlhaus et al., 1994), Boylov et al. decided to introduce an approximated algorithm, which can resolve iteratively some sub-problems until convergence. (Kolmogorov & Zabih, 2001) have continued the works of Boykov trying to ameliorate the energy function to represent more explicitly the occlusion. The sub-optimal approach has a wider application spectrum than the one proposed by Roy and Cox or by Ishikawa and Geiger. However, it is iterative and sub-optimal, and then its convergence speed and the quality of the obtained minimum must be supervised.

In spite of their good results, methods based on graph cuts are disadvantaged by several limitations. Firstly, because of using the bidimensionnal continuity constraint, these methods can cause an excess of regularization (over smoothing), which can appear as the effects of a local filter. Secondly, as the penalty function (cost of attributing different disparities to neighbour pixels) is not always convex, this returns the minimization of energy function to be a NP-complete problem (Kolmogorov & Zabih, 2001), and in consequence, the obtained solution is only an approximation of the exact one.

3. Minimum cut problem

3.1 Graph definition

A graph is a set of vertices connected by edges. A binary relation between vertices, called adjacency relation, defines the connection. A graph **G** be defined as a pair **(V,E)**, where **V** is a set of vertices, and **E** is a set of edges between the vertices **E** = {(u,v) | u, v ∈ **V**}. If the graph is undirected (see figure 1), the adjacency relation is symmetric, or **E** = {{u,v} | u, v ∈ **V**} (sets of vertices rather than ordered pairs). In a directed graph (cf. figure 2), the edges are ordered, the edge (u,v) is different from the edge (v,u). When an edge e=(u,v) connects the vertices u and v, these two vertices are adjacent, and called the extremities of the edge e. Two edges are called adjacent if they have in common one vertex. A weighted graph is a graph having a weight, or a number, associated with each edge. Some algorithms require all weights to be non-negative, integral, positive, etc.

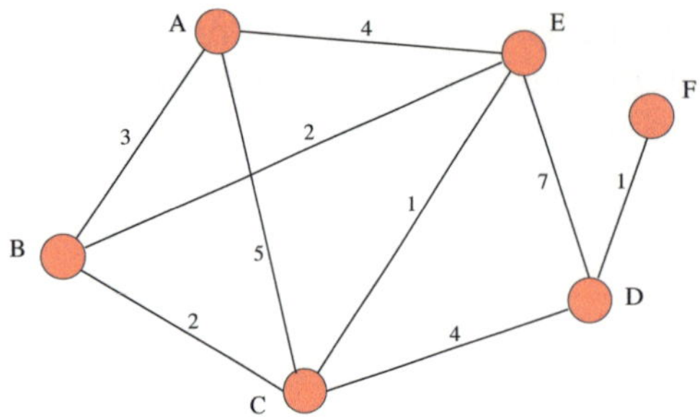

Fig. 1. Weighted Undirected Graph

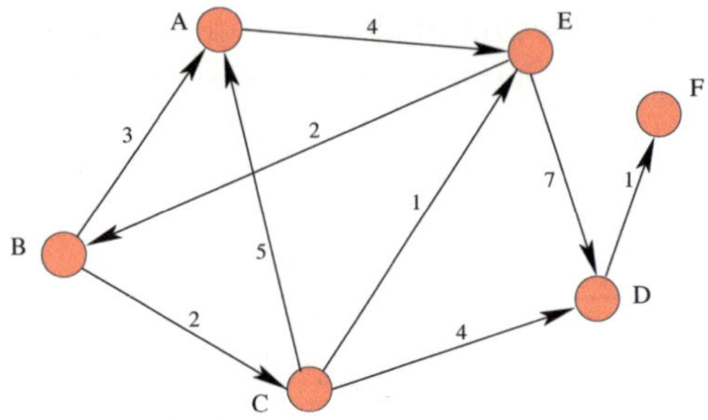

Fig. 2. Weighted Directed Graph

3.2 Graph representation

A graph can be represented by two data structures: adjacency-matrix or adjacency-list.

Adjacency-matrix Representation:

A directed graph with v vertices can be represented by v x v matrix, where the entry at (i,j) is 1 if there is an edge from vertex i to vertex j; otherwise the entry is 0. A weighted graph may be represented using the weight as entry. An undirected graph may be represented using the same entry in both (i,j) and (j,i) or using an upper triangular matrix. Figure 3 illustrates a representation of the directed graph of the figure 2 by an adjacency-matrix.

Adjacency-list Representation:

A directed graph with v vertices can be represented using a set of v lists of vertices. List i contains vertex j if there is an edge from vertex i to vertex j. A weighted graph may be represented with a list of (vertex, weight) pairs. An undirected graph may be represented by having vertex j in the list for vertex i, and vertex i in the list for vertex j. Figure 4 illustrates a representation of the directed graph of the figure 2 by an Adjacency-list. The arrow (->) means a link in a list. In general, the adjacency-list representation is more compact for a sparse graph.

	A	B	C	D	E	F
A	0	0	0	0	1	0
B	1	0	1	0	0	0
C	1	0	0	1	1	0
D	0	0	0	0	0	1
E	0	1	0	1	0	0
F	0	0	0	0	0	0

Fig. 3. Adjacency-Matrix Representation

A → E

B → A → C

C → A → D → E

D → F

E → B → D

F

Fig. 4. Adjacency-List Representation

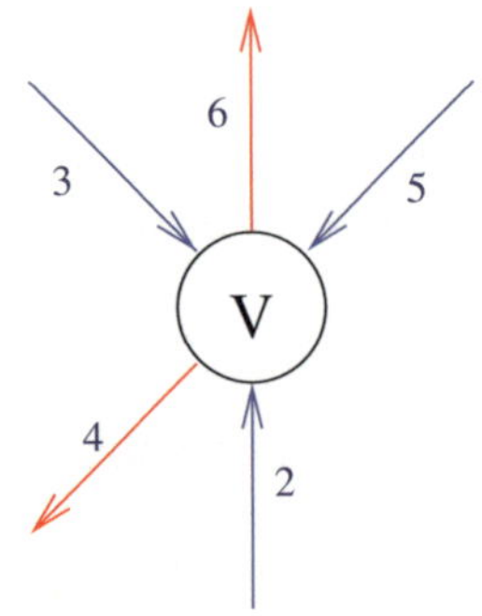

Fig. 5. Entering edges (blue) and leaving edges (red).

3.3 Graph cut definition

Let **G** = (**V**,**E**) denotes a graph, with **V** the set of vertices and **E** the set of edges. A cut is formed by partitioning the nodes of a graph into two mutually exclusive sets **S** and **T**. An edge between a node in one set and a node in the other is said to be crossing the cut and called a **cut edge**. In weighted graphs, the weight, or the capacity, of a cut is the sum of weights of the edges that have exactly one endpoint in each of the two sets. The problem of finding the minimum weight cut in a graph is called Minimum Cut Problem.

3.4 Minimum s-t cut problem

In a directed graph **G** = (**V**,**E**), the edges having the same beginning vertex v are called leaving edges of v, and the sum of their capacities (weights) is called leaving degree of vertex v. The edges having the same arriving vertex v are called entering edges, and the sum of their capacities (weights) is called entering degree of vertex v. See figure 5.

$$\text{entering degree of v} = \sum_{(u,v)\in E} c(u,v) \tag{1}$$

$$\text{eaving degree of v} = \sum_{(v,w)\in E} c(v,w) \tag{2}$$

We call a **source** s a vertex in a directed graph with entering degree equal to zero, and a **sink** t is a vertex with zero leaving degree. The s-t cut is the problem of finding a cut in the graph to separate the source s from the sink t. A minimum s-t cut is a cut with a minimum weight separating the vertices into two sets, the source s will be in one of them and the sink will be in the other.

3.5 Flows in graph

It is frequently common to introduce flows in a graph by the example of water conduit in pipe networks.

Water flow in networks:

Suppose we have a water source, a sink and a network of pipes relating them. We accept that the source can give (and the sink can receive) an unlimited amount of water. Find the maximum flow is the research for the maximum amount of flow capable to pass from the source into the sink via the network. Supposing the unlimited capacity for the source and the sink, the problem is only constrained by the capacity of the network. The capacity of each pipe is proportional to its diameter. In this network, there will be obstructions in some pipes. To pass from to the source to the sink, water must absolutely take one of these obstructed pipes. In an intuitive case, if all the pipes are full, the flow is equal to the sum of their capacities. In particularly, the obstruction corresponds to a set of pipes, which separates the source from the sink, and the sum of its capacities is equal to the maximum flow capable to pass from the source to the sink. Once the flow is maximal, we are in the situation where all the pipes of obstruction are full (saturated).

Mathematical definition:

A flow in a weighted graph **G**=(**V**,**E**), where each edge e=(i,j) has a real positive capacity c(e), is a map Φ between the edges set **E** and the set of real numbers.

A flow $\Phi : VxV \rightarrow R$

- $\Phi(i,j) = -\Phi(i,j)$
- $\sum_{i\in V-\{s,t\}} \Phi(i,j) = 0$: Flow conservation: what goes into a vertex is equal to what goes out of it (except for the source and the sink)
- $\Phi(i,j) \leq c(i,j)$ for all $(i,j) \in E$: Flow in an edge is less or equal to its capacity

- c(i,j) = 0 for all $(i,j) \notin E$

An edge is saturated if the flow in it is equal to its non-zero capacity:

$$e=(i,j) \text{ is saturated } \Longleftrightarrow (c(i,j) = \Phi(i,j) \text{ and } c(i,j)>0)$$

3.6 Residual graph

Let Φ be a flow in the graph **G=(V,E)**. The residual graph $\mathbf{G_r} = (\mathbf{V}, \mathbf{E_r})$ is constructed in the following way:

$\forall\ (i,j) \in \mathbf{E}$:

$$\text{If } \Phi(i,j) < c(I,j) \text{ then: } (i,j) \in \mathbf{E_r} \text{ and } c_r(i,j) = c(i,j) - \Phi(i,j)$$

$$\text{If } \Phi(i,j) \geq 0 \text{ then: } (j,i) \in \mathbf{E_r} \text{ and } c_r(j,i) = \Phi(i,j)$$

Thus, for each edge in the graph **G** with non-zero capacity c(i,j) and with a flow Φ (i,j), there are two edges in the residual graph $\mathbf{G_r}$ with capacities $c_r(i,j) = c(i,j) - \Phi(i,j)$ and $c_r(j,i) = \Phi(i,j)$. Figure 6 represents a maximum flow in a graph, and the construction of the residual graph illustrated in the figure 7.

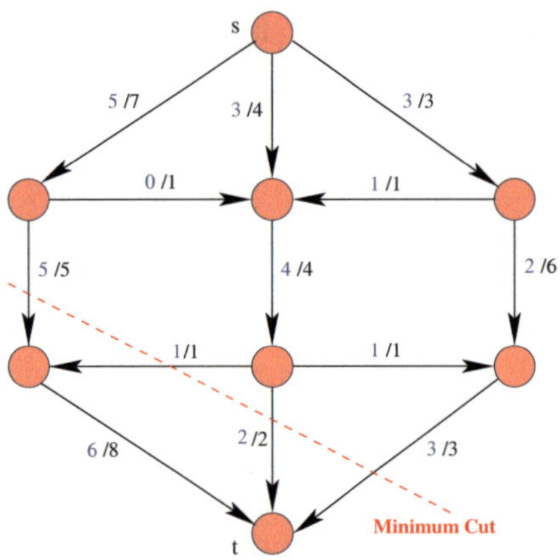

Fig. 6. Minimum Cut in a graph

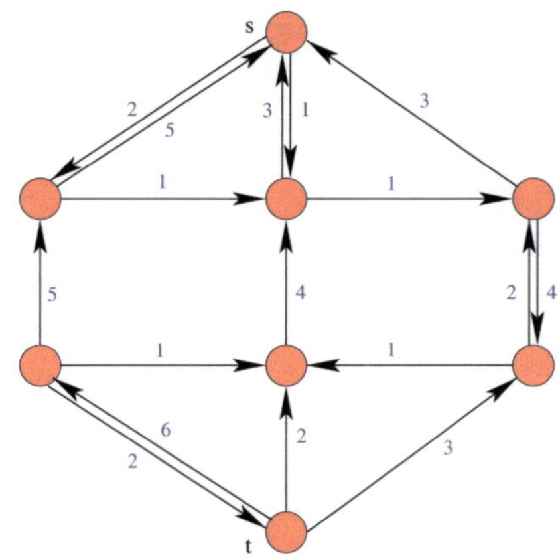

Fig. 7. Residual Graph

Augmenting Path:

An Augmenting Path in the residual graph $\mathbf{G_r}$ is a path from the source s to the sink t composed of residual edges with non-zero residual capacities.

3.7 Solutions of the minimum s-t cuts problem

The minimum s-t cut problem is traditionally resolved using the maximum flow algorithms in networks. Let s be the source and t the sink in a graph **G**, the maximum flow – minimum cut theorem says:

$$\text{Max Flow }(s,t) = \text{ Min Cut }(s,t)$$

This theorem says that the maximum value of a flow in a graph is equal to the minimum value of an s-t cut. (Ford & Fulkerson, 1962) have proved that the flow from the source s to the sink t will cause the saturation of a set of edges dividing the vertices into two sets **S** and **T**, with s in **S** and t in **T**. In other words, the theorem states that the maximum flow in a network is

dictated by its bottleneck. Between any two nodes, the quantity of material flowing from one to the other cannot be greater than the weakest set of links somewhere between the two nodes. This theorem links the value of the maximum flow with the value of the minimum cut, but it does not specify any relation between the minimum cut value and the set of cut edges. However, once we have a maximum flow, we can find the set of edges forming the minimum s-t cut. The algorithm 1 illustrates a procedure to obtain the minimum s-t cut.
The figure 6 illustrates a minimum cut in a weighted directed graph. The weights of edges are in black, and the flow values are in blue. The dashed red line crosses the edges of the minimum cut. In this example, we notice that the value of the maximum flow is 11, and the weight of the minimum cut is 11.

1. Resolve the Maximum s-t Flow problem.
2. Find the set of vertices **S** reachable from the source s in the residual graph.
3. Find the set of vertices **T** non-reachable from the source s in the residual graph.
4. Consider the edges starting from a vertex in **S** and ending in a vertex in **T**.
   The set of these edges forms the minimum s-t cut.

Algorithm 1. An algorithm of minimum s-t cut.

3.8 Maximum flow algorithm

We find in the literatures two approaches for solving the maximum flow problem (Cormen et al., 2001). The first is the augmented path algorithm proposed by (Ford & Fulkerson, 1962), and the second is the Push-Relabel algorithm.

Augmented path algorithm:

(Ford & Fulkerson, 1962) have developed a solution of the maximum flow problem based on the augmented path. In fact, if there is an augmented path in the residual graph $\mathbf{G_r}$, then the flow is not maximal. The algorithm commences by finding an augmenting path, next, it adds the capacity of this path to the flow, and then updates the residual graph $\mathbf{G_r}$. This operation is repeated until there is no augmented path.

```
G=(V,E) : a graph, V is the set of vertices, and E is the set of edges.
          s is the source, t is the sink
∀ (u,v) ∈ E :
     Φ(u,v) ← 0
     Φ(v,u) ← 0
Construct the residual graph Gr
while (there is an augmenting path p in the residual graph Gr between s and t)
begin
     cΦ(p) = min{ cr(u,v) : (u,v) ∈ p}  : the capacity of the augmented path.
     for each (u,v) ∈ p
     begin
          Φ(u,v) ← Φ(u,v) + cΦ(p)
          Φ(v,u) ← Φ(u,v)
     end
     Update the residual graph Gr
end
```

Algorithm 2. Maximum Flow: the augmented path algorithm.

Push-Relabel algorithm:

The idea of the Push-Relabel algorithm can be clarified by the following example. Suppose that the graph can be represented as follow: the source s is on the top and the sink t is at the height zero (bottom). We send water from the source down to the sink. In each vertex, water in excess will flow to the vertices at smaller height, constrained with the capacity of the edge relating the two vertices. After each operation, water will be in a lower vertex, and the excess is reduced in the upper vertex. We track water until arriving to the sink t. The algorithm stops when there is no excess in any vertex.

The **Height** of a vertex: We observe that the longest path from the source to the sink contains at maximum V vertices. Thus, we can assign a height to each vertex in the following way:

- Height(s) = V
- Height (t) = 0
- $\Phi(u,v)>0$ ==> Height (u) > Height (v)

In other words, the heights of nodes (except s and t) are adjusted, and flow is sent between vertices, until all possible flow has reached t. Then we continue increasing the height of internal nodes until all the flow that went into the network, but did not reach t, has flowed back into s. In order to introduce the notion of excess, we define first the preflow.

Preflow Definition:

A preflow is similar to a flow with the exception that the total amount which can flow into a vertex can be bigger than the flow out of the vertex (the principle of flow conservation is no longer respected). The notion of preflow is introduced by (Karzanov, 1974). Let **G=(V,E)**be a graph. A preflow is a map verifying:

Un preflow $P : VxV \rightarrow R^{+}$:

- $\forall (u,v) \in$ **E** $: P(u,v) \leq c(u,v)$
- $\forall (u,v) \in$ **E** $: P(v,u) = -P(u,v)$
- $\forall u \in V\text{-}\{s\} : \sum_{(w,u)\in E} P(w,u) - \sum_{(w,u)\in E} P(u,w) \geq 0$

Let $\epsilon(u) = P(V,u)$, $\epsilon(u)$ is called the excess of vertex u, which is the difference between the in and out flows of the vertex u. A vertex $u \in V\text{-}\{s,t\}$ is called overflowed if $\epsilon(u) > 0$. The algorithm 3 gives a general implementation of the Push-Relabel algorithm, which uses the Push and the Relabel operations defined below (Goldberg & Tarjan, 1988).

```
G=(V,E) : a graph, V is the set of vertices, and E is the set of edges.
          s is the source, t is the sink
while (there is a vertex v having an excess (є(v) > 0))
begin
     Select the vertex v.
     do a Push operation (Push(v))
     or do a legal Relabel operation (Ralabel(v))
end
```

Algorithm 3. Maximum Flow: General algorithm of Push-Relabel.

Push operation:

A push from u to v means sending a part of the excess flow from u into v. Three conditions must be met for a push to take place:

- $\epsilon(u) > 0$: there is an excess in the vertex u, which means that the flow into this vertex is bigger than the flow out of it.
- $c(u,v) - \Phi(u,v) > 0$: the edge (u,v) is not saturated.
- Height(u) > Height(v). The vertex u is higher than the vertex v.

When these conditions are verified, we send a flow equal to $\min(\epsilon(u), c(u,v) - \Phi(u,v))$.

Relabel operation:

Relabelling a vertex u means increasing its height until it is higher than at least one of the vertices it has available capacity to them. Conditions for a relabel:

- $\epsilon(u) > 0$: There must be an excess in the vertex in relabelling.
- Height(u) ≤ Height(v) for all v such that $c(u,v) - \Phi(u,v) > 0$. The only vertices we have available capacity to are higher.

When we re-label a vertex u, we adjust its height to be equal to the smallest value verifying Height(u) > Height(v) for all the vertices v having $c(u,v) - \Phi(u,v) > 0$.

Complexity of the maximum flow algorithms:

- The complexity of the augmenting path algorithm is O(e*MaxFlow) (Ford & Fulkerson, 1962), where e is the number of edges in the graph, and MaxFlow is the value of the maximum flow.
- The Push-Relabel algorithm is one of the most efficient algorithms to compute a maximum flow. The general algorithm complexity is $O(v^2e)$ (Cormen et al. 2001), where v is the number of vertices in **V** and e is the number of edges in **E**. An implementation with FIFO vertex has a complexity of $O(v^3)$ (Cormen et al. 2001). The theoretical complexity of the algorithm proposed by Goldberg is $O(v\, e \log(v^2/e))$ (Goldberg & Tarjan, 1988).

In our work, we use an implementation of Push-Relabel algorithm proposed by (Goldberg, 1985), which is included in *Boost Graph Library*[1].

Note that (Boykov & Kolmogorov, 2004) have proposed a new algorithm, and they have proved that for typical graphs in vision, their algorithm is about 2 to 5 times faster than the other algorithms including the approach Push-Relabel. Their algorithm belongs to the group of algorithms based on augmenting path. The algorithm starts by building search trees for detecting augmenting paths. In fact, the good performance is due to the reuse of these trees and never restarts building them from scratch.

4. Stereo matching using graph cuts

4.1 Labelling problem

Many problems in vision can be formulated as a labelling problem. In such a problem, we have a set of sites and a set of labels. Sites represent features extracted from the image (pixels, segments, etc.), for which we want to estimate some quantity. Labels represent the quantities associated to these sites: intensity, disparity, region number, etc. Let P={1,2,...,n} be a set of n sites, and $L=\{l_1, \ldots, l_k\}$ be a set of k labels. Labelling is a map from P into L:

$$f : P \dashrightarrow L : s_p \dashrightarrow f_p = f(s_p) = l_i \quad (3)$$

---

[1] http://www.boost.org/libs/graph/doc/index.html

Thus, $f(P) = \{f_1,\ldots, f_n\}$. We assign an energy function to the labelling map: a general form of energy function can be written:

$$E(f)=E_{data}(f)+\lambda E_{prior}(f) \tag{4}$$

The first term $E_{data}(f)$ represents the intrinsic data energy, which translates the constraints of associating labels to data. The second term $E_{prior}(f)$ aggregates the extrinsic energies (*prior energy*) which translate the constraints defined by the prior information. The constant λ can control the relative importance of the two terms; bigger values of λ give more importance to prior information.

The energy function $E_{data}(f)$ associates an important cost for association data/label, which are less pertinent.

$$E_{data}(f) = \sum_{p \in P} D_p(f_p) \tag{5}$$

Where $D_p(f_p) \geq 0$ measures the cost of assigning the label f_p to the site p.

The prior energy function $E_{prior}(f)$ will assign an important cost to the associations f_p not compatibles with prior information. The choice of this function depends on the type of the problem, but in general, a type of prior energy expresses the smoothing constraint. This constraint is very famous in computer vision, and it is well adapted when the estimated quality changes slowly everywhere or nearly everywhere: in 3D, this corresponds to the hypothesis that the world is continuous piecewise. This hypothesis is considered by taking prior information of smoothing type E_{smooth}.

To formulate the smoothing energy, we need to model how the pixels interact between them: often it is sufficient to express how a pixel interacts with its neighbours. Let N_p be the set of neighbour pixels to the pixel p, and N be the set of neighbour pairs {p,q}: N is named neighbourhood system. Smoothing energy can be written:

$$E_{smooth}(f) = \sum_{\{p,q\} \in N} V_{\{p,q\}}(f_p, f_q) \tag{6}$$

Where $V_{\{p,q\}}(f_p,f_q)$ is the neighbourhood interaction function: this function attributes high penalties to the pair {p,q} if the pixels p and q have different labels. The form of $V_{\{p,q\}}(f_p,f_q)$ determines the type of prior smoothing. With these notations, the global smoothing energy is the sum of neighbourhood interaction functions of all neighbour pixels. Often we choose $V_{\{p,q\}}(f_p,f_q)$ as:

$$V_{\{p,q\}}(f_p,f_q)=u_{\{p,q\}} V(f_p,f_q) \quad \text{with } u_{\{p,q\}} \in R^+ \tag{7}$$

Where V is a homogeneous potential, and $u_{\{p,q\}}$ is a multiplying term depending on the pair of pixels. The figure 8 shows a linear potential, $V(f_p,f_q)= | f_p - f_q |$.

In stereo vision, the term $u_{\{p,q\}}$ is often a decreasing function of the norm of the gradient between the sites p and q, which favours the coincidence of disparity discontinuities with the contours of the reference image. This choice is translated by the map $u_{\{p,q\}} : u_{\{p,q\}} = U(|I_p - I_q|)$. The term $u_{\{p,q\}}$ represents the penalty of assigning different disparities for neighbour pixels p and q. The value of this penalty must be small for a pair {p,q} which is on a contour, and therefore with a large difference of intensity $|I_p - I_q|$. In practice, we use a decreasing empiric function U(.). The figure 9 illustrates the function $u_{\{p,q\}}$.

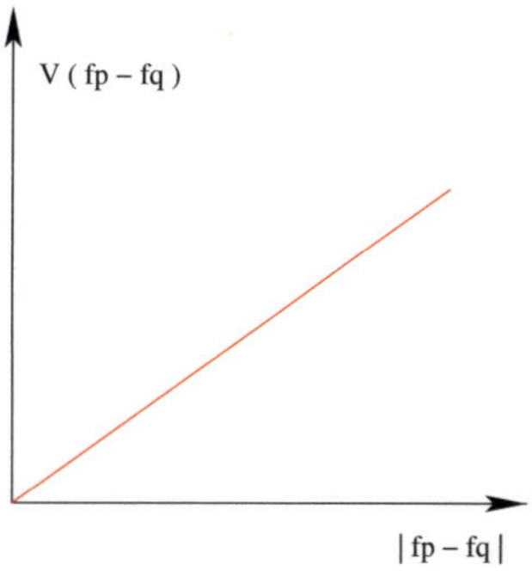

Fig. 8. Function of linear potential

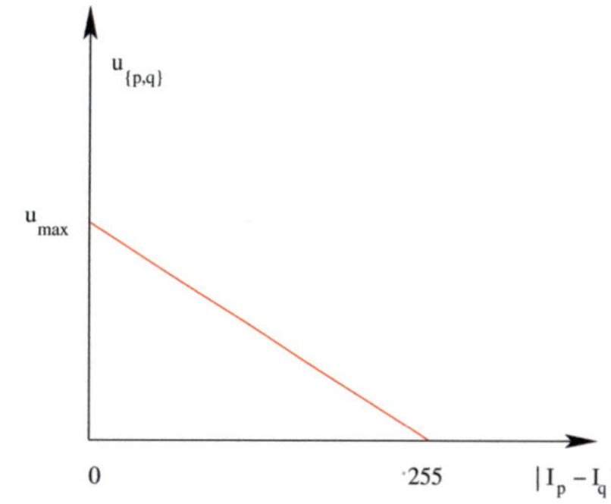

Fig. 9. The function $u_{\{p,q\}}$.

4.2 Energy and graph

The principle of graph cuts methods is to create a graph in such a way a cut represents a function and the weight (cost) of this cut represents the energy value of this function. The value of the minimum cut will be equal to the minimum of the energy function.
Starting from the general formulation of labelling problem, let us consider the linear potential map:

$$V_{\{p,q\}}(f_p,f_q) = u_{\{p,q\}} \; | f_p - f_q | \tag{8}$$

This gives the global energy:

$$E(f) = \sum_{p \in P} D_p(f_p) + \lambda \sum_{\{p,q\} \in N} u_{\{p,q\}} |f_p - f_q| \tag{9}$$

We will show how to construct a graph associated to an energy function, and then how to obtain the minimum value of this energy function by finding the minimum cut of the graph. This method is due to (Roy & Cox, 1998), but it profits of reformulations proposed by (Veksler, 1999) who put in evidence the energy to be minimized. Our description of the graph will be based on this formulation. First, we will explain the construction of a full graph, and then we explain our contribution in reducing its size to accelerate the algorithm.

4.3 Building a complete graph

We aim to construct a graph **G=(V,E)**, where the set of vertices **V** has two special vertices: a source s and a sink t, and in which a minimum cut will represent a minimization of the stereo matching cost. The graph is constructed as follow:

- Let k be the number of possible matches (in stereo vision, k is given by the range of disparity, which is related to the minimum and maximum depths in the scene). For example, for a disparity range [0,40] , the value of k is 41.
- To each pixel p in the image, we associate a chain composed of k-1 nodes (sites), noted p_1, p_2, …, p_{k-1}. These nodes are connected by edges called *t-links* or disparity edges, and noted $\{t_1^p, t_2^p, \ldots, t_k^p\}$, where $t_1^p = (s, p_1)$, $t_j^p = (p_{j-1}, p_j)$, and $t_k^p = (p_{k-1}, t)$. The t-links of a pixel are a set of edges linking the source to the first node, and then link this node to the next one, until linking the last node in the chain to the sink. For a disparity range of k values, and for each pixel p, we will have k-1 nodes in the chain and k-2 t-

links between these nodes, and two other t-links, the first with the source and the other with the sink. Thus, we have in total k t-links. The capacity (weight) of each of these t-links is equal to the cost of matching of the pixel to a corresponding one in the other image with a disparity equal to the zero-index of the t-link in the chain.

- The capacity of the t-link number j is set to $K_p+D_p(l_j)$, with $D_p(l_j)$ is the cost of matching of the pixel p with a disparity l_j, and K_p is a constant satisfying the constraint 10.

$$K_p > (k-1) \sum_{q \in N_p} u_{\{p,q\}} \tag{10}$$

In fact, the constant Kp is chosen to be bigger than the sum of capacities of penalty edges (n-links, cf. below) which have one extremity in the chain of nodes associated to the pixel p.

- To each pair of neighbour pixels p and q, we link the corresponding chains with edges called *n-links* or penalty edges. The n-link edge at the level j $\in \{1, 2 \ldots k\text{-}1\}$, is noted $\{p_j,q_j\}$, has a capacity equals to $u_{\{p,q\}}$. The figure 10 illustrates the nodes related to one pixel, the t-links (which are coloured in blue) and the n-links (in green).

The capacity of an s-t cut of a graph is the sum of capacities of all cut edges. Due to the construction method of this graph, a cut capacity will be composed of two parts: the first is the sum of capacities of the cut t-link edges, and the second is the sum of capacities of cut n-link edges. In fact, the addition of the constant K_p allows us to assure the uniqueness of the cut of each t-link chain, see (Roy & Cox, 1998) for a proof.
Another method can assure the uniqueness of cutting the chain of t-links in exactly one edge is proposed by (Ishikawa, 2000). Their graph is very similar but it is oriented, and they add to the chain of pixel p another inverse chain with infinite capacity.

$$E(cut) = \sum_{p \in Image} (K_p + D_p(f_p)) + \lambda \sum_{\{p,q\} \in N} u_{\{p,q\}} | f_p - f_q | \tag{11}$$

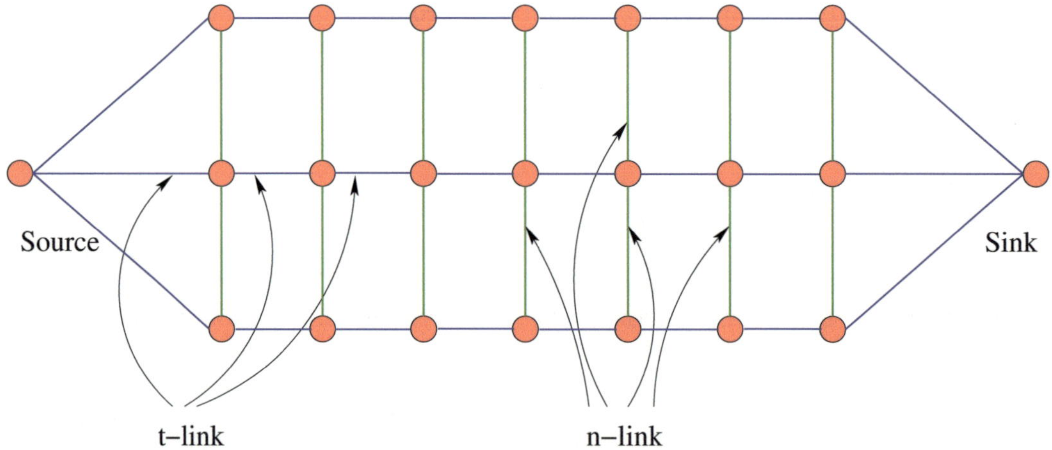

Fig. 10. The t-links and the n-links.

A graph cut consists in splitting the graph into tow parts. The cut t-link edges form the surface of searched depth, and this can allow the assignment of a disparity to each pixel. The problem of graph cut can be solved using the maximum flow (as seen before). To determine the disparity of a pixel p, the minimum cut will pass through one (and only one) t-link of the chain associated to the pixel p. The index (zero-based) of this edge in the chain allows finding the disparity of this pixel. Hence, for a disparity range $[d_{min}, d_{max}]$, a t-link of index j corresponds to a disparity of $(d_{min} + j)$.

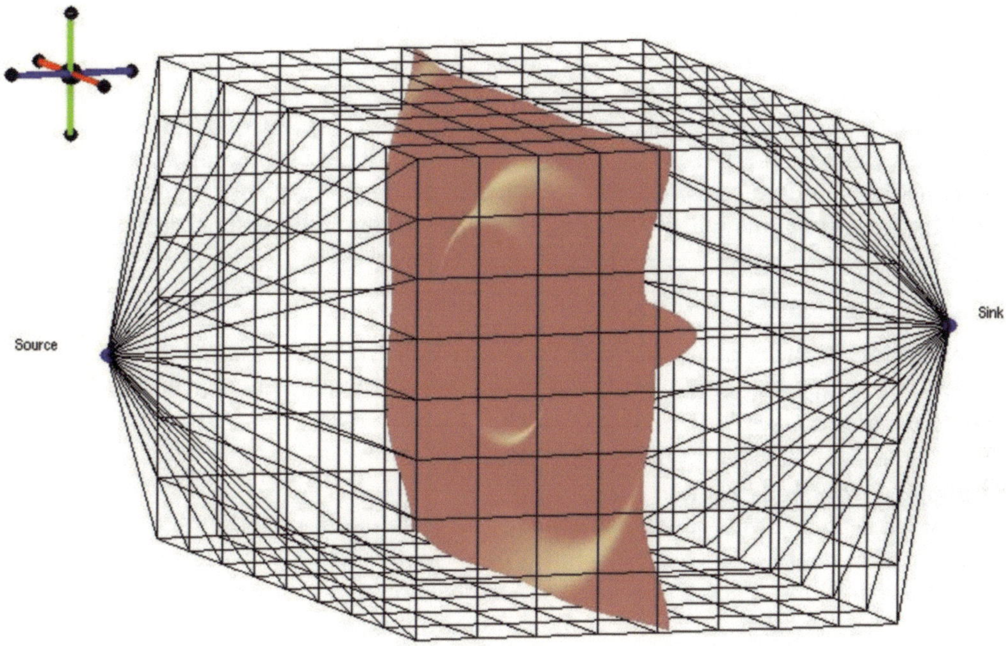

Fig. 11. The disparity surface correspondent to a minimum cut of the graph, as proposed by (Roy & Cox, 1998)

This method enables us to attribute a disparity to each pixel in the image. The disparity surface can be visualized as a surface traversing all the cut t-links in the graph. The figure 11 illustrates a minimum cut and the corresponding disparity surface. We note that this method allow us to obtain integer disparity only, and cannot deal with real disparities (sub-pixel). In fact, the construction of the complete graph requires that the zero-index of a t-link in the chain corresponds to a value in the disparity range $[d_{min}, d_{max}]$, and in consequence, non-integer (real) values of disparity are not acceptable by this method. We will see in the next section (by the reduced graph) how to overcome this problem.

Requirements of this method:

We will calculate the total number of vertices and edges in the complete graph in order to put in evidence two constraints: the constraint of needed memory, and the constraint of execution time.

Let W and H be the width and the height of the image, $[0,d_{max}]$ the disparity range. Let v the vertices number in the graph, and e the number of edges. For each pixel, the number of possible matches is $k=d_{max}+1$. For each pixel, the corresponding chain has $k-1=d_{max}$ nodes. Thus, the total number of nodes in the graph is (the term +2 is added for the source and the sink):

$$v = W\ H\ d_{max} + 2 \quad (12)$$

In each chain (except at the boundaries), each node is linked with its 6 neighbours by edges. The number of edges in the graph is:

$$e = 6\,W\ \,H\,d_{max} - 2\,d_{max}\,(\,W + H) \tag{13}$$

The term 2 d_{max} (W + H) is added because there are fewer edges at the boundaries of the graph. To have an idea about the total number of vertices and edges in the complete graph, see the table 1, which is calculated for an image of size 512x512 pixels and for different disparity ranges. In this table, and for a disparity range [0, 40], there are more than 10×10^6 vertices and more than 60×10^6 edges. This example illustrates the necessity to have huge memory (RAM). Indeed, because the complexity of maximum flow algorithm (as seen before), the execution time will be very big, especially for large disparity ranges.

	$d_{max} = 30$	$d_{max} = 40$	$d_{max} = 50$
Vertices Number v	7 864 320	10 485 760	13 107 200
Edges Number e	47 124 480	62 832 640	78 540 800

Table 1. Vertices and edges numbers in the complete graph for an image of size 512x512, for different disparity ranges.

4.4 Building a reduced graph

The major problems of the complete graph method presented in the previous section are its greediness to memory and execution time. To overcome these problems, we propose to construct a reduced graph. This reduced graph will contain a reduced number of vertices and edges. This allows us to decrease the amount of needed memory, and to lighten the execution time (Zureiki et al., 2007).

In the reduced graph: for each pixel in the image, we keep only some potential disparity values, resulting from a local method of stereo matching. However, in the method of the complete graph, we keep for each pixel all possible values of disparity (as seen in previous section).

The construction steps of the reduced graph are similar to the complete graph, with some differences:

- By mean of a local matching method (based on local similarity measurement, as SAD for example), we calculate for each left pixel p, the matching costs for all possible values in the disparity range [d_{min}, d_{max}].
- Then, we choose the N best values (for illustration purposes, we choose N=4 without lack of generality). The choice may be done according to different criteria, for example, with a classic ZNCC score; we keep the disparity values around the peak, or the N best local maxima (if they exist), etc. We will note our N selected disparities for the pixel p as {$d_{1,p}$, $d_{2,p}$, ... $d_{N,p}$}, and the corresponding costs of matching as $\{ D_{p,d_{1,p}} ,\ D_{p,d_{2,p}} ,\ \dots ,\ D_{p,d_{N,p}} \}$.
- To reduce the size of the graph, for each pixel p of the image, instead of keeping a chain of k-1 nodes and k edges, we remove all the nodes and edges except N-1 nodes (and thus, N edges).
- Thus, to each pixel p we associate a novel chain of nodes $\{ p_{L_1} ,\ p_{L_2} ,\ \dots,\ p_{L_{N-1}} \}$. These nodes are connected by edges (t-links) noted $\{ t_1^p ,\ t_2^p ,\dots,\ t_N^p \}$ with $\{ t_1^p = (s, p_{L_1}) ,$ $t_2^p = (p_{L_1}, p_{L_2}) ,\ \dots ,\ t_N^p = (p_{L_{N-1}}, t) \}$.

- The capacity of the t-link edge i is $C + D_{p,d_{i,p}}$, where C is a constant satisfying the constraint 14. The addition of this constant assures the uniqueness of the cut of the t-links chain in exactly one edge.

$$C > N * \max_{\{p,q\} \in N} (u_{\{p,q\}}) * | d_{\max} - d_{\min} | \tag{14}$$

- For each adjacent pixels p and q, the corresponding chains are related by edges (n-links) at the N-1 levels, with a capacity as in equation 15.

$$\text{n-link capacity at level i} = u_{\{p,q\}} (| d_{i,p} - d_{i,q} | +1) \tag{15}$$

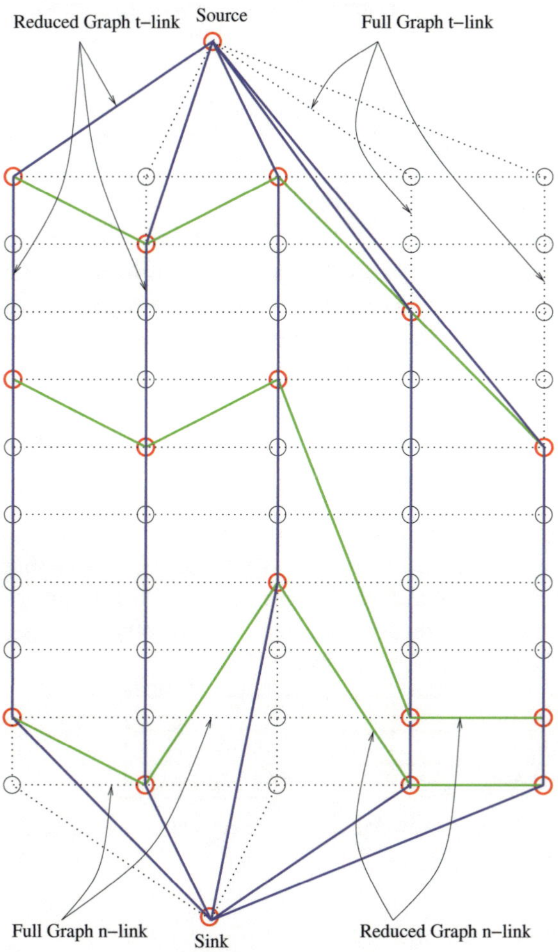

Fig. 12. Construction of the Reduced Graph.

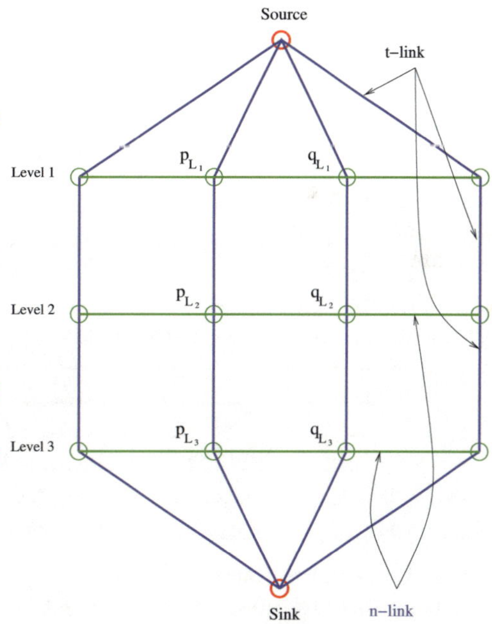

Fig. 13. The Reduced Graph.

The figure 12 illustrates the construction of the reduced graph. The dashed edges are for the complete graph (previous section). The not removed nodes are in red, the new t-links are in blue and the new n-links are in green. The figure 13 illustrates a frontal projection of the reduced graph.

Minimized Energy by the Reduced Graph:

We have seen that the minimized energy by the complete graph contains two terms. The first term represents the intrinsic data energy, which translates the costs of attributing disparities to pixels of the image. The second term aggregates the constraint of continuity (adjacent pixels have similar disparities). The constant λ can control the relative importance of the two terms. Hence, the prior energy appears in the weights associated to the n-link edges in the complete graph. In our reduced graph, we do distinguish between two types of prior information, the first translates the information acquired by the local method and acts in the choice of the selected nodes, while the second (smoothing) interferes in the penalties associated to the n-link edges. Thereby, we exploit the prior knowledge that the disparity of a pixel p has only N possible values (the most probable) in a new way. In fact, we consider that removing non-potential nodes as a novel form of representing this prior knowledge. The minimized energy can be written as:

$$E(f) = \sum_{\substack{p \in P \\ f_p \in \{d_{p,1}, \ldots, d_{p,N}\}}} D_p(f_p) + \lambda \sum_{\substack{\{p,q\} \in N \\ f_p \in \{d_{p,1}, \ldots, d_{p,N}\} \\ f_q \in \{d_{q,1}, \ldots, d_{q,N}\}}} u_{\{p,q\}} \, |f_p - f_q| \tag{16}$$

Requirements of this method:

We will calculate the total numbers of vertices and edges in the reduced graph, using the same notation as previously for W, H, d_{max}, k, etc. Let N be the number of chosen disparities among all the possible values in the disparity range. We can obtain the number of vertices and edges by replacing d_{max} by (N-1) in the equation for the complete graph:

$$v = W\,H\,(N-1) + 2 \tag{17}$$

$$e = 6\,W\,H\,(N-1) - 2\,(N-1)\,(W + H) \tag{18}$$

The table 2 gives the numbers of vertices and edges in the reduced graph for an image of size 512x512, and for any value of disparity range, while using N=4 or N=5. We clarify that the disparity range does not any more influence the size of the graph, but the number of pre-selection N. For N=4, we obtain less than 0.8×10^6 vertices and less than 0.160×10^6 edges. This example illustrates clearly the immense gain in needed memory (RAM).

	N=4	N=5
Vertices Number v	786 432	1 048 576
Edges Number e	153 600	1 564 672

Table 2. Vertices and edges numbers in the reduced graph for an image of size 512x512 with the number of pre-selected disparities N= 4 or 5.

Advantages of the Reduced Graph:

The benefits of reduced graph construction are:

- The graph becomes less voluminous. The vertices number is divided for example by 10 for a disparity range of width 40 and with the number of pre-selection N=4.
- The size of the graph is no more dependent of the disparity range. This allows us to use larger disparity ranges.
- The index of the edge in the t-link chain associated to a pixel p in the complete graph corresponds to a value in the disparity range. This prevents using non-integer

disparities. On the contrary, in the reduced graph, the value associated to the edge is added as an attribute, and there is not any constraint to be an integer or real value. Hence, we can use real value for disparity, and especially for sub-pixel disparities.

- Due to the size of the reduced graph in numbers of vertices and edges, and knowing that the maximum flow algorithms are proportional (of order 2 or 3) to these numbers, the execution time to find the minimum cut in the reduced graph is extremely smaller than the one for the complete graph. Table 3 gives some experimental results for execution time.

5. Implementations and results

In our work, images are acquired by a pre-calibrated stereo rig. In addition, our stereo matching algorithms considers that the left and right images are rectified, which means that epipolar lines are horizontal and have the same line index in the left and right images. Thus, the rectified images come from the pre-calibrated cameras are the input of the graph cuts algorithms. We suppose that the rectification is exact, which means that the disparity depends only on the column index of the pixel: the pixel (u,v) in the left image is matched to the pixel (u, v-d) in the right image. The sub-pixelic resolution is applied only to the value d. Graph constructions and processings, are designed using the Boost Graph Library (BGL)[2]. Boost is an open source library distributed under Boost Software License[3]. The BGL is a C++ library built by using the principles of generic programming for the construction of data structures and for the implementation of the different algorithms on graphs. For more details, see (Siek et al., 2001). Among the many algorithms for manipulating graphs, BGL makes available in particular an implementation of the algorithm Push-Relabel Maximum Flow proposed by (Goldberg, 1985). This algorithm has a complexity of $O(v^3)$, with v is the vertices number. In fact, this implementation is not the best one in performance, but it allows us to evaluate the execution time for both the complete and reduced graphs. Indeed, for the reduced graph, the influence of algorithm is already reduced because the graph is less voluminous.
Algorithms of construction and of minimum cut have been implemented first on the complete graph, as described in section 4.3. Next, these algorithms of construction and of minimum cut have been adapted for the reduced graph (section 4.4). To evaluate the performance of these methods, we have used the image *sawtooth* (Scharstein & Szeliski, 2002) illustrated on figure 14. For this image, the exact disparity (round truth) is given on figure 15. For the reduced graph, the selection of a limited set of disparities on every pixel is provided by a local matching method based on SAD with window size equals to 7 pixels.
The test is done on a P4 with 3GHz and 512MB of RAM. Table 3 gives the execution time for the image sawtooth. We note that for the size 434x380 and 20 levels of disparities, we could not test the method of the complete graph (memory explosion). On the contrary, with the reduced graph, we obtain the disparity image in less than one minute. Further, on, we tested the two methods with a half-size image (217x190), we can notice a factor of 40 between the complete and reduced graphs in execution time (150 seconds against 4 seconds). In addition, we remark that for the reduced graph, the pre-selection number N influence explicitly the

---

[2] http://www.boost.org/libs/graph/doc/index.html
[3] http://www.boost.org/LICENSE_1_0.txt

execution time for larger sizes of images, but its influence is nearly neglected for small images. Thus, this table put in evidence the enormous gain in execution time provided by the reduced graph approach. We notice that the complete graph cannot be manipulated on normal machines (with ordinary memory), whereas the reduced graph can be built and analyzed in acceptable time. Therefore, the reduced graph algorithm is faster, in spite of a supplementary phase of calculation of local costs (which can be neglected compared with the execution time of graph cut algorithm).

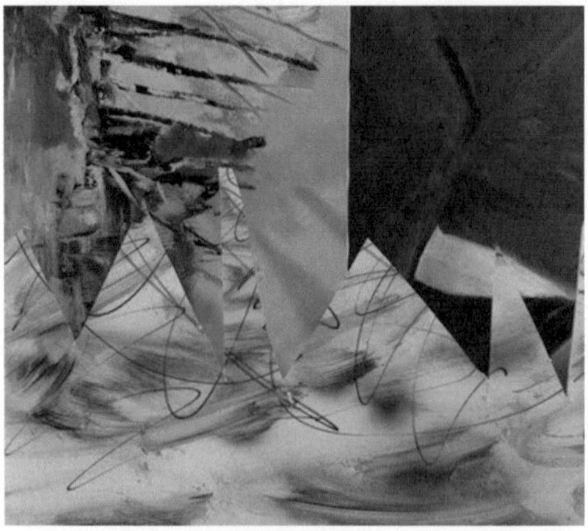

Fig. 14. The sawtooth image.

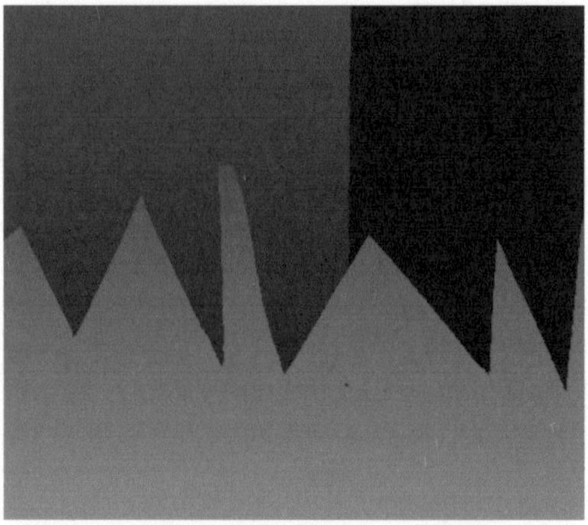

Fig. 15. The sawtooth image: true disparity.

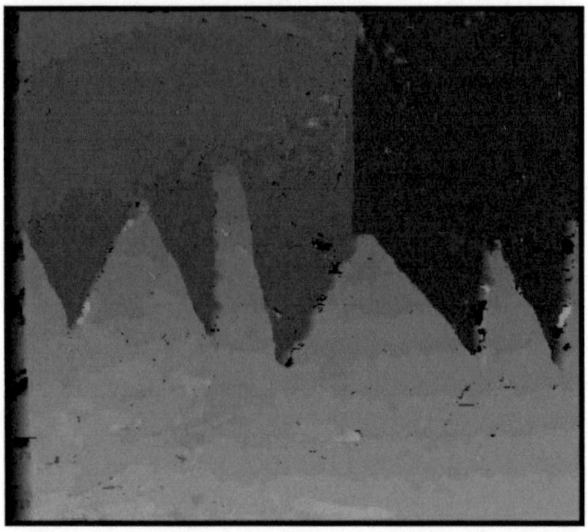

Fig. 16. Disparity image by reduced graph.

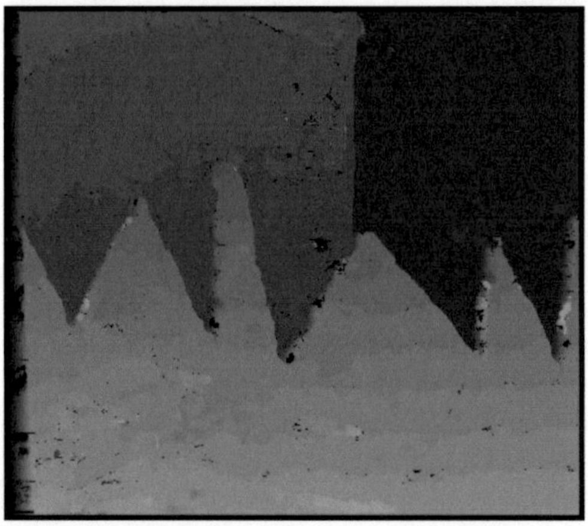

Fig. 17. Disparity image by complete graph.

Image Size	Complete Graph	Reduced Graph	
		N=4	N=5
434x380	NA	15 s	50 s
217x190	150 s	4 s	5 s

Table 3. Execution time in seconds for stereo matching using graph cuts in Complete graph and Reduced graph, for a disparity range [0, 20], and with the number of pre-selected disparities N= 4 or 5.

Figure 16 represents the disparity image obtained by the reduced graph approach, and figure 17 shows results by the complete graph. We can visually appreciate the quality of the obtained disparity images.

Fig. 18. The Flowerpots image.

Fig. 19. The Flowerpots image: true disparity.

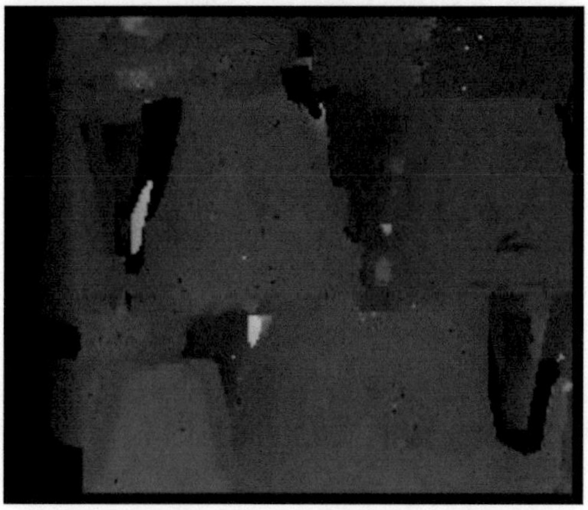

Fig. 20. Disparity image by reduced graph.

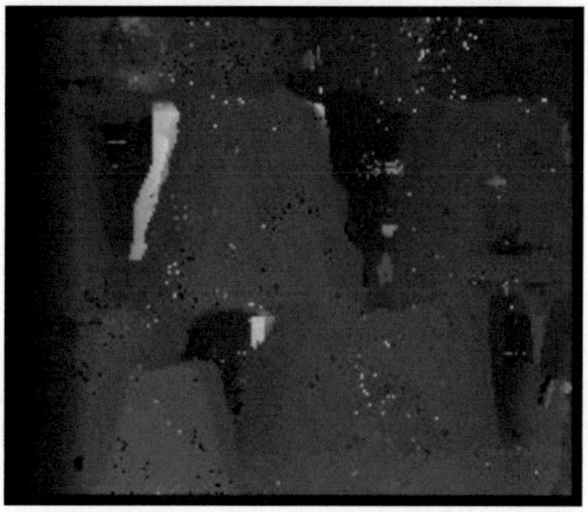

Fig. 21. Disparity image by SAD matching.

To illustrate the improvement of the reduced graph with respect to the local matching method, let us use the image *Flowerpots* [4] (Hirschmüller & Scharstein, 2007) shown in the figure 18. The true disparity of this image is given in figure 19. We used the SAD local method with a window size 7, this gives the disparity image of figure 21. When using our reduced graph method to calculate the disparity image, the result is shown in figure 20. We notice that many false matches in the local method were removed by the global method. In spite of these ameliorations, there still exist some errors in the disparity image. These errors (see figure 20), are in general in zones with very poor texture (or even mono coloured zones). In fact, these are among the most difficult problems for stereo matching.

Although the reduced graph matching method provides several ameliorations, it has some weak points, summarized in three main drawbacks:

[4] http://vision.middlebury.edu/stereo/

- The reduced graph method inherits some problems from the used local method. In fact, if the local method supplies N erroneous values of a pixel disparity, the resulting disparity for this pixel will be absolutely one of these erroneous values.By now, the global step cannot correct errors introduced by the local step.
- There are smoothing effects (like all global methods) due to the regularization. The smoothing will be more important when using bigger value for the penalty cost.
- In spite of its relative good execution time, the reduced graph method is not yet adapted to be used in real-time applications, like robotics. A parallel version of the maximum flow algorithm could be proposed to overcome this problem, taking advantage of a dedicated architecture.

6. Conclusion and perspectives

We described in this chapter, our evaluation of global stereo correspondence methods based on graph cuts. The combination of a local method, able to select a reduced set of possible matches for each pixel, and a global method, based on the graph cuts algorithm, let us to achieve two goals

- Sensibly ameliorate the quality of disparity images obtained only by a local method.
- Avoid the combinatorial explosion of the Graph Cuts method executed without preliminary reduction of the graph.

Some optimizations of our algorithm are currently studied. Note that we work always on pre-rectified images, and we produce an integer disparity image: we will study how to adapt this algorithm in order to find stereo matching on non-rectified images.

The reduced graph approach allows us to limit the execution time and to use larger disparity ranges. We write the code with the aim to obtain results and not to optimize the performance. In addition, we did not use the best maximum flow algorithm. In the next period, we could very easily improve performances, in order to satisfy as much as possible, real time constraints.

Section 4.4 has shown that the reduced graph is able to deal with sub-pixel disparities (real values). Testing this property and obtaining sub-pixel disparity images will be our main objective in our near future works.

7. References

Belhumeur, P. N. (1996). A bayesian approach to binocular stereopsis. *International Journal of Computer Vision*, 19(3):237-260.

Bellman, R. (1957). *Dynamic programming*, Princeton University Press.

Bobick, A. F. & Intille, S. S. (1999). Large occlusion stereo. *International Journal of Computer Vision (IJCV)*, 33(3):181-200.

Boykov, Y. & Kolmogorov, V. (2004). An experimental comparison of min-cut/max-flow algorithms for energy minimization in vision. *In IEEE Transactions on Pattern Analysis and Machine Intelligence*, pp. 1124-1137.

Boykov, Y., Veksler, O. & Zabih, R. (1998). Markov random fields with efficient approximations. *In Proceedings of the IEEE Computer Society Conference on Computer Vision and Pattern Recognition (CVPR)*, pp. 648-655.

Boykov, Y., Veksler, O. & Zabih, R. (1999). Fast approximate energy minimization via graph cuts. *In Proceedings of International Conference on Computer Vision (ICCV)*, volume 1, pp. 377-384.

Chen, Q. & Medioni, G. (1999). A volumetric stereo matching method : Application to image based modeling. *IEEE Computer Society Conference on Computer Vision and Pattern Recognition (CVPR)*.

Cormen, T. H., Leiserson, C. E., Rivest, R. L. & Stein, C. (2001). *Introduction to Algorithms, Second Edition*, The MIT Press.

Cox, I. J. (1992). Stereo without disparity gradient smoothing: a bayesian sensor fusion solution. *In proceedings of British Machine Vision Conference (BMVC)*, pp. 337-346.

Cox, I. J., Hingorani, S. L., Rao, S. B. & Maggs, B. M. (1996). A maximum likelihood stereo algorithm. *Computer Vision and Image Understanding*, 63(3):542-567.

Dahlhaus, E., Johnson, D. S., Papadimitriou, C. H., Seymour, P. D. & Yannakakis, M. (1994). The complexity of multiterminal cuts. *SIAM Journal on Computing*, 23(4):864-894.

Faugeras, O. (1993). *Three-Dimensional Computer Vision : a Geometric Viewpoint*, MIT press.

Faugeras, O., Vieville, T. et al. (1993). *Real-time correlation-based stereo: algorithm, implementations and applications*. Rapport technique RR-2013, INRIA-Sophia Antipolis.

Ford, L. & Fulkerson, D. (1962). *Flows in Networks*, Princeton University Press.

Goldberg, A. V. (1985). *A new max-flow algorithm*. Rapport technique MIT/LCS/TM-291, Massachusetts Institute of Technology, Cambridge, MA.

Goldberg, A. V. & Tarjan, R. E. (1988). A new approach to the maximum-flow problem. *Journal of the Association for Computing Machinery (JACM)*, 35(4):921-940.

Grandjean, P. & Lasserre, P. (1995). Stereo Vision Improvments. *In IEEE International Conference on Advanced Robotics*, Barcelona (Spain).

Hirschmüller, H. (2001). Improvements in real-time correlation-based stereo vision. *In Proceedings of the IEEE Workshop on Stereo and Multi-Baseline Vision (SMBV)*.

Hirschmüller, H. and Scharstein, D. (2007). Evaluation of cost functions for stereo matching. *In IEEE Computer Society Conference on Computer Vision and Pattern Recognition (CVPR)*, Minneapolis, MN, pp. 1-8.

Horn, B. (1986). *Robot Vision*, MIT Press.

Ishikawa, H. (2000). *Global optimization using embedded graphs*. PhD Thesis, New York University. Adviser : Davi Geiger.

Ishikawa, H. & Geiger, D. (1998). Occlusions, discontinuities, and epipolar lines in stereo. *In Proceedings of the 5th European Conference on Computer Vision (ECCV)*, pp. 232-248. Springer-Verlag.

Julesz, B. (1962). Towards the automation of binocular depth perception. *In IFIP Congress*, pp. 439-444.

Karzanov, A. V. (1974). Determining the maximal flow in a network by the method of preflows. *Soviet Mathematics Doklady*, 15:434-437.

Kolmogorov, V. & Zabih, R. (2001). Computing visual correspondence with occlusions using graph cuts. *In Proceedings of the International Conference on Computer Vision (ICCV)*, pp. 508-515.

Kolmogorov, V. & Zabih, R. (2002a). Multi-camera scene reconstruction via graph cuts. *In Proceedings of the 7th European Conference on Computer Vision (ECCV)*, pp. 82-96. Springer-Verlag.

Kolmogorov, V. & Zabih, R. (2002b). What energy functions can be minimized via graph cuts? *In Proceedings of the 7th European Conference on Computer Vision (ECCV)*, pp. 65-81. Springer-Verlag.

Ohta, Y. & Kanade, T. (1985). Stereo by intra- and inter-scanline search using dynamic programming. *In IEEE Transactions on Pattern Analysis and Machine Intelligence*, pp. 139-154.

Roy, S. (1999). Stereo without epipolar lines: A maximum-flow formulation. *International Journal of Computer Vision*, 34(2-3):147-161.

Roy, S. & Cox, I. (1998). A maximum-flow formulation of the n-camera stereo correspondence problem. *In Proceedings of the International Conference on Computer Vision (ICCV)*, pp. 492-499.

Sára, R. (2002). Finding the largest unambiguous component of stereo matching. *In Proceedings of the 7th European Conference on Computer Vision (ECCV)*, pp. 900-914. Springer-Verlag.

Scharstein, D. & Szeliski, R. (2002). A taxonomy and evaluation of dense two-frame stereo correspondence algorithms. *International Journal of Computer Vision*, 47.

Siek, J. G., Lee, L.-Q. & Lumsdaine, A. (2001). *Boost Graph Library, The : User Guide and Reference Manual*. Addison-Wesley Professional.

Veksler, O. (1999). *Efficient graph-based energy minimization methods in computer vision*. PhD Thesis, Cornell University. Adviser : Ramin Zabih.

Zureiki, A., Devy, M. & Chatila, R. (2007b). Stereo matching using reduced-graph cuts. *In IEEE International Conference on Image Processing (ICIP)*, San Antonio, Texas (USA).